International Conference on Aerospace and Control Engineering (ICoACE 2022)

Journal of Physics: Conference Series Volume 2489

Guangzhou, China
23-25 December 2022

ISBN: 978-1-7138-7303-7
ISSN: 1742-6588

Printed from e-media with permission by:

Curran Associates, Inc.
57 Morehouse Lane
Red Hook, NY 12571

Some format issues inherent in the e-media version may also appear in this print version.

Institute of Physics
Dirac House, Temple Back
Bristol BS1 6BE UK

Phone: 44 1 17 929 7481
Fax: 44 1 17 920 0979

techtracking@iop.org

Additional copies of this publication are available from:

Curran Associates, Inc.
57 Morehouse Lane
Red Hook, NY 12571 USA
Phone: 845-758-0400
Fax: 845-758-2633
Email: curran@proceedings.com
Web: www.proceedings.com

TABLE OF CONTENTS

Journal of Physics: Conference Series **2489** (2023) 011001 doi:10.1088/1742-6596/2489/1/011001

Preface

The purpose of the three-day 2022 International Conference on Aerospace and Control Engineering (ICoACE 2022), held via virtual form on December 23rd-25th, 2022 in Guangzhou, China, was to create a timely forum for multi-disciplinary discussions related to the recent developments on control engineering triggered by aerospace.

As usual, the program of the meeting consisted of keynote speeches, oral presentations, the former including speeches delivered by six sophisticated professors from different areas. The speakers covered three main topics: i) Several Key Issues for High-Fidelity Simulation Requirement of Modern Aeroengine; ii) Aircraft Power Supply System Development: From Multi-Electric Aircraft to Electric Propulsion Aircraft; iii) Based on IoTs Parking Navigation With Continuous Information Potential Field Research. Among them, the first report mainly introduced the current situation of high fidelity simulation technology of aeroengine, the application of high fidelity simulation technology in Aeroengine development process, and the future demand of the digital twin applied in the aeroengine high fidelity simulation technology.

In the framework of experimental and theoretical approaches, the conference gathered together 100 delegates from all over the world and addressed a number of highly relevant aspects of aerospace and control engineering. Covering Materials Surface Physics, Electric Automation, Distributed Control System, Physics of Gases, Plasma Physics, the selected contributions of this Special Issue provide guidance for future interdisciplinary developments. In this way, the multi-scale aspects of material-based aerospace, mechanics and control engineering should be considered.

We received various manuscripts of invited papers and contributions presented at the Seminar. All of these have been peer reviewed and are collected in this volume. We would like to thank all the anonymous colleagues who have acted as referees to assess the suitability of the various articles for publication in the Journal of Physics: Conference Series. We are confident that the high quality of both invited and contributed papers contained in these Proceedings will be appreciated by the aerospace community.

We would like to express our thanks to all the authors for their time and genuine efforts, and to the reviewers for their fruitful comments during the preparation of this volume. We also acknowledge the support provided in various ways by Nanchang Institute of Technology who acted as host to the Conference.

The Committee of ICoACE 2022

Committee member

Conference General Chairs
Prof. Xiaolin Qiu, Nanchang Institute of Technology, China
Prof. Devinder Yadav, University of Nottingham Ningbo, China

Academic Committee Chair
Prof. Xianyu Li, Nanchang Institute of Technology, China

Publication Chair
Prof. Ruixiang Bai, Dalian University of Technology, China

Local Committee Chairs
Prof. Yanbin Fan, Nanchang Institute of Technology, China
Prof. Ning Sun, Nankai University, China

Technical Program Committee Chair
Prof. Gang Zhang, Harbin Institute of Technology, China

Technical Program Committees
Prof. Enwei Chen, Hefei University of Technology, China
Prof. Ozgur Koray, Biruni University, Turkey
Prof. Hossein Hosseinkhani, Innovation Center for Advanced technology, Matrix, Inc., Canada
Prof. Minoru Sasaki, Gifu University, Japan
Prof. Predrag Ivanis, University of Belgrade, Serbia
Prof. N.Rajathi, Kumaraguru College of Technology, India
Prof. Subhendu Kumar Pani, Orissa Engineering College, India
Assoc. Prof. Muhyiddine Jradi, University of Southern Denmark, Denmark
Assoc. Prof. SANDIP KUNAR, Aditya Engineering College, India
Assoc. Prof. Sahil Verma, Lovely Professional University, India
Assoc. Prof. Rajeev Tiwari, University of Petroleum and Energy Studies, India
Dr. GUELAILIA Ahmed, Satellite Development Center, Algerian Space Agency, Oran, Algeria

Academic Committees
Prof. Naim L Braha, University of Prishtina, Kosovo
Prof. Badrul Hisham, Universiti Teknikal Malaysia Melaka (UTEM), Malaysia
Prof. Mohammad Reza Safaei, Florida International University, America
Prof. Marjan Goodarzi, Lamar University, America
Prof. Zhilong Chen, Nanchang Institute of Technology, China
Prof. Lushen Wu, Jiangxi Province Engineers Federation, China
Assoc. Prof. Shuhua Bai, Nanchang Institute of Technology, China
Researcher Yubao Wang, Nanchang Institute of Technology, China

Local Committees
Prof. Hongyi Yuan, Nanchang Institute of Technology, China
Prof. Xiaoli Wan, Nanchang Institute of Technology, China
Prof. Quanxin Zhu, Hunan University, China
Prof. Weidong Zhang, Shanghai Jiao Tong University, China
Prof. Mingwei Qin, Southwest University of Science and Technology, China
Dr. Siting Chen, South China University of Technology, China
Assoc. Prof. Chao Chen, Taiyuan University of Technology, China
Assoc. Prof. Ke Li, Nanchang College of Technology Jiangxi Province Electromechanical Experimental Teaching Demonstration Center, China
Assoc. Prof. Jiuming Xie, Tianjin Sino-German Universities of Applied Sciences-Mechanical Engineering College, China
Assoc. Prof. Liangxi Ding, Nanchang Institute of Technology, China
Assoc. Prof. Zhengkai Sun, Nanchang Institute of Technology, China
Engineer Yelin Li, Jiangxi Province Engineers Federation, China

Peer Review Statement

All papers published in this volume have been reviewed through processes administered by the Editors. Reviews were conducted by expert referees to the professional and scientific standards expected of a proceedings journal published by IOP Publishing.

- **Type of peer review:** Single Anonymous
- **Conference submission management system:** Morressier
- **Number of submissions received:** 101
- **Number of submissions sent for review:** 95
- **Number of submissions accepted:** 52
- **Acceptance Rate (Submissions Accepted / Submissions Received × 100):** 51.5
- **Average number of reviews per paper:** 2
- **Total number of reviewers involved:** 30
- **Contact person for queries:**
  Name: Xuexia Ye
  Email: xx.ye@keoaeic.org
  Affiliation: AEIC - AEIC

Study on the aerodynamic performance of inlet pipe different bending radius of ram air system

Xiaohong Liu[1a*] and Fuming Xia[2b*]

[1]Department of Manufacturing Support Engineering, Shanghai Aircraft Design and Research Institute, Shanghai 201210, China

[2]Center of Process Equipment, Shanghai Aircraft Manufacturing Company, Shanghai 201100, China

[a*]liuxiaohong@comac.cc, [b*]xiafumin@comac.cc

Abstract—In this paper, the bending radius deviation of the intake duct is studied. The computational fluid dynamics (CFD) method is used to evaluate the performance of the inlet by the coupling calculation of the internal and external flow of aircraft. Firstly, the CFD method is used to calculate the theoretical numerical model. The results show that when the pipeline flow is constant and the pipeline R/D decreases, such as from 1.84 to 1.3, the total pressure recovery decreases, especially when the pipeline R/D is less than 1.5, the total pressure recovery decreases rapidly. At the same time, the drag force of the pipeline does not change much, and the flow field of the inlet pipeline is analyzed to analyze the internal reasons for the total pressure recovery and flow rate change with R/D, which has important guiding significance for the application of batch production in future.

1. Introduction

In the production of a certain type of aircraft in the ram air system, there exist some problems about out of tolerance problems of inlet pipe bending radius, to assess the different inlet pipe bending radii varies how to affect the performance of the pipeline and judge whether this pipe can be accepted. It is carried out by the CFD analysis [1] method for integration with the aircraft's inflow and outflow calculation, at a first simplified model of fluid is developed. The conclusion will be developed according to the influence of the total pressure recovery, Mach number, resistance analysis, flow analysis, bending pipe wall pressure analysis, and flow field analysis of different radius R/D changes on the inlet pipe outlet. The performance of different bending pipes is finally confirmed, which provides an important reference value for aircraft production about out of tolerance of pipe bending radius.

2. Ram-air system design requirements and performance definition

Ram air system design requirements mainly include ram air inlet pressure: [-20, 300] mbar, flow rate: 32-250kg/min, and total pressure recovery coefficient($C=P_t/P_{t\infty}$) not less than 0.792. Pt displays the total pressure of the outlet of the inlet pipe, $P_{t\infty}$ displays the total pressure of far-field gas; the drag coefficient is C_X ($X/(0.5\rho_\infty A\ V_\infty V_\infty)$), ρ_∞ displays density, V_∞ displays velocity, A displays flow area, and X displays drag.

3. Numerical models and grids

3.1 Simplification of model

In this simulation calculation, firstly, the aircraft model is simplified. The principles of simplification are as follows: 1) The non-aerodynamic area of the aircraft model is repaired by CATIA software and the surface with problem topology is closed [3]; 2) The problems of aircraft models such as small gaps and overlapping surfaces are simplified and dealt with; 3) Considering that wingtip winglet far away from fuselage has little influence on the flow field around ram-air intake [2]. According to experience, calculation ignores the influence of wingtip winglets on the intake of the ram air system such as in Fig. 1.

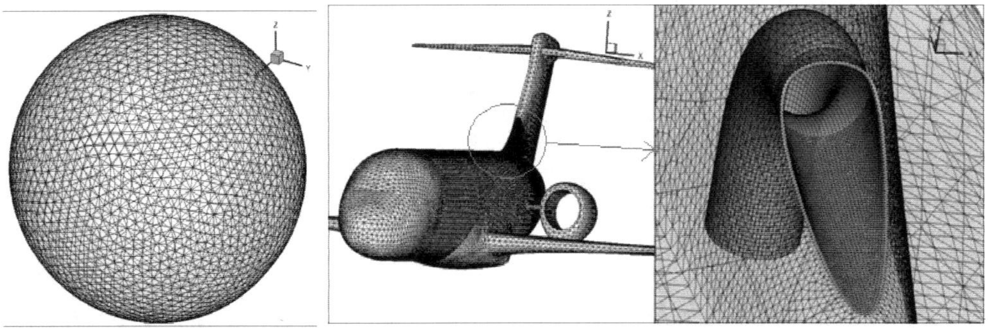

Fig. 1. Computational domain and ram air inlet pipe grid layout

3.2 The boundary conditions

The boundary conditions are mainly: the initial condition, far field condition, outlet boundary condition, and solid wall boundary condition [4]. (1) Initial conditions: The far-field flow condition is used as the initial condition of the flow field. (2) Far-field conditions: You can only get a position at a certain distance from the object. When the flow is subsonic, the disturbance wave will propagate to infinity in all directions. At this time, the boundary should be divided far away from the object as far as possible. Such a boundary is the far-field boundary. (3) Outlet boundary conditions: When different back pressure values are given, different working states of pipelines can be obtained. (4) Solid wall boundary conditions: when the wall is solid, the fluid velocity is zero, that is, there is no slip condition. The turbulence model is the k-ε model which is widely used in the aviation field. The residual of this calculation is set as 0.00001. It takes about 8000 steps to calculate with more than 1.5 million grids. The total pressure recovery curve converges.

4. Analysis of the influence of manufacturing deviation of the ram-air system inlet line

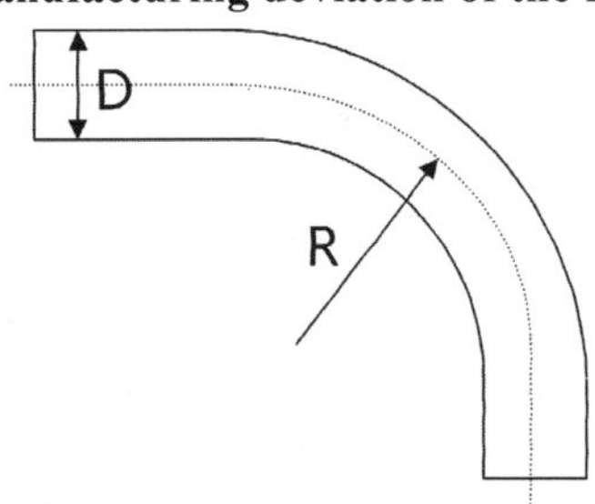

Fig. 2 Ram air inlet pipe different R/D center line

Due to the influence of the outflow field and the fuselage, it is difficult to evaluate the influence of the general calculation on the ram air system. The numerical simulation method is used for this kind of situation. The R/D of the theoretical pipeline is 1.84, and the R/D of the most serious over-error model pipeline is 1.75. To make a more detailed analysis, this paper tries to find out the internal rules and provide a direct basis for pipeline out-of-tolerance treatment. In this calculation, R/D was selected as 1.84, 1.75, 1.51, and 1.3 corresponding to the four different radii in Fig. 2.

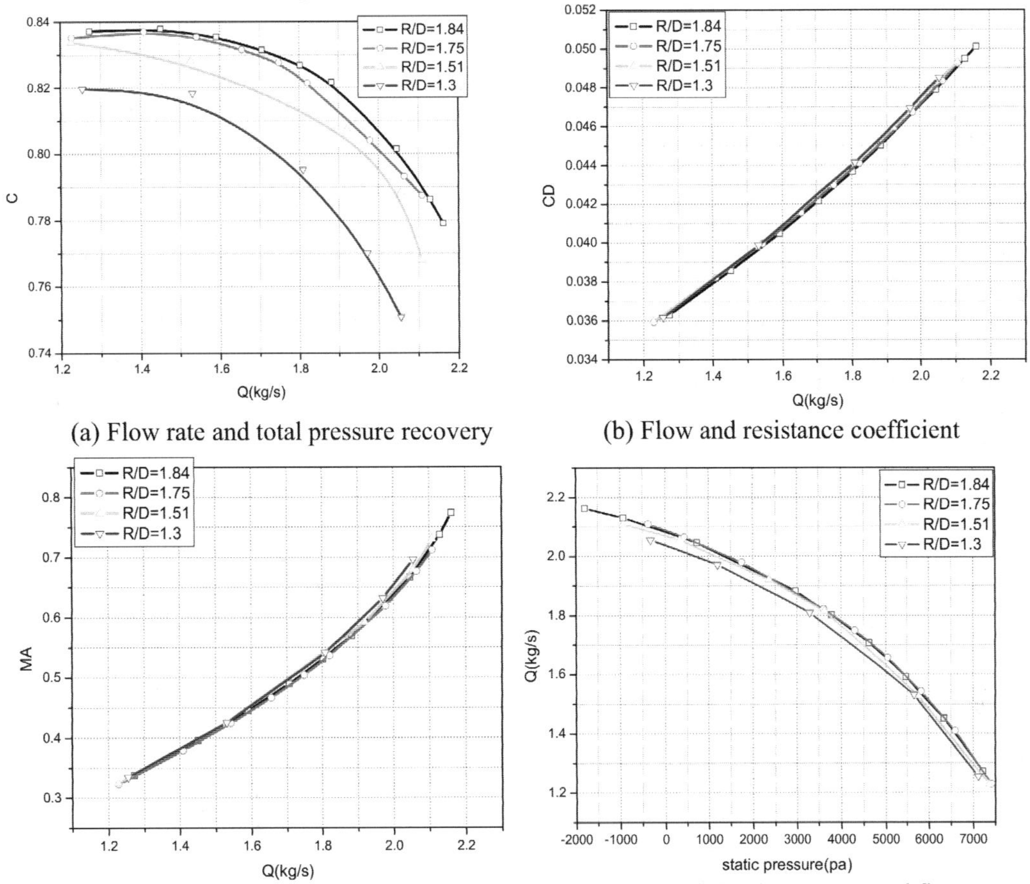

(a) Flow rate and total pressure recovery (b) Flow and resistance coefficient

(c) Flow and Mach number (d) Static pressure and flow

Fig. 3 Aerodynamic performance of Ram air inlet pipe with different R/D center line

4.1 Curve of total pressure recovery with ram air inlet flow

When the pipeline R/D is constant, the total pressure recovery decreases with the increase of the flow rate, which is caused by the increase of the flow rate and the increase of the resistance in the pipeline. At the same time, when the flow rate is less than about 1.6kg/s, the total pressure recovery changes gently, and when the flow rate exceeds 1.6kg/s, the total pressure recovery decreases rapidly. When the pipeline flow rate is constant, the total pressure recovery decreases as the pipeline R/D decreases, such as from 1.84 to 1.3. Especially when the pipeline R/D is less than 1.5, the total pressure recovery decreases rapidly. According to the supplier's documents, the total pressure recovery requirement is not less than 0.792. When H=35000ft and Mach number MA=0.78 flow value, the maximum flow value is now 2kg/s, and it is the worst working condition. Therefore, when pipeline R/D is less than 1.5, the total pressure recovery loss is unacceptable. When the pipe R/D is greater than or equal to 1.5, the total pressure recovery loss meets the design requirements. When pipeline R/D=1.84, the total pressure recovery is about 0.805. When pipeline R/D=1.75, the total pressure recovery is about 0.8. When pipeline R/D=1.51, the total pressure recovery is about 0.793. When pipeline R/D=1.3, the total pressure recovery is about 0.76 such as in Fig. 3(a).

4.2 Curve of drag coefficient with ram air inlet flow rate
When the pipeline R/D is constant, the resistance coefficient increases with the increase in flow rate. This is because the flow rate increases, the speed increases, and the impact force on the pipe increases, leading to an increase in resistance. When the pipeline flow rate is constant, the resistance coefficient changes with R/D unchanged such as in Fig. 3(b).

4.3 Curve of each number with ram air inlet flow rate
When the pipeline R/D is constant, the inlet Mach number of the ram air pipeline increases with the increase of the flow rate. When the pipeline flow rate is constant, the inlet Mach number of the ram air pipeline is unchanged with the change of R/D such as in Fig. 3(c).

4.4 Ram-air inlet flow rate varies with back pressure
When the pipeline R/D is constant, the flow rate of the ram air pipeline decreases with the increase of the inlet back pressure of the ram air pipeline. When the pipeline flow is constant and R/D is greater than 1.51, the ram air inlet pipeline flow is constant. When R/D=1.3, the ram air inlet pipeline flow decreases such as in Fig. 3(d).

4.5 Variation law of pipe wall pressure distribution with back pressure

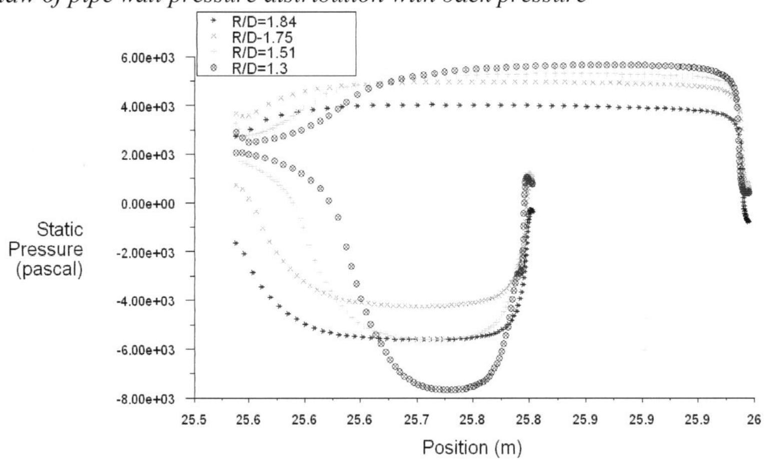

(a) Back pressure -4000Pa

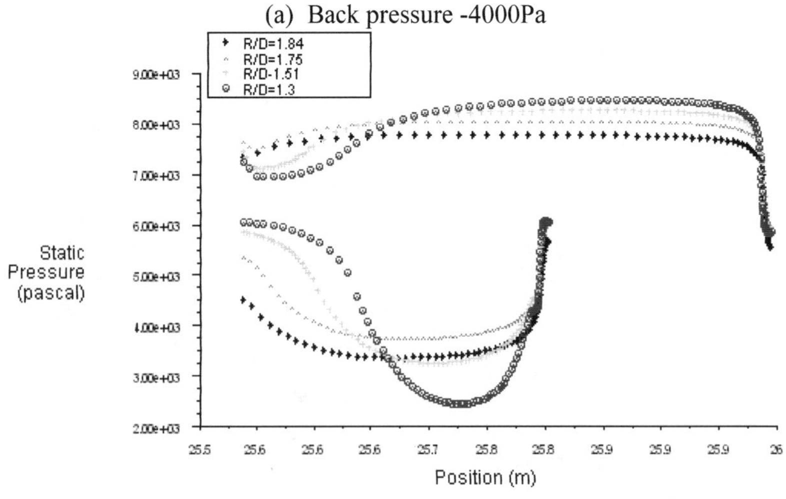

(b) Back pressure 5000Pa

Fig.4 Variation curve of the static pressure of pipe wall with different R/D

Analysis of the pressure distribution law of the X=0 section of different R/D pipelines is shown in Figs. 4(a) and (b). When the pipeline is bent, the fluid will cause local loss when passing through the turning part, including eddy current loss, collision loss, and "secondary flow" loss. For outside the pipe wall, about before Y = 25.65 m, outer pipe wall pressure increases. When R/D=1.75 and 1.84, the pressure increases gently, while when R/D=1.3 and 1.51, the pressure increases sharply and the pressure gradient is larger than that of the former. At about Y=25.65m-25.9m, the pressure keeps a constant value, and when R/D increases from 1.3 to 1.84, the constant value of pressure decreases successively. After Y=25.9m, no matter how big the R/D value is, the pressure drops sharply; Before Y=25.6m, no matter how big the R/D value is, the pressure changes sharply. When R/D=1.51 and 1.3, the pressure decreases sharply, the minimum value is even lower than R/D=1.84, and the positive pressure gradient increases. When Y exceeds 25.8m, no matter how big R/D is, the pressure value increases sharply. The presence of secondary flow in the elbow increases the local loss, and the energy loss caused by secondary flow is related to R/D. The smaller R is, the larger the pressure difference between inside and outside, the larger D is, and the larger the range of secondary flow is. Vortex and secondary flow are the sources of energy loss in curved pipe flow.

4.6 The flow field analysis

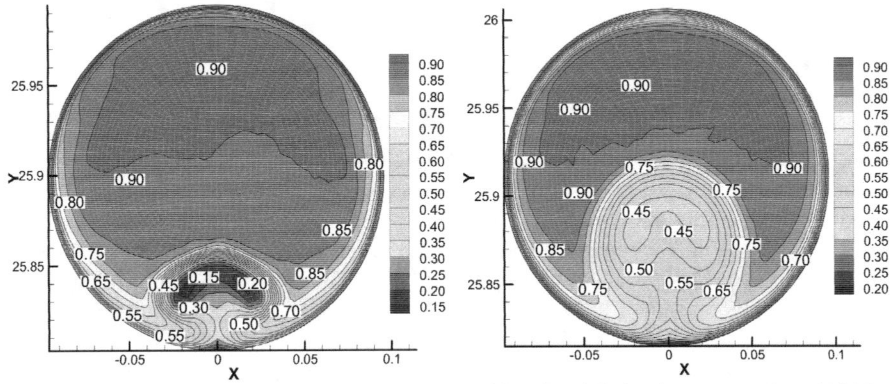

(a) R/D=1.84 back pressure P= -4000Pa (b) R/D=1.3 back pressure P= -4000Pa

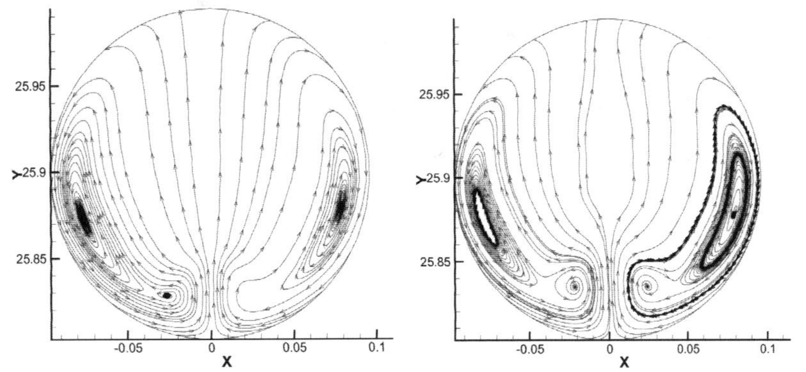

(c) R/D=1.84 back pressure P= -4000Pa (d) R/D=1.3 back pressure P= -4000Pa

Fig. 5 Total pressure recovery distribution and streamlined distribution at the ram air outlet

There is a low-pressure area on the inner side of the ram-air inlet due to the thickening of the boundary layer after the turn of the airflow. When the back pressure is P=-4000Pa and R/D is from 1.84 to 1.3, the secondary flow loss at the ram air inlet increases, and the single vortex structure is transformed into a double vortex structure on one side [5]. At the same time, the low-pressure area of the total pressure recovery coefficient increases according to Fig. 6. When R/D is constant and the back pressure increases (P=-4000Pa), the secondary flow loss decreases, and the total pressure recovery increases such as in Fig. 5.

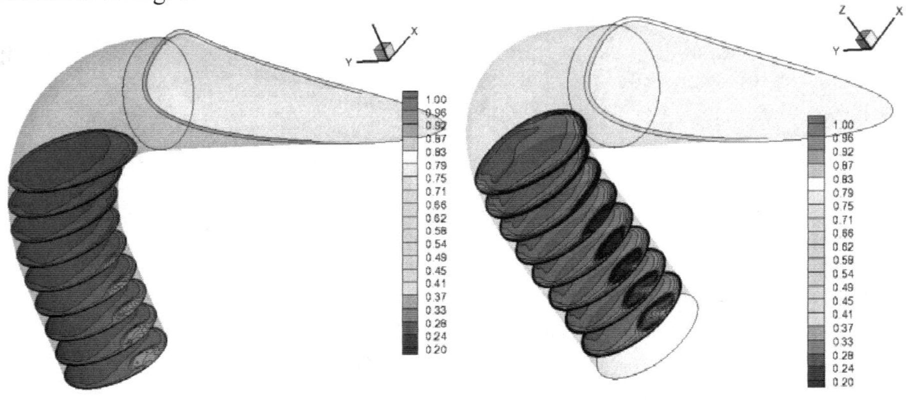

(a) R/D=1.84 (b) R/D=1.3

Fig. 6 Total pressure recovery distribution along the ram air inlet pipe

5. Conclusion

Based on the results and discussions presented above, the conclusions are obtained as below:

(1) The total pressure recovery decreases with the increase of the flow, and the total pressure recovery decreases as the pipeline R/D decreases. When R/D is less than 1.5 total pressure recovery is unacceptable, it must be greater than or equal to 1.5 for pipe R/D.

(2) The drag coefficient increases with the increase of the flow when the pipeline flow rate is constant, and the resistance coefficient changes with R/D unchanged.

(3) When the pipeline R/D is fixed, the flow rate of the ram air pipeline decreases with the increase of the back pressure at the inlet of the ram air pipe. When R/D is greater than 1.51, the flow of the ram air inlet pipe does not change much. When R/D=1.3, the flow of ram air inlet pipe decreases rapidly.

(4) When R/D is greater than 1.51, the pressure gradient of the inner wall is relatively gentle. With the increase of R/D, the pressure of the inner wall increases. When R/D=1.3, the pressure decreases sharply, and the pressure gradient increases, which leads to the increase of the pressure difference between the outer wall and the inner wall and the enhancement of the secondary flow loss.

(5) There is a low-pressure area on the inner side of the ram air inlet. When R/D is changed from 1.84 to 1.3, the secondary flow loss of the ram air inlet increases, and the single vortex structure is transformed into a double vortex structure on one side.

(6) In the future, the elbow Angle of the ram air intake pipe will be studied, and the optimal pipeline shape is designed, which has relatively large total pressure recovery and less separation, and improves the aerodynamic performance of the pipe. It provides an important reference value for the design of ram air pipelines for civil aircraft

References

[1] CFD Analysis of Engine Air Intake for A Pusher Type Light Transport Aircraft [O]. Vinay CA, Bhaskar Chakravarthy, Dutt HNV 2013.

[2] G. C. Paynter. CFD Status for Supersonic Inlet Design Support. AIAA 1994-465,1994.

[3] A. M. Kirk, A. Kumar, J. I. Gargoloff. Numerical and Experimental Investigation of a Serpentine Inlet Duct. AIAA 2007-842, 2007

[4] J. J. Korte, D. J. Singh, A. Kumar, A. H. Auslender. Numerical Study of the Performance of Swept, Curved Compression Surface Scramjet Inlets. Journal of Propulsion and Power. 1994, 10(6)

[5] J. J. Seddon, E. L. Goldsmith. Intake Aerodynamics. Second Edition. London Blackwell Science, 1999

Volatility analysis of the flight block time based on the stochastic volatility model

Ya Sun[1a*], **Meiyi Wang**[1b] **and Hua Xie**[1c]

[1]Nanjing University of Aeronautics and Astronautics

[a*]sy1999.1016@nuaa.edu.cn, [b]wangmeiyi@nuaa.edu.cn, [c]marcuse@163.com

Abstract. To effectively predict the volatility of flight block time, this paper constructs a stochastic volatility model based on actual flight block time data, solves the model parameters by the Markov chain Monte Carlo method, and uses the standard stochastic volatility (SV-N) model and thick-tailed stochastic volatility (SV-T) model to characterize the volatility of flight block time. The results show that the thick-tailed stochastic volatility model is better than the standard stochastic volatility model in describing the volatility of the segment runtime, and the thick-tailed stochastic volatility model is chosen to predict the volatility of the flight block time. Predicting the flight block time volatility in real time can provide a theoretical basis for traffic traveler planning.

1. Introduction

The flight block time is the total time from the removal of the wheel block before the aircraft slides to the placement of the wheel block after the aircraft lands and stops, which is the sum of the aircraft's ground taxiing time and air flight time [1]. Due to the random changes in traffic, stormy weather, and other uncontrollable factors, the transformation of the flight block time has the characteristics of time-varying, irregular and random and presents more complex fluctuation characteristics. Many studies have been conducted at home and abroad on flight block time. Still, there needs to be more research on the volatility of the flight block time and the prediction of the flight block time volatility of air transportation systems. Stochastic volatility (SV) models have been used in financial data to build many models to fit and analyze the data to make reasonable forecasts and estimates. However, the SV models contain latent variables, making their likelihood functions extraordinarily complex and challenging to solve for their assessments. To solve this problem, in 2002, JACQULER et al. [2] combined Bayesian theory with the SV model and used Markov Chain Monte-Carlo (MCMC) method to estimate, and found that its estimation effect is more effective than the likelihood method; however, in the field of transportation, the SV model is less applied to analyze data. Among them, Wei-Feng Li et al. [3] involved the SV model in showing that it also has good feasibility in explaining road travel time volatility. Therefore, this paper analyzes the volatility of the flight block time by constructing the SV model, which in turn can reveal the characteristics of the fluctuation structure of the flight block time and has practical significance for exploring the prediction of the traveler's flight block time. Therefore, this paper constructs the SV model to analyze the volatility of the flight block time and then can reveal the characteristics of the fluctuation structure of the flight block time, which is of practical significance for exploring the prediction of the flight block time.

2. Review of SV model and Markov Monte Carlo

2.1. Thick-tailed stochastic volatility model

In the SV model, volatility is assumed to wander randomly and obey geometric Brownian motion, whereas the expected return of the underlying asset, $\phi(t)$ is the instantaneous volatility of the underlying asset, and $d\omega_0(t)$ is the standard Brownian sign [4]. In the SV model, volatility variation depends on unobservable stochastic processes. In the mid-1980s, Taylor (1986) [5] in the United States first proposed the standard stochastic volatility model (SV-N), which has the following model form:

$$y_t = \exp(h_t / 2)\varepsilon_t, \varepsilon_t \sim i.i.d.N(0,1) \tag{1}$$

$$h_{t+1} = \mu + \phi(h_t - \mu) + \eta_t, \eta_t \sim i.i.d.N(0, \sigma^2) \tag{2}$$

where time $t = 1, 2, \cdots, T$, y_t is the log return of the asset at the time t; the disturbance term is independently and identically distributed and follows a normal distribution; h_t is the implied volatility of the return at the time t. The implied log volatility obeys the Gaussian AR(1) process, ϕ is the persistence parameter of volatility, which can reflect the current state on the degree of future volatility, and when $|\phi| < 1$, the model is covariance-smooth; μ is the average volatility level parameter; η_t is used to measure the degree of perturbation of fluctuations, independently and identically distributed in N (0, σ), and are independent of each other. In the SV-N model, the errors are assumed to be normally distributed and smooth, and the parameters to be estimated are the mean μ , the persistence parameter ϕ , and the perturbation level σ^2 .

The thick-tailed stochastic volatility (SV-T) model can be obtained by replacing the distribution in Equation (1) with the student t distribution, which has the following model form:

$$y_t = \exp(h_t / 2)\varepsilon_t, \varepsilon_t \sim i.i.d.t(\omega) \tag{3}$$

$$h_{t+1} = \mu + \phi(h_t - \mu) + \eta_t, \eta_t \sim i.i.d.N(0, \sigma^2) \tag{4}$$

In Eqs. (3) and (4), the significance of the parameters is consistent with the SV-N model except that ε_t obeys the t distribution. It is usually assumed that $\varepsilon_t \sim N(0,1)$ or $\varepsilon_t \sim t(\omega)$. Based on the advantage of the MCMC method in solving the parameters of the SV model, the basic theory of MCMC is given below.

2.2. MCMC estimation method

The static Monte Carlo method is slow and inefficient in dealing with high-dimensional data, so the Markov process is introduced into the Monte Carlo simulation to obtain the dynamic Monte Carlo method, i.e., MCMC, which is widely used for sampling high-dimensional random vectors [6]. The basic idea of the MCMC method is the following three steps [7]:

(1) Select the appropriate Markov chain in X such that the transfer kernel is, where appropriate means the smooth distribution corresponding to X.

(2) Generate a sequence from a point X0 in X using the Markov chain in Step (1).

(3) For some m and a relatively large n, the expectation of any function is estimated as.

According to the above three steps of the MCMC method, the parameters in the SV model can be solved. Based on the aforementioned SV-N model, the parameters to be estimated are μ , ϕ , σ^2 and the potential fluctuation series h_t and $t = 0, \cdots, T$ is added to the parameters to be estimated so that the total number of parameters is 4+T. Therefore, the joint prior density is [8]:

$$P(\mu, \phi, \sigma^2, h_0, h_1, \cdots, h_T) = P(\mu, \phi, \sigma^2) P(h_0 | \mu, \phi, \sigma^2) \prod P(h_{t+1} | h_t, \mu, \sigma_n^2) \tag{5}$$

The expressions for the functions of $\eta_t \sim i.i.d.N(0,\sigma^2)$, $P(h_{t+1}|h_t,\mu,\phi,\sigma^2)$ can be easily introduced. The likelihood function of the model is $L(\mu,\phi,\sigma^2,h_{0:T})$, and the conditional distribution of y_t gives the likelihood function as:

$$P(y_1,\cdots,y_T)=\prod_{t=1}^{T}P(y_t|h_t) \tag{6}$$

Based on this, the joint posterior distribution of the parameters is proportional to the product of the likelihood function, the joint prior distribution, based on Bayes' principle:

$$P(\mu,\phi,\sigma^2,h_0,h_1,\cdots h_t|y_1,\cdots,y_T)\propto$$

$$P(\mu)P(\phi)P(h_0\mu,\phi,\sigma^2)\prod_{t=1}^{T}P(h_{t+1}|h_t,\mu,\phi,\sigma^2)\prod_{t=1}^{T}P(y_t|h_t) \tag{7}$$

The Gibbs sampling method can be used here to sample the SV model so that the 4+T parameters in the model can be estimated. Before the construction of Gibbs sampling, it is assumed that the density function of X is π (X), which is difficult to achieve in practice, but this does not affect the use of the method. In application, Gibbs sampling can be repeated for $i=1,\ldots,n$. Typically, the results of the final iteration converge to a π distribution by the distribution. The exact steps of sampling are as follows [9][10].

(1) Determine the initial point $(x_1^{(0)},\cdots,x_n^{(i)})$ and let i = 0;

(2) Draw $x_1^{(i+1)}$ from distribution $\pi(x_1|x_2^{(i)},x_3^{(i)},\cdots,x_n^{(i)})$;

(3) Set $i=i+1$ and go to step (2).

Denote $x^t=(x_1^t,x_2^t,\cdots,x_n^t)$, then c is the realized value of the Markov chain. Typically, methods such as ACF discrimination and Geweke's test are used to determine whether the sampling converges to a stable distribution. The MCMC method, by using repeated sampling, enables the final posterior distribution of Bayesian estimates of each parameter to be obtained.

3. Empirical analysis

The data of flights from Beijing to Guangzhou and from Guangzhou to Beijing are selected for the period from January 1, 2018, to December 31, 2018, with each data interval of 1 hour, totaling 19,362 pieces of data. The two models will be trained with the flight block time data in June, July, and August, so the data can be analyzed for volatility based on the model-solving parameters. Then the models can be selected and proposed to forecast the flight block time one by one. Here, the first-order logarithmic difference of the flight block time is defined as the volatility of the flight block time and is calculated as:

$$y_t=(\ln T_t-\ln T_{t-1})\times 100\% \tag{8}$$

where T_t is the flight block time at time t.

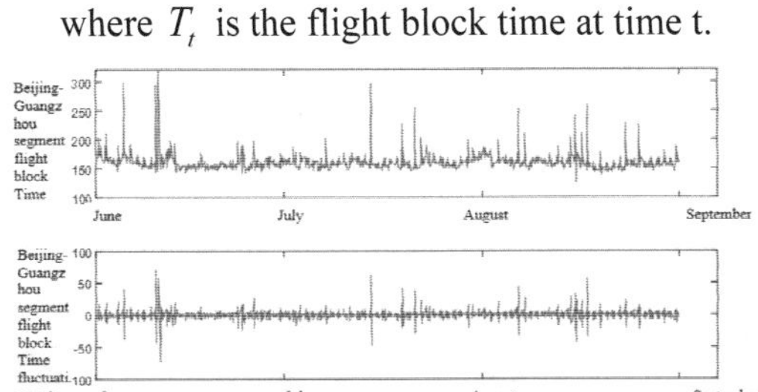

Figure 1. Beijing-Guangzhou segment flight block Time and the fluctuation rate chart.

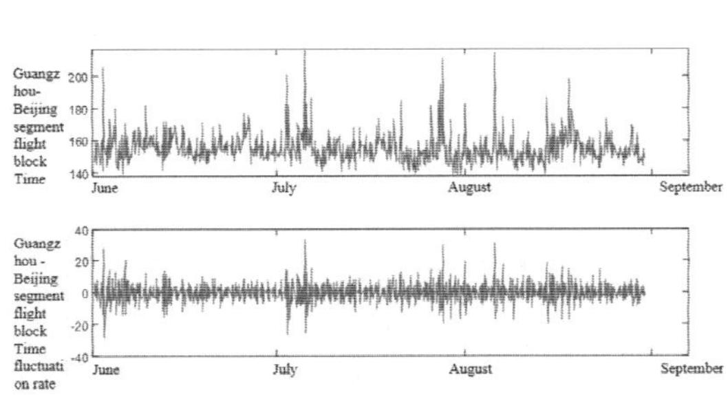

Figure 2. Guangzhou- Beijing segment flight block Time and the fluctuation rate chart.

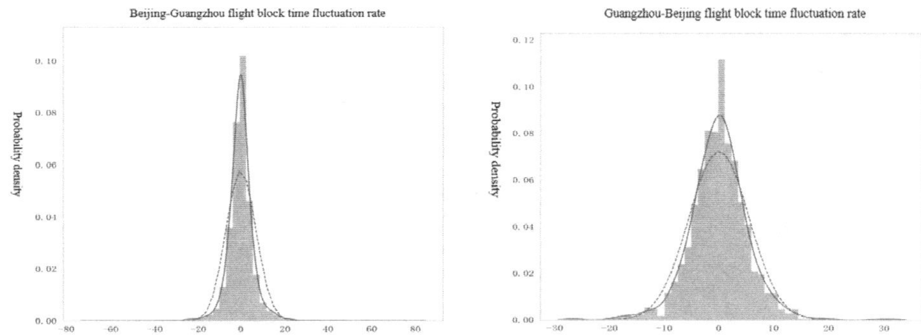

Figure 3. Kernel density estimation curve of flight block time

Table 1. Table of flight time fluctuation rate parameters

Flight segments	Mean value	Standard deviation	Skewness	Peakedness
Beijing-Guangzhou	0.0010728	6.8874	0.9710	24.8016
Guangzhou-Beijing	0.003716	5.5184	0.1730	4.3437

To briefly describe the basic characteristics of the Beijing-Guangzhou and Guangzhou-Beijing segment runtimes, the flight block time diagrams and the volatility timing diagrams of the flight block times for the two segments are given, as shown in Figures 1 and 2. As can be seen from Figures 1 and 2, the size of the trip time varies from June to August 2019, with the Beijing-Guangzhou segment showing significant ups and downs in early June, mid-to-late July, and mid-to-late August, resulting in different mains for the two. As can be seen from Figure 3, there is a significant volatility time-variability in the volatility time series of both trip times, i.e., pileups occur in both high and low volatility phases.

Based on the flight block time in the example, the frequency distribution histogram is used to estimate the kernel density curve of the flight block running volatility and analyze the essential characteristics of the data. The results are shown in Figure 3. The red dashed line is the standard distribution probability density curve, and the solid gray line is the kernel density estimation curve. The temporal distribution of the runtime volatility of the two segments has much more values near the mean than the normal distribution, showing spikiness; the values distributed on both sides of the histogram are also more than the normal distribution, showing thick-tailedness. The results are shown in Table 1, which shows that the volatility time series of both segments are concentrated around the mean value; the kurtosis of the Beijing-Guangzhou segment and Guangzhou-Beijing segment are 24.8016 and 4.3437 respectively, which are both greater than 3, reflecting the spikes and thick tails of both segments again.

3.1. MCMC estimation for SV-N model

MCMC may require several iterations before the sequence approaches the target distribution. The results of estimating each parameter of the SV-N model for the Beijing-Guangzhou segment and Guangzhou-Beijing were estimated using MCMC iterations 10,000 times. This is shown in Figure 4 and Figure 5. From the marginal distribution in Figure 6 and Figure 7, we can estimate the means and standard errors of the parameters, and the results are shown in Table 2. The estimated results of each parameter of the SV-N model for the Beijing-Guangzhou segment and the Guangzhou-Beijing segment are shown in Table 3. In Table 3, the estimated values of the fluctuation level parameter μ for the Beijing-Guangzhou segment and the Guangzhou-Beijing segment are 0.76583 and 0.94056, respectively. Compared with the Guangzhou-Beijing segment, the absolute value of the fluctuation level parameter for the Beijing-Guangzhou segment is more minor, indicating that the trip time volatility of the Guangzhou-Beijing segment is higher. From the perspective of the fluctuation persistence parameter ϕ, the Beijing-Guangzhou segment is 0.68456 and the Guangzhou-Beijing segment is 0.9357, both of which are less than 1, indicating that both sectors are smooth and have strong fluctuation persistence, and the fluctuation persistence of the Guangzhou-Beijing sector is longer. The level of volatility perturbation σ_v^2 is greater than 0.3425 from 0.7357, again indicating a higher level of volatility perturbation in the Guangzhou-Beijing sector than in the Guangzhou-Beijing sector. From the model degrees of freedom estimation, the Beijing-Guangzhou segment is 26.389 and the Beijing-Guangzhou segment is 29.4352, again indicating that the volatility distribution is non-normal for both segments.

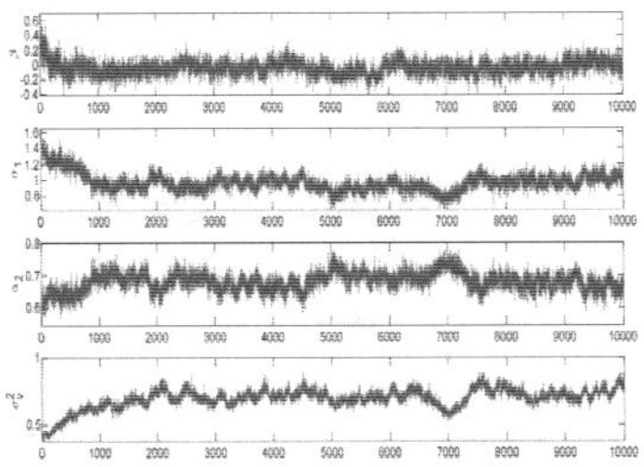

Figure 4.
MCMC iterative process for Beijing-Guangzhou

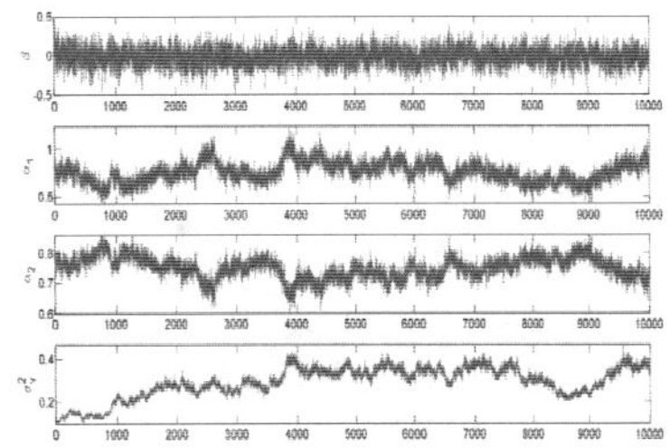

Figure 5.
MCMC iterative process for Guangzhou-Beijing

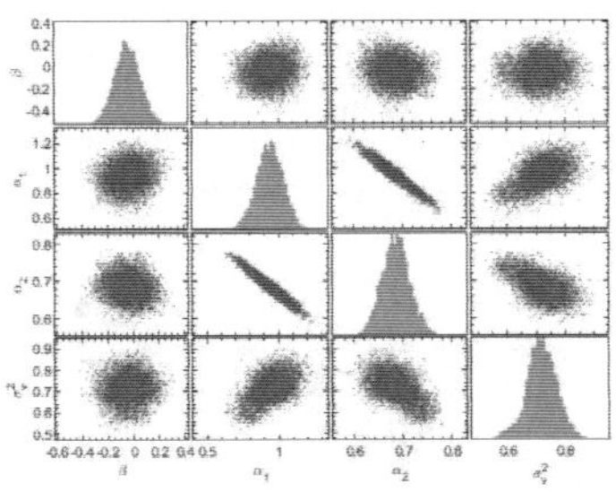

Figure 6.
Scatterplot and histogram of the joint Beijing-Guangzhou distribution of parameters

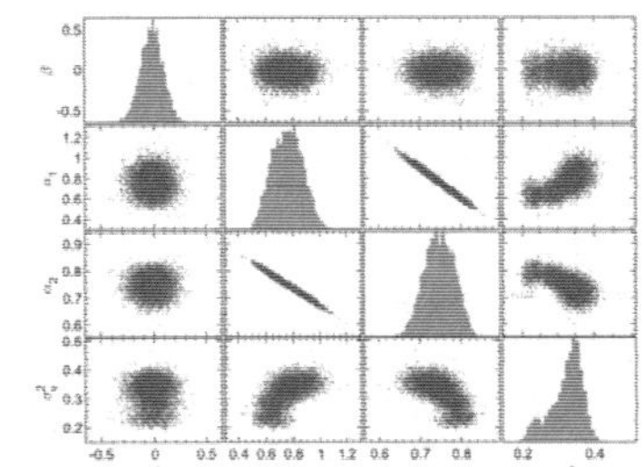

Figure 7.
Scatterplot and histogram of the joint Guangzhou-Beijing distribution of parameters

Table 2. Estimation results of various parameters of the SV-N model

Flight segments	Parameters	Mean value	Standard deviation	5% quantile	50% quantile	95% quantile
Beijing-Guangzhou	ω	26.389	24.565	3.47	23.4678	82.784
	μ	0.7536	0.09134	0.7810	0.8740	1.0827
	ϕ	0.68456	0.0290	0.6376	0.6652	0.7344
	σ_v^2	0.3425	0.0485	0.2631	0.3522	0.3960
Guangzhou-Beijing	ω	29.4352	26.1348	4.445	27.5672	85.4573
	μ	0.7647	0.103	0.5877	0.6750	0.9280
	ϕ	0.9357	0.09112	0.7780	0.8631	1.0234
	σ_v^2	0.7357	0.0555	0.6251	0.6974	0.8147

3.2. MCMC estimation for SV-T model

The results of estimating each parameter of the SV-T model for the Beijing-Guangzhou segment and Guangzhou-Beijing were estimated using MCMC iterations 10,000 times. This is shown in Figure 8 and Figure 9. From the marginal distribution in Figure 10 and Figure 11, we can estimate the means and standard errors of the parameters, and the results are shown in Table 3. In Table 3, the estimated values of the fluctuation level parameter for the Beijing-Guangzhou and Guangzhou-Beijing sectors are 0.76583 and 0.94056, respectively, and the absolute values of the fluctuation level parameter for the Beijing-Guangzhou sector are more minor compared with those for the Guangzhou-Beijing segment, indicating that the trip time volatility of the Guangzhou-Beijing segment is higher. From the perspective of the fluctuation persistence parameter ϕ, the Beijing-Guangzhou segment is 0.68508 and the Guangzhou-Beijing segment is 0.74489, both of which are less than 1, indicating that both segments are smooth and have strong fluctuation persistence, and the fluctuation persistence of the Guangzhou-Beijing segment is longer. The volatility perturbation levels are greater from 0.71675 to 0.32761, which also indicates that the volatility perturbation level of the Guangzhou-Beijing segment is higher than that of the Guangzhou-Beijing segment. From the model degrees of freedom estimation, the Beijing-Guangzhou segment is 28.563 and the Beijing-Guangzhou segment is 29.4632, again indicating that the volatility distribution is non-normal for both segments.

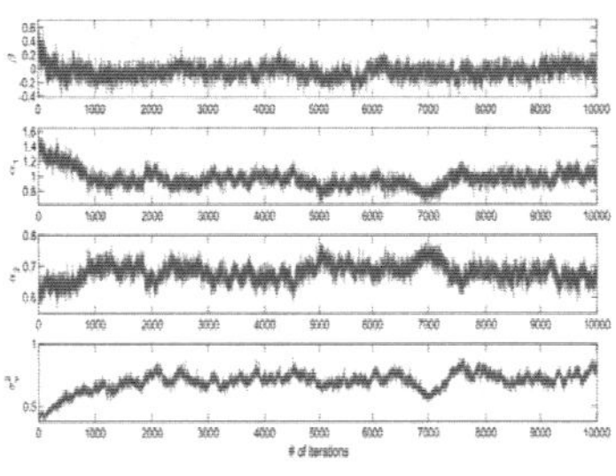

Figure 8.
MCMC iterative process for the Beijing-Guangzhou segment

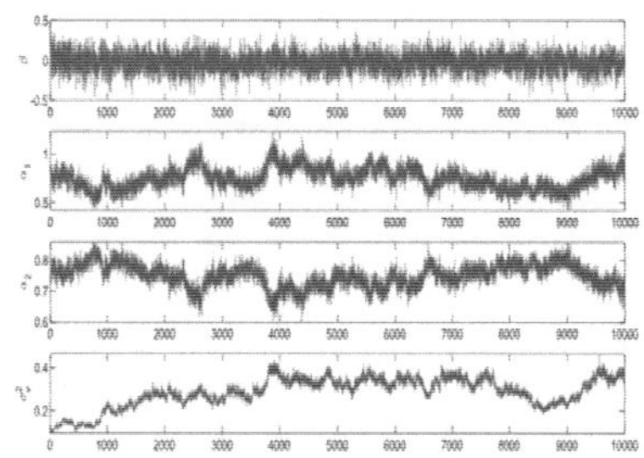

Figure 9.
MCMC iterative process for the Guangzhou-Beijing segment

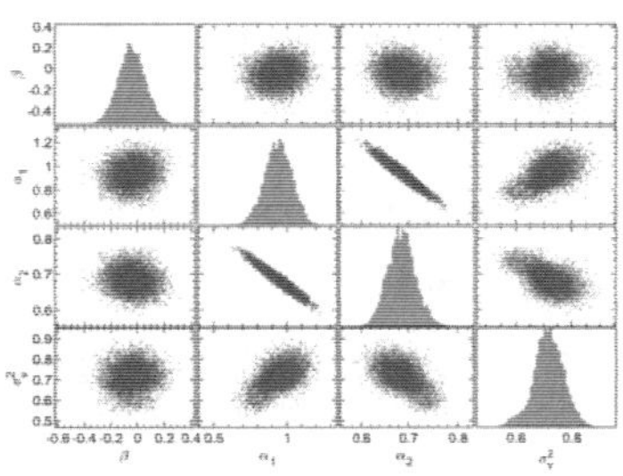

Figure 10.
Scatterplot and histogram of the joint Beijing-Guangzhou distribution of parameters

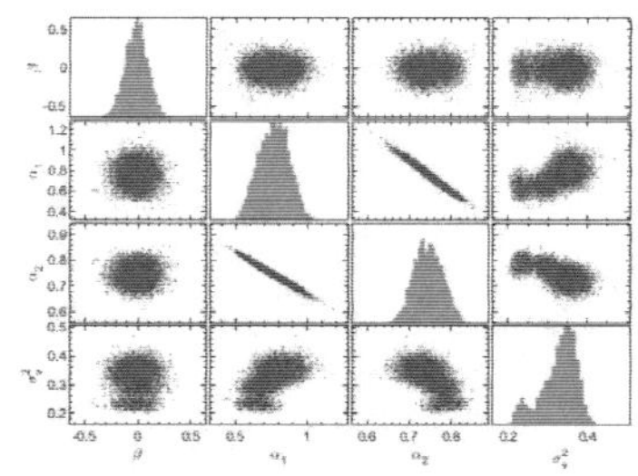

Figure 11.
Scatterplot and histogram of the joint Guangzhou-Beijing distribution of parameters

Table 3. Estimation results of various parameters of the SV-T model

Flight segments	Parameters	Mean value	Standard deviation	5% quantile	50% quantile	95% quantile
Beijing-Guangzhou	ω	28.563	25.4655	3.67	24.5634	83.324
	μ	0.76583	0.092881	0.7826	0.8780	1.0937
	ϕ	0.68508	0.034772	0.6898	0.7202	0.8025
	σ_v^2	0.32761	0.045328	0.2334	0.3016	0.3872
Guangzhou-Beijing	ω	29.4632	26.2343	4.575	27.46758	86.5743
	μ	0.76583	0.10586	0.5898	0.6880	0.9370
	ϕ	0.94056	0.092881	0.7826	0.8780	1.0937
	σ_v^2	0.71675	0.054885	0.6204	0.6827	0.8054

3.3. Model Selection

The estimation results of the model parameters in Table 2 (SV-N) and Table 3 (SV-T) are combined to compare the effects of the SV-N model and the SV-T model in fitting the example data from a single segment analysis. For the estimated values of the fluctuation levels μ, both the Beijing-Guangzhou segment and the Guangzhou-Beijing segment exhibit larger absolute values than the estimated values of the SV-T model, indicating that both the Beijing-Guangzhou segment and the Guangzhou-Beijing segment exhibit more substantial volatility under the SV-T model than the SV-N model. In terms of the estimated value of the sustainability parameter ϕ, the estimated value of the Beijing-Guangzhou segment is around 0.68 and the Guangzhou-Beijing segment is around 0.93, indicating that the Guangzhou-Beijing segment has more volatility aggregation effect than the Beijing-Guangzhou segment; on the whole, the estimated value of ϕ in the SV-T model is more significant than that in the SV-N model, indicating that the SV-T model can better characterize the volatility of the two segments than the SV-N model. Volatility persistence than the SV-N model. For the parameter σ_v^2, which evaluates the disturbance level, the greater the sustainability parameter ϕ and the more minor then σ_v^2, the more predictable the course of the fluctuations. In both Table 2 and Table 3, the σ_v^2 values of the SV-T model are smaller than those in the corresponding SV-N model, indicating that the SV-T model is a better fit than the SV-N model.

3.4. Flight block time volatility forecast

From the above analysis of the volatility of the Beijing-Guangzhou segment and Guangzhou-Beijing segment, the SV-T model can better describe the volatility of the flight block time of the Beijing-Guangzhou segment and Guangzhou-Beijing segment. Therefore, the SV-T model is selected to predict the volatility of the flight block time. Figure 6 shows the prediction results of the flight block time (June, July, and August) of the Beijing-Guangzhou and Guangzhou-Beijing segments based on the SV-T model. Figure 12 and Figure 13 show that the SV-T model has excellent prediction results, and when applied to volatility prediction, the SV-T model can provide confidence in the prediction.

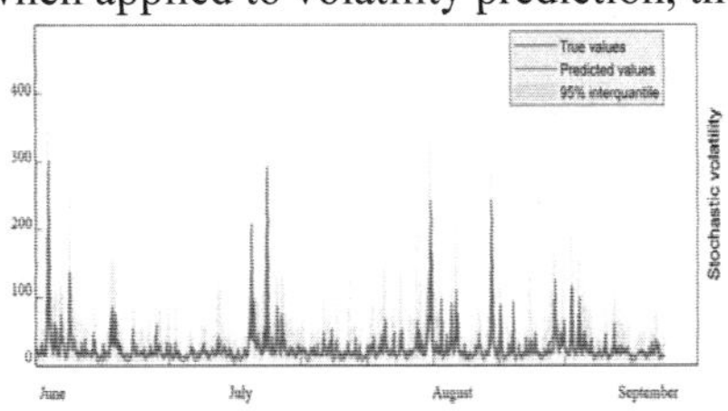

Figure 12. A posteriori means of random volatilities after 4000 tests on the Beijing-Guangzhou segment. Confidence intervals between the 95th percentile of the stochastic volatility are indicated in red.

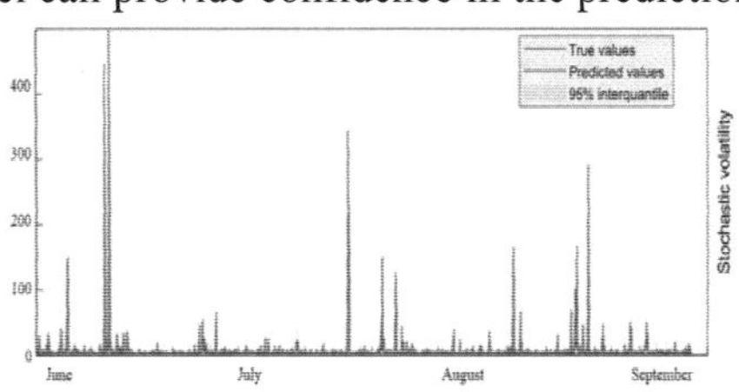

Figure 13.A posteriori means of random volatilities after 4000 tests on the Guangzhou-Beijing segment. Confidence intervals between the 95th percentile of the stochastic volatility are indicated in red.

4. Conclusion

In this paper, we select the flight block time data of the Beijing-Guangzhou and Guangzhou-Beijing segments in 2018, construct the SV-N model and SV-T model, respectively, and solve the model parameters using the MCMC method to compare the model fitting effect. The results show that: the flight block time of both segments shows the characteristics of spikes and thick tails; the fluctuation level parameters show that the flight block time of the Guangzhou-Beijing segment is significantly more volatile, while the fluctuation of the Guangzhou-Beijing segment is persistent; from the perspective of disturbance fluctuation level, the flight block time of Guangzhou-Beijing segment is relatively higher. The example analysis confirms that the SV-T model is better than the SV-N model in characterizing the fluctuation of volatility. The SV-T model was chosen to predict the volatility of the flight block time of 678 months one by one, and the prediction effect of the two segments was more accurate.

Acknowledgments

This paper is supported by the National Natural Science Foundation of China (NSFC) under grant No.71971112.

References

[1] Civil Aviation Administration of China. Civil Aviation Comprehensive Statistical Report System (2015) [E]. Beijing: Civil Aviation Administration of China, 2015.12.
[2] JACQUIER E, POLSON N G, ROSSI P E. Bayesian analysis of stochastic volatility models[J]. Journal of Business & Economic Statistics,2002,20(1):69－87.
[3] Li, Wei-Feng, Duan, Zheng-Yu, Yang, Dong-Aid. Analysis of fast road travel time volatility[J]. Computer Engineering and Applications,2015(4):260-265.
[4] TSE Y K, Yu J, CHANG X B. Estimation of Hyperbolic Diffusion using MCMC Method[J]. 2002(9): 1-15.
[5] TAYLOR S J. Modeling Stochastic Volatility: A Review and Comparative Study Modeling[J]. Mathematical Finance,1994,4(2):183-204.
[6] Zhang Bo, Shang Hao. Applied stochastic processes[M]. Fourth edition. Beijing: People's University of China Press,2016: 202-217.
[7] Wang Yiming. (2009). A comparison of return volatility models under different distributions (Master's thesis, East China Normal University). https://kns.cnki.net/KCMS/detail/detail.aspx?dbnam-e=CMFD2009&filename=2009187464.nh.
[8] KASTNE R G, F R UHWI R TH-SCHNATTE R S. Ancillarity-Sufficiency Interweaving Strategy (ASIS) for Boosting MCMC Estimation of Stochastic Volatility Models[J].Computational Statistics & Data Analysis, 2014, 76:408-423.
[9] Yu Ranchun. (2014). An empirical study of stochastic volatility model based on MCMC Bayesian approach (Master's thesis, Shanghai Normal University). https: //kns. cnki. net/KCMS/detail/detail.as-px?dbname=CMFD201501&filename=1014334179.nh.
[10] Hung, C. W. (2019). Bayesian estimation of stochastic fluctuation models based on integrated nested Laplace approximation methods (Master's thesis, Shandong University). https://kns.cnki.net/K-CMS/detail/detail.aspx?dbname=CMFD201902&filename=1019043634.nh.

A micro-force measurement torsion pendulum design with dual laser detection

Yujie Zhao[1,2,3], Bojian Wu[1,2,3], Yuandong Jia[4], Zongfeng Li[1,3]*

[1] Key Laboratory of Space Utilization, Technology, and Engineering Center of Space Utilization, Chinese Academy of Sciences, Beijing 100094, China;

[2] University of Chinese Academy of Sciences, Beijing 100049, China;

[3] Technology and Engineering Center for Space Utilization, Chinese Academy of Sciences, Beijing 100094, China;

[4] Lanzhou University, Lanzhou 730000, China.

Corresponding Author e-mail: lzfeng@csu.ac.cn

Abstract: Aiming at the ground test and evaluation of the suspended test bench of the high microgravity scientific experiment cabinet in the Tianhe core module of the space station to obtain its exact thrust resolution, a micro-force measurement torsion device with dual laser detection is designed. The main configuration of the device is the suspension fiber torsion pendulum, and tungsten wire with a length of 400 mm and a diameter of 800 μm is used as the suspension fiber. Under micro thrust, the suspension fiber is twisted, which drives the mirrors at both ends of the scale arm to rotate along the suspension fiber. The displacement detection of the reflector is carried out, and the magnitude of the micro thrust is calculated in combination with the theorem of momentum. Experimental results indicate that the design of the torsion of micro-force measurement provides an effective test method for the ground test of the suspension test bench. The device can be used to measure and evaluate the micro-propulsion force of the air jet port. After replacing the tungsten fiber with a diameter of 1mm, it can also replace the air floating platform to carry out the air jet closed-loop control verification work, which can effectively and accurately complete the ground test and evaluation of the suspension test bench.

1. Introduction

With the successful docking between the Tianzhou-4 cargo spacecraft and the core module of the space station, it marks the official entry into the construction stage from the key technology verification stage of the Chinese space station. This is a symbol of our more profound understanding of space. There are many experimental phenomena and theories discovered in microgravity environments.

However, the microgravity state of the Chinese space station is not absolute microgravity. The resultant force of the spacecraft is equal to the centripetal force of the spacecraft to orbit around the Earth, so its microgravity level is about 10^{-3} g to 10^{-5} g. The researchers used a double-layer suspension technology to increase the microgravity level by two orders of magnitude to 10^{-7} g [1]. The suspension test bench is a part of the high microgravity scientific experiment cabinet. To ensure its normal operation, experiments should be designed to test and verify it on the ground. The injection micro thrust is also measured to verify the injection closed-loop control algorithm.

The commonly used microforce measurement modes include direct measurement schemes and indirect measurement schemes [2]. In the direct measurement scheme, the thrust is directly acted on by the force sensor to obtain the thrust. Commonly used indirect measurement modes include lever amplification mode, electromagnetic balance mode, and torsion pendulum mode [3].

In this paper, a torsion pendulum device for micro force measurement with double laser detection is designed. The high precision and large-range laser displacement sensor are selected as the sensing systems. Combined with the appropriate torsional stiffness, a large range of force measurements was achieved, and the micro-force measurement accuracy of the μN level was achieved. The measuring device has the advantages of high precision, fast response, and a large range, which fully meets the thrust measurement requirements for ground tests and evaluation of suspension test bench. The diversity of micro force measurement is enriched.

In addition, the torsion pendulum designed in this paper can also be used in ground-based experiments of inertial sensors for space gravitational wave detection. It can not only test the influence of the interference force generated by the locking release operation on the inspection quality but also can calibrate and measure the electrostatic force generated by the test mass, which has a high practical application value.

2. Scheme Design and Principle of Torsion Pendulum

2.1. The scheme design of the torsion pendulum

The torsion pendulum system for micro force measurement is based on torsion design. It uses high-purity tungsten wire as the torsion fiber, and the torsion fiber is connected with the lower side structure through the chuck. The lower side is mainly the scale rod and the tooling plate. Calibration balls were arranged on the scale rod, and reflectors corresponding to laser displacement sensors were installed at both ends. The counterweight threaded sliding table is installed on the tooling plate, and the counterweight is added according to actual needs. The payload fixed on the tooling plate is provided with a jet thruster. The magnetic damping device is composed of a flange threaded sleeve and two ring magnets and is installed on the torsion pendulum bracket between the first torsion fiber and the second torsion fiber. The three-dimensional diagram is shown in Figure 1.

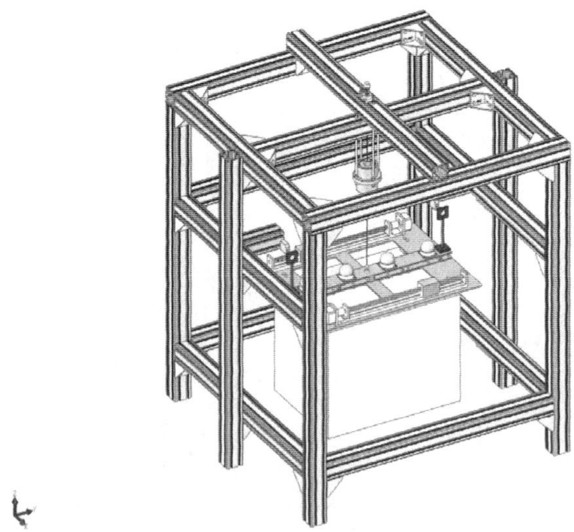

Figure 1. Micro-force measurement torsional pendulum system

2.2. Principle of torsion pendulum measurement

The principle of the torsion pendulum for micro force measurement is to use the suspension fiber to hang the object to be measured in the air. The micro thruster on the object to be measured generates thrust, which makes the object twist under the action of micro thrust. According to the moment of momentum theorem, the equation of motion of the torsion pendulum under the action of micro thrust [4] can be expressed as:

$$\ddot{\theta}(t) + \beta\dot{\theta}(t) + \Gamma\theta(t) = T(t) \quad (1)$$

Formula (1): J is the rotational inertia of the pendulum, β is the dissipation term (damping force), Γ is the torsion stiffness constant of the suspension fiber, and *T*(t) is all possible torsional torque. When no torque is applied by the torsion pendulum, $T(t) = T_{th,n}(t)$, and $T_{th,n}(t)$ is the thermal noise torque of the suspension fiber [5].

When a small thrust is applied to one end of the scale rod, the scale rod twists under the thrust to drive the torsion fiber. Then the displacement is measured by a laser displacement sensor. Micro thrust can be the electromagnetic force of a voice coil motor or an electrostatic comb or the force of gravity exerted by a tiny weight. The numerical value of the micro force is combined with the measured displacement to calculate the torsion coefficient of the torsion fiber. Then, the torsion pendulum turns during the suspension test bench jet. Displacement is also measured by the displacement sensor. At this time, the thrust is directly calculated from the obtained displacement and the known bearing torsion coefficient. J, β, Γ can be regarded as a constant when the torsion pendulum oscillates slightly. It is difficult to measure the angular displacement (*θ*) of the torsion pendulum in practice. Displacement sensors are dexterously used to measure the displacement of the pendulum end and then convert the angle of the swing. In a small Angle, the torsion angle can be expressed as:

$$\theta \approx \mathrm{tg}\theta = \frac{x_1+x_2}{L} \quad (2)$$

Where *θ* is the torsion angle, for mirror displacement, x_1 is the displacement of the No.1 reflector, x_2 is the displacement of the No.2 reflector, and L is the distance between the center points of the two reflectors.

Since this experiment is a static test, both the angular velocity terms and the angular acceleration terms are 0, then the magnitude of thrust torque can be expressed as:

$$T = \Gamma \cdot \Delta\theta \quad (3)$$

Thus, the size of the micro thrust can be written as:

$$F = \Gamma\frac{x_1+x_2}{bL} \quad (4)$$

Where *F* is the torsion angle, and *b* is the size of the micro thrust arm. Then the error sources of the results are analyzed. The results are analyzed and corrected to obtain more accurate thrust value data. Thus, the micro thrust value is measured. The block diagram of its working test process is shown in Figure 2:

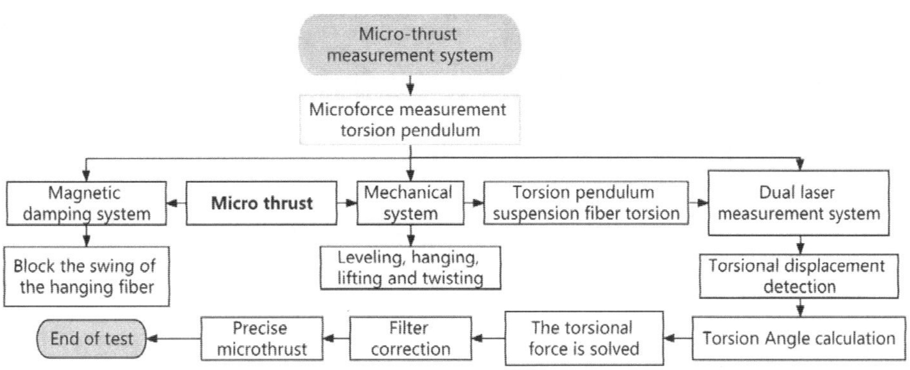

Figure 2. Block diagram of micro force measurement torsion pendulum system

2.3. Standard force-generating device

This system can use the voice coil motor, electrostatic comb, and weight as the micro force generation device [6-7]. Weights are used as preliminary calibration tools.

The gravity method[8] was adopted to calibrate the torsion pendulum system of microforce measurement. High-precision weights with mg level were selected for testing, as shown in Figure 3.

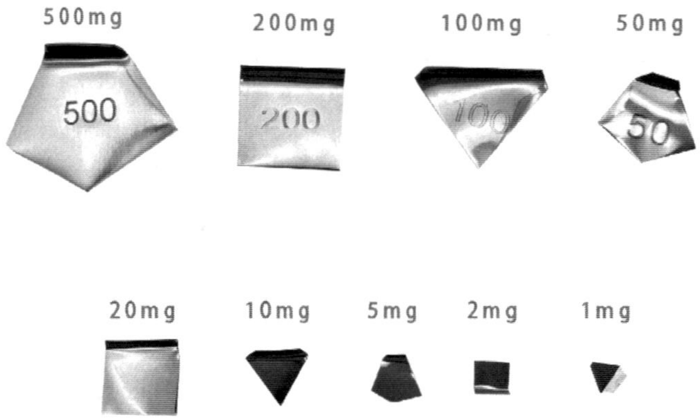

Figure 3. Weight used for the calibration

The weight installation layout is shown in Figure 4:

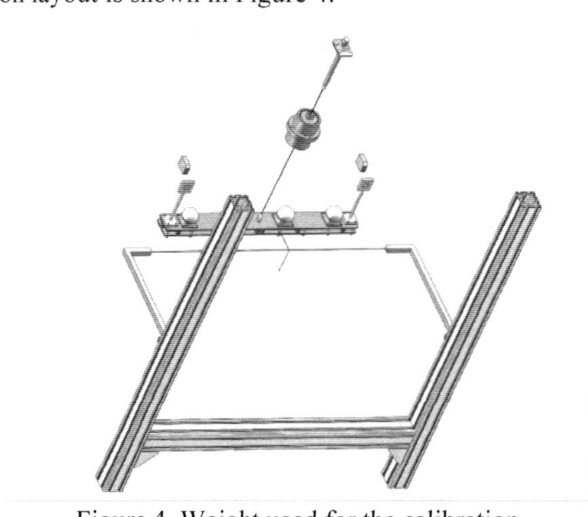

Figure 4. Weight used for the calibration

At this point, the micro force of the twist is:

$$F_W = mg - f \quad (5)$$

Where F_w is the force of the weight on the twisting pendulum, m is the mass of the weight, g is the gravity acceleration, and f is the friction drag. Different microforce values can be generated according to the selection and combination of weights with different masses.

3. Simulation analysis of the system

3.1. Simulation of mechanics

The main structure of the torsion pendulum is mainly composed of torsion fiber, scale rod, calibration ball, and workpiece. Torsion fiber bearing capacity 50kg sets thrust as 1mN and moment arm as 0.35m. Then a torque of $3.5 \times 10^{-4} \mathrm{N} \cdot \mathrm{m}$ is added to the torsional fiber to simulate the thrust torque of the thruster acting on the torsional fiber. The main body of the torsion pendulum is further simplified, and finite element analysis is carried out with the ANSYS tool, as shown in Figure 5.

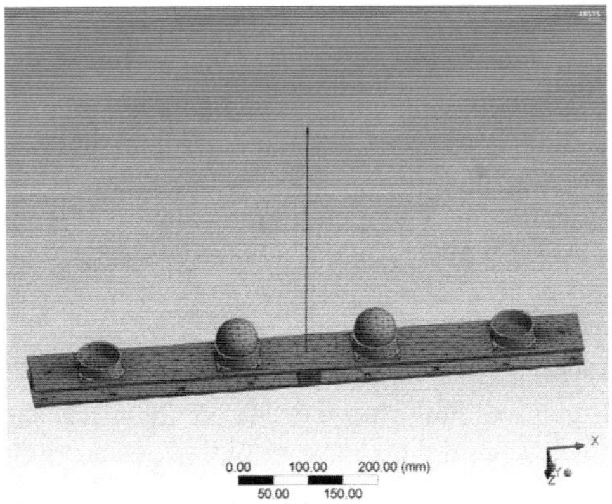

Figure 5. Grid division of the simplified torsion pendulum model

The stress and strain diagram of the torsion model after 1mN is shown as Figure 6:

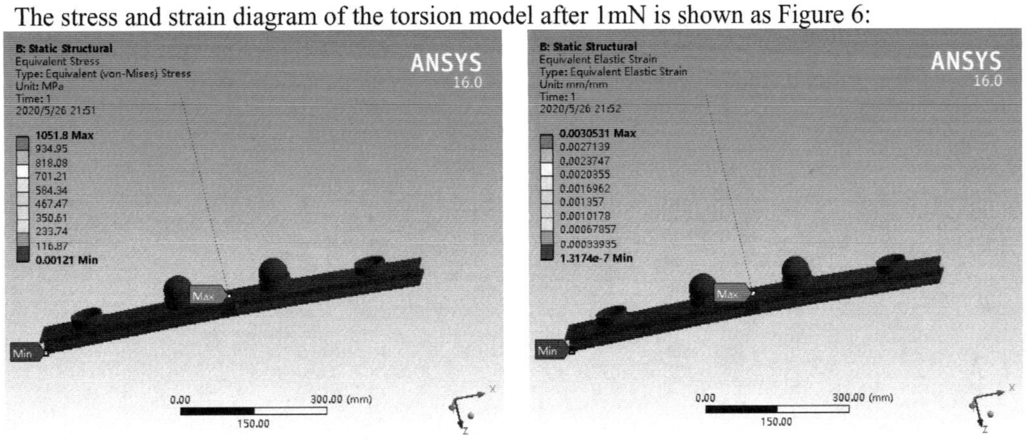

(a) Stress simulation results (b) Strain simulation results

Figure 6. Stress and strain diagram of simplified torsion pendulum model F=1mN

Similarly, when the micro thrust is 1μN, other conditions remain the same. Finite element analysis was conducted on the simplified model again, and the analysis results were shown as Figure 7:

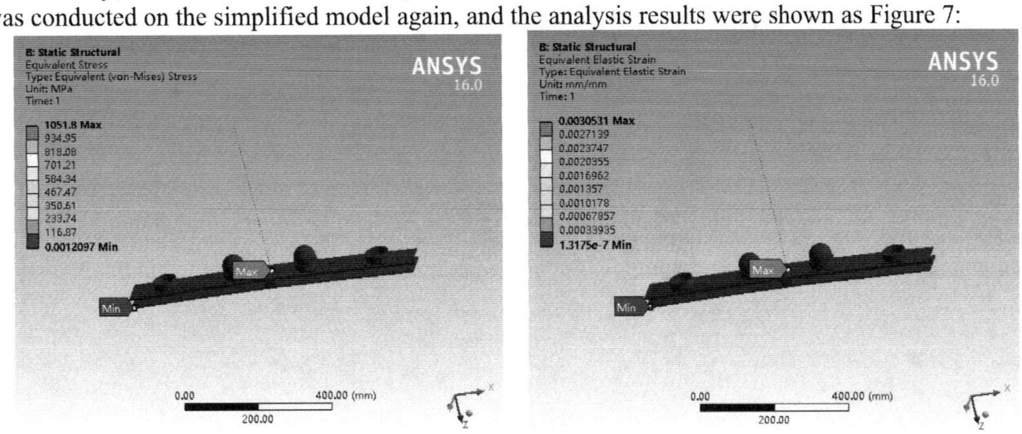

(a) Stress simulation results (b) Strain simulation results

Figure 7. Stress and strain diagram of simplified torsion pendulum model F=1μN

Due to the large bearing load, the twisted fiber is the key to the device. As can be seen from the Figure 6 and Figure 7, the stress concentration point is at the connection between the torsion fiber and the scale rod during the stress process, the maximum strain of the simplified model is 0.003053mm, the stress concentration occurs on the torsion fiber, and the maximum stress is 1051.8Mpa, which is much less than the tensile yield strength of the torsion fiber of 1700Mpa. So the torsion fiber can carry the workpiece to achieve the torsion deformation under micro thrust. Therefore, the mechanical analysis results of the main body show that the experiment is feasible when bearing 50kg and a micro thrust of 1mN.

3.2. Simulation of magnetic damping system

Magnetic damping system takes advantage of the eddy current effect in a magnetic field[9]. The horizontal movement of the internal aluminum disk is prevented, thus preventing the swing of the torsion pendulum system, so the torsion fiber can only do torsion deformation. It is ensured that the torsion is always perpendicular to the scale rod, that is, perpendicular to the force being applied, so that the measurements are accurate.

The magnetic field between the damper ring magnets was simulated and analyzed by Maxwell. Rubidium ferroboron ring magnet has an outer diameter of 58mm, an inner diameter of 34mm, and a thickness of 4mm. The 3D model and meshing are shown in Figure 8 and Figure 9:

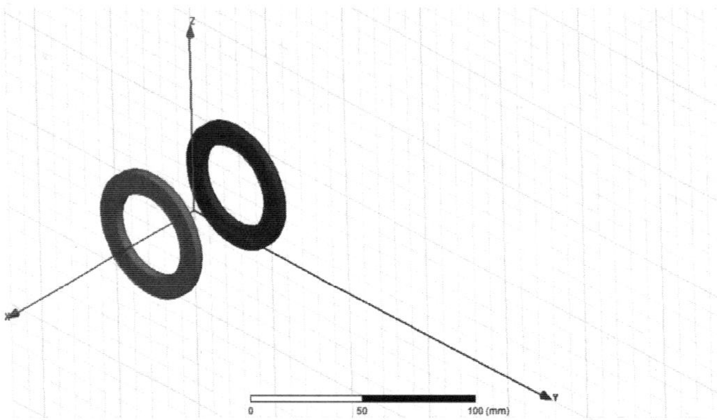

Figure 8. Model of Ring Magnet

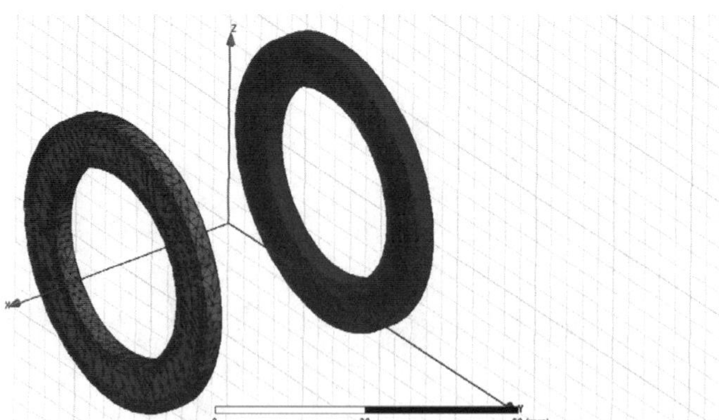

Figure 9. Model Mesh Dividing Diagram

The N30 rubidium ferroboron magnet was selected to simulate the magnetic field of the ring magnet. By setting the parameters of the ring magnet, the magnetic field distribution and magnetic induction distribution of the ring magnet surface are obtained.

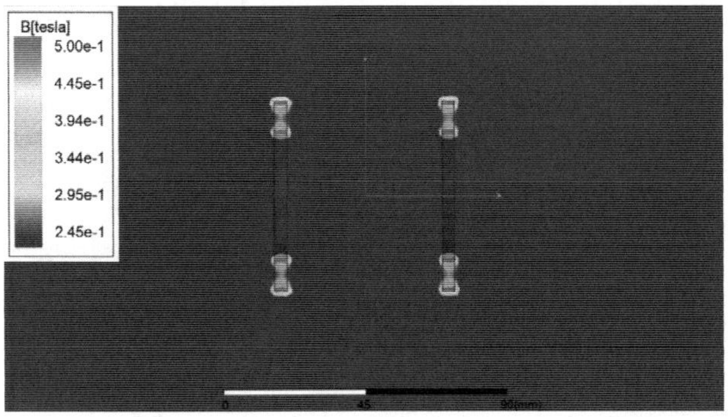

Figure 10. Distribution of magnetic field on the magnet surface

As can be seen from the Figure 10, the strongest magnetic field is B=0.5T when the magnet is 50mm apart. Different magnet materials and brands correspond to different magnetic field strengths. To further understand the damping mechanism of the magnetic damper on the aluminum disk, the distribution of magnetic sensing lines between the ring magnets is drawn as Figure 11:

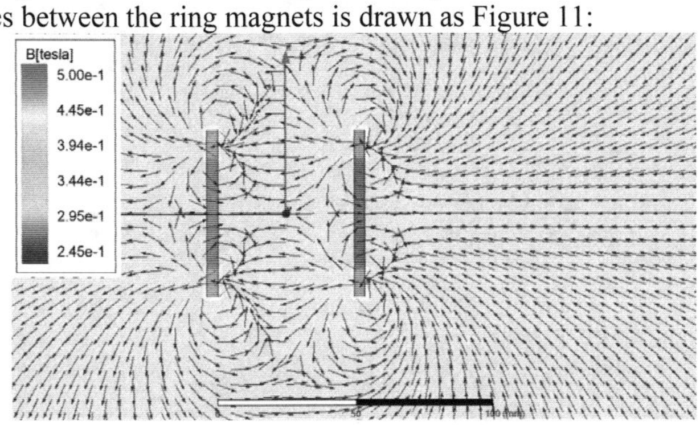

Figure 11. Distribution of magnetic inductance of ring magnet

Because the aluminum disk sits in the middle of the ring magnet, as seen from the Figure 11, the ring magnet produces a magnetic sensing line perpendicular to the aluminum disk. The magnetic sensing lines are dense in the center and sparse away from the center. The horizontal movement of the aluminum disk will cause a change in magnetic flux, resulting in a damping force, which prevents the aluminum plate from swinging horizontally.

3.3. Analysis of the open loop system

The function representation is used to deal with the equation of motion of the torsion pendulum system. Thus, the input torque is represented by $f_i(t)$, and the corresponding output torsion angle is represented by $\theta_o(t)$ [10].

According to the motion Equation (1) of the torsion pendulum system under the action of micro forces, we can see that its mathematical model is a second-order linear differential equation with constant coefficients:

$$J\frac{d^2}{dt^2}\theta_o(t)+\beta\frac{d}{dt}\theta_o(t)+\Gamma\theta_o(t)=f_i(t) \tag{6}$$

The block diagram of the transfer function of the open-loop system is shown as Figure 12 :

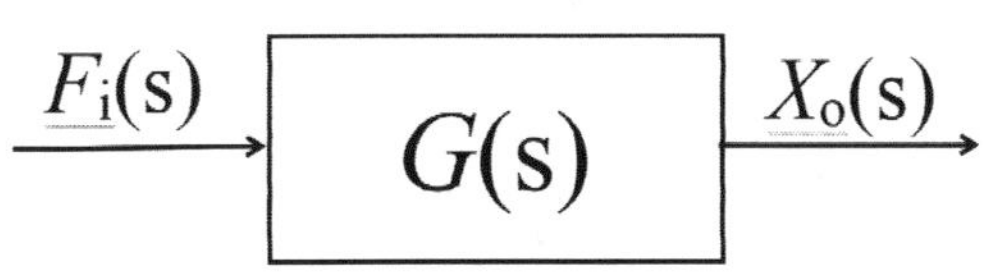

Figure 12. Block of torsion pendulum transfer

Where $X_o(s)$ and $F_i(s)$ are the Laplace transform of output $\theta_o(t)$ and input $F_i(t)$, respectively. Then the Laplace transform of this equation of motion is:

$$Js^2X_o(s) + \beta sX_o(s) + \Gamma X_o(s) = F_i(s) \tag{7}$$

By comparing the output function Xo(s) with the input function Fi(s), the transfer function of the torsion pendulum system can be written:

$$G(s) = \frac{X_o(s)}{F_i(s)} = \frac{1}{Js^2+\beta s+\Gamma} \tag{8}$$

In the torsion pendulum system, the moment of inertia $J = 3.71kg \cdot m^2$ and the torsional stiffness coefficient $\Gamma = 0.013536N \cdot m / rad$ are substituted into the transfer function. The system has slight damping, and the damping coefficient is set as 0.02. The Rous array is shown as Table 1:

Table 1. List of Rous array of the torsion pendulum system

parameter	numeric value	
S2	3.71	0.0135364
S1	0.02	0
S0	0.0135364	0

It can be seen from the table that the coefficients in the first column of the Rous array are all positive. The system is stable according to Rous's criterion.

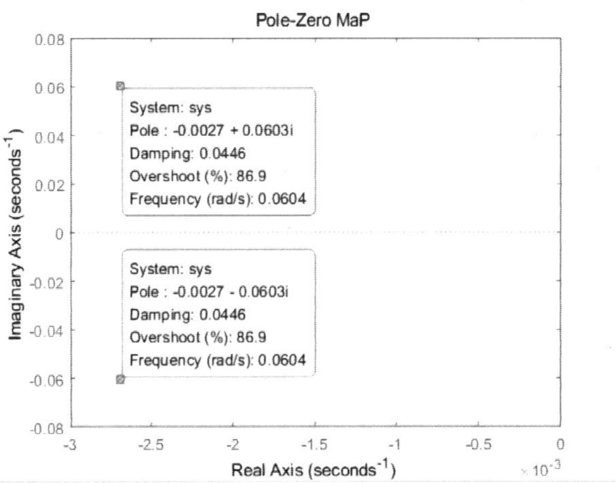

Figure 13. Pole-Zero Map of the torsion pendulum system

According to the stability of the open-loop system, the real parts of the zero poles of the characteristic polynomial are less than zero, so the open-loop system is stable. It can be seen from the Figure 13 that the transfer function has only two poles, and the real part is less than zero, so the system is stable.

Through the stability analysis of the torsion pendulum open-loop system, the stability of the system is supported by the theoretical data and the results of the chart.

4. Experiments and Calibration

4.1. Calibration of micro force generating device

The microforce-generating device needs to be calibrated. High-precision Mettler balance can be used to calibrate it.

Mettler precision balance is a commonly used micro force calibration and detection device [11,12]. The balance model is Mettler: XPR5045 (Figure 14), measuring range; 510g, resolution: 0.1mg. The instrument needs to be zeroed before use. And then, we measure the mass of the weight. According to the local gravitational acceleration g=9.8015m/s^2 in Beijing, the value of the applied microforce can be obtained.

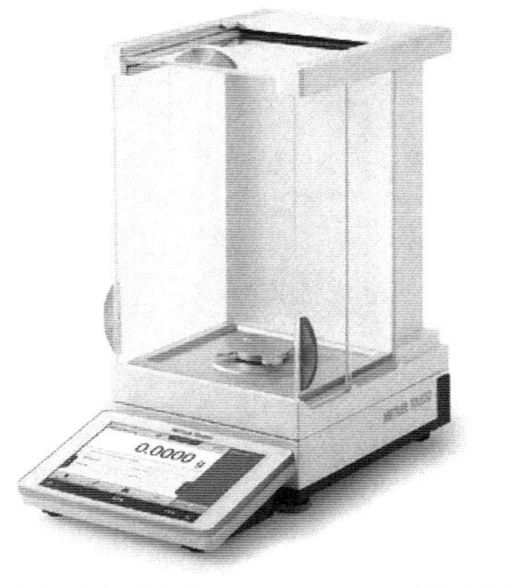

Figure 14. Mettler Precision Balance

The actual mass of the used weight obtained through calibration is shown in the Table 2:

Table 2. Weight calibration data sheet

Serial Number	Standard Mass /mg	Actual Mass /mg
1	100	97.8
2	150	147.5
3	200	201.4
4	250	249.6
5	300	297.7
6	500	499.6

4.2. Calibration of Torsion Pendulum

The torsional stiffness coefficient of the torsion pendulum system was calibrated based on the gravity method. The corresponding relation between torsion angle and micro thrust can be obtained. Thus, the conversion between the torsion displacement of the torsion pendulum and micro thrust is realized [13].

First, the torsion pendulum is adjusted to the horizontal posture, so the torsion pendulum is only subjected to the horizontal force. At the same time, because the weight is too light, the friction resistance of the rope and pulley will be too obvious. Therefore, a larger weight is selected for calibration measurement. The weights of 100mg, 150mg, 200mg, 250mg, 300mg, and 500mg were selected for the test. The corresponding tension forces of weights are 0.981mN, 1.4715mN, 1.962mN, 2.4525mN, 2.943mN, and 4.905mN, respectively. The experimental model is shown as Figure 15:

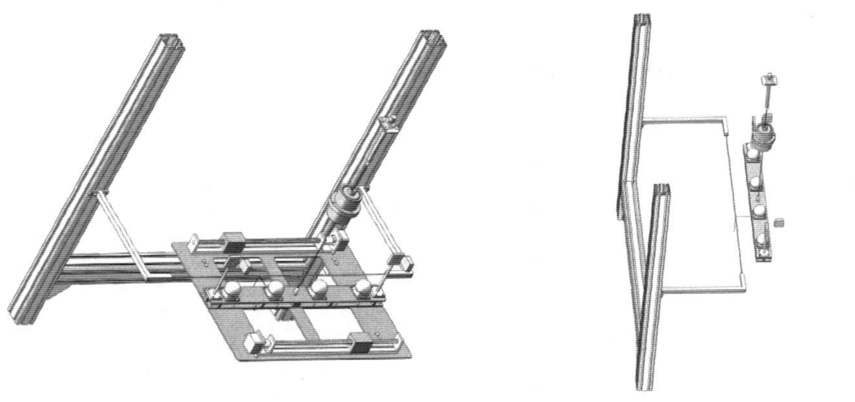

Figure 15. Model diagram of gravity calibration

Because the weight mass is relatively heavy, the corresponding tension is larger. Therefore, the torsional thrust arm was shortened to reduce the torsional moment. The string 85mm is hung from the center. Its position is shown in the Figure 16:

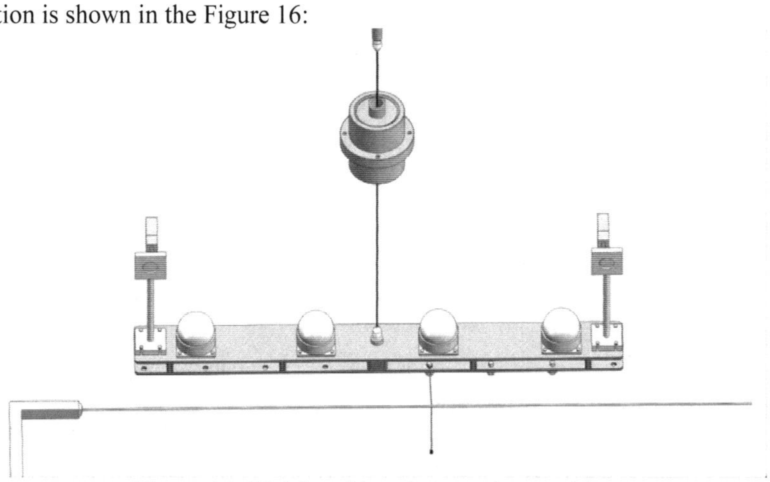

Figure 16. Weight with string traction position

The test results are shown as Figure 17:

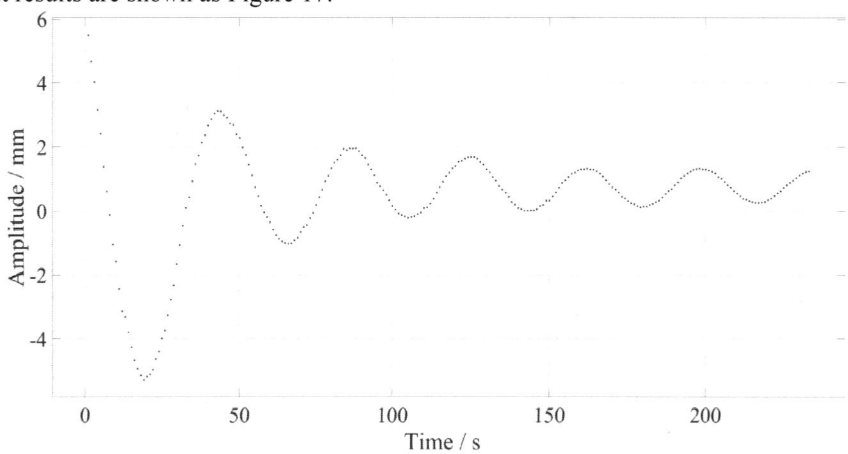

Figure 17. Calibration data fitting curve

By analyzing the experimental data, the calculated torsional stiffness coefficient of the torsion pendulum system is $\Gamma = 0.014105 N \cdot m / rad$. The theoretical value of the calibrated torsional stiffness coefficient of the torsion pendulum system is $\Gamma_t = 0.013536 N \cdot m / rad$. The difference between the actual value and theoretical value is 4.03%.

According to the torsional stiffness coefficient and thrust moment arm, the relationship between micro thrust and torsion angle (Figure 18) can be calculated as follows:

$$F = \frac{\Gamma \cdot \Delta\theta}{a} = 0.0352625 \cdot \Delta\theta \tag{9}$$

Where, F is the force exerted by the torsion pendulum, $\Delta\theta$ is the torsion angle, and a is the size of the thrust moment arm.

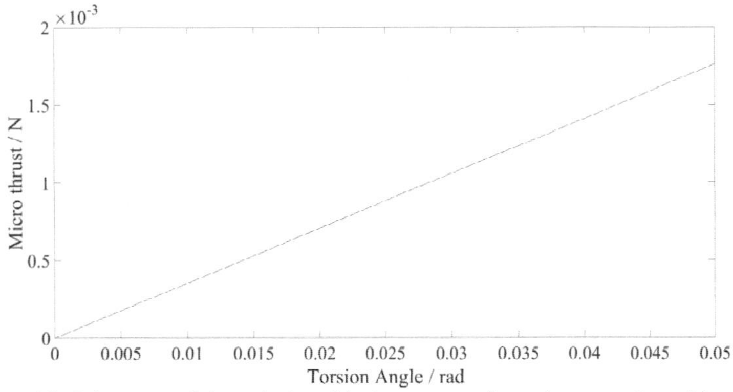

Figure 18. Diagram of the relationship between micro thrust and torsion angle

5. Error analysis

5.1. Temperature influence

When the torsion pendulum measures the micro thrust value, the main error comes from the temperature difference. This has a fundamental influence on the measurement accuracy of the torsion pendulum in microforce measurement. When measuring the micro thrust, the influence of temperature is mainly reflected in the change of torsional stiffness of the fiber. Because of the large bearing capacity of twisted fiber, the influence of a large load on the stiffness of twisted fiber should be considered. Torsion fiber stiffness is:

$$\Gamma = G\frac{\pi \cdot r^4}{2l} + \frac{Mgr^2}{2l} \tag{10}$$

Where G is the shear modulus of the twisted fiber, M is the bearing mass of the twisted fiber, r is the radius of the twisted fiber, and l is the distance between the reflector and the twisted fiber. The shear modulus G of twisted fiber can be expressed as:

$$G = \frac{E}{2(1+\mu)} \tag{11}$$

Where μ is Poisson's ratio, and E is twisted Young's elastic modulus. At this time, the torsion angle can be expressed as:

$$\Delta\theta = \frac{4Fbl(1+\mu)}{\pi r^4 \cdot E + 2Mgr^2(1+\mu)} \tag{12}$$

The elastic modulus of a body is related to temperature and other factors. The secondary torsion fiber diameter is 0.8mm. Under the influence of the thermal effect, the torsional stiffness coefficient of the torsion pendulum changes when the micropulse force is applied to it. Therefore, it is necessary to strictly control the temperature of the torsion pendulum experiment environment to 25.0°C. Moreover, the temperature difference of the experimental environment should be ensured within ±1.0°C to weaken the influence of the thermal effect on measurement results [14]. In addition, thermal noise interferes with the torsion pendulum system of this set of micro force measurements as a high-frequency signal [15]. The interference of thermal noise can be eliminated by installing a low-pass filter on the laser measuring controller.

5.2. Installation error

Torsion pendulum system parts have form error and fit tolerance. Laser sensor installation may also have deviations, resulting in measurement errors. The measuring range of the torsion pendulum is 1μN~1mN, and the measuring accuracy is high. Therefore, the installation requirements of the torsion pendulum for micro force measurement are higher. The torsion fiber should be located in the center of the scale when the torsion pendulum is installed. Otherwise, the center of gravity will be offset and cannot accurately balance the weight. Thus, the precise measurement of micro thrust value is affected.

By ensuring high machining accuracy of parts, the installation error is reduced. It is ensured that the surface roughness is 1.6μm and the symmetry is good. At the same time, the verticality deviation of the torsion pendulum is less than $\pm 2\times 10^{-4}$rad should be ensured so that the installation error is within the allowable range [16].

5.3. Measurement instrument error

The measurement deviation is mainly caused by the measurement accuracy and installation accuracy of the laser displacement sensor. The accuracy of the laser displacement sensor ensures the measuring range and resolution, the measuring fault tolerance Angle, and the tolerance focal length. The accuracy of the measurement results is guaranteed.

The micro-force measurement should ensure that the surface roughness of the mirror is not greater than 1.0μm [16]. The verticality deviation between the laser displacement sensor and the mirror is less than $\pm 1.47\times 10^{-3}$rad.

According to the general matrix error relation, the error caused by the position deviation of the laser displacement sensor and reflector in the installation process is introduced [17], according to the following system of linear equations:

$$Ax = b \tag{13}$$

Where, A is the coefficient matrix, x is the solution vector, and b is the right-end vector. It is reflected in the matrix A in the formula:

$$||\delta x|| \leq \text{cond}(A)\frac{||\delta A||}{||A||}||x|| \tag{14}$$

Where δx is the error solution vector, and δA is the error coefficient matrix. When the installation error of the laser displacement sensor with a resolution of 10μm is 100μm, the calculation accuracy is ±30μm; that is, the torsion angle deviation is $\pm 6.5\times 10^{-7}$rad. When the torsion force equation is used, it is found that the deviation is 10^{-7}N, which is less than the minimum resolution of the torsion pendulum system.

6. Conclusion

In this paper, the ground test and evaluation of the high microgravity science experiment cabinet in the core module of the Space station are carried out. A torsion pendulum device for micro force measurement with double laser detection is designed. The measuring principle and structure of the device are introduced. The standard thrust generator is calibrated first, and then the torsion pendulum system is calibrated. The main error sources are analyzed. Finally, the following conclusions are drawn:

(1) Use small weights to generate standard micro forces. A high-precision Mettler balance was used to calibrate and verify the actual weight. The value of the tiny standard force is obtained indirectly. Precision balance for force calibration can be used as a standard force measurement method. It is also suitable for testing and calibration of other calibration structures such as electrostatic comb.

(2) Choose high-purity tungsten wire with high stiffness as suspension fiber. The torsional stiffness coefficient is 0.014105N·m/rad, and the micro-force coefficient of the torsional pendulum system is 0.0352625. The high stiffness of tungsten wire is shown to be suitable for torsion tests with a large load and small angle. In addition, with the development of material science, non-metallic fibers with higher stiffness can be considered substitutes for micro-force torsion pendulum suspension fibers. It can get rid of the limitations of stiffness and strength of existing materials and further improve the accuracy of measurement.

(3) The torsion pendulum system uses high precision and large-range laser displacement sensor. The laser displacement sensor has a range of 25 mm and a resolution of 1μm. The measuring positions were distributed at both ends of the beam, which had the effect of microforce amplification. Combined with torsional stiffness, the range of force measurement can be realized, including 1μN~1mN. The thrust measurement requirements for the ground test and evaluation of the suspension test platform are fully satisfied.

References

[1] Epic. Doing experiments in the sky and appearing in "Tiangong Class" is a key role [N]. Science and Technology Daily,2022-03-31(005).

[2] Wanlong Liu, Haowei Zhu, Shujiang Sun, et al. Development status of domestic micro thrust test technology [J]. Journal of Rocket Propulsion, 2015, 41(05): 7-11.

[3] Xiong Mo. Research on Micro-Ox Thrust Measurement Technology [D]. University of Electronic Science and Technology of China,2019.

[4] Haonan Zhang, Buren Duan, Lizhi Wu, Development of a steady-state micro thrust measurement stand for micro spacecraft, Measurement, Volume 178, 2021.

[5] Daren Yu, Xiang Niu, Taibu Wang, et al. Research progress of micro-propulsion technology for space gravitational wave detection mission [J]. Journal of Sun Yat-sen University (Natural Science Edition),2021.

[6] Yanji Hong, Xing Jin, et al. Micro Thrust and Micro Impulse Measurement Method [M]. Beijing: National Defense Industry Press, 2014.

[7] Hao Wu, Fujun Zhang, Zhenyu Zhang, et al. A Simulation Study of Static Electromagnetic Characteristics of Voice Coil Motor Injector, IFAC-Papers Online, Volume 54, Issue 10, 2021.

[8] Wang Dapeng, Jin Xing, Zhou Weijing, et al. Generation and Characteristic Analysis of High-precision Electromagnetic Calibration Force [J]. Chinese Journal of Scientific Instrument, 2017.

[9] Physics-Applied Physics; Reports from Shanghai University Describe Recent Advances in Applied Physics (Optimum Design of an Eddy Current Damper Considering the Magnetic Congregation Effect) [J]. Journal of Physics Research, 2020.

[10] Qingqing Wang, Junwe Jiai, Yujing Wu, et al. Theoretical Design and Simulation Analysis of Sensing Unit of Torsion Pendulum Thrust Rack [J]. Measurement Technology,2022.

[11] Karadag B, Cho S, Funaki I. Precision balance for sub-millinewton resolution direct thrust measurement. Rev Sci Instrum. 2018.

[12] Yanji Hong, Xiaohui Feng, Haichao Cui, Sibo Wang. Thrust Uncertainty Calibration Method for micro thrust Measurement System [J]. Propulsion Technology,2022.

[13] Chao Yang, Jianwu He, Qi Kang, et al. Design and experimental study of submicron Bovine thrust measurement system [J]. Chinese Optics, 2019.

[14] S M Merkowitz, P G Maghami, A Sharma, et al. A μNewton thrust stands for LISA[J]. Classical and Quantum Gravity, 2002, 19(7): 1745-1750.

[15] Chao Yang, Jianwu He, Chu Zhang, et al. Micro Impulse Measurement Method, and Experimental Research Based on Torsion Pendulum [J]. Chinese Journal of Theoretical and Applied Mechanics, 2022.

[16] Weiyi Wang. Research on Thrust Measurement Technology of Milli union-class ion Thruster [D]. University of Electronic Science and Technology of China,2021.

[17] Yueqiao Ma, Zongfeng Li. (2021). Test mass poses measurement using the laser below the sensor. The 229-235. 10.1109 / CMMNO53328.2021.9467611.

Semiconductor Bridge Temperature Change Study under Different Excitation Methods and Sizes

Nianrong HUANG[1], Feng Feng[2], Jianing DUAN [3], Bingdong TAN[4]

[1]School of Mechanical Engineering, Nanjing University of Science and Technology, 210094, Nanjing, People's Republic of China.

[2]School of Mechanical Engineering, Nanjing University of Science and Technology, 210094, Nanjing, People's Republic of China.

[3]School of Mechanical Engineering, Nanjing University of Science and Technology, 210094, Nanjing, People's Republic of China.

[4]Sichuan Navigation System Engineering Research Institute, 610100, Sichuan, People's Republic of China.

e-mail: 15838176881@163.com

Abstract. The temperature of the semiconductor bridge increases due to the Joule heat effect in the way of capacitive discharge voltage excitation and current excitation, but the temperature distribution and the effect of different excitation methods and different sizes on the temperature change are not very clear; aiming at this problem, the temperature distribution of the semiconductor bridge after electrical energy excitation is simulated by finite element simulation. The effects of excitation mode and bridge area size on the temperature rise process of the bridge were studied by the control variable method; finally, the high-temperature area of the bridge body is only distributed in the bridge area, the temperature of the bridge body is positively correlated with electrical energy but negatively correlated with the area of the bridge area, and the current excitation method is better than the effect of voltage excitation. In addition, the conclusion is that the appropriate increase in excitation energy and the reduction of bridge size can help the temperature rise to improve the efficiency of the semiconductor bridge in practical applications.

1. Introduction

The concept of semiconductor bridge fireworks originated in the 1960s, but it was not until the mid-1980s that it was really entered into people's vision and began to be valued by countries around the world until the Sandri Laboratories in the United States conducted in-depth research and patented in 1987 [1–3]. Semiconductor bridge pyrotechnic products are a new type of fireworks that use microelectronics manufacturing technology to use a thin heavy doped silicon bridge on a silicon substrate or sapphire substrate as the ignition and detonation unit [4].

It is characterized by its small size, low energy, and rapid response, but also has good anti-static safety performance, so it can be used in extremely demanding launch systems and detonation systems [5]. The principle of semiconductor bridge ignition is to use electric energy to make the silicon bridge rapidly undergo multiphase changes in solid, liquid gas, and plasma states in a short period of time to produce high-temperature plasma and ignite the gunpowder through plasma [6,7]. There are two ways to generate plasma by rapidly heating up the electrokinetic semiconductor bridge, namely current excitation and voltage excitation suitable for capacitor discharge. However, the temperature distribution of the bridge body is not very clear, and there are few studies on the difference between the two ways of excitation at home and abroad. Therefore, this paper intends to use simulation software to simulate and analyze the process of heating up the bridge body under the two excitation methods, and study the influence of different sizes, voltages, and currents on the heating of the bridge body.

2. Model building with basic assumptions

A semiconductor bridge consists of a substrate, an insulating layer, a semiconductor material layer, and an electrode layer. The substrate is mainly used to support other components and can also play a role in dissipating heat to prevent accidental fire in the bridge, which is usually composed of sapphire or monocrystalline silicon material [1]. The material of the insulating layer is silicon dioxide, which prevents the additional current of the material layer from flowing to the base layer [8]. The semiconductor material layer is generally monocrystalline silicon, polycrystalline silicon, or other semiconductor materials with negative temperature resistance characteristics, which is the most important part of the semiconductor bridge structure [1,8]. The electrode layer is usually made of aluminum. Based on the above analysis of the semiconductor bridge materials, the model diagram shown in Fig 1 is established by the 3D modeling software.

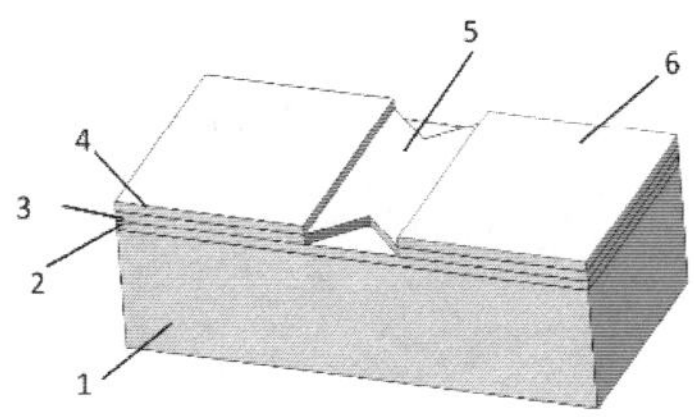

Fig 1. 3D model diagram of the semiconductor bridge

Notes: 1-base; 2-silica layer; 3-semiconductor material layer; 4- electrode; 5 -Bridge District; 6-electrode

Since polysilicon usually vaporizes at a temperature of about 3000 to produce plasma and detonate gunpowder, only the temperature change of the bridge body before vaporization is simulated in the actual simulation. The following assumptions are made before the finite element simulation:

- The material of respective materials of the internal structure of the semiconductor bridge is uniform, and the physical parameters such as the specific heat capacity, density, and thermal conductivity of the respective are fixed.
- Regardless of the heat dissipation of the semiconductor bridge during the heating process.
- The temperature of each surface of the model is 300 K room temperature.

3. Numerical simulation

According to the electric heating effect, the heating process of the semiconductor bridge is an electrothermal conversion process, which follows the Joule heat law:

$$Q = J \cdot E \tag{1}$$

formula: Q is the Joule heat generated (W/m^3), J is the current density (A/m^2), and E is the electric field strength (V/m).

Fig 2 is a cloud map of the current density distribution of the semiconductor bridge during the heating process of electric energy excitation, and it can be seen from the figure that the current density at the core area of the bridge body is higher than that of other parts, especially at the V angle. This is mainly because the current flowing from the edge of the positive electrode to the negative electrode must pass through the bridge area, and at the V-angle, the current per unit area is largely due to the small area, resulting in the maximum current density at the V-angle.

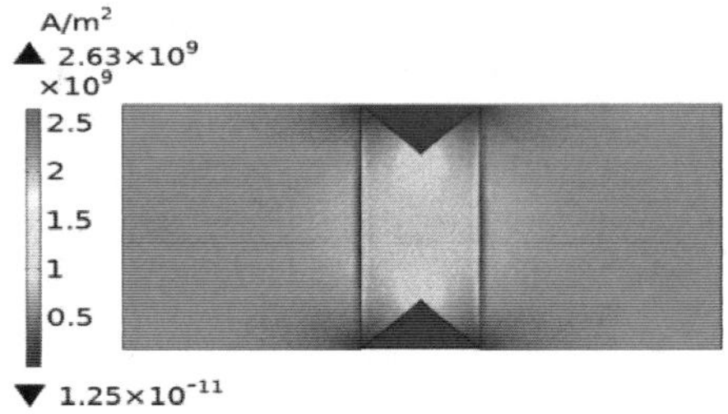

Fig 2. Semiconductor bridge current density distribution cloud map

Fig 3 shows the electric field strength distribution cloud map of the semiconductor bridge, which shows that there is a large electric field strength at the edge of the two angles in the bridge area. At the V-angle, there is only a certain electric field strength. According to the Joule Heat Law (1), the Joule heat generated after the bridge body is energized is determined by the product of the current density and electric field strength, so only the bridge area can meet the conditions for the generation of Joule heat.

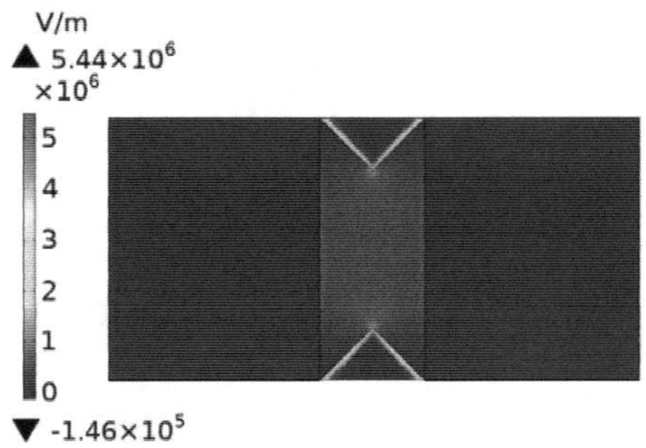

Fig 3. Semiconductor bridge electric field intensity distribution cloud map

Fig 4 shows a cloud plot of the temperature distribution after the semiconductor bridge is powered up. As can be seen from the figure, the bridge body temperature is highest at the V-shaped sharp angle and diffuses around the sharp corner in the bridge area, which is consistent with the results of analyzing the current density and electric field strength cloud map.

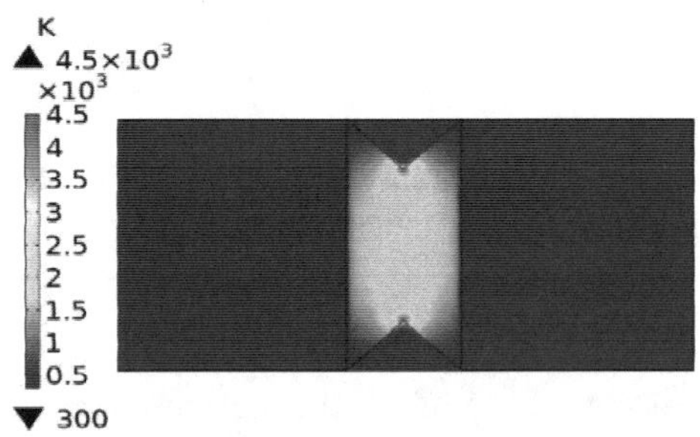

Fig 4. Cloud map of semiconductor bridge temperature distribution

The straight-line section with the tip of the two V angles is used as the monitoring line to study the change of the temperature of the bridge area along the width direction, and the obtained curve is shown in Fig 5. As can be seen from the figure, the curved trend of temperature change along the width direction is consistent with the temperature cloud map shown in Fig 4. The entire temperature curve is V-shaped, with the temperature highest at the lower tip, it begins slowly downward at the lowest in the middle, then rises and is highest at the upper tip.

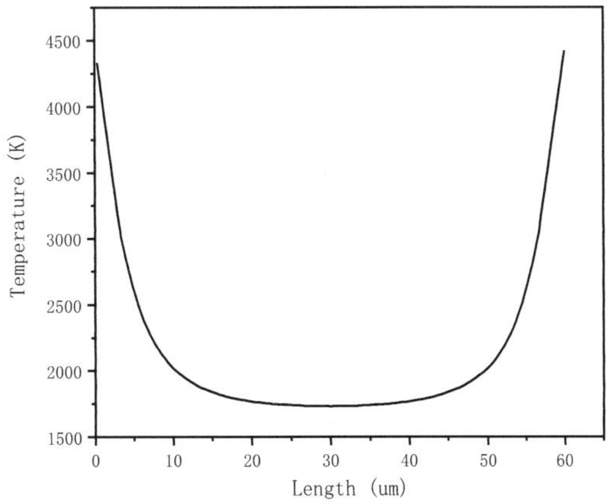

Fig 5. The temperature distribution curve of the bridge body along the width direction

In the semiconductor material layer, a vertical line is taken from the V angle to make a temperature monitoring line in the altitude direction, and the result is shown in Fig 6. As can be seen from the figure, the outer surface temperature of the bridge area is the highest, and the temperature is getting lower and lower as it goes down. This is mainly caused by the gradual decrease in current density and electric field strength in the area from the outer surface to the inner surface.

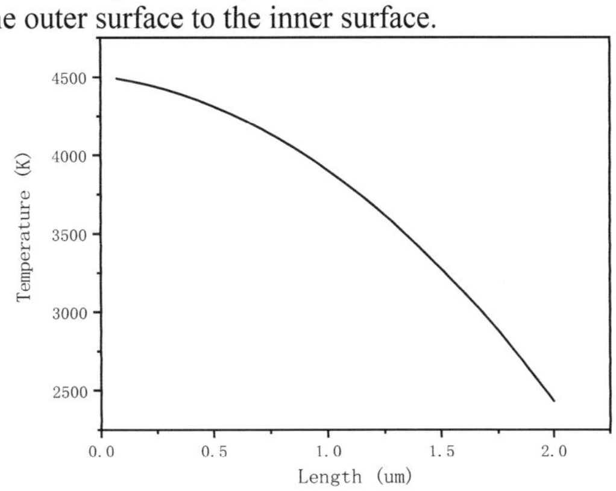

Fig 6. The temperature distribution curve of the bridge body along the height direction

4. Analysis of temperature changes of bridge bodies under different excitation energy and sizes

According to the results of the analysis study in the previous section, under the stimulation of a certain amount of electrical energy, the semiconductor bridge will heat up rapidly under the action of Joule Heat's law to reach the temperature of plasma ignition. This section will study the temperature change of semiconductor bridges under different excitation electrical energy and multiple sizes.

4.1. Simulation of fire and heating up of the bridge body under voltage excitation

Under the method of capacitor discharge voltage excitation, the influence of different excitation voltages and different bridge zone sizes on the temperature change of the bridge body is simulated and analyzed by using the control variable method. The operating conditions are shown in Table 1 and Table 2.

Table 1. Condition 1

Bridge body size /(um)	Bridge area size /(um)	Voltage/(v)
		18
100 x 200	100 x 40	20
		21

Table 2. Condition 2

Voltage/(v)	Bridge body size /(um)	Bridge area size /(um)
	100 x 200	100 x 40
20	110 x 250	110 x 45
	120 x 260	120 x 50

According to the above set working conditions, when the bridge size is fixed, the temperature change of the semiconductor bridge under different excitation voltages is simulated, and the obtained results are shown in Fig 7; when the voltage is fixed, the temperature change under the condition of changing the size of the bridge is shown in Fig 8.

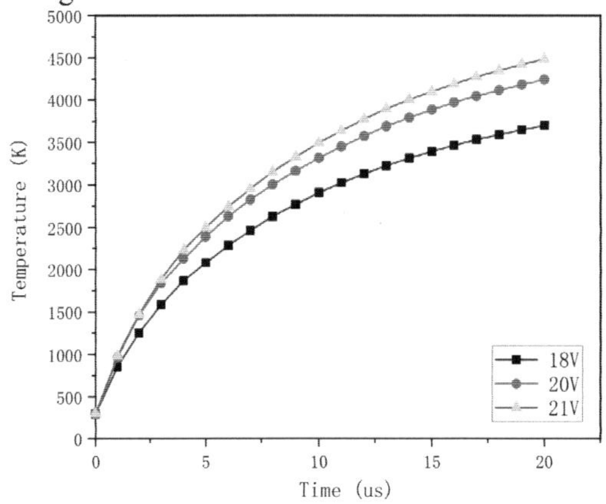

Fig 7. The temperature change curve of the bridge body under different voltages

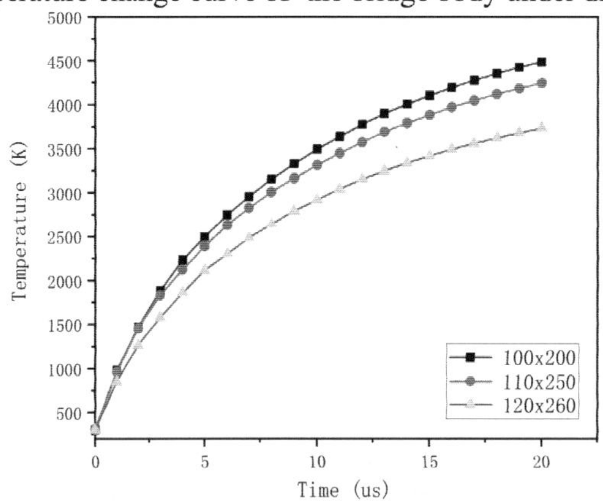

Fig 8. The temperature change curve of the bridge body under different sizes

As can be seen from Fig 8, the bridge temperature increases with the increase of the excitation voltage. Because a larger voltage can provide more current subs, both the current density and the electric field strength are improved, and combined with Joule Heat's law, the temperature will also increase. According to Fig8, when the excitation voltage is unchanged, the increase in the bridge area leads to a decrease in both the current density and the electric field, resulting in a slower temperature increase and a lower temperature value, making it difficult to achieve the temperature required for ignition.

4.2. Simulation of fire and heating up of the bridge body under current excitation

This section examines the law of temperature change of the bridge body under different input currents and different bridge zone sizes. The operating conditions are shown in Table 3 and Table 4.

Table 3. Condition 3

Bridge body size /(um)	Bridge area size /(um)	Current/(A)
		0.14
100 x 200	100 x 40	0.15
		0.16

Table 4. Condition 4

Current/(A)	Bridge body size /(um)	Bridge area size /(um)
	100 x 200	100 x 40
0.16	110 x 250	110 x 45
	120 x 260	120 x 50

According to the above set working conditions, when the bridge size is fixed, the temperature change of the semiconductor bridge at different currents is simulated, and the result is shown in Fig 9; when the current is fixed, the temperature change after changing the condition of the bridge area size is shown in Fig 10.

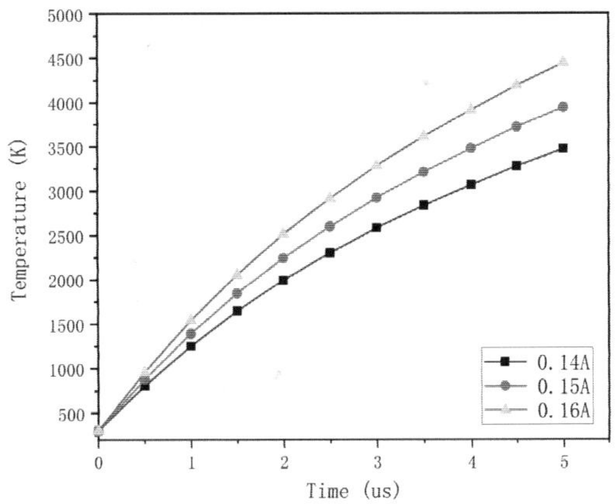

Fig 9. The temperature change curve of the bridge body under different currents

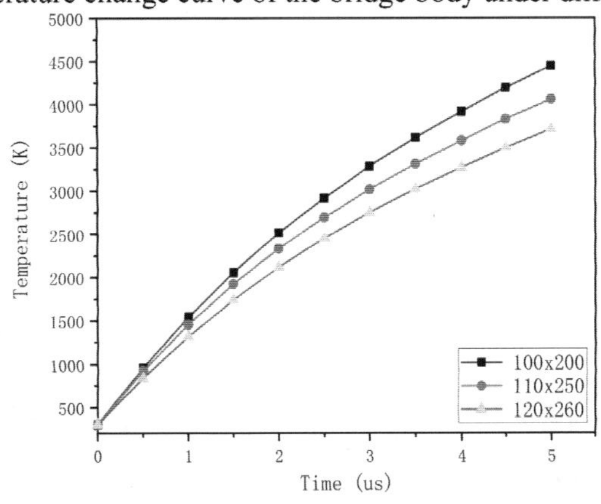

Fig10. The temperature change curve of the bridge body under different sizes

From Fig 10, the bridge body temperature is positively correlated with the magnitude of the current. According to the calculation formula of the current density ($J = I/S$, J is the current density; I is the current;

S is the area of the current flowing through). Whether it is increasing the current or reducing the area of the bridge area, the current density can be greatly improved, and the electric field strength will also increase a little, resulting in a more obvious electric heating effect. This explains the positive correlation between bridge temperature and the magnitude of the current, which also makes a reasonable explanation of the negative correlation between the bridge temperature change and the bridge area shown in Fig10.

5. Conclusions

The temperature of the semiconductor bridge rises due to the Joule heating effect under the mode of capacitive discharge voltage excitation and current excitation, but the temperature distribution and the effect of different excitation methods and different sizes on temperature changes are not very clear. Based on this phenomenon, it is extremely important to discuss the law of temperature distribution of the bridge body and the relationship between temperature and electric energy, the mode of electrical energy excitation, and the size of the bridge body during the energization process of the semiconductor bridge. Therefore, in this paper, the temperature change of a V-shaped semiconductor bridge after power-up is simulated, and the results show that the high-temperature region of the bridge body is mainly distributed in the bridge area, and the temperature is highest at the V angle at both ends. In addition, the comparative analysis of the temperature change of semiconductor bridges under different excitation methods and multiple sizes is made, and the main conclusions are as follows:

- Under the excitation mode of the two electrical energies, the increased rate of the temperature in the bridge area is positively correlated with the size of the excitation energy.
- The rate of temperature increase in the bridge area is negatively correlated with the area.
- No matter what kind of excitation method is used to input energy to the semiconductor bridge, the higher the input electrical energy, the higher the temperature of the bridge body.
- When the bridge temperature is met to meet the ignition conditions, the effect is better by current excitation.

According to the simulation results, under the premise that the V-type semiconductor bridge meets the safety of ignition, the appropriate input of large excitation energy to the bridge body and the reduction of the size of the bridge body help to improve its heating change, which provides an effective reference value for the practical application of semiconductor bridge fireworks. In the future, it is hoped that more scholars can study the influence of different sizes of V angles and other bridge shapes on temperature changes during energization.

References

[1] Benson, D.A., Renlund, A. M. (1987) Semiconductor bridge: A plasma generator for the ignition of explosives[J]. Journal of Applied Physics, 62(5):1622-1632.
[2] Bickes, R.W., Schwarz, A.C. (1987) Semiconductor bridge (SCB) igniter: US, US4708060 A[P].
[3] Zhu, F.C., Xu Z.X., Zhou, B. (2003) Research and Development of SCB Pyrotechnics[J]. Blasting Equipment, 01: 18-23.
[4] Hu, J.S., Jiao, Q.J. (2006) A preliminary study on the electrothermal properties of semiconductor bridges[J]. Engineering Blasting Science, 01: 90-91+47.
[5] Ren, W., Zhou, Z., Liu, J.P. (2007) Design and research of semiconductor bridge igniter[J].Pyrotechnics, 06:43-46.
[6] Liu, M.F. (2012) Research on intelligent ignition and internal ballistic performance control of semiconductor[D]. Nanjing University of Science and Technology.
[7] Zvulun, E., Toker, G.R., Gurovich, V.T., (2014) Shockwave generation by a semiconductor bridge operation in water. IEEE International Conference on Plasma Sciences.
[8] Park, M.I., Choo, H.T., Yoon, S.H. (2004) Comparison of plasma generation behaviors between a single crystal semiconductor bridge (single-SCB) and a polysilicon semiconductor bridge (poly-SCB)[J]. Sensors & Actuators A Physical, 115(1):104-108.

Temperature Field and Thermal Stress Analysis of Rocket Engine Composite Structure Nozzles

Yunpeng Liu[1]*, Feng Feng[2], Xiaodong Shen[3] and Chao Chen[4]

[1]School of Mechanical Engineering, Nanjing University of Science and Technology, 210094, Nanjing, People's Republic of China.

[2]School of Mechanical Engineering, Nanjing University of Science and Technology, 210094, Nanjing, People's Republic of China.

[3]School of Mechanical Engineering, Nanjing University of Science and Technology, 210094, Nanjing, People's Republic of China.

[4]Anhui Shenjian Technology Co., Ltd, 230000, Hefei, People's Republic of China.

E-mail: 1007978676@qq.com

Abstract. As one of the core components of the rocket engine, the working conditions of the nozzle are very bad. Ensuring the nozzle works normally is an important prerequisite for the normal operation of the rocket engine. To ensure the safety of the solid rocket motor nozzle in use, the temperature distribution and thermal stress of the nozzle must be preliminarily analyzed. Therefore, the three-dimensional model of an L-shaped nozzle is established. The temperature field and stress field of the structure is numerically simulated by FLUENT and Static Structural. The influence of gas temperature and material thermal conductivity on temperature distribution and thermal stress of solid structures are analyzed. The results show that the throat has the highest temperature and maximum thermal stress. When the gas temperature is 2600K, the maximum temperature of the throat is 1400K, and the thermal stress is 158MPa. When the thermal conductivity of the insulation layer is 2.8 W/(m·K), the thermal stress on the throat is 170 MPa. The conclusion can provide an important reference for the thermal protection design of the nozzle.

1. Introduction

The nozzle design is directly related to the performance of the whole solid rocket motor [1]. The nozzle is subject to the erosion of high-temperature and high-pressure gas during working, and its structural strength decreases with the increase in temperature. The resulting thermal protection problems bring hidden dangers to the safety of the rocket engine. Composite ablation-resistant materials are often used as thermal protection materials for nozzles in engineering [2]. Carbon/phenolic materials, as the representative of them, have the characteristics of good ablation resistance and low density. It is widely used in the ablation-resistant layer of the nozzle. In addition, the protective layer of the nozzle is composed of materials with different thermophysical parameters. Composite polymer materials are used in the convergence section and expansion section, and graphite is used in the throat.

Analysis of temperature and thermal stress distribution is the main content of nozzle thermal analysis and design [3]. In this paper, the thermal structural coupling analysis of the temperature field and stress field of an L-shaped gas passage and a composite structure nozzle is carried out. At the same time, the influence of different factors on the temperature and stress distribution of the nozzle insulation layer is analyzed, which provides an important reference for the nozzle thermal protection design.

2. Thermal structure coupling model

2.1. Physical model and grid division

The structure of the L-shaped gas passage and the composite nozzle is shown in Figure 1. The throat diameter is 6.4mm, and the outlet diameter is 10mm. The calculation model is divided into two parts: the fluid domain and the solid domain. The gas flow area is fluid, and the throat insert, thermal insulation layer, and shell are solid areas. Structured grid division shall be carried out, with axial densification near the throat and radial densification near the inner wall [4]. The mesh size of the first layer at the fluid boundary is 1×10-6m. The three-dimensional structural grid of the model is shown in Figure 2.

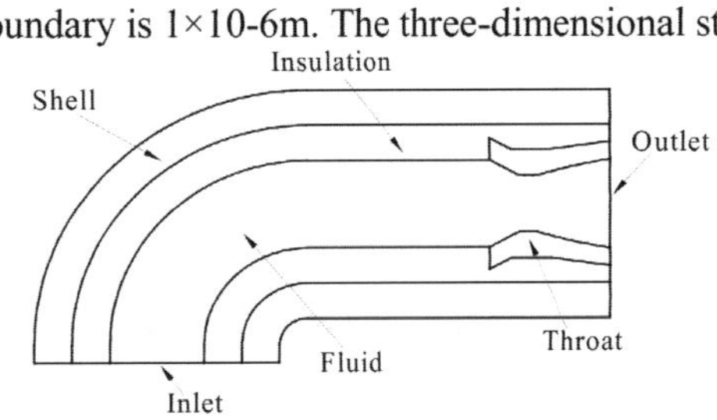

Figure 1. Composite nozzle structure diagram

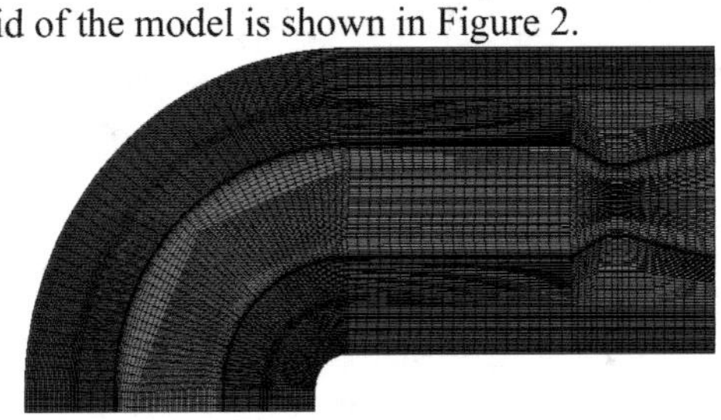

Figure 2. Nozzle grid diagram

The shell is made of structural steel, the thermal insulation layer is made of carbon/phenolic composite thermal insulation material, and the throat insert is made of graphite resistant to high-temperature scouring. Table 1 lists the physical performance parameters of various materials.

Table 1. Material Property Parameters

Parameter	Structural steel	Carbon/phenolic	Graphite
Density /($kg{\cdot}m^{-3}$)	8030	22	7
Thermal conductivity/$W{\cdot}(m{\cdot}K)^{-1}$	16.3	5	151
Specific heat capacity/$J{\cdot}(kg{\cdot}K)^{-1}$	502.8	871	710
Coefficient of linear expansion	1.2×10^{-5}	4×10^{-6}	4.8×10^{-6}
Elastic modulus/Pa	2×10^{11}	1.66×10^{10}	1.2×10^{10}
Poisson's ratio	0.3	0.33	0.425

2.2. Boundary conditions

In this paper, combining high-precision numerical simulation of the flow field with a calculation of the temperature field of solid structure, the structure of the solid region is analyzed using the fluid-solid coupled heat transfer model [5]. The inlet is a pressure inlet with a total pressure of 9MPa and a total temperature of 2500K; The outlet is a pressure outlet with a temperature of 300K and a pressure of 101325Pa.

First, the wall between fluid and solid is set as the thermal insulation surface; After the gas flow is stable, the wall conditions are changed and transient calculations are conducted. The temperature field is imported into the static analysis module for thermal stress analysis.

3. Numerical simulation Thermal stress results of composite structure nozzle

3.1. Results of composite structure temperature field

Fluent is used to conduct numerical calculations. Figure 3 shows the curve of Mach number and temperature along the central axis in the steady flow field. Then the wall conditions between the fluid domain and the solid domain are changed and transient heat transfer calculations are conducted. Figure 4 shows the radial temperature curve of the throat at different times. Figure 4 shows us that by comparing three different materials, the temperature change of the graphite throat insert is the largest. In 0.2s~1s, the temperature at the throat inserts increased by 380K. When t=0.6s, the wall temperature of the metal shell starts to rise. When t=1s, the temperature of the throat insert directly contacting with high-temperature gas is the highest, reaching 1360K, and the temperature difference between the internal and external walls of the insulation layer is about 900K. In addition, the temperature curve at the material interface shows a sharp drop. The carbon/phenolic material has low thermal conductivity and good thermal insulation performance, which makes its temperature change very obvious in the radial direction.

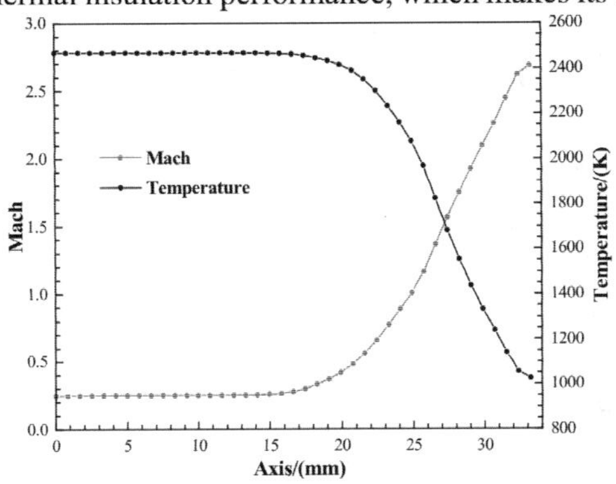

Figure 3. Flow parameter diagram on axis

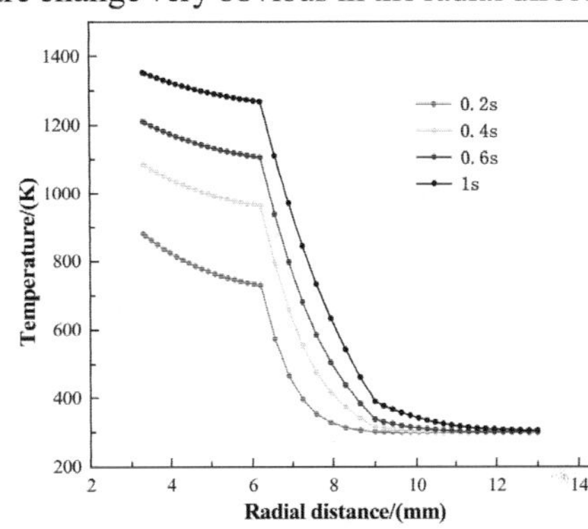

Figure 4. Radial temperature distribution of throat

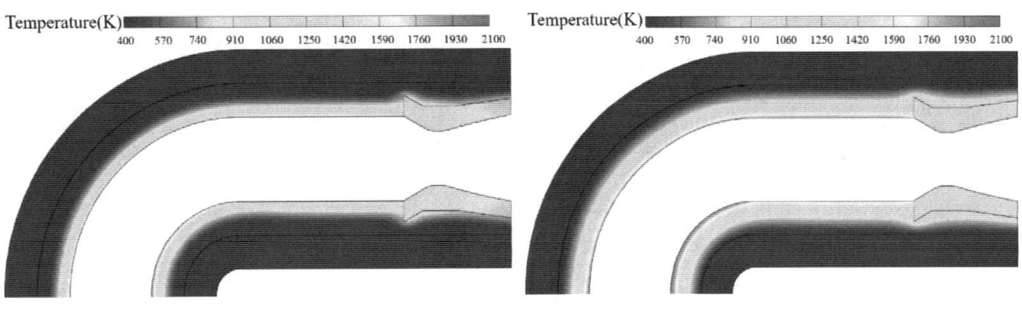

(a) t=0.4s (b) t=1s

Figure 5. Temperature distribution at different times

The temperature distribution of the channel and nozzle wall is shown in Figure 5. When t=0.4s, the temperature of the throat insert is significantly higher than that of the carbon/phenolic insulation layer due to the good thermal conductivity of graphite material. When t=1s, the temperature of the throat insert is between 1280K-1400K, the maximum temperature of the insulation layer protected by graphite is 1150K, and the temperature gradually decreases along the radial direction.

3.2. Thermal stress calculation results

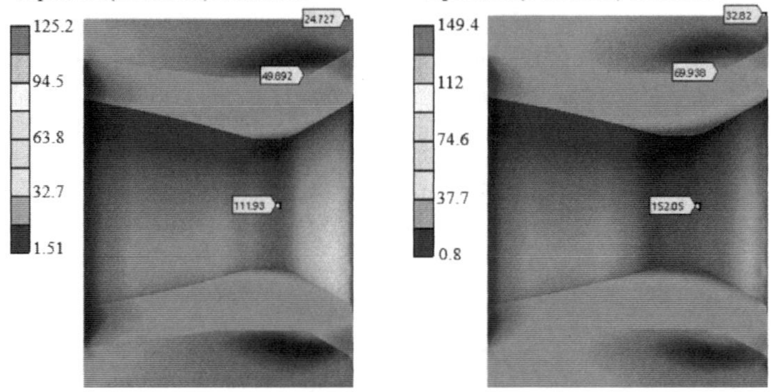

(a) t=0.4s (b) t=1s

Figure 6. Thermal stress distribution at different times

The temperature results are imported into the Static Structural for analysis. The thermal stress distribution of nozzles with different materials is mainly analyzed. Figure 6 shows the stress contours of the throat insert and insulating layer.

As shown in Figure 6, the thermal stress of the throat inserts and insulation layer increases gradually with the increase of time. When t=0.4s, the thermal stress has been generated in the throat insert and at the junction of the throat insert and the carbon/phenolic insulation layer. The maximum thermal stress of the throat insert is located near the throat, with a size of 110MPa. The temperature of the throat is the highest in the whole throat insert, and there is a large temperature gradient, so the thermal stress is the highest here. When t=1s, the maximum thermal stress of the throat reaches 152MPa.

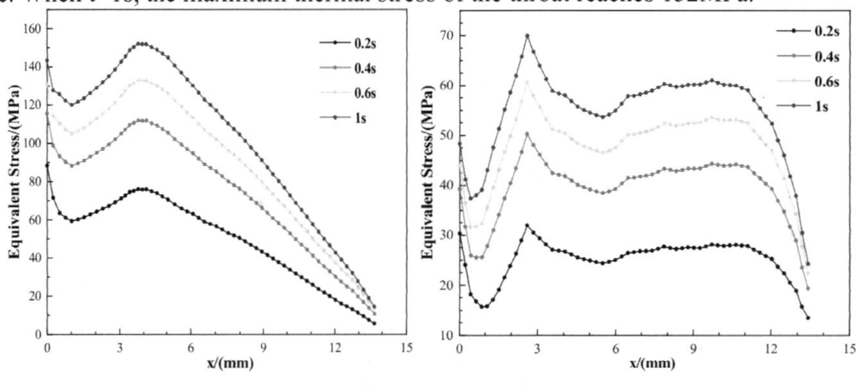

(a) Inner wall surface (b) Outer wall surface

Figure 7. Thermal stress curves of throat insert at different times

Figure 7 shows the thermal stress curve of the inner and outer wall surfaces of the throat insert along the two axial paths at different times. As shown in Figure 7, the thermal stress on the inner wall first increases and then decreases, and the stress at the front end of the throat reaches the maximum. There is an obvious stress wave peak on the outer wall, which corresponds to a large turning point of the throat insert back wall, where the surface is not rounded, and a stress concentration occurs.

4. Analysis Influence of different factors on the thermal stress of nozzle structure

4.1. Influence of gas temperature on heat transfer and thermal stress

To analyze the influence of gas temperature on the working process of the nozzle, 2200K, 2400K, and 2600K gas temperatures are set to analyze the solid temperature field and stress field. Figure 8 shows the temperature curve of the throat along the radial direction when t=1s under different gas temperatures.

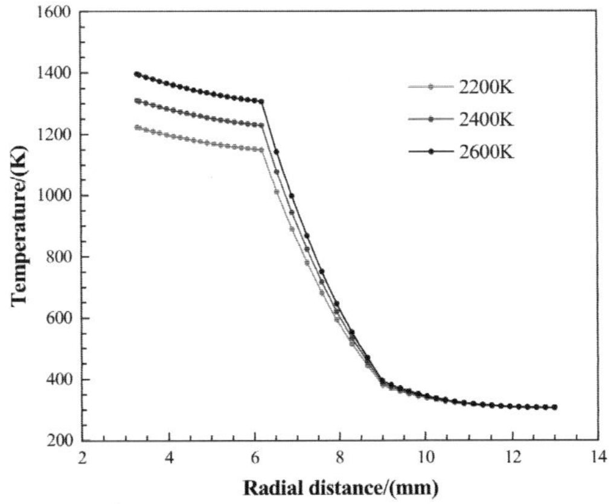

Figure 8. Radial temperature curve under different gas temperatures

As shown in Figure 8, under different gas temperatures, the radial temperature distribution in the solid area is the same. Changing the gas temperature has the greatest influence on the temperature change of the graphite throat insert. When the gas temperature is 2200K, the temperature range of the throat insert is about 1120K~1220K. When the gas temperature rises to 2600K, the maximum temperature at the throat reaches about 1400K.

The temperature fields of three conditions are imported into Static Structural for thermal stress calculation. The thermal stress curve of the inner and outer wall surfaces of the throat insert along the two axial paths is shown in Figure 9. The curve of thermal stress on the inner wall first increases and then decreases along the axial direction. With the increase of gas temperature, the peak value of thermal stress on the inner wall increases from 132 MPa to 158 MPa; At the stress concentration point of the outer wall, the thermal stress increases from 60MPa to 78MPa.

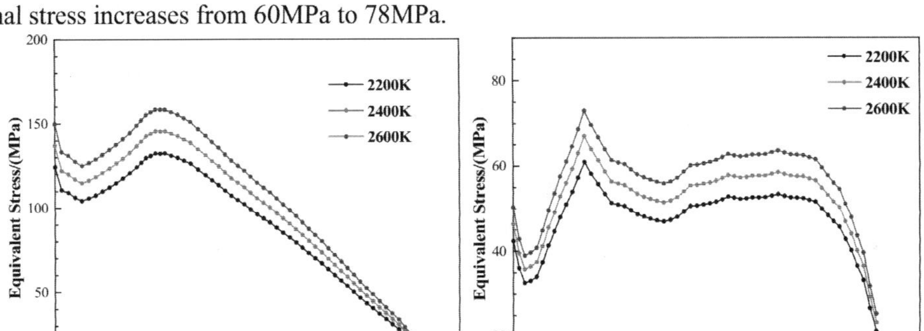

(a) Inner wall surface (b) Outer wall surface

Figure 9. Thermal stress curves at a different gas temperature

4.2. Influence of Carbon/Phenolic thermal conductivity on heat transfer and thermal stress

Figure 10 shows the temperature curve of the throat along the radial direction with different thermal conductivity of carbon/phenolic materials at 1s. The thermal conductivity range of carbon/phenolic materials is set as 2.8 W/(m·K) to 5.8 W/(m·K), with four groups.

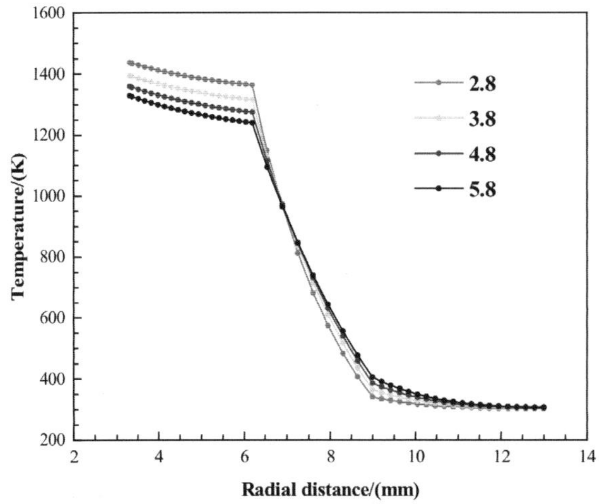

Figure 10. Temperature curve under different thermal conductivity of insulation layer

As shown in Figure 10, before the radial distance is 7mm, the larger the thermal conductivity is, the lower the radial temperature is; After the radial distance is 7mm, the situation is the opposite. This is because the lower the thermal conductivity is, the better the thermal insulation performance is, and the faster the temperature drops.

According to temperature field data, the thermal stress analysis is carried out by Static Structural. The thermal stress curves on the inner and outer wall surfaces of the throat insert under different thermal conductivity of carbon/phenolic materials are obtained, as shown in Figure 11.

The distribution of thermal stress on the inner and outer wall surfaces of the throat insert is the same under different thermal conductivity. With the decrease in thermal conductivity of the insulation layer, the thermal stress on the inner and outer walls increases gradually. When t=1s, the thermal stress at the throat insert on the inner wall increases from 145MPa to 169MPa as the thermal conductivity decreases; At the stress concentration point of the outer wall, the thermal stress increases from 67MPa to 81MPa.

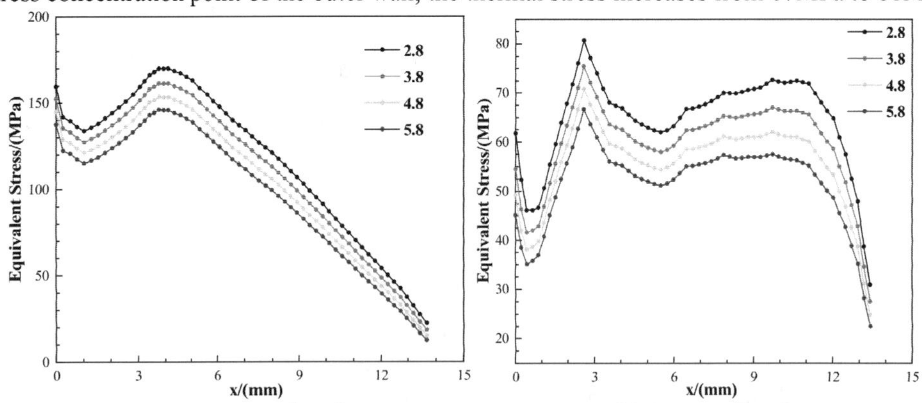

(a) Inner wall surface (b) Outer wall surface

Figure 11. Thermal stress curve under different thermal conductivity of insulation layer

5. Conclusions

In this paper, the L-shaped nozzle is numerically simulated to analyze the temperature field and thermal stress distribution in the solid area, and the following conclusions are drawn:

- After the nozzle works for 1s, the temperature of the inner wall of the nozzle shell is about 390K, meeting the thermal protection requirements.

- When the gas temperature rises from 2200K to 2600K, the throat temperature reaches about 1400K, and the thermal stress increases from 132MPa to 158MPa. When the thermal conductivity of carbon/phenolic material is 2.8 W/(m·K), the temperature of the outer wall of the throat insert reaches 1360 K, and the thermal stress on the throat reaches 170 MPa.
- When designing the nozzle, the thermal conductivity of the insulation layer can be appropriately increased to make the heat transfer to the outside as soon as possible if the working time of the nozzle is short.
- The analysis conclusion of the temperature field and stress field in this paper can reduce the test cost, speed up the test process, and provide an important reference for nozzle thermal protection design.

The later research can consider the density change of the insulating layer and the generation of pyrolysis gas, and analyze the volume ablation of the insulating layer when the nozzle works.

References

[1] Zhou, C.S., Ju, Y.T., Zhu, F.Y. (2005) Design theory of rocket. M. Beijing Institute of Technology Press.
[2] Wang T.B., Xue T.S., Zhou C.S.(2012) Numerical simulation of temperature field and stress field of compound structure nozzle. J. Journal of Ballistics,24:88-91.
[3] Huang H.Y., Wang Q.(2008) Coupled numerical study of the flow field and heated structure for axisymmetric nozzles. J. Journal of Propulsion Technology,29:194-198.
[4] Zhang, X.G., Wang, C.H., Liu Y. (2011) Coupled analysis of flow field and thermal structure of solid rocket motor throat insert. J. Journal of Solid Rocket Technology, 34:579-583.
[5] Hu, H.F., Bao F.T., Wang Y.J. (2011) Numerical simulation and experimental study on flow heat structure sequential coupling of nozzle separated flow. J. Journal of Astronautics, 32:1534-1541.

Investigation of the Design Conformity of Civil Aircraft Battery System to Airworthiness Items

Shanyong Wu

Electrical Integration Department, Shanghai Aircraft Design and Research Institute, Shanghai, China

wushanyong@comac.cc

Abstract—Due to its material, function, safety, and other characteristics, there are also special requirements in terms of the design and verification of civil aircraft batteries. This paper studies the airworthiness compliance design method of the civil aircraft battery system. First, it introduces the battery technology used in mainstream civil aircraft, obtains specific design requirements by analyzing the airworthiness clauses applicable to battery systems, and then proposes the key issues that need to be considered in the design process of civil aircraft battery systems based on these design requirements. That is to say, the design of a civil aircraft battery system must fully consider the requirements of function, performance, and installation, which are defined in airworthiness regulations, and then it can be used safely on a civil airplane.

1. INTRODUCTION

At present, the mainstream civil aircraft battery system is mainly composed of batteries and their controllers. The design requirements of the battery system refer to SAE ARP 1870 [8], MIL-STD-464C [7], MIL-STD-704F [9], and DO-160G [10]. When the AC power fails, the battery can supply power for important DC electrical equipment that affects flight safety, and can also supply power for single-phase AC important electrical equipment through a converter. In addition, the battery can also provide electrical power for starting the auxiliary power unit (APU). The battery controller automatically controls the charge and discharge of the battery according to the working state of the electrical power system.

The battery plays an extremely important role as the emergency power supply equipment of the aircraft's electrical power system. At the same time, due to the battery containing chemical substances, many factors need to be considered in the design and installation of aircraft batteries. Therefore, based on the above background and reasons, it is necessary to study the design of civil aircraft battery systems and obtain a design method that meets the requirements. CCAR 25 [1], as a guiding document for the design of civil aircraft in China, puts forward requirements for the design and installation of the battery system in Clause 25.1353(b) [2][3]. This article studies the design of the battery system of civil aircraft based on this clause.

2. TYPES OF CIVIL AIRCRAFT BATTERIES

At present, the commonly used civil aircraft batteries mainly include the following types.

Lead-acid battery: a battery whose electrodes are mainly made of lead and its oxides, and whose electrolyte is a sulfuric acid solution. In the discharge state of the lead-acid battery, the main component of the positive electrode is lead dioxide, and the main component of the negative electrode is lead; in the charging state, the main component of the positive and negative electrodes is lead sulfate. Lead-acid batteries have the advantages of stable quality, high reliability, simple maintenance, and low cost of use. But at the same time, the electrolyte (liquid) of lead-acid batteries is highly corrosive, so it is not environmentally friendly, and the low energy density leads to large weight and volume and short life.

Nickel-cadmium battery: it refers to an alkaline battery that uses metal cadmium as the negative active material and nickel hydroxide as the positive active material. Nickel-cadmium batteries have the advantages of small internal resistance, high current charge and discharge, high energy density, overcharge resistance, simple maintenance, and more charge and discharge cycles. At the same time, cadmium-nickel batteries also have shortcomings such as cadmium emissions that will have a great impact on the ecological environment and serious memory effects.

Lithium battery: it is a type of battery that uses lithium metal or lithium alloy as the negative electrode material and uses a non-aqueous electrolyte solution. Lithium batteries have the advantages of high energy density, long service life, and high-power handling capacity. At the same time, lithium batteries also have disadvantages such as poor safety, prone to thermal runaway, poor high-current discharge capability, complex charge, and discharge control circuit, and high cost.

The operational performance standards of the batteries mentioned above can be referred to as RTCA/DO-293 [6]. At present, the batteries used in mainstream civil aircraft are still nickel-cadmium batteries, and some lead-acid batteries are also used. Due to the advantages of high energy density and lightweight, lithium batteries have attracted more and more attention from the civil aviation industry. With the continuous development of lithium battery technology, lithium batteries have been used in a new type of aircraft, such as the Boeing 787.

3. Requirements of Civil Aircraft Airworthiness Clauses on Battery System Design and Installation

Clause (b) of 25.1353 of CCAR 25 has special requirements for the design and installation of batteries. The details of the terms are as follows.

Storage batteries must be designed and installed as follows:

3.1.Safe cell temperatures and pressures must be maintained during any probable charging or discharging condition. No uncontrolled increase in cell temperature may result when the battery is recharged (after the previous complete discharge)-

- At maximum regulated voltage or power;
- During a flight of maximum duration; and
- Under the most adverse cooling condition likely to occur in service.

3.2.Units Compliance with Paragraph (b)(1) of this section must be shown by test unless experience with similar batteries and installations has shown that maintaining safe cell temperatures and pressures presents no problem.

3.3.No explosive or toxic gases emitted b any battery in normal operation, or as the result of any probable malfunction in the charging system or battery installation, may accumulate in hazardous quantities within the airplane.

3.4.No corrosive fluids or gases that may escape from the battery may damage surrounding airplane structures or adjacent essential equipment.

3.5.Each nickel-cadmium battery installation must have provisions to prevent any hazardous effect on the structure or essential systems that may be caused by the maximum amount of heat the battery can generate during a short circuit of the battery or individual cells.

3.6.Nickel-cadmium battery installations must have

- A system to control the charging rate of the battery automatically to prevent battery overheating;
- A battery temperature sensing and over-temperature warning system with a means for disconnecting the battery from its charging source in the event of an over-temperature condition; or
- A battery failure sensing and warning system with a means for disconnecting the battery from its charging source in the event of battery failure.

Through further analysis of the airworthiness clause 25.1353, the airworthiness clause mainly puts forward three requirements for the design and installation of the battery system, that is, the function, performance, and installation design of the battery system. The details are as follows:

- The functional requirements of the battery system, mainly include the charge and discharge control function of the battery, the function of the starting auxiliary power unit, and the detection and protection functions of the battery state;
- The performance requirements of the battery system mainly refer to the ability of the battery cells and components to withstand the pressure and temperature under the expected charge and discharge state, as well as the explosion of the battery which is caused by a large current impact caused by a short circuit or thermal runaway;
- The installation requirements of the battery system mainly refer to the installation of the battery. Measures should be taken to avoid the impact of the thermal runaway state of the battery on the adjacent airframe structure and airborne equipment when the battery fails.

4. COMPLIANCE DESIGN OF BATTERY SYSTEM TO AIRWORTHINESS CLAUSES

4.1.The function of the design for the battery system

The functional design of the battery system should cover the following aspects.

4.1.1.The function of providing electric power for onboard equipment and auxiliary power unit (APU) startup.

Civil aircraft batteries can supply power to important DC electrical equipment through the power supply buses on the aircraft, and can also supply power to single-phase AC important electrical equipment through static inverters. At present, the APU of civil aircraft uses a DC motor for electric starting, and its starting power usually comes from the onboard battery, and the battery is required to provide the power which is enough to start the APU three times. The block diagram of the battery-starting APU is shown in Figure 1, and the current curve of the battery-starting APU is shown in Figure 2. Before the battery was installed on the aircraft, a series of identification tests are carried out, including the APU start-up test, backup power supply, emergency power supply, charging stability, cycle life, and other tests would be carried out to prove its compliance with the design. After the battery was installed on the aircraft, a series of onboard function tests would be carried out to prove its compliance with the design.

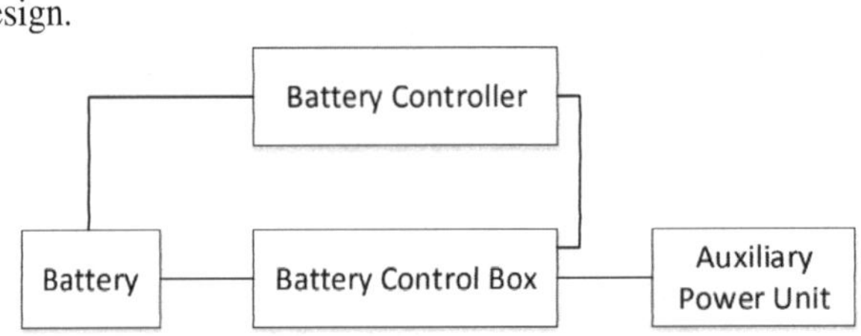

Figure 1. The principle block diagram of battery starting APU

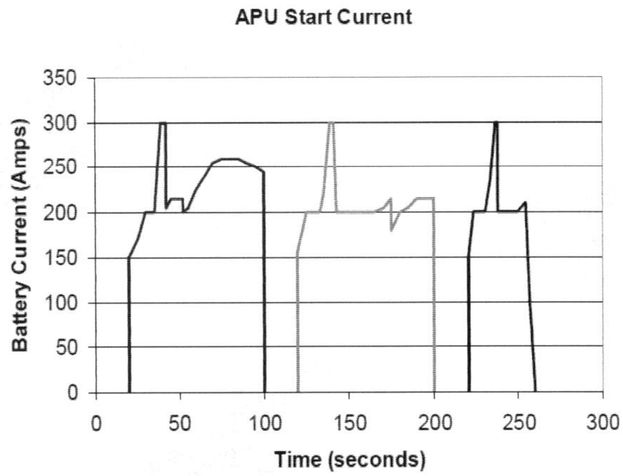

Figure 2. The current curve of battery starting APU

4.1.2.The function of battery status detection and display.
The status detection and display of the battery include two situations: normal and abnormal. Under normal conditions, when the battery is working normally, the voltage, current, temperature, charge, and discharge status of the battery are detected, and the battery controller transmits the detection data to the avionics system, and the avionics system displays it on the multi-function display screen of the aircraft.

When the battery system is abnormal, such as when the battery is overheated or in an emergency, the battery enters the discharge mode, and when the aircraft is powered on, but the battery cannot be charged, the corresponding over-temperature alarm, discharge alarm, and battery disconnect alarm will appear on the multi-function display in the cockpit of the aircraft. After the battery was installed on the aircraft, a series of onboard function tests would be carried out to verify the function of battery status detection and display.

4.1.3.The function of battery manual disconnect.
The top panel of the aircraft cockpit is equipped with a battery switch. When the battery has a temperature alarm, the battery can be manually disconnected from the onboard buses by pressing the battery switch on the top panel, to protect the battery and prevent the battery from overheating and other dangerous situations. After the battery was installed on the aircraft, a series of onboard function tests would be carried out to verify the function of the battery manual disconnect.

4.2.Battery system performance design
The performance of the battery system should meet the following requirements.

4.2.1.The normal charge and discharge performance of the battery.
In the predictable state, the charge and discharge of the battery are automatically controlled by the battery controller, and the charge and discharge control logic of the battery is shown in Figure 3. Before the battery is installed on the aircraft, a series of identification tests are carried out, including an APU start-up test, backup power supply, emergency power supply, charging stability, cycle life, and other tests. These tests proved that the battery can withstand temperature and pressure under normal charging and discharging conditions.

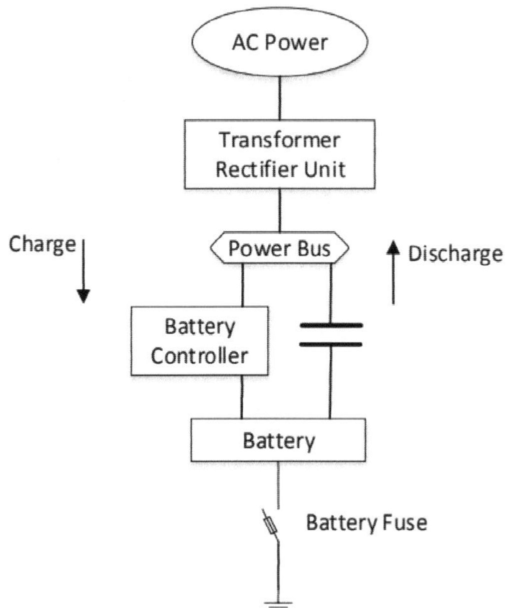

Figure 3. Schematic diagram of a battery charging and discharging

4.2.2.External short-circuit performance of the battery.

The design of the battery for civil aircraft shall be able to withstand the current of short-circuit tests without being damaged when it is fully charged. For example, a battery with a capacity of 27Ah and a rated voltage of 24V was short-circuited, the short circuit resistance was 2mΩ, and the battery discharged for 60 seconds in this condition (the test conditions are more severe than the short-circuit condition when a battery is installed on an aircraft because the resistance value of any place on the aircraft that causes a short-circuit of the battery are greater than 2mΩ), the test data are as follows:

- The initial short-circuit current of the battery is 2923A;
- After the 60s, the short-circuit current of the battery is 597A;
- The maximum temperature of the battery is 118°C;
- There is no leakage of any electrolyte;
- The battery is not damaged after the test is completed.

It can be seen from the above test that the battery itself can withstand the high current impact caused by the external short circuit, and can also withstand the high temperature caused by the short circuit, but it is still necessary to carry out short circuit protection for the battery. A usual method is to set a fuse inside the battery, such as an NB-400 fuse, with a rated current is 400A. When the current is twice the rated current, the fusing time is 2s~20s, and in the above test, the duration of the short-circuit current which is higher than 800A is much longer than the 20s. Therefore, when the battery has an external short circuit, the battery fuse can effectively protect the battery.

4.3.Installation design of battery system

The installation design of the battery system mainly considers the requirements of the following aspects.

4.3.1.The installation design of the battery can help Achieve the expected function.

For example, in a certain type of civil aircraft, the main battery is installed in the electronic and electrical equipment cabin with many important DC power equipment, and the APU battery is installed in the cabin which is close to the APU, which is conducive to the laying of the APU starting cable. It can greatly shorten the power distribution cable between the battery and the electrical equipment, not only can reduce the weight of the cable, but also can effectively reduce the voltage drop of the line and improve the power supply performance.

4.3.2.The installation design of the battery must consider the emission of flammable and explosive gases generated when the battery is overcharged.

When the battery is overcharged, the electrolysis of water will occur in the battery cell to produce hydrogen and oxygen, and the hydrogen and oxygen will easily explode when they reach a certain ratio in the air. Therefore, the battery box should be designed as a sealed body. To ensure that the gas will not leak into the cabin where the battery is installed, at the same time, the battery box should also be designed to exhaust the flammable gas produced by the chemical reaction inside the battery, using the pressure differences between the cabin and the outside of the aircraft to eliminate. Explosive gas is exhausted out of the airframe through the exhaust device so that it will not accumulate inside the aircraft and reach a dangerous level. Before the battery was installed on the aircraft, its venting function would be verified by a venting function test.

4.3.3.The installation of the battery must consider the internal short-circuit heat radiation of the battery.

There may be a short circuit inside the battery (the short-circuit protection function of the battery itself is invalid in this case), which will cause a large amount of heat to accumulate, generate heat radiation and even cause an explosion. So, when the battery is designed, the battery box should have an explosion-proof function, which can ensure that when the battery explodes due to heat accumulation caused by an internal short circuit, the box will not burst and damage other equipment. And it is also necessary to keep enough safety distance to prevent heat radiation from the battery. Before the battery was installed on the aircraft, the battery short-circuits test and explosion containment test would also be carried out to prove its compliance with the design.

4.3.4.Battery installation must consider internal corrosive electrolyte leakage.

Usually, because the electrolyte inside the battery is highly corrosive, and the cover of the battery box is sealed with airtight rubber, the liquid inside the battery will generally not leak out of the battery. However, in the installation design of the battery, it should still be fully considered and prevented from causing harm to the aircraft structure and airborne equipment. Therefore, no equipment should be arranged within the range of battery electrolyte leakage, including the same level as the battery and below the battery, and the structure within the leakage range should be treated with anti-corrosion technology to prevent damage caused by battery electrolyte leakage.

In addition, as a kind of airborne equipment, the battery must meet general requirements such as mechanical strength, equipment identification, and maintainability during equipment installation, and they won't be repeated here.

5. Conclusion

Through the analysis of the airworthiness clauses applicable to the airborne battery system and the compliance design of the battery system to the airworthiness clauses, the battery is a special electrical device on the aircraft, and its design must fully consider the requirements of function, performance, and installation, which are defined in airworthiness regulations. This paper mainly studies the airworthiness compliance design of nickel-cadmium and lead-acid battery systems which are widely used by current civil aircraft. In the future, the airworthiness compliance design of lithium batteries will be studied to meet the development requirements of the new generation of civil aircraft.

References

[1] Civil Aviation Administration of China. CCAR25 China Civil Aviation Regulations Part 25 Airworthiness Standards for Transport Airplanes [s]. Beijing: Civil Aviation Administration of China, 2011.11.7
[2] CCAR 25.1353 Electrical equipment and installations.
[3] CCAR 25.1353(b) Storage batteries must be designed and installed as follows.
[4] Ma Shuxun. Aircraft Design Manual Volume 16: Electrical System Design [M]. Beijing: Aviation Industry Press, 1999.
[5] Xie Shaojun, Chen Yong. Development of a general aviation battery charger [S]. Journal of Nanjing University of Aeronautics and Astronautics, 2003, 35(1): 96-100.
[6] RTCA/DO-293, Minimum Operational Performance Standards for Nickel-Cadmium, Nickel Metal-Hydride, and Lead Acid Batteries, December 2, 2009.
[7] MIL-STD-464C, Electromagnetic Environmental Effects, Requirements For Systems, December 1, 2010.
[8] SAE ARP 1870, Aerospace Systems Electrical Bonding and Grounding for Electromagnetic Compatibility and Safety, April 1, 1999.
[9] MIL-STD-704F, Aircraft Electric Power Characteristics, March 12, 2004.
[10] DO-160G, Environmental Conditions and Test Procedures for Airborne Equipment Section 16, December 8, 2010.

Research on the influence of pilot workload on local route path planning of helicopters in low-altitude penetration

Shen Li and Pan Li

National Key Laboratory of Rotorcraft Aeromechanics, NUAA, Nanjing 210016, China

Email:lipan@nuaa.edu.cn

Abstract. Targeting the problem of high pilot workload caused by trajectory planning of low altitude penetration, this paper proposed a trajectory planning method considering the pilot workload. Firstly, by decomposing the three-dimensional trajectory into horizontal and vertical directions, the trajectory tree method based on the dynamic programming technique is used for horizontal trajectory planning, and the comprehensive slope-curvature limitation method is used for altitude trajectory planning. Then, a method was proposed to quantify the control inputs and pilot workload based on trajectory tracking simulation. The pilot workload cost, together with flight altitude cost and route length cost, was added to the local planning of a reference trajectory. To solve the problem of dimension explosion in trajectory tree search, a segmented optimal route tree algorithm is proposed. Finally, the simulation results show that the proposed method can effectively consider the pilot workload in trajectory planning, which is helpful to improve the flight ability of the planning trajectory and reduce the pilot workload at the same time.

1. Introduction

For the trajectory planning of helicopter in low altitude penetration, the general trajectory planning mainly focus on the trajectory length (or fuel consumption) cost, flight altitude cost, terrain shielding cost, and enemy threat cost. The factors related to pilot workload are often not considered. This problem can lead to heavy workloads on the pilot when executing the mission on the planned trajectory, which in turn makes the mission difficult to complete, especially for helicopters with poor flying quality. Therefore, it is necessary to consider the pilot workload in trajectory planning for low-altitude penetration.

Trajectory planning, as the core technology of low-altitude penetration of helicopters, has always been a hotspot research issue. Trajectory planning is a multi-constrained optimization problem that requires a lot of information to be processed simultaneously. In recent years, many researchers have proposed many new trajectory planning algorithms or combined several different planning algorithms to obtain the optimal trajectory. Szczerba R J et al. [1] simplify the UAV constraints and introduce them into the cost function, using the sparse A* algorithm to reduce the search space and improve the search efficiency. Fu Y et al. [2] present an improved particle swarm optimization algorithm that can be used to solve continuous function optimization problems in trajectory planning with faster convergence and stronger robustness. Wang Q [3] combines the artificial potential field method with the simulated annealing algorithm to solve the local minimum problem in the traditional artificial potential field and the influence of parameters on the escape effect of the search route point.

The military helicopter flight quality specification based on ADS-33E and GJB-902B specifies that the flight quality of the helicopter is determined by the capability level of the helicopter to complete tasks and the workload of the pilot [4]. The evaluation of pilot workload is a difficult part of helicopter flight quality research in recent decades, for which a lot of research has been done at home and abroad. Padfield et al. [5] used the wavelet analysis method to extract information from the pilot's control input. After the whitening processing of the pilot's control input data, the wavelet coefficient cloud is obtained by using the discrete wavelet transform. The number of local peaks in the cloud is counted to indicate the pilot's substantial workload. Tritschler [6] used the time-frequency amplitude diagram to analyze the pilot's operations and used it as a tool to evaluate the pilot workloads. The results show that the method can be used for quantitative analysis.

At present, there is research on low-altitude penetration trajectory planning of helicopters considering pilot workload. This paper studies a sectional optimal trajectory tree algorithm, analyzes how to quantify the pilot workload cost, and incorporates it into local trajectory planning appropriately. Finally, a comprehensive trajectory planning method considering the pilot workload is obtained.

2. Comprehensive optimal trajectory

The comprehensive optimal trajectory for low altitude penetration is a three-dimensional ideal flight path calculated by optimizing performance indicators, which can satisfy the maneuvering performance requirements of the helicopter based on the reference route provided by the mission plan, using terrain data and according to tactical requirements (such as maximum lateral deviation from the reference route, minimum ground clearance, pilot workloads, etc.), as shown in Figure 1.

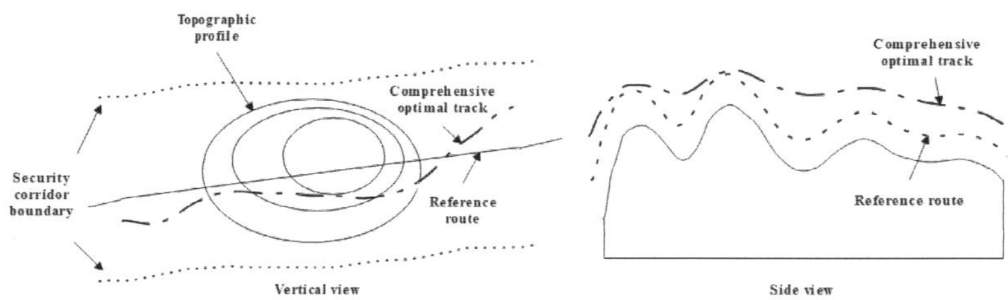

Figure 1. Schematic diagram of the comprehensive optimal track

2.1. Performance metrics

The main factors that affect the penetration mission are flight altitude and route length (flight time). High flight altitudes and long route lengths (long flight time) would easily expose the helicopter itself. However, if the trajectory is too complex, it will bring a heavy workload to the pilot. So, it is necessary to consider pilot workload in route planning. The comprehensive performance metrics used in this paper are pilot workload cost, flight altitude cost, and route length cost as Equation (1) shows.

$$J = \omega_1 J_{workload} + \omega_2 J_{height} + \omega_3 J_{length} \tag{1}$$

where ω_1, ω_2 and ω_3 are the weight coefficients of pilot workload cost, flight altitude cost, and route length cost respectively. The workload cost represents the amplitude and frequency of the pilot's joystick control within a period. The flight altitude cost represents the height of the helicopter off the ground within a period. And the cost of route length represents the degree of deviation from the reference route or fuel consumption of the helicopter within a period.

The order of magnitude of each item in the cost function varies greatly, so the order of magnitude should be unified. In this paper, all costs are unified to within 1. The proportion of the weight coefficient determines the helicopter to select terrain tracking or terrain avoidance or reduce pilot workload. Weight coefficients are set according to low altitude penetration mission.

2.2. Performance constraints
Considering the performance of the helicopter, the flight path planning mainly meets the following constraints:

2.2.1 Limitation on maximum climb/descent angle. The climb angle is the included angle between the helicopter's climb or descent trajectory and the horizontal level. Due to the performance constraints of the helicopter itself, the climb angle and descent angle cannot be too large at a certain flight speed, γ_{max} is the maximum climb(descent) angle, the space coordinate of the point i is (x_i, y_i, z_i), then it satisfied

$$\frac{|z_{i+1} - z_i|}{\sqrt{(x_{i+1} - x_i)^2 + (y_{i+1} - y_i)^2}} \leq \tan \gamma_{\max} \quad (2)$$

2.2.2 Limitation on the maximum rate of change of climb/descent angle. Due to the limitation of the normal load factor, the rate of change of the climb/descent angle of the helicopter in the vertical plane is limited.

$$-\Delta\gamma_{\max\_n} \leq \gamma(t) - \gamma(t-1) \leq \Delta\gamma_{\max\_p} \quad (3)$$

where $\Delta\gamma_{\max\_p}$ and $-\Delta\gamma_{\max\_n}$ are the limits of positive and negative track inclination angles in unit time interval respectively.

2.2.3 Limitation on minimum turning radius. The flight speed and roll angle should also be taken into account when turning, so there exists a minimum turning radius. The turning radius must be larger than the minimum turning radius r_{min}.

$$r > r_{\min} \quad (4)$$

3. Three-dimension trajectory planning
To improve the efficiency of calculation and satisfy the requirement of real-time, the three-dimensional trajectory can be decomposed into horizontal plane and vertical plane respectively.

3.1. Horizontal trajectory generation algorithm
Horizontal trajectory planning adopts a combination of tree search and dynamic planning methods. In the process of trajectory planning, according to the helicopter's horizontal motion equation, the trajectory tree search determines all possible trajectory node states of the helicopter in the horizon plane under the constraint conditions. However, this kind of trajectory tree has the problem of dimension explosion, so the dynamic programming method is used to optimize the end nodes of a certain section of the planning, to reduce the number of possible trajectories and reduce the amount of calculation.

3.1.1 Trajectory tree search. Considering the lateral motion of the helicopter during coordinated turning, as the helicopter rolls, the lift component can generate lateral acceleration. Figure 2(a) shows the forces on the helicopter during lateral movement, while Figure 2(b) shows the geometric relation during the coordination turning process.

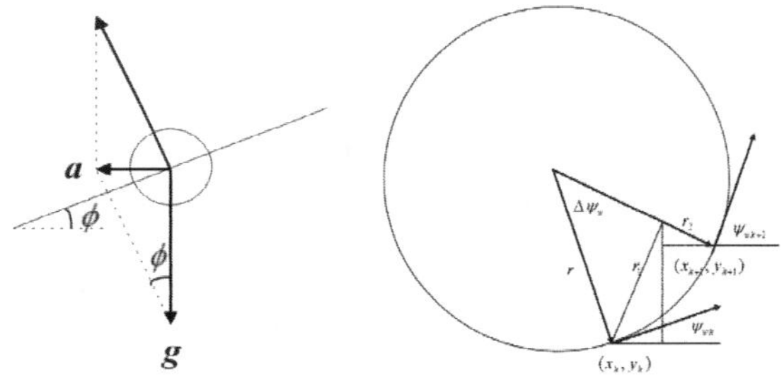

(a) (b)

Figure 2. Lateral force of helicopter and geometric relation during turning

Assuming that the speed of the aircraft is constant, the centripetal acceleration during turning is

$$a = g\tan\phi = \frac{V^2}{r} = \rho V^2 \tag{5}$$

where V: flight speed, ϕ: roll angle, r: turning radius, ρ: turning curvature. Formula (5) shows that the turning radius is determined only by the roll angle when the flight speed is constant.

Let the coordinate position of the helicopter at the current moment be (x_k, y_k), and let the trajectory angle be ψ_{w_k}. Then increment of trajectory angle after experiencing one cycle Δt is

$$\Delta\psi_w = \rho V \Delta t \tag{6}$$

The helicopter's trajectory angle at the next sampling time is

$$\psi_{w_{k+1}} = \psi_{w_k} + \Delta\psi_w \tag{7}$$

Based on the geometric relation, the coordinates of the aircraft at the next sampling time can be calculated as

$$\left.\begin{aligned} x_{k+1} &= x_k + r_1\cos(\psi_{w_k} + \Delta\psi_w) + r_2\cos(\frac{\pi}{2} - \psi_{w_k} - \Delta\psi_w) \\ &= x_k + r\cos\psi_{w_k}\sin\Delta\psi_w - r(1-\cos\Delta\psi_w)\sin\psi_{w_k} \\ y_{k+1} &= y_k + r_1\sin(\psi_{w_k} + \Delta\psi_w) - r_2\sin(\frac{\pi}{2} - \psi_{w_k} - \Delta\psi_w) \\ &= y_k + r(1-\cos\Delta\psi_w)\cos\psi_{w_k} + r\sin\psi_{w_k}\sin\Delta\psi_w \end{aligned}\right\} \tag{8}$$

From the analysis above, when the flight speed V and command cycle Δt of the helicopter are fixed, different coordinate points at the next sample time can be obtained according to the different roll angles at the current moment. To calculate the horizontal path by trajectory tree search, the roll angle needs to be discretized, assuming that the maximum roll angle of the helicopter is ϕ_{max}, which is discretized into

$$\phi = 0, \pm\frac{1}{n}\phi_{\max}, \pm\frac{2}{n}\phi_{\max}, \ldots, \pm\phi_{\max} \tag{9}$$

Through Equations (5) ~ (9), using the discretized roll angle, a trajectory tree can be obtained as Figure 3 shows.

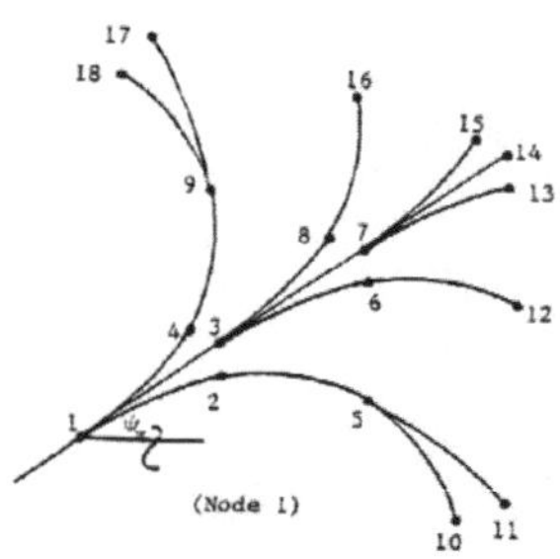

Figure 3. Schematic diagram of a trajectory tree

Considering the limitation of roll acceleration, the variation of roll angle in one command cycle needs to be limited, i.e., the roll angle between the parent node and child nodes could not exceed a certain value. For any node k of the trajectory tree, the following status needs to be stored:

$$S_k = \begin{cases} 1.position(x_k, y_k) \\ 2.track\ yaw\ angle\text{:}\ \ \psi_{w_k} \\ 3.parent\ node \\ 4.\cos t\ values \\ 5.roll\ angle\ of\ node\ k\text{:}\ \ \phi_k \end{cases} \tag{10}$$

3.1.2 Dynamic programming. The problem of dimension explosion in trajectory tree search can be found above. If a large-scale trajectory tree is used directly to search from the starting point to the endpoint, enormous computing effort will be paid. Therefore, we generate a certain generation number of trajectory trees one by one and prune them by dynamic programming to search for a suboptimal trajectory. The process of dynamic programming is as followed:

(1) From the root node, generate the next generation of child nodes along the initial trajectory angle;
(2) Compute the coordinates and cost of the next generation of child nodes for all the children of the current generation;
(3) If the generation number of the trajectory tree does not reach the set value, return to repeat Step (2);
(4) According to the cost of all nodes of the last generation, find the lowest cost node, and search its parent node in opposite direction, all the way back to the root node, to obtain an optimal two-dimensional flight path;
(5) Regarding the end node of the previous tree as the new root node, repeat Steps (1) ~ (4) until the aircraft is near the goal point.

3.2. Vertical trajectory generation algorithm
As long as there is a horizontal trajectory, there must be an optimal vertical trajectory corresponding to it. In this paper, a comprehensive slope-curvature limited smoothing algorithm [7] is used to generate a vertical trajectory.

3.2.1 Slope limited smoothing algorithm. Based on the horizontal trajectory planning, a terrain elevation sequence $h_i (i = 1,2, \dots, N)$ is obtained. The next step is to limit the gradient of this set of sequences. If the gradient of h_i is greater than the maximum gradient k_{max}, then raise h_i to the height of the maximum gradient. If the gradient of h_i is less than the maximum negative gradient $-k_{max}$, then raise h_{i+1} to the height of the maximum negative gradient. The slope limit is shown in Figure 4.

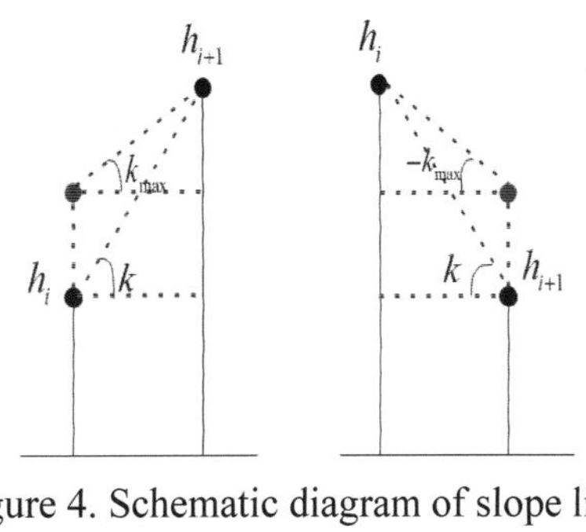

Figure 4. Schematic diagram of slope limit

Calculations are traversed from the first point through to the last point. Since raising one of these points will affect the points previously traversed, it is necessary to traverse again and again until no height adjustment is made. The calculation procedure is shown in Figure 5.

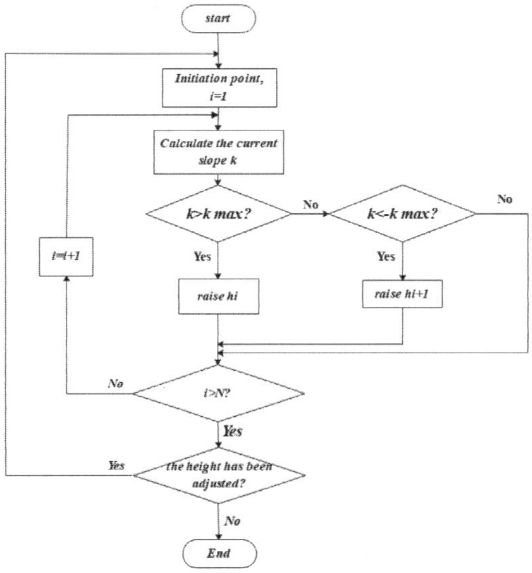

Figure 5. Flow diagram of slope limit algorithm

3.2.2 Curvature limited smoothing algorithm. Although the slope limit algorithm solves the limit of the maximum climb/descent angle of the helicopter, however, the maximum normal load factor is not considered. The relationship between trajectory curvature ρ_V in a vertical plane and the normal load factor n_z is

$$\rho_V = \frac{n_z - g}{V^2} \tag{11}$$

When the flight speed V is constant, limit curvature ρ_V could also limit the normal load factor. For safety, only the height of the corresponding point is raised and the curvature of the trajectory point is adjusted.

3.2.3 Comprehensive terrain smoothing algorithm. Combining the slope-limited smoothing algorithm with the curvature-limited smoothing algorithm to smooth the entire trajectory in the vertical plane is the comprehensive terrain smoothing algorithm. The steps are as followed:

(a) Use the slope-limited smoothing algorithm;

(b) Use the curvature-limited smoothing algorithm;

(c) After curvature limitation is finished, judge again whether the slope limitation meets the condition, and end the algorithm if it satisfies, otherwise return to (a).

For instance, assuming that the gradient limit is 0.15, the normal load factor limit is [0,2g], the flight speed is 40 meters/sec, and the minimum safety clearance is 50 meters, the result of the vertical trajectory can be obtained by comprehensive terrain smoothing algorithm, as shown in Figure 6.

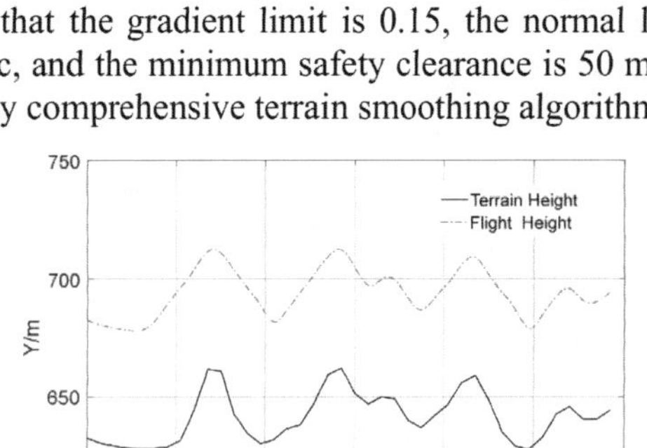

Figure 6. Result of vertical route

4. Trajectory tracking simulation and pilot workload assessment

Through the tree search in the two-dimensional horizontal plane and the comprehensive terrain smoothing algorithm in the vertical plane, we can get a three-dimensional trajectory tree. Since the costs defined are the route length cost, flight altitude cost, and pilot workload cost, to get the costs of all sub-nodes, we need to realize trajectory tracking simulation and discuss how to add and define the pilot workload costs.

4.1. Trajectory tracking

Taking the helicopter's three-axis velocity increment Δu, Δv and Δw, three-axis angular velocity p, q and r, and three-axis angle increment $\Delta\phi$, $\Delta\theta$ and $\Delta\psi$ as state variables, and taking the lateral cyclic pitch angle increment $\Delta\delta_{lat}$, longitudinal cyclic pitch angle increment $\Delta\delta_{lon}$, tail rotor pitch angle increment $\Delta\delta_{ped}$, and collective angle increment $\Delta\delta_{col}$ as control variables. Linearizing the UH-60 helicopter [8-9] from its trim state, we can obtain system Matrix $\boldsymbol{A}$ and control Matrix $\boldsymbol{B}$, and the state-space equation is

$$\Delta \dot{\boldsymbol{X}} = \boldsymbol{A}\Delta\boldsymbol{X} + \boldsymbol{B}\Delta\boldsymbol{U} \tag{12}$$

where $\Delta\boldsymbol{X} = [\Delta u\ \Delta v\ \Delta w\ p\ q\ r\ \Delta\phi\ \Delta\theta\ \Delta\psi]$, $\Delta\boldsymbol{U} = [\Delta\delta_{lat}\ \Delta\delta_{lon}\ \Delta\delta_{ped}\ \Delta\delta_{col}]$.

4.1.1 Trajectory description. The trajectory described here is the three-dimensional trajectory generated in Section 3. What the mission needs to achieve is that the helicopter tracks the two-dimensional trajectory in the horizontal plane with a fixed flight speed, and tracks the height of the vertical plane at the same time. Therefore, it is only necessary to describe the trajectory angle command as a function of time $\psi_{s\_cmd} = \psi_{s\_cmd}(t)$ and the height command (or climb rate command) as a function of time $h_{cmd} = h_{cmd}(t)$. Since the node's trajectory angle information has been retained in the trajectory tree, and the arrival time t of each node $t_k = k\Delta t$ is also known. The trajectory angle can be interpolated by third-order Hermite interpolation, and the height can be interpolated by Spline interpolation.

4.1.2 Tracking control. The track has been described previously. What we need to do now is to track the trajectory angle command and the height command while keeping the flight speed *V*=40m/s and zero sideslip angle. The trajectory tracking control structure is shown in Figure 7.

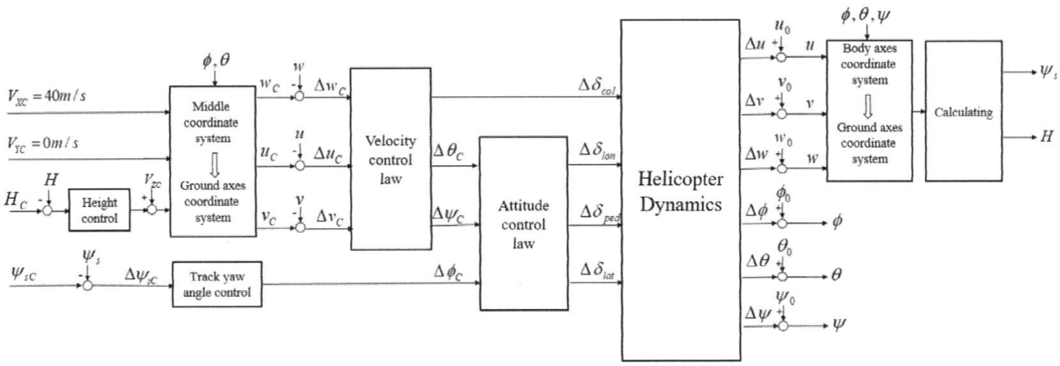

Figure 7. The control structure of trajectory tracking

The control law in the control structure uses proportional-integral feedback control. To eliminate the side slip, the yaw channel needs to add a coordinated turning command.

To track all trajectories, a sufficient closed-loop stability margin is required for each channel. The tracking results of a trajectory are shown in Figure 8.

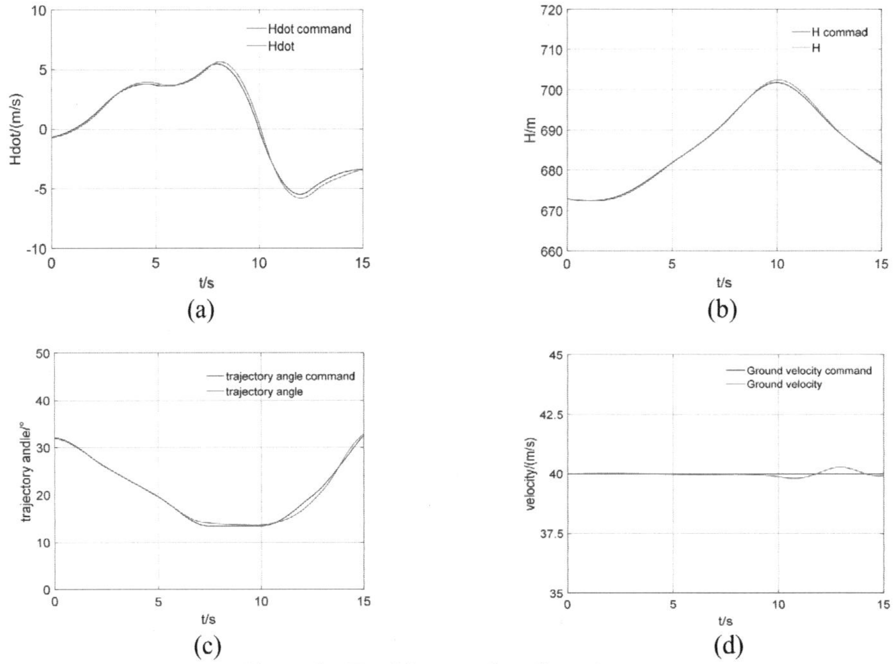

Figure 8. Tracking results of a trajectory

4.2. Pilot workload

The piloted workload [9-10] is measured by calculating the control attack parameter, which is defined as the ratio of the peak rate of control displacement $\dot{\eta}_{pk}$ to the magnitude of change in the controlled displacement $\Delta\eta$

$$A_{\eta} = \frac{\dot{\eta}_{pk}}{\Delta\eta} \tag{13}$$

Figure 9 shows an example of two ideal control inputs; decisive and quick, and smooth and slow, having ramps of different slopes.

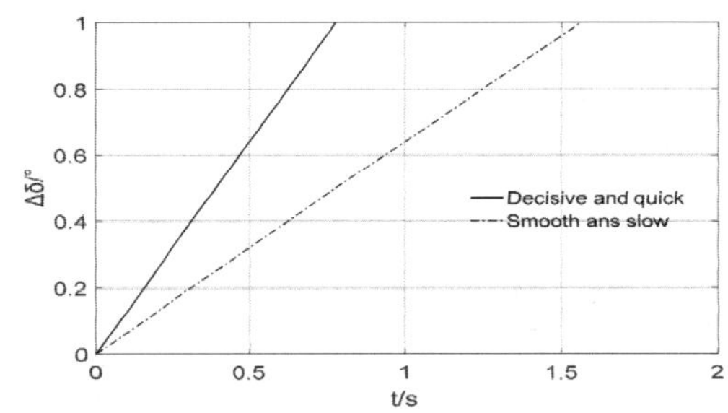

Figure 9. Ideal control inputs showing two ramps of different slopes

Using the control attack concept, the Attack Number $A_{\eta N}$ and Attack Activity Rate $A_{\eta R}$ metrics are defined as [11]:

(a) Attack Number $A_{\eta N}$: the total number of times that the pilot displaces a particular control having a magnitude higher than a specified threshold value.

(b) Attack Activity Rate $A_{\eta R}$: the ratio of the total number of $A_{\eta N}$ to the time length ΔT.

Here, the feathering angle is used to replace the pilot joystick displacement. Taking the lateral cyclic pitch angle as an example, the threshold value is 0.375° (2.5% of maximum travel range), and the Attack Number and Attack Activity Rate can be obtained by calculating as Figure 10 shows.

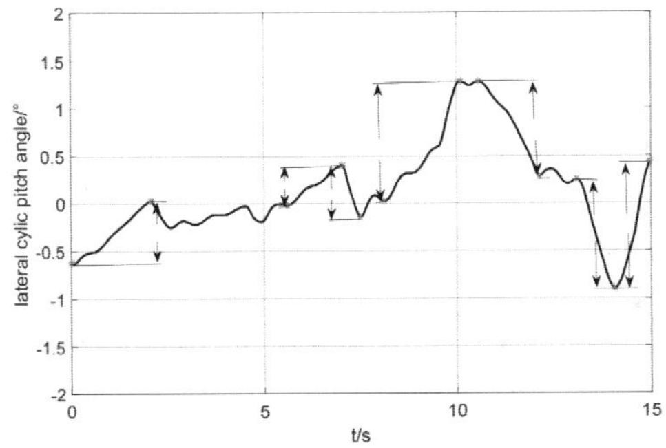

Figure 10. Attack number of lateral cyclic pitch

To obtain the comprehensive workload, the four channels need to be comprehensively weighted:

$$A_{\eta RC} = \sum_{i=1}^{4} \frac{A_{\eta Ni}}{A_{\eta NTot}} A_{\eta Ri} \tag{14}$$

Taking $A_{\eta RC}$ as the pilot workload cost in this period, trajectory planning can be carried out together with flight altitude cost, route length cost, and pilot workload cost.

5. Comprehensive local trajectory planning algorithm

The 3D trajectory planning generation algorithm and the definition of pilot workload have been discussed in Section 3 and Section 4. Now it is necessary to unify the algorithm process. Since dynamic planning is adopted in the form of sectional optimization, and the flight altitude cannot be adjusted directly in sections, the helicopter must take off from the starting point each time. Table 1 shows the procedural flow gram of the local trajectory planning algorithm.

Table 1. Procedural flow gram of the local trajectory planning algorithm

Program: Comprehensive local trajectory planning algorithm

Begin:
- Initialization;
- Call in the digital map, set the maximum roll angle and discretize it, define the comprehensive cost, define the starting point and fly along the initial heading for a small distance, and set it as the optimal trajectory;
- While (The target point has not been reached)
  - Using the terminal node of the optimal trajectory as the root node to generate the trajectory tree
  - While (Not reaching the scale of one trajectory tree)
    - Generate trajectory tree;
  - End: Generate trajectory tree
  - Track from the starting point, calculate workload cost and other costs, and find the best node of the last generation;
  - While (Not Search backward to the root node)
    - Reverse search trajectory tree to obtain optimal sub-route
  - End: Reverse search
  - Optimal trajectory =Optimal trajectory +Optimal sub-route;
- End: (main loop)

End: Comprehensive local trajectory planning algorithm

6. Simulation and analysis

Comprehensive optimal trajectory planning for a certain helicopter in low altitude penetration is carried out in this section. Assuming the flight speed V=40m/s and the level number of the trajectory tree is 6, the other limitations are as follows: the maximum climb/slip angle is 15 degrees and the normal load factor limit is [0.5g,2g].

Example 1: The maximum roll angle of the helicopter is ± 30 °, the number of discrete roll angles is 7 (that is, the maximum difference between the roll angles of the parent node and the child node is 10 °), and the command cycle Δt=2.5s, the results are shown in Figure 11.

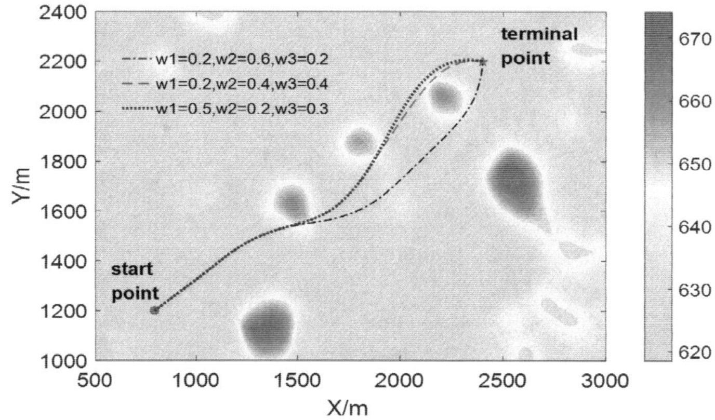

Figure 11. The comprehensive optimal trajectory of e.g., 1

It can be seen that different trajectory planning results can be obtained by adjusting the cost weights. Table 2 shows three kinds of cost values.

Table 2. Cost values under different weights of e.g., 1

ω_1	ω_2	ω_3	Workload cost	Height cost	Distance cost
0.2	0.6	0.2	0.2406	0.2681	0.1162
0.2	0.4	0.4	0.2663	0.2761	0.0525
0.5	0.2	0.3	0.2399	0.2723	0.0702

Example 2: The initial heading is offset from the reference trajectory by 10 °, the maximum roll angle of the helicopter is ± 35 °, and the number of discrete roll angles is 5 (i.e., the maximum difference between the roll angles of the parent node and the child node is 17.5 °), and the command period Δt=3.5s. The results are shown in Figure 12 and cost values are shown in Table 3.

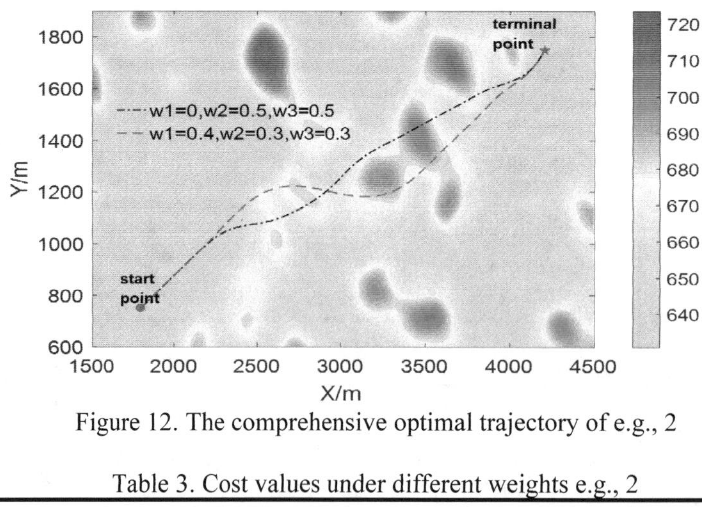

Figure 12. The comprehensive optimal trajectory of e.g., 2

Table 3. Cost values under different weights e.g., 2

ω_1	ω_2	ω_3	Workload cost	Height cost	Distance cost
0	0.5	0.5	0.3003	0.3914	0.0323
0.4	0.3	0.3	0.1895	0.3805	0.1083

When the height cost weight and the distance cost weight are 1:1, it is more difficult for the pilot to complete the flight mission without considering the workload. But when the workload is considered, the pilot workload can be reduced a lot, so the pilot could complete the mission of low altitude penetration more easily.

7. Conclusion

This paper aims at the problems of flight difficulty and the heavy workload of pilots in the planned trajectory of a helicopter in low-altitude penetration. The pilot workload is quantized by Attack Number $A_{\eta N}$ and Attack Activity Rate $A_{\eta R}$. A comprehensive optimal trajectory planning algorithm based on subsection optimization is developed. The results of simulation and analysis show that the method can consider the factors of pilot workload, reduce the pilot's burden in trajectory planning, and improve the success rate in low-altitude penetration missions.

References

[1] Szczerba R and Galkowski P 2000 Robust algorithm for real-time route planning, *IEEE Transactions on Aerospace and Electronic Systems*, **36**(3): 869-878.

[2] FU Y, Ding M and Zhou C 2012 "Phase angle-encode and quantum-behaved particle swarm optimization applied to three-dimensional route planning for UAV". *IEEE Transactions on Systems Man and Cybernetics – Part A Systems and Humans*, **42**(2): 511-526.

[3] Wang Q, Zhang A and Wu Z 2014 UAV path planning based on improved artificial potential field and simulated annealing. *Fire Control & Command Control*, **39**(08): 70-73.

[4] Ford and Terry 1989 Helicopter Handling Qualities. *Aircraft Engineering & Aerospace Technology*, **61**(1): 2-4.

[5] Jones J G and Padfield G D 1999 Wavelet analysis of pilot workload in helicopter low-lever flying tasks. *Aeronautical Journal*, **103**(1019): 55-63.

[6] Tritschler J and John k 2016 Use of Time-Frequency Representations for Interpreting Handling Qualities Flight Test Data. *Journal of Guidance, control, and Dynamics: A Publication of the American Institute of Aeronautics and Astronautics Devoted to the Technology of Dynamics and Control.*

[7] Hu Z Z, Xu K H and Shen C L 2000 Smoothing Approach to Digital Terrain for Low Altitude Penetration. *Journal of Nanjing University of Aeronautics& Astronautics*, **05**: 493-498.

[8] Hibert K B 1984 A Mathematical Model of the UH-60 Helicopter. NASA-TM-85890.

[9] Li P and Chen R 2010 Formulation and Validation of a Helicopter Model for Pull-up Maneuverer Simulation. *Acta Aeronautical et Astronautical Sinica*,**31**(12):2315-2323.

[10] Padfield G D, Jones J P and Charlton M T 1994 Where Does the Workload Go when Pilots Attack Manoeuvres? An Analysis of Results from Flying Qualities Theory and Experiment. In Proceedings of the 20th European Rotorcraft Forum, Amsterdam, The Netherlands, 4–7 October.

[11] Perfect P, Gubbels A W, and White M D 2013 Rotorcraft Simulation Fidelity: New Methods for Quantification and Assessment. *Aeronaut. J*, **117**: 235–282.

Structure design and analysis of upper limb wearable passive power-assisted exoskeleton robot

Yeyan Ning[1]*, Yunji Li[1], Zhenyu Zhang[1], Kaixuan Liu[1], Jianwen Liu[1], Jun Cai[1], Chunlei Wang[1]

[1] Robot Center, The 21st Research Institute of China Electronics Technology Group Corporation, Shanghai, Shanghai, 200233, China

*Corresponding author's e-mail: ningyeyan@126.com

Abstract. The upper limb exoskeleton robot has been a key research direction at home and abroad in recent years. Still, the existing upper limb exoskeleton robot has poor coupling with the human body, and the power assistance effect could be better. In this paper, the strength of each muscle group of the upper limb when the human body is carrying objects is measured as reflected by the surface EMG signal of the human body, and the human kinematics is analyzed to guide the design of the passive exoskeleton robot of the upper limb. First, the spring tension and compression system model is established for the upper limb of the human body. The model is statically analyzed, the forces and moments on each joint are calculated, and each muscle's actual output data when carrying objects is estimated. Based on the muscle kinematics model, a passive upper limb exoskeleton robot is designed. The robot is divided into three modules, namely, the arm module, shoulder module, and back module, using ergonomics and modular design ideas. The energy storage module and joint degree of freedom of the exoskeleton are analyzed to ensure that the exoskeleton meets the requirements of ergonomics, and the key components are checked and simulated. The results show that the designed upper limb wearable passive power-assisted exoskeleton robot meets the strength requirements, which verifies the rationality of the structure design.

1. Introduction

The exoskeleton robot of upper limbs combines human physical strength and mechanical structure energy ingeniously and has been a key research direction at home and abroad in recent years[1-3]. It combines bionics, machinery, sensing, control and algorithm, and other comprehensive technologies, becoming one of the intelligent devices to improve work efficiency and living standards[4-6].

The design of upper limb exoskeleton robots requires interdisciplinary integration involving biomechanics, ergonomics, mechanical engineering, and other knowledge[7-9] and has played a role in industrial lifting and sports rehabilitation. In the field of industrial handling and lifting, in the manufacturing and maintenance workshops, workers often need to use tools and lift them over the top to complete screw removal, chassis inspection, maintenance, etc. Over time, muscle fatigue is easily due to physical decline, affecting work efficiency, causing labor injuries, and even causing potential safety hazards. The use of upper limb exoskeleton robots can help the shoulders assist during lifting, reduce shoulder pressure, reduce labor intensity and improve the working environment of workers. In sports rehabilitation, upper limb motor dysfunction caused by accidents and diseases affects patients' normal life, and they cannot complete the training actions. The upper limb exoskeleton robot can help patients carry out auxiliary rehabilitation training, restore muscle strength, compensate for upper limb motor function, improve physical fitness, and enhance patients' happiness.

In 2017, Yu proposed a new 7-DOF upper limb robot exoskeleton[10]. Each arm includes force and torque sensors to detect human motion, but the balance between high-intensity load and lightweight design still needs to be solved; in 2018, Ma designed an exoskeleton robot driven by steel wires to achieve motion and torque transmission[11]. The flexible mechanism design of joints can be achieved by adjusting the stiffness of the rope drive system, but the reliability and stability of the equipment need to be strengthened; in 2020, Xu used the rotation angle to solve the redundant problem of robot inverse kinematics online[12], laying a foundation for further research on robot kinematics, control and human-computer interaction strategies, but the equipment flexibility needs to be further improved; in 2020, Chen proposed an upper limb exoskeleton rehabilitation robot based on pneumatic muscle actuator[13], which can plan the rehabilitation trajectory through curve fitting, and use fuzzy synovial controller (FSMC) for rehabilitation control, but how to ensure the stable storage of gas is a problem to be solved; the wearable exoskeleton upper limb rehabilitation robot designed by J in 2020 has solved the problem of large size of supporting upper limb rehabilitation robot[14], but the comfort of wearing effect needs to be improved.

The existing upper limb exoskeleton robots have poor coupling with the human body, poor power assist effect, and can only achieve simple functions. To improve the comfort of the exoskeleton robots, the involved upper limb exoskeleton robots incorporate the ergonomic design concept, and the shoulder motion center coincides with the human shoulder joint rotation center, which has a good matching effect with human motion.

2. Static analysis of the human upper limb muscle model

Electromyography (EMG) is the superposition of many muscle fibers' motor unit action potential (MUAP) in time and space. The human muscle is modeled and statically analyzed as a tension and compression system similar to a spring, as shown in Figure 1.

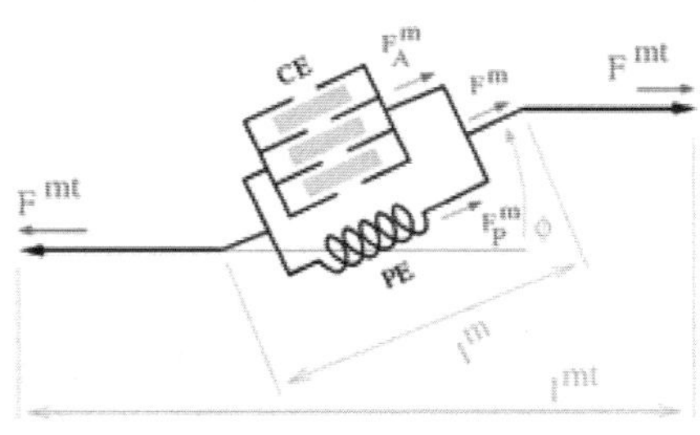

Figure 1. Upper limb muscle model.

The model includes two parts: the contraction unit generates active muscle F_A^m and the parallel elastic element generates passive force F_P^m:

$$F^m = F_A^m + F_P^m \tag{1}$$

The force of the contraction unit is driven by the muscle a (u). The maximum isometric force F of the optimal muscle length F_0^m (the force generated under the condition of equal length, that is, the muscle has no length change) and the active force length function $f_A(\tilde{l}^m)$ are multiplied to get:

$$F_A^m = f_A(\tilde{l}^m) F_0^m a(u) \tag{2}$$

$$\tilde{l}^m = \frac{l_m}{l_0^m} \tag{3}$$

$$u(t) = L(|e(t) - \bar{e}(t)|) - u_0 \tag{4}$$

$$a_{exp}(u) = \frac{e^{AuR^{-1}} - 1}{e^A - 1} \tag{5}$$

where L represents the low-pass filter, R represents the maximum estimated value of signal u(t), and A represents the nonlinear shape factor that defines the function curvature, -5<A<0. Normalized muscle fiber length $\tilde{l}^m$ and muscle fiber length $l_m$divided by the optimal muscle fiber length l_0^m. The driven force is the maximum equidistant force F_0^m and normalized passive force length curve f_P:

$$F_P^m = f_P(\tilde{l}^m)F_0^m \quad (6)$$

This model will be simplified in this experiment, and the parallel elastic element will generate the passive force. F_P^m will be ignored, i.e.:

$$F^m \approx F_A^m \quad (7)$$

And when $\tilde{l}^m$After m normalization is:

$$0 < a_{exp}(u) < 1 \quad (8)$$

3. Design of passive exoskeleton robot for upper limbs

3.1. Overall structure design of the robot

There are 27 degrees of freedom for human upper limbs, as shown in Figure 2, including 20 degrees of freedom for palms, 2 degrees of freedom for wrist joint bending and extension, adduction and abduction, 2 degrees of freedom for elbow joint bending and extension, internal rotation and external rotation, and 3 degrees of freedom for shoulder joint bending and extension, adduction and abduction, and internal rotation and external rotation.

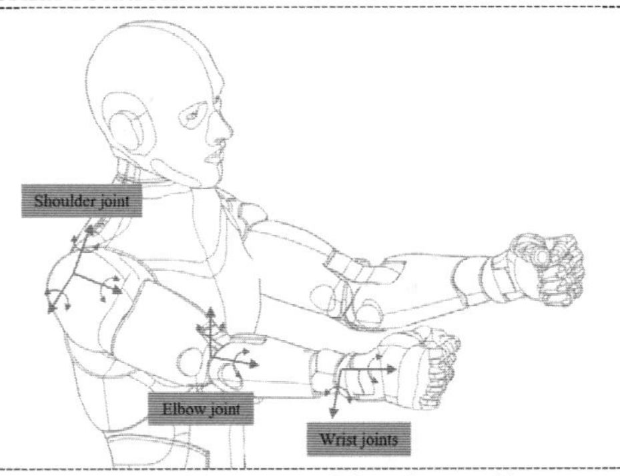

Figure 2. Distribution of degrees of freedom of human upper limbs.

Referring to the distribution of degrees of freedom of human upper limbs, the upper limb passive exoskeleton robot includes several modules: arm, shoulder, and back support. As shown in Figure 3, the structural dimensions of each module refer to the body shape design of normal adult males, meet the ergonomic requirements, and make wearing as comfortable as possible.

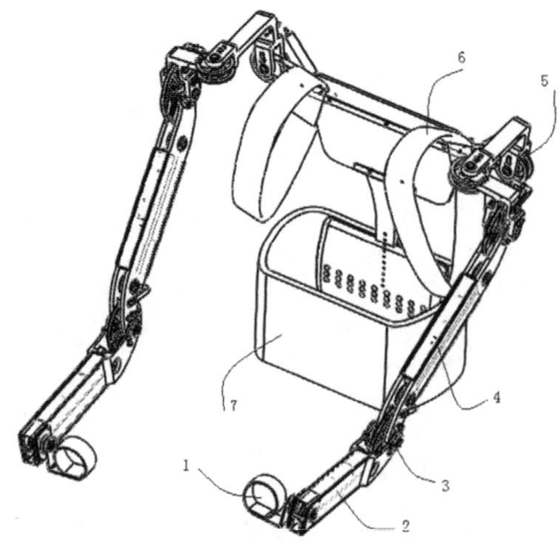

Figure 3. The overall structural layout of the exoskeleton robot.

The fixing straps 1, 6, and 7 are used to fix the wrist, shoulder, and waist. The arming module includes forearm arm frame 2, elbow joint 3, and upper arm frame 4. The shoulder module is 5.

The arming module includes a wrist joint, arm frame, reversing pulley, friction pulley, wire rope, and spring. The arm module structure is flexible with the human arm, which is used to receive the torque output from the spring energy storage module and transmit it to the human arm to assist in lifting. When installing the spring, we tension the wire rope to give it the initial potential energy. When straightening and bending, the arm will produce vertical displacement in the sagittal plane. At this time, the wire rope drives the spring to stretch and store energy. When the arm is bent at 90 degrees, the displacement is minimum. The spring releases energy to give the lifting force. When the arm is straightened vertically, we store the gravitational potential energy of the downward movement of the arm itself. At this time, the arm displacement is maximum. The wire rope drives the spring to stretch to complete the energy storage process. It is released when the human arm bends upward to realize the circular movement.

The back support module includes a back support plate that fits the back of the human body and a flexible strap that contacts the back of the human body. The back support plate is worn on the back of the human body and connected to the shoulder exoskeleton module. The male and female ends of the buckle are set at the end of the flexible bandage to facilitate wearing and disassembling with the human body. The flexible bandage shares the weight of the exoskeleton equipment with the back of the human body, reducing the burden on the shoulders and arms. The fixed connection function of the back and waist transfers the heavy objects lifted and carried by the wearer to the waist position through the exoskeleton robot's forearm, upper arm, and shoulder. The back plate of the upper limb exoskeleton robot has a modular mechanical interface, which can be combined with the lower limb exoskeleton robot.

The shoulder module includes the shoulder frame, pulley, wire rope, adapter block, and other structures. The connection between the back support plate and the shoulder module is rotationally connected, which can realize the internal and external rotation, adduction and extension, and internal flexion and extension of the shoulder joint. The rotation axis is located close to the center of the human shoulder joint to ensure that the axis of the exoskeleton shoulder is coaxial with the axis of the human shoulder joint. This rotation center is the same axis of the human shoulder joint for adduction and abduction, internal and external rotation, and forward flexion and backward extension as shown in the figure. So, when the exoskeleton rotates, it can imitate the human shoulder joint movement, follow the shape of the human body movement, and achieve a man-machine integration effect, improving the flexibility of exoskeleton movement at the shoulder joint and the convenience and comfort of users. The end of the wire rope is fixed to the inside of the shoulder frame by bolts.

3.2. Structural design of joint degrees of freedom

The upper limb passive exoskeleton robot has eight degrees of freedom, as shown in Figure 4. The left and right passive and active degrees of freedom are symmetrically distributed, which can meet the requirements of left and right shoulder joint passive internal rotation and external rotation, left and right shoulder joint passive adduction and extension, left and right shoulder joint beidon flexion and extension, and left and right elbow joint active bending and extension. The cooperation of multiple joint degrees of freedom can improve the coupling between the robot and the human body and optimize human-computer interaction. The upper limb passive exoskeleton robot can assist in the bending and stretching of the left and right elbow joints and assist in completing daily work and life tasks. The passive exoskeleton of the upper limb is fixedly connected with the wearer at the wrist joint of the arm by a bandage.

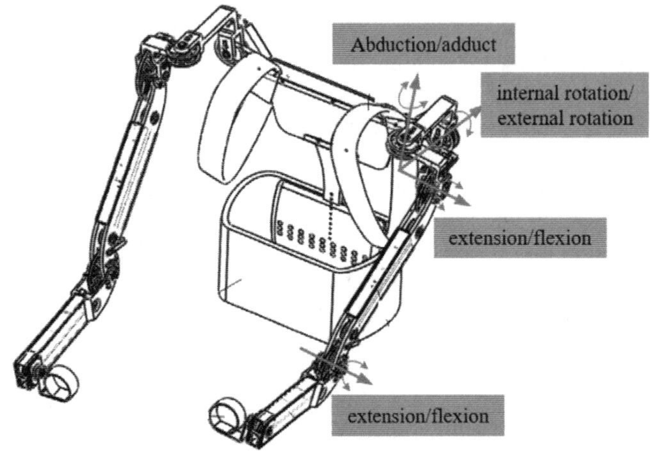

Figure 4. Joint configuration of the degree of freedom for upper limb passive exoskeleton robot.

4. Check and analyze of force on parts

The exoskeleton robot's big arm and small arm are important force-bearing parts of the entire robot and important carriers for installing energy storage devices. They occupy the largest volume and weight in the entire equipment. Therefore, lightweight design is very important for reducing the weight of the entire equipment. Figure 5 shows the three-dimensional model of the jib. As a whole, the structure is hollow and thin-walled. The cross-section of the jib is I-beam type. The flange is wide, the lateral stiffness is large, and the bending resistance is strong. The two planes are parallel to each other, which is convenient for processing and manufacturing. The maximum stress in the middle is free of force. The middle of the upper and lower edges of the I-beam is hollowed as far as possible, reducing the weight and facilitating the installation of the spring energy storage device, making the whole structure compact. The advantages of high space utilization are highlighted. At the same time, the forearm is designed with an arc transition at the elbow joint for a beautiful appearance, increasing the joint's range of motion. During force analysis, we add 200 N pressure in the vertical direction at the end of the jib to fix the end hole. The added 200 N pressure is equivalent to a weight of 20 Kg, which can simulate the ultimate weight lifted under normal use conditions to ensure the rationality of the structure.

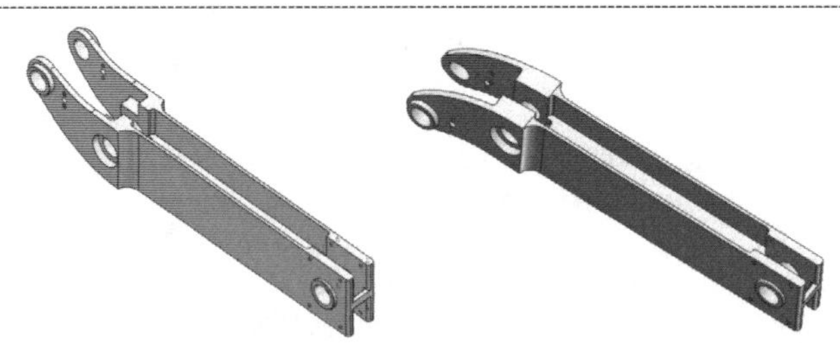

Figure 5. 3D model of the jib.

The jib is made of hard aluminum alloy, the new national label number is 2A12, and the yield strength is 325 Mpa. When 200 N force is applied, according to the stress analysis in Figure 6, the maximum stress is 151.6 Mpa, and the maximum deformation is 1.13 mm, which is completely within the material safety permission range.

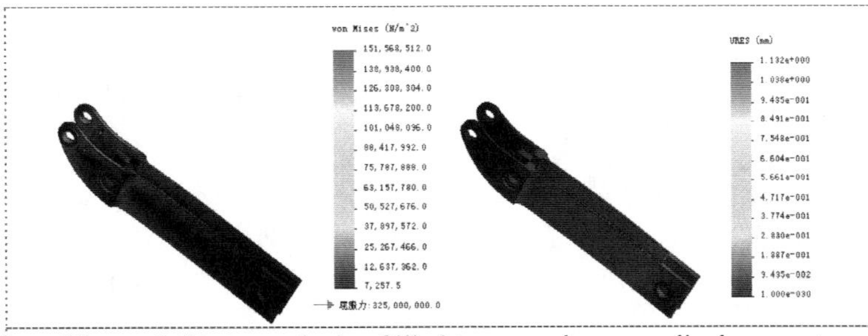

Figure 6. Stress analysis of jib (stress nephogram, displacement nephogram).

As shown in Figure 7, the three-dimensional mechanism model of the exoskeleton robot's big arm is shown. The structural design of the big and small arms is much the same, and the I-beam type structural design is also adopted, which also ensures the structural strength and the installation space position of the energy storage device. The boom is also made of 2A12 hard aluminum alloy with a yield strength of 325 Mpa. One end is fixed, and the other is applied with 200 N pressure and 60 Nm torque, as shown in Figure 8. The maximum stress is 136.9 Mpa, and the maximum displacement is 0.56 mm, all in line with the design requirements of structural strength.

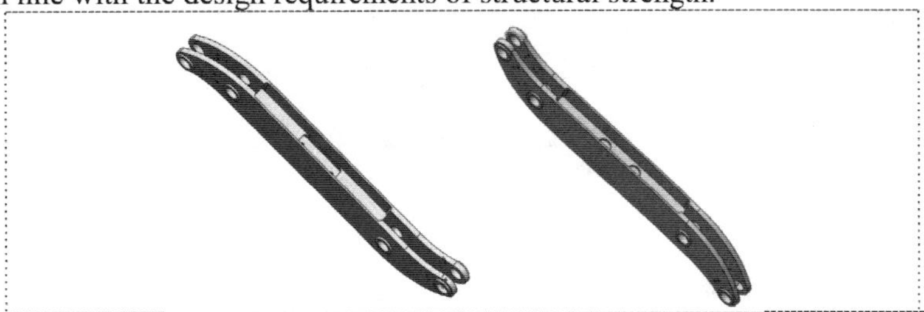

Figure 7. 3D model of the boom.

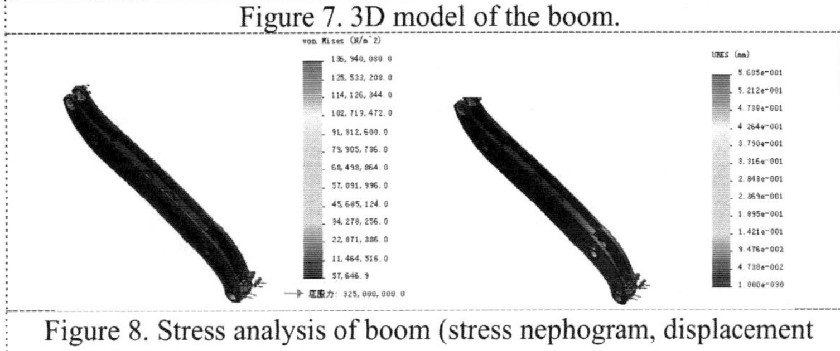

Figure 8. Stress analysis of boom (stress nephogram, displacement nephogram).

5. Conclusion and prospect

Based on the ergonomics design concept, this paper establishes a spring tension and compression system model for human upper limb muscles and conducts a static analysis of the muscle model. The obtained joint data guide the design of a passive upper limb exoskeleton robot. To ensure the rationality of the design, the comfort of the contact position between the robot and the human is fully considered, and the motion analysis of the spring energy storage module is carried out, as well as the verification and simulation of key components. The results show that the robot design is reasonable. On the premise of meeting ergonomics, achieving the balance and unity between the power-assisted effect and lightweight design is the next research direction.

References

[1] Ren, J. L., et al. "Deep Learning based Motion Prediction for Exoskeleton Robot Control in Upper Limb Rehabilitation." 2019 International Conference on Robotics and Automation (ICRA) 2019.
[2] Li, Z., et al. "Design and Analysis of Joint-Assisted Exoskeleton Control System of Upper Limb in Active Spacesuit." 2020 39th Chinese Control Conference (CCC) 2020.
[3] Qingcong, and Hongtao. "Development, Dynamic Modeling, and Multi-Modal Control of a Therapeutic Exoskeleton for Upper Limb Rehabilitation Training. " Sensors (Basel, Switzerland) (2018).
[4] Chen, L., et al. "Construction and analysis of muscle functional network for exoskeleton robot." Shengwu Yixue Gongchengxue Zazhi 36.4(2019):565-572.
[5] Zheng, J., P. Shi, and H. Yu. "A Virtual Reality Rehabilitation Training System Based on Upper Limb Exoskeleton Robot." 2018 10th International Conference on Intelligent Human-Machine Systems and Cybernetics (IHMSC) 2018.
[6] Ning, L., et al. "Bio-inspired wearable soft upper-limb exoskeleton robot for stroke survivors." 2017 IEEE International Conference on Robotics and Biomimetics (ROBIO) IEEE, 2017.
[7] Li, N., et al. "Bio-inspired upper limb soft exoskeleton to reduce stroke-induced complications." Bioinspiration & Biomimetics (2018).
[8] Leiyu Zhang, Jianfeng Li, Shuting Ji, Peng Su, Chunjing Tao, Run Ji. Design and human–machine compatibility analysis of Co-Exos II for upper-limb rehabilitation[J]. Assembly Automation, 2019, 39(4).
[9] Wang, W., et al. "A novel fatigue detection method for rehabilitation training of upper limb exoskeleton robot using multi-information fusion:" International Journal of Advanced Robotic Systems 17.6(2020):379-386.
[10] Ho Yu, et al. "Development of a upper-limb exoskeleton robot for refractory construction." Control Engineering Practice 72. (2018). doi:10.1016/j.conengprac.2017.09.003.
[11] Ma, R., et al. "Mechanism Design and Kinematics Analysis of an Original Cable-driving Exoskeleton Robot." 2018 2nd IEEE Advanced Information Management, Communicates, Electronic and Automation Control Conference (IMCEC) IEEE, 2018.
[12] Xu, N., et al. "Modeling and Kinematics Analysis of a Novel 5-DOF Upper Limb Exoskeleton Rehabilitation Robot." 2020 39th Chinese Control Conference (CCC) 2020.
[13] Chen, C. T., et al. "Implementation of an upper-limb exoskeleton robot driven by pneumatic muscle actuators for rehabilitation." MDPI AG 4(2020).
[14] Guo, J., P. Li, and S. Guo. "Design and Analysis of a Wearable Exoskeleton Upper Limb Rehabilitation Robot." 2020 IEEE International Conference on Mechatronics and Automation (ICMA) IEEE, 2020.

High-fidelity Modelling Approach for Investigating the Tightening Torque Effect in Thread Connection

Hui Zhang, Fei Yu and Houbin Wang

Department of Composite Structures, Aircraft Strength Research Institute, Xi'an, 710075, China

zhangh226@avic.com

Abstract. Accurate replication of the torque-tightening effect is important when simulating the mechanical behavior of composite joints using the finite element (FE) approach. In this study, a parametric mesh construction strategy was developed to automatically build a high-fidelity 3D FE model with lift angle to simulate bolt thread interactions. An experimental approach was proposed to calibrate the FE model. The relation between tightening torque and the axial force was studied by the FE method, experiment, and engineering algorithm. Results showed that traditional empirical formula overestimate the clamping force. By using the proposed model to calculate the axial force of the bolt, the accuracy of the axial force can be significantly improved when analyzing the composite bolted joints. Finally, parametric simulations were performed to investigate the influence of coefficients of friction at thread interfaces and the nut/washer interfaces. The pre-tightening force was significantly influenced by the frictional coefficients at these interfaces. Results indicated that increasing the coefficient of friction (CoF) at the nut/washer interface from 0.1 to 0.2 can lead up to 28% reduction in the clamping force, while this increase was 23% for the threads interfaces.

1. Introduction

The mechanical connection is simple and convenient to assemble and disassemble, which is an important form of composite joint. For aviation composite structures, bolts mainly bear a shear load, and the shear strength of the structure, which can be affected by many factors, such as the orientation of the ply, the thickness of the plates, the friction coefficient of the interface and the pre-tightening torque, etc. A number of studies have been devoted to investigating above-influencing factors[1–7]. Increase in the tightening torque typically results in higher clamping force, which can transfer a part of the shear force through friction between the plate, thus improving the strength of the composite joint. The accuracy of final failure load prediction is therefore closely related to the accuracy of the axial force applied in the finite element (FE) model.

Accurate prediction of the stiffness and strength of the joint requires correctly modelling the tightening torque in FE models. The common approaches for defining the axial force caused by tightening torque are typically temperature, displacement, section forces, etc. However, most of these methods are based on empirical approaches. For instance, Xie[8] compared two empirical methods for calculating clamping force and found that the difference between them is about 15%. Experimental study implied that the difference between the real axial force and the force derived from an empirical formula is significant[9]. The detailed FE model considering the lift angle on threads can obtain an accurate torque-axial force relation, resulting in improved accuracy for tightening torque. Whilst the detailed geometry of threads can be conveniently discretised using tetrahedral elements, this may lead to serious convergence problems and even questionable results in some cases. Therefore, it is necessary to use hexahedral finite elements for the threads, in order to avoid potential convergence problems. The commonly used construction method for threaded hexahedrons is mainly based on the slicing method[10–13], but manually meshing is extremely tedious.

In this paper, the equation describing thread section was derived based on the slicing method, enabling a parametric mesh construction strategy to build the high-fidelity FE model for the bolt-joint structure. This model was subsequently use to numerically establish the relation between the tightening torque and the axial force for a single bolt-joint, showing a good agreement with experimental data.

2. Meshing strategy

2.1. Thread section equation

The method to construct the hexahedral mesh for the thread is mainly based on the following acknowledges:

- The profile of the slicing plane remains the same for different height positions along the axis of the bolt.
- The profile of the slicing plane rotates as increasing its height position. This rotation θ is defined as a function of the height position.

$$\theta = \frac{\Delta_h}{p} \times 2\pi \tag{1}$$

Where p denotes the axial distance between two adjacent threads and Δ_h is the height between two slice planes.

The detailed mesh can be constructed by repeating the rotation of the profiles. Here, taking the triangular thread of GB/T 196-2003 as an example, its thread profile is shown in Figure 1.

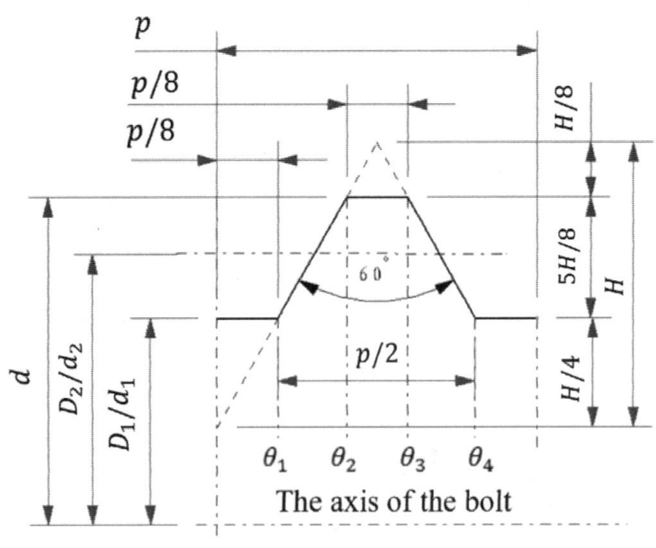

Figure 1. The profile of triangular thread by GB/ 196-2003

The thread profile can be expressed as below.

$$r(\theta)=\begin{cases}\frac{d}{2}-\frac{5}{8}H, & 0\le\theta<\theta_1\\ \frac{d}{2}-\frac{5}{8}H+k(\theta-\theta_1), & \theta_1\le\theta<\theta_2\\ \frac{d}{2}, & \theta_2\le\theta<\theta_3\\ \frac{d}{2}-\frac{5}{8}H-k(\theta-\theta_3), & \theta_3\le\theta<\theta_4\\ \frac{d}{2}-\frac{5}{8}H, & \theta_4\le\theta<2\pi\end{cases} \quad (2)$$

Where H is the height of the original triangle and $H=p\sqrt{3}/2$; d is the nominal diameter. $\theta_1-\theta_4$ represent the corners of the profile and $\theta_1=\pi/4$, $\theta_2=7\pi/8$, $\theta_3=9\pi/8$, $\theta_2=7\pi/4$. $k=\sqrt{3}p/2\pi$. D and d represent the diameters of male thread and female thread respectively, and subscripts 1 and 2 denote pith diameter and minor diameter respectively. P is the pitch of the thread.

Due to the tolerance of machining and fitting, there will be a small gap between the male thread and the female thread in the natural occlusion state. In order to avoid convergence problem, an offset distance δ was assigned between the thread tips and the trough on nuts. Therefore, the equation above needs to be revised in consider this offset distance.

For male thread:

$$r(\theta)=\begin{cases}\frac{d}{2}-\frac{5}{8}H-\frac{\delta}{2}, & 0\le\theta<\theta_1\\ \frac{d}{2}-\frac{5}{8}H+k(\theta-\theta_1)-\frac{\delta}{2}, & \theta_1\le\theta<\theta_2\\ \frac{d}{2}-\frac{\delta}{2}, & \theta_2\le\theta<\theta_3\\ \frac{d}{2}-\frac{5}{8}H-k(\theta-\theta_3)-\frac{\delta}{2}, & \theta_3\le\theta<\theta_4\\ \frac{d}{2}-\frac{5}{8}H-\frac{\delta}{2}, & \theta_4\le\theta<2\pi\end{cases} \quad (3)$$

where $\theta_1=\frac{\pi}{4}-\frac{\delta}{2k}$, $\theta_2=\frac{7}{8}\pi-\frac{\delta}{2k}$, $\theta_3=\frac{9}{8}\pi+\frac{\delta}{2k}$, $\theta_2=\frac{7}{4}\pi+\frac{\delta}{2k}$

For female thread:

$$r(\theta)=\begin{cases}\frac{d}{2}-\frac{5}{8}H+\frac{\delta}{2}, & 0\le\theta<\theta_1\\ \frac{d}{2}-\frac{5}{8}H+k(\theta-\theta_1)+\frac{\delta}{2}, & \theta_1\le\theta<\theta_2\\ \frac{d}{2}+\frac{\delta}{2}, & \theta_2\le\theta<\theta_3\\ \frac{d}{2}-\frac{5}{8}H-k(\theta-\theta_3)+\frac{\delta}{2}, & \theta_3\le\theta<\theta_4\\ \frac{d}{2}-\frac{5}{8}H+\frac{\delta}{2}, & \theta_4\le\theta<2\pi\end{cases} \quad (4)$$

where $\theta_1=\frac{\pi}{4}+\frac{\delta}{2k}$, $\theta_2=\frac{7}{8}\pi+\frac{\delta}{2k}$, $\theta_3=\frac{9}{8}\pi-\frac{\delta}{2k}$, $\theta_2=\frac{7}{4}\pi-\frac{\delta}{2k}$.

The coordinates of nodes on the profile can be determined using the equations above.

2.2. Program implementation

For versatility and scalability, this paper constructs a general meshing algorithm to adapt different solvers. Mesh seeds are uniform allocated in the circumferential direction and biased in the radial direction. For the center part of the bolt, the wedge elements are used, and the rest are all hexahedron elements. Figure 2 shows the program structure. The node coordinates are automatically calculated by a three-layers loop, and the loop parameters i, j and k respectively correspond to the number of slice layers, the number of radial seeds, and the number of circumferential seeds. Each node is mapped to the loop parameters i, j and k.The structure for generating elements is similar to that of nodes, and the mapping relationship between element number and node number is established. The main program gets the corresponding element and node definition rules by parsing the solver template. According to the rules, the executive program will calculate the node number and its coordinate, the element number, and the node array. Finally, the nodes number and coordinates, elements number, and the nodes of itself will be written into the corresponding output file according to the rules.

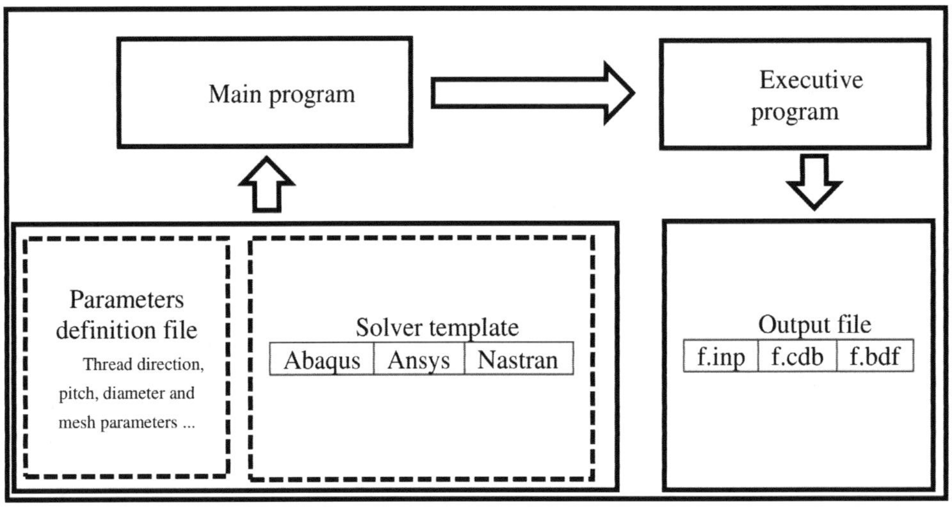

Figure 2. Program structure

Figure 3 shows the FE meshes of bolt and nut automatically generated by the program. For the generation of different shapes of threads, the detailed mesh processing of the corresponding threads can be completed only by deducing their specific parametric equations. For the convenience of engineering, the program also supports the generalized definition of section profile, which can be directly fitted by the given interpolation points. Their interpolation points can be directly derived from CADtools, thus improving its usability.

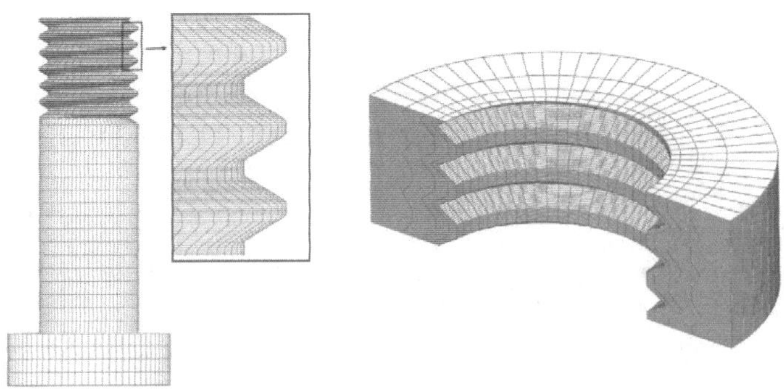

(a) bolt mesh (b) nut mesh
Figure 3. Mesh Examples

3. FE model and experimental verification

3.1. FE model

Considering the nonlinearity of contact analysis, this paper takes Abaqus for solver as an example. First, the program above was used to automatically generate the mesh of bolt and thread, as shown in Figure 3. Then the input files of the bolt and nut were imported into Abaqus for assembly and establish the FE model of a single bolt as shown in Figure 4. The pre-tightening force is mainly balanced by the friction between the thread pairs and the friction between the nut and the washer. The corresponding contact relationship is established for bolt-nut and nut-washer respectively, where the contact interaction was defined using the Abaqus in-built hard contact algorithm with a penalty control. All friction coefficients for this model were set to 0.15.

The finite sliding formula based on surface-to-surface is adopted in the contact property as a significant rotation of the nut is accompanied by the pre-tightening process. Tie constraints are applied between the bolt head and plate, plate interface, washer and plate, so the bolt will not rotate in the process of Pre-tightening. Clearance fit is adopted between the bolt and the plate hole, in which the shank diameter was set to 8mm and the diameter of the hole was set to 8.05mm. The friction between the bolt shank and plate hole is not considered as a clearance fit, which has been applied.

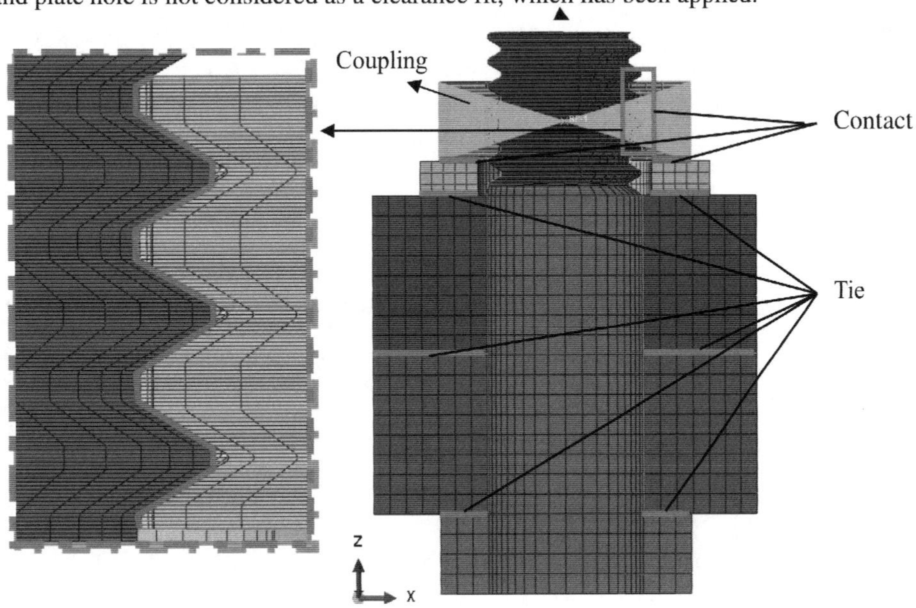

Figure 4. The FE model of the bolt joint

To alleviate the convergence problem, the load was applied to the nut in a form of displacement: the side surface of the nut is coupled with a reference point in the center as shown in Figure 4, and the corresponding axial rotation was applied. The section force of the bolt and reaction moment for RF3 was required then the corresponding torque-axial force curve can be obtained. Except for the C3D6 element in the center of the bolt, C3D8I elements were used in the rest of the whole model. The model contains 197680 elements, in which 158328 for bolts, 41400 for nuts, and 8784 for plates and washers.

3.2. Experimental verification

An instrumented bolt for measuring axial force by the material of 30CrMnSi was manufactured as shown in Figure 5. The bolt was designed specially on the bolt head to embed two pairs of electric resistance strain gauges, by which the axial force of the bolt was reversed.

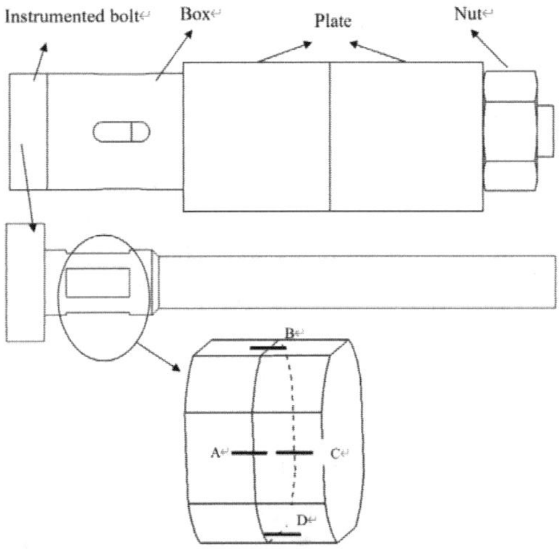

Figure 5. Diagram of the instrumented bolt

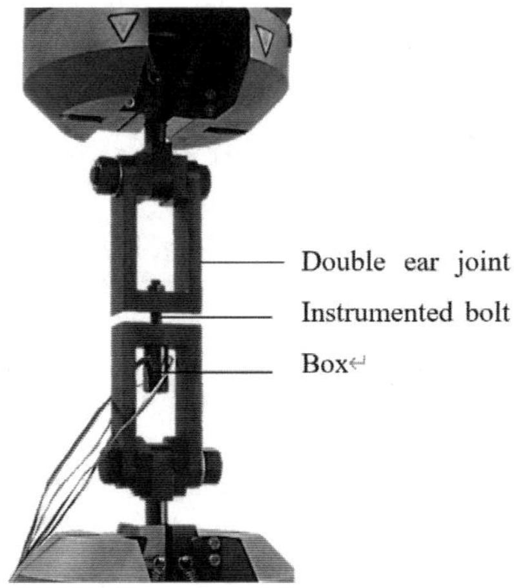

Figure 6. Diagram of pure tension test

The axial tension was carried out on the Instron-8801 mechanics machine to calibrate the curve of load-strain. The experiment method was shown in Figure 6. The average of all strains was taken as the result to reduce the error caused by the sticking precision of the strain gauge or flatness of the bolt, which may contribute to uneven strains. The calibration results were shown in Figure 7. More details about the experiment and instrumented bolt can be referred to scholars[9,14].

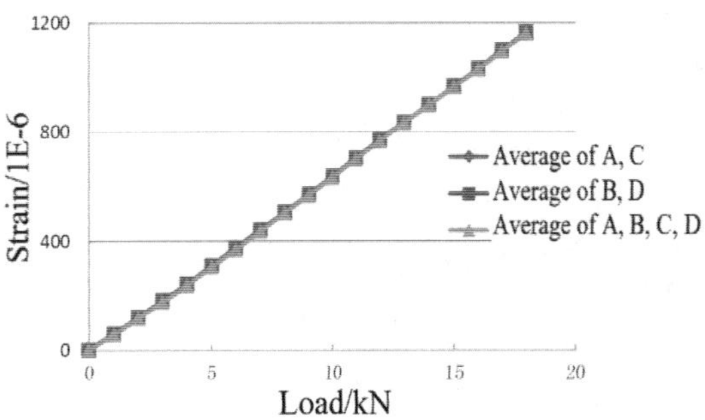

(a) Average of the strains

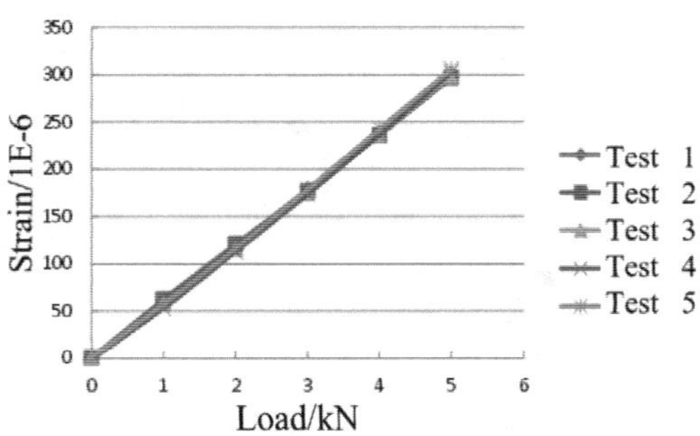

(b) Test results for 5 times

Figure 7. result by experiment

It can be seen that the linearity of curve for the axial force-average axial strain of the test is quite good, and the load-strain curves obtained by the 5 tests are consistent as shown in Figure 7. The dispersion coefficient of slope for the curves is 1.73%, which can meet the engineering requirements.

4. Analysis

4.1. Experiment result

Torque was applied on the bolt instrument with a torque wrench to get the axial strain for each tightening step. The corresponding axial force-torque curve is shown in Table 1.

Table 1. Axial force by the experiment

Torque/ N·m	0	5	10	15	20	25
Axial force /kN	0	1.86	5.03	7.74	10.07	11.49

It can be seen from Table 1 that the axial force-torque curve is nonlinear under low and high tightening torque. It can be explained by the poor surface fitting under low tightening torque, while the thread becomes locally plastic under high tightening torque. To eliminate the nonlinear influence, the slope of the linear section of the axial force-torque curve was used to represent the axial force generated by the unit torque, as shown in Figure 8.

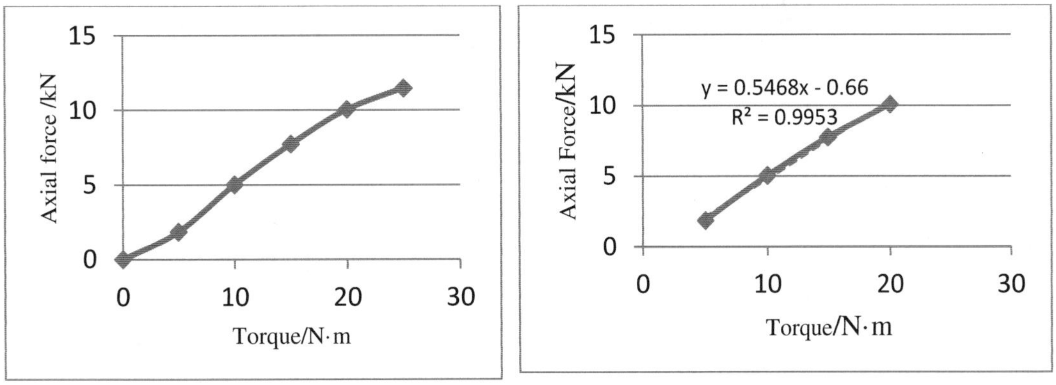

(a) Curve for axial force-torque (b) Fitted curve for axial force-torque

Figure 8. Curve for axial force-torque by experiment

4.2. FE result

The stress distribution by the FE method is shown in Figure 9(a) and stress distribution for bolt is shown in Figure9(b). When the torque is 9.162N·m, the total force on the section of the bolt is 5431N, and the average axial stress is only 108Mpa. However, the maximum stress at the front end of the thread reaches 606Mpa, which is 6 times the average. The typical stress at the root of the thread reaches 300Mpa, which is 3 times the average axial stress. The analysis result shows that the average axial stress of the bolt is far lower than the yield strength of the material, and part of the thread reaches a local plastic state. This is quite consistent with the phenomenon observed by the experiment, which indicates that the tightening torque must be reasonably limited in engineering and a high tightening torque will lead to the damage of thread.

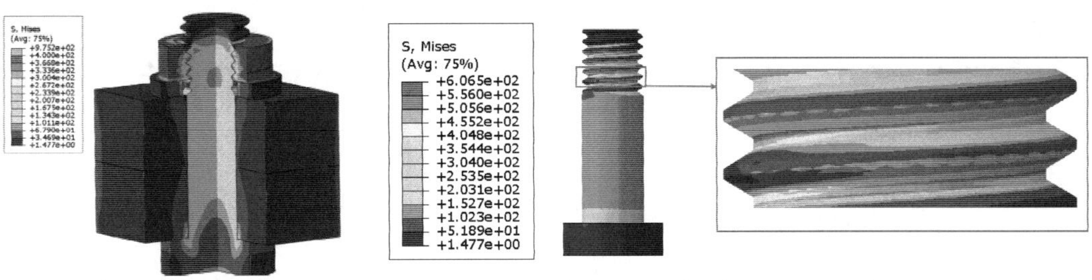

(a) Stress distribution (b) Stress distribution of the bolt

Figure 9. Stress distribution by FE model

The longitudinal profile stress distribution of the bolt is shown in Figure 10. As it can be seen from Figure 10(a), the stress of the shank is larger on both sides and smaller in the center. There is a significant stress concentration at the thread and the original bolt head. The first and second thread bear seriously, and the stress concentration is more obvious. Figure 10(b) shows the contact stress distribution of the nut, indicating that the first and the second thread bear most of the contact force. The above analysis results are consistent with the fracture phenomenon of bolts in engineering, which confirms the accuracy of the model. And the curve for the axial force torque by the FE model is shown in Figure 11.

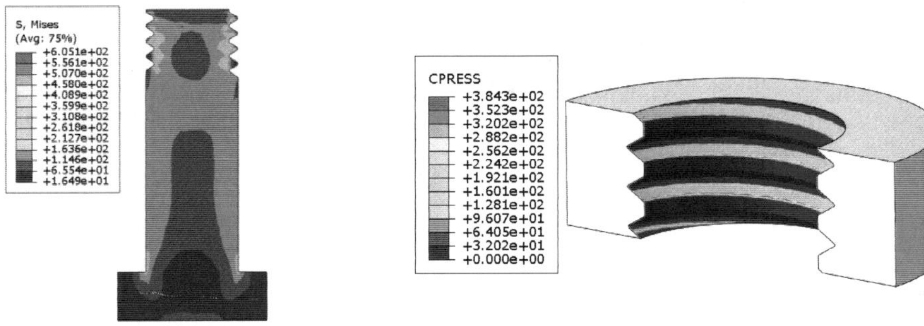

(a) Stress distribution by longitudinal profile

(b) Contact pressure of the nut

Figure 10. Stress distribution for bolt and nut

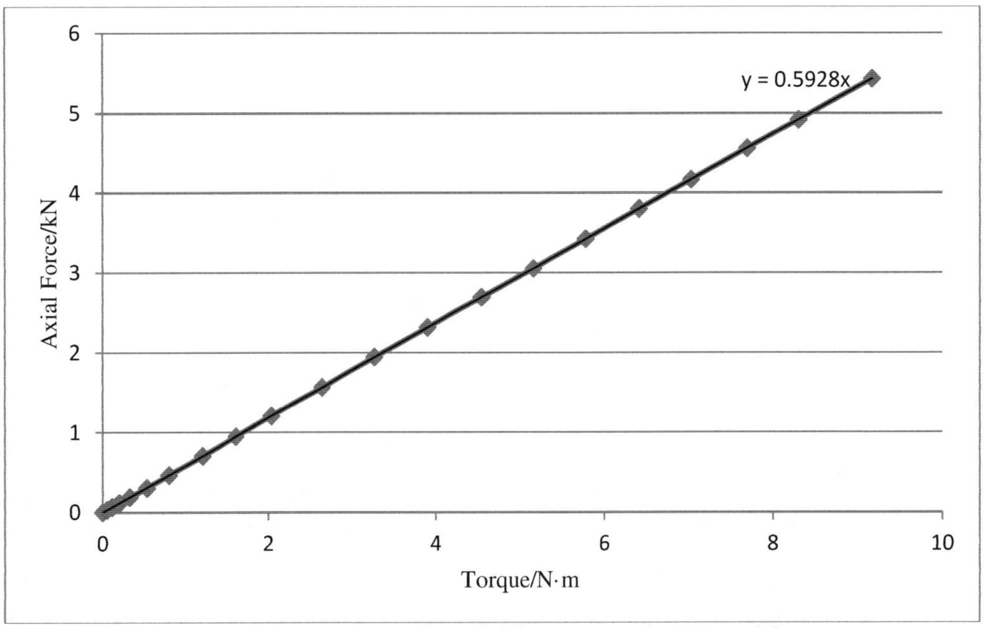

Figure 11. Curve for axial force-torque by FE model

5. Discussions

In engineering, empirical formulas are usually used to calculate the corresponding axial force or lateral restraint stress, for example [13]:

$$P_t = \frac{2\pi T}{p + 2\pi\mu d} \tag{5}$$

Where p is the pitch and μ represents the friction coefficient between thread interfaces, which is usually set to 0.15. The axial forces of the bolt under unit torque obtained by experiment, FE method, and empirical formula are shown in Table 2. The result illustrates that the FE method is closer to the experimental results, and the axial force obtained by the traditional empirical formula is significantly higher than the experiment.

Table 2. The axial force under unit torque

Experiment	FE model		Empirical formula	
	Axial force value/ N	Relative error	Axial force value/ N	Relative error
547	593	8.4%	715	30.7%

The friction f_1 between the thread pairs and f_2 between the nut and washer are the main factors affecting the axial force during the process of tightening torque. Traditional empirical formula usually ignores these frictions, for example, empirical formula in this paper only considers the friction f_1 and ignores the friction between the nut and the washer. In fact, due to the rotation of the nut, the friction between the nut and the washer will transfer a considerable proportion of pre-tightening force.

Friction coefficients of f_1 and f_2 ranged from 0.1 to 0.2 with a space of 0.05 are studied in this paper. The torque-axial force curves under different friction coefficients are shown in Figure 12(a) and Figure 12(b). When the friction coefficient f_1 is 0.15, and f_2 for 0.1, 0.15, 0.2 respectively, the axial force corresponding to the unit tightening torque is 711N, 593N, and 509N as shown in Figure 12(a). The clamping force decreases by 28% when f_2 variety from 0.1 to 0.2. When the friction coefficient f_2 is 0.15, and f_1 for 0.1, 0.15, 0.2 respectively, the axial force corresponding to the unit tightening torque is 681N, 593N, and 525N as shown in Figure 12(b). The clamping force decreases by 23% when f_1 variety from 0.1 to 0.2.

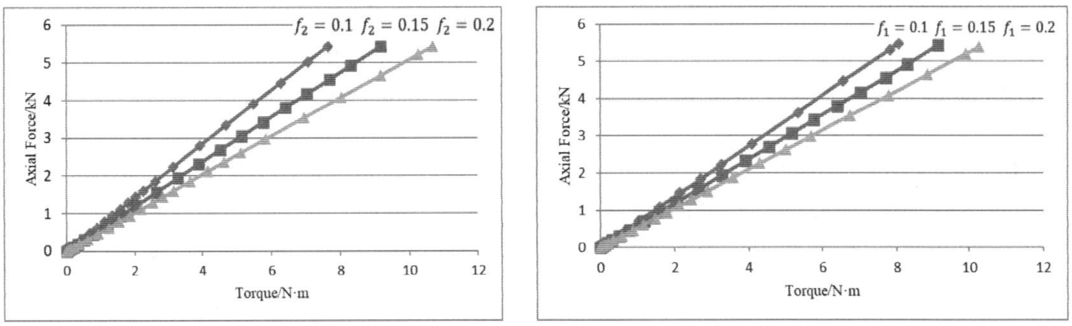

$(a) f_1 = 0.15$ $(b) f_2 = 0.15$

Figure 12. Effect of axial force-torque on different friction coefficients

All the above analysis shows that the friction coefficient between thread interfaces and the washer-nut interface have an obvious influence on the transformation of pre-tightening. For the instrumented bolt described in this paper, the influence of the friction coefficient between the washer-nut interfaces is a litter greater than that of the friction coefficient between thread interfaces on the transformation of tightening torque, which also indicates the necessity of the detailed model.

6. Conclusion

The main work of this paper lies in:

- The thread section equation based on the slice method was derived. A parametric mesh construction strategy was developed to automatically build a high-fidelity 3D FE model with lift angle, which can be widely used for detailed modelling for different shapes of thread. As the mesh generation algorithm is not relate to the solver, the program only needs to parse the input file of a specific solver to adapt the solution, which makes the program in this paper widely applicable.
- The detailed 3D FE model for a single bolt was established and calibrated by experiment using an instrumented bolt. The relation between tightening torque and the axial force was studied by FE method, experiment, and empirical formula. The results indicate that the axial force of the

bolt obtained by the engineering algorithm is larger, and the axial force obtained by the detailed FE model is consistent with the experiment.

- Different friction coefficients between the thread interfaces and washer-nut interfaces were analysed by parameters study, which shows that both of these coefficients have significant influence on the conversion of pre-tightening force to axial force. The clamping force decreases by 28% when f_2 varies from 0.1 to 0.2 and 23% when f_1 varies from 0.1 to 0.2
- Traditional empirical formula ignores the friction transferred between the nut and the washer, which overestimates the clamping force of the bolt. By using the detailed model described in this paper to calculate the axial force of the bolt, and then applying it to the simplified model, the accuracy of the axial force can be significantly improved.

References

[1] McCarthy C T, Gray P J. An analytical model for the prediction of load distribution in highly torqued multi-bolt composite joints[J/OL]. Composite Structures, 2011, 93(2): 287–298. DOI:10.1016/J.COMPSTRUCT.2010.09.017.

[2] Sharos P, Egan B, Mccarthy C. An analytical model for strength prediction in multi-bolt composite joints at various loading rates[J/OL]. Composite Structures, 2014, 116: 300–310. DOI:10.1016/j.compstruct.2014.05.021.

[3] Lawlor V P, McCarthy M A, Stanley W F. An experimental study of bolt-hole clearance effects in double-lap, multi-bolt composite joints[J]. Composite Structures, 2005, 71(2): 176–190. DOI:10.1016/j.compstruct.2004.09.025.

[4] Nezhad H Y, Egan B, Merwick F, et al. Bearing damage characteristics of fiber-reinforced countersunk composite bolted joints subjected to quasi-static shear loading[J/OL]. Composite Structures, 2017, 166: 184–192. DOI:10.1016/J.COMPSTRUCT.2017.01.029.

[5] Gray P J, O'Higgins R M, McCarthy C T. Effect of thickness and laminate taper on the stiffness, strength and secondary bending of single-lap, single-bolt countersunk composite joints[J/OL]. Composite Structures, 2014, 107: 315–324. DOI:10.1016/j.compstruct.2013.07.014.

[6] Fengrui Liu, Jianyu Zhang, Libin Zhao, et al. An analytical joint stiffness model for load transfer analysis in highly torqued multi-bolt composite joints with clearances[J/OL]. Composite Structures, 2015, 131: 625–636[2022–08–28]. DOI:10.1016/j.compstruct.2015.06.003.

[7] Tianliang Qin, Libin Zhao, Jianyu Zhang. Fastener effects on mechanical behaviors of double-lap composite joints[J/OL]. Composite Structures, 2013, 100: 413–423[2022–08–28]. DOI:10.1016/j.compstruct.2013.01.008.

[8] Mingjiu X. Joint for Composites Materials[M/OL]. Shanghai Jiao Tong University Press, 2016.

[9] Houbin Wang. Study on Axial Force Measurement Method for Bolt of Joint [J/OL]. Aircraft Manufacturing Technology, 2018, 61(22): 82–86.

[10] Liu Z, Yan X, Zheng M, et al. The effect of tightening again on bolt loosening under transverse load: experimental and finite element analysis[J/OL]. Structures, 2022, 44: 1303–1311[2022–12–06]. DOI:10.1016/j.istruc.2022.08.049.

[11] Liu Z, Wang B, Li Y, et al. Analysis of self-loosening behavior of high strength bolts based on accurate thread modeling[J/OL]. Engineering Failure Analysis, 2021, 127: 105541[2022–12–06]. DOI:10.1016/j.engfailanal.2021.105541.

[12] Fukuoka T, Nomura M. Proposition of helical thread modeling with accurate geometry and finite element analysis[J/OL]. Journal of Pressure Vessel Technology, 2008, 130(1)[2022–11–11]. DOI:10.1115/1.2826433.

[13] Yu Q, Zhou H, Wang L. Finite element analysis of the relationship between tightening torque and initial load of bolted connections[J/OL]. Advances in Mechanical Engineering, 2015, 7(5): 1687814015588477[2022–05–14]. DOI:10.1177/1687814015588477.

[14] WANG Houbing, WEI Jingchao, CHENG Li'nan, LI Xinxiang, ZHAO Rong. Force sensor for axial force and shear force of bolts in joints[J]. ACTA AERONAUTICAET ASTRONAUTICA SINICA, 2021, 42(3): 424459-424459

Study on the combustion process of the ramjet combustion chamber

Tingliang, Zhang[1*], Hui Zhao[1], Lantao He[2] and Tianyu Wang[2]

[1]Engineering College of Aeronautics, Air Force Engineering University,

[2]Unit 94106, Xi'an, China

[*]Corresponding author's e-mail: 942632944@qq.com

Abstract: To explore the ignition performance of the ramjet combustion chamber, this paper combined the specific structure and working principle of the ramjet to model the ramjet combustion chamber, and carried out numerical simulation research on the flow field characteristics and combustion process of the stable operation of the combustion chamber. The results show that for the blunt-body of the flame stabilizer, the overall flame stability zone is relatively large at low inlet velocity, but the strength of the return zone at low inlet velocity is smaller than that at high speed. The flame stability zone formed at high inlet velocity is more stable and has stronger resistance to external disturbance. When the ratio of oil and gas mixtures is in the flammability limit range, the selection of ignition position has an important influence on the success of ignition. When the high-temperature zone formed by the initial ignition core can be diffused into the reflux zone and maintain sufficient temperature, the ignition can be successful.

1. Foreword

At present, with the rapid development of the hypersonic vehicle, an increasing thrust-to-weight ratio of aero-engine is increasingly needed. As the core component of ramjet, the reliable ignition and stable combustion of the ramjet combustion chamber are of great significance to the stable operation of ramjet. In the past, the development of ram-combustor is mainly based on experiments, but there are some problems in the experiment, such as complicated equipment, high cost, and inconvenient measurement. With the improvement of computing power and the development of CFD numerical simulation software, the numerical simulation method is more and more applied in the process of combustion chamber design and analysis.

As the core component of ramjet, the ramjet combustion chamber plays a pivotal role in its ignition reliability and combustion stability [1]. Therefore, experts and scholars at home and abroad have carried out a lot of research on the ramjet combustion chamber. In terms of ignition process research, many basic theoretical studies have been conducted abroad, including forced ignition of premixed gas [2], the establishment of a numerical simulation model of ignition phenomenon [3-5], flame diffusion mechanism research [6-9], etc. Zetterstrom [10] studied the liquid fuel ramjet with different intake modes and found that the unsteady vortex system in the dual intake combustor could be improved by using appropriate injection modes. Zhang Bin [11] used the EDC combustion reaction model to conduct a numerical simulation study on the ignition process and characteristics of the aero-engine main combustion chamber. Tano et al. [12] conducted an experimental study on the combustion characteristics of kerosene in scramjet engines. Wang Xiaodong [13] found through research that the axial vortex generated in the downstream flow field of the nozzle in the ramjet combustor is the key to improving the combustion efficiency, which provides a thought for improving the combustion efficiency in the ramjet combustor. However, there are no relevant studies on how the flow field changes after the flow passes through the blunt body evaporative flame stabilizer in the stamping combustor and how the flame forms and develops after the blunt body evaporative flame stabilizer in the open literature.

Due to the high flow speed in the ramjet combustor, the complexity of the gas combustion process, and the influence of experimental conditions and monitoring means, it is difficult to obtain the temperature field distribution and velocity field distribution in the combustor through experimental methods, which limits the experimental study of the ramjet combustor to a certain extent. Therefore, this paper combined the specific structure and working principle of the ramjet to model the ramjet combustion chamber, the flow field characteristics of the ramjet combustion chamber and the combustion process were numerically simulated, and the flow field and the flame development process were analyzed. It provides a reference for the experiment, design, and improvement of the ramjet combustion chamber.

2. Model Establishment

2.1 Geometric Model

The object studied in this paper is the ramjet combustion chamber, whose overall structure is shown in Figure 1. The blunt body evaporative flame stabilizer is adopted in the stamping chamber. Compared with the ordinary blunt-body flame stabilizer, the internal ring cavity of the blunt-body evaporative flame stabilizer is provided with a fuel distribution pipe, and 36 pairs of oil injection holes are arranged on the distribution pipe for oil injection. An air slot for cooling is provided on the front wall of the blunt body flame stabilizer; The igniter is inserted into the flame stabilizer at an Angle perpendicular to the cut plane of the flame stabilizer.

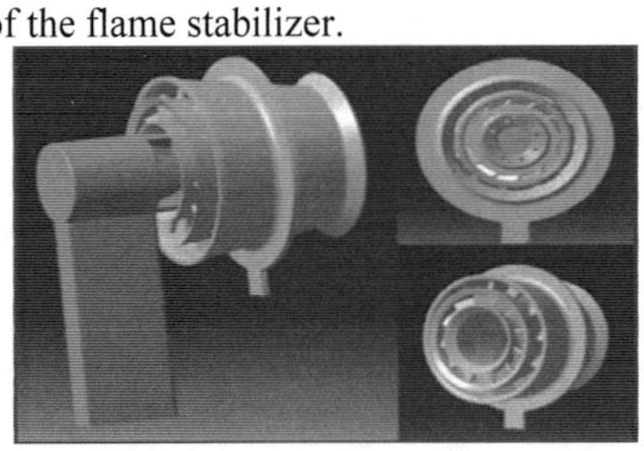

Figure 1. Overall geometric model of stamping combustor

2.2 Grid Division

The structure of the ram-combustor is relatively simple and symmetrical. The structure of the stabilizer, distribution tube, and igniter in the front section of the ram-combustor are relatively precise, and the discrete phase of fuel droplets is involved here. In this paper, the method of block grid division is used to encrypt the precise parts of the structure. To obtain more accurate calculation results for the divided grids, grid independence verification was carried out before calculation. Under the same setting conditions, the position and shape of the reflux area and exit parameters were taken as a comparison basis to calculate the models with different mesh numbers and compare the results. Considering the amount of computation and the reliability of the results, the total number of grid cells for calculation is about 2 million. The grid of the calculation domain is shown in Figure 2.

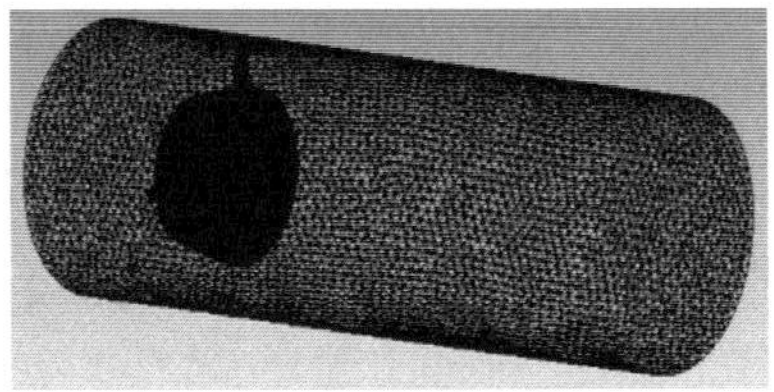

Figure 2. Stamping combustors two-dimensional surface computing domain grid

2.3 Simulation model

Due to the complexity and particularity of the structure of the evaporative flame stabilizer used in the stamping combustor, the actual oil-gas mixing process is very complicated, involving the atomization, crushing, evaporation of the fuel itself, as well as gas-liquid interaction, droplet and wall interaction, droplet interaction and so on. Considering the limitation of the calculation amount, it is necessary to simplify it reasonably. In the simulation, the discrete phase model (DPM) was used to simulate the oil-gas mixing process in the stamping combustion chamber, and the cold flow field after oil-gas mixing in the combustion chamber was taken as the initial flow field. The EDC combustion reaction model in the finite rate model was opened, and the chemical reaction was carried out by simulating different ignition modes through a high-temperature jet or setting a local high-temperature zone.

2.4 Boundary conditions

The inlet boundary of the ram burner is set as the velocity inlet, and the outlet boundary is set as the free outlet. The oil-gas mixed flow field was taken as the initial flow field of the hot flow field, and the EDC chemical reaction model was started based on the setting of the cold simulation model. The intake speed is set at 200 m/s, the intake temperature is 500K, the (relative) intake pressure is 5000Pa, and the flame stabilizer wall temperature is 1000K. The fuel nozzles of the ram combustor are respectively located in front of each intake nozzle of the evaporative flame stabilizer, and the nozzles are provided at the corresponding positions (injection). The particle size range of fuel oil droplets is set as 30-80 microns, the average particle size is set as 50 microns, and the droplet distribution is rosin-rambler distribution. The interaction between the droplet and wall surface was set as reflection type, and the oil supply of each nozzle was 0.015kg/s.

3. Result Calculations

3.1 The influence of different inlet velocities on the flow field in the ram-combustor

Before the ram-combustion chamber ignition, the inlet velocity ranges from Mach 0.15-0.28, about 50m/s to 100m/s. The cold flow field is calculated with 10m/s as the interval, and the boundary of the return area is shown in the same figure for comparison, as shown in Figure 3. It can be seen from the figure that the change of intake velocity within this range has little influence on the boundary of the return zone of the evaporative flame stabilizer, and the contour lines are overlapping.

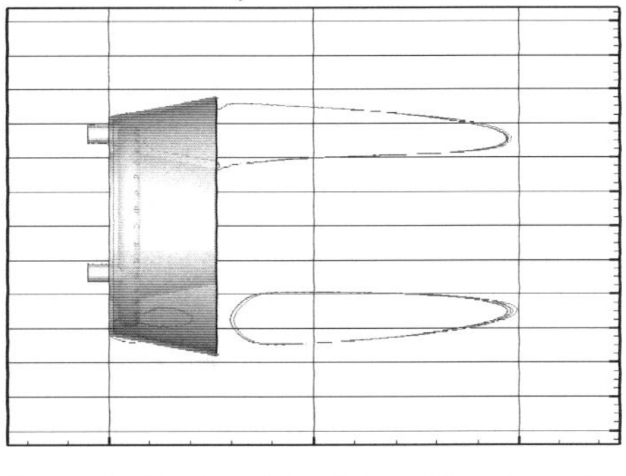

Figure 3. 50m/ s-100m /s backflow boundary

From the principle of steady flame in the combustion chamber, the flame surface can be stable because the flame propagation speed is equal to the flow speed. If the flow speed in the flow field is less than the flame propagation speed, the flame can be maintained here and not blown away. To further explore the influence of intake velocity on the flame stability characteristics of the flame stabilizer, the low-speed isolines near the reflux area are shown in Figure 4. To facilitate the observation of the results, the isolines with the absolute value of axial velocity within the range of 10m/s are made.

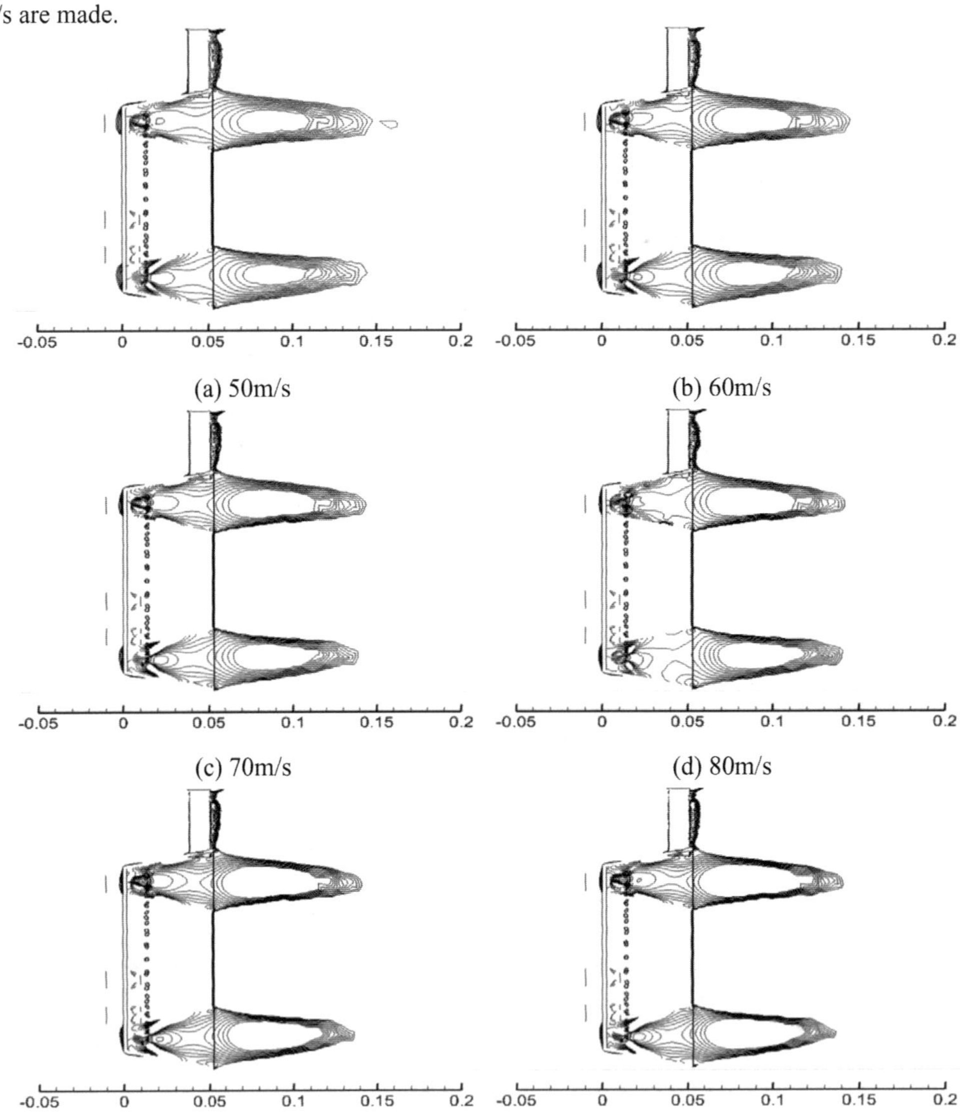

(a) 50m/s (b) 60m/s

(c) 70m/s (d) 80m/s

(e) 90m/s (f) 100m/s

Figure 4. Axial velocity contour at different intake velocities

As can be seen from Figure 4, contour lines are distributed evenly. With the increase of intake velocity, the outermost low-speed boundary shortens and becomes thinner, while the inner low-speed boundary expands, and the low-speed area around the reflux area shrinks. Since the area that can stabilize the flame only needs to meet the flow velocity less than the flame propagation velocity, for the blunt-body of the flame stabilizer, the overall stable flame area is relatively large when the intake velocity is low. However, it is also found that the strength of the return flow area at low inlet velocity is not as strong as that at high inlet velocity, and there is a phenomenon of oscillation and separation of the steady flame area. Therefore, the steady flame area formed at high inlet velocity is relatively small but more stable, with stronger resistance to external disturbance.

3.2 Calculation results and analysis of cold flow field
The proper cold flow field before ignition in the stamping combustion chamber plays an important role in successful ignition. Through the calculation of the cold flow field, the velocity field, pressure oil, and gas distribution in the stamping combustion chamber can be obtained, and the position of the reflux area and the total pressure loss of the combust chamber flow can be analyzed. The cold flow field shown in Figure 5 is obtained by calculation. Figure 5 (a) is the local flow diagram after the blunt body of the evaporative flame stabilizer. It can be seen from the streamlined path shape that a stable backflow area is formed behind the blunt-body of the flame stabilizer. Figure 5 (b) is the cloud map of axial velocity. The red solid line is the boundary line of the return flow area with an axial velocity of 0. According to the velocity cloud image, a convergent channel is formed between the blunt-body of the flame stabilizer and the shell of the combustion chamber, and the velocity of the subsonic flow through the convergent channel increases.

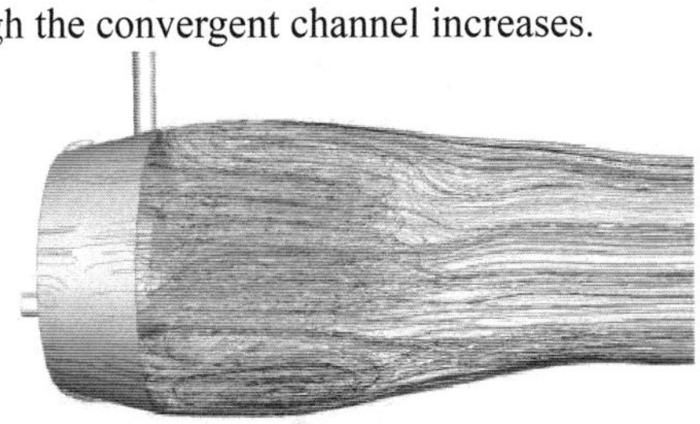

(a) Local streamline diagram behind the flame stabilizer

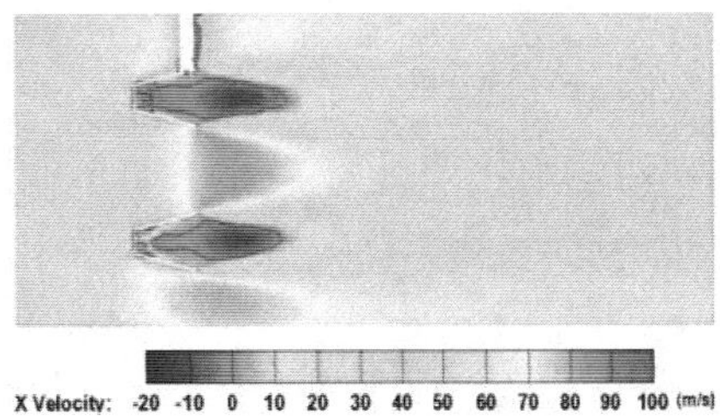

(b) Axial velocity nephogram

Figure 5. Three-dimensional cold flow field

The calculation results of the oil-gas mixed flow field with a discrete phase model are shown in Figure 6. Figure 6 shows the movement trajectory of fuel droplets, and the color of the droplets represents the mass flow fraction of fuel here. Through the analysis, it is found that the fuel concentration near the intake nozzle is low, while the fuel concentration between the intake nozzle is relatively high. The reason is that the air concentration near the intake nozzle is relatively high, the airflow is fast, and the air flows out directly from the small hole of the distribution pipe close to the intake nozzle, which makes the mass fraction of the fuel in the oil and gas mixture low.

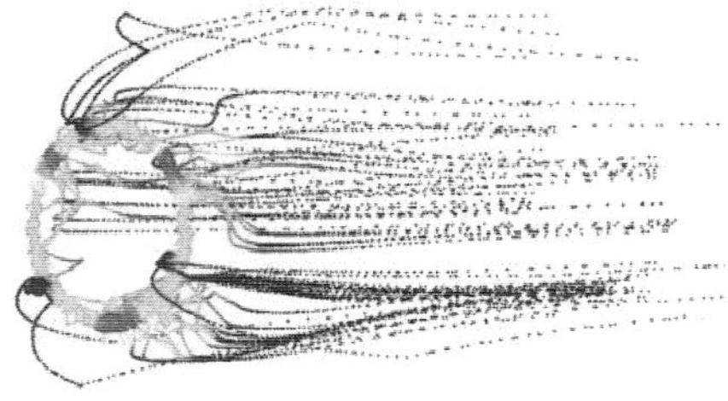

Figure 6. The trajectory of fuel droplets

3.3 Comparison of different ignition methods
Different ignition positions and ignition modes have a crucial impact on the success of ignition. In the model, high-temperature zones with different positions and shapes are set to simulate the initial ignition core region formed by different ignition positions and ignition modes. The temperature in the high-temperature zone is set at 2000K and the ambient temperature in the combustion chamber is 350K. Nine ignition schemes are set as shown in Figure 7.

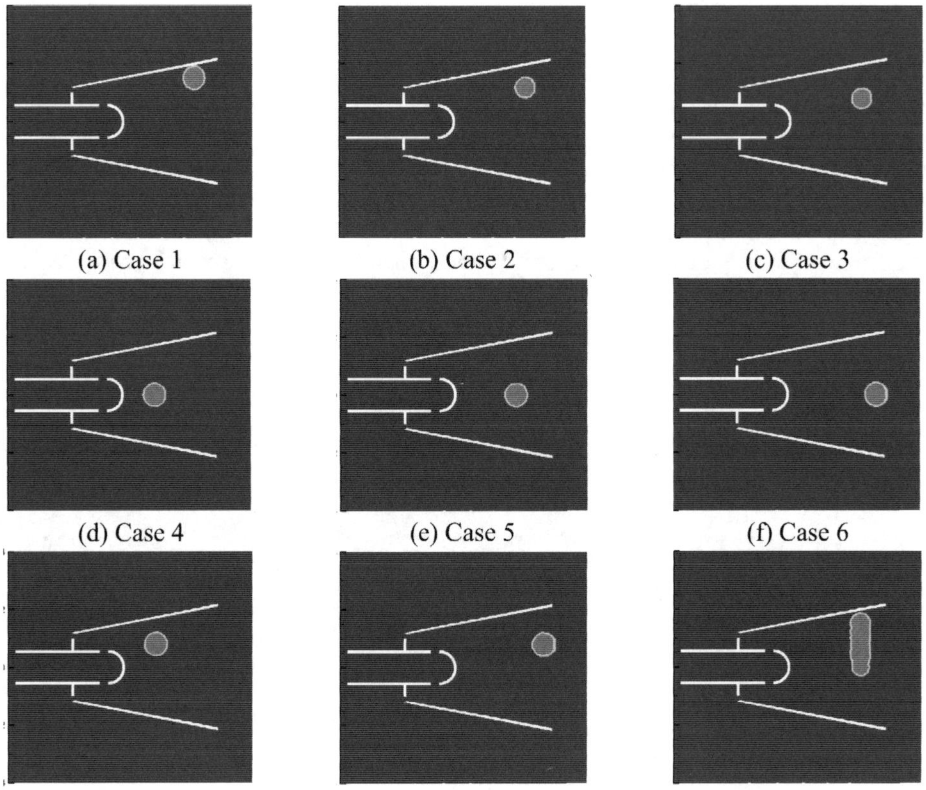

(a) Case 1 (b) Case 2 (c) Case 3

(d) Case 4 (e) Case 5 (f) Case 6

(g) Case 7 (h) Case 8 (i) Case 9

Figure 7. Ignition scheme

The long high-temperature zone in Case 9 simulates the high-temperature jet column formed by the hot jet ignition. Case 1, Case 2, and Case 3 respectively represent different insertion depths of the igniter; Case 4, Case 5, and Case 6 respectively represent ignition positions with different coaxial distances from the distribution tube axis in the flame stabilizer; The ignition positions of Case 3, Case 7, and Case 8 are set at different positions on the boundary of the return zone of the cold flow field. Although the ignition methods are different, the steady-state flow field finally achieved under the successful ignition condition is the same due to the same boundary conditions. Whether the ignition is successful or not is judged by the temperature cloud map, as shown in Figure 8. Where (a) is the temperature field cloud diagram of ignition failure. The high-temperature area of ignition fails to ignite the oil and gas mixture successfully, and the temperature decreases and is blown downstream under the action of low-temperature incoming flow; (b) is the temperature field cloud map of successful ignition. The temperature rapidly rises to nearly 3000K, and the high-temperature area reaches the root of the reflux area behind the distribution tube, gradually forming a stable ignition source and continuously igniting the subsequent oil and gas mixture. Therefore, if the temperature of the root of the reflux zone reaches more than 2000K after fire proliferation, the ignition is successful. The flame morphology is uncertain due to the disturbance of the unsteady flow behind the blunt body.

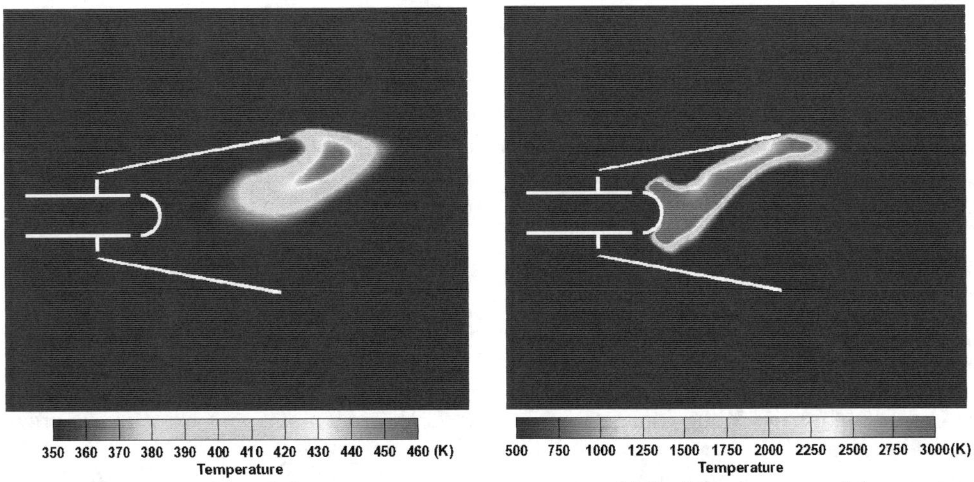

(a) Ignition failure (b) Ignition is successful

Figure 8. Temperature cloud image after ignition

A steady-state solution was carried out for the nine ignition schemes under the same conditions, and the ignition success was judged according to the temperature at the root of the reflux area. The ignition results of each ignition scheme were shown in Table 1.

Table 1. Ignition results of different ignition schemes

Mode of ignition	Ignition success or not
Case 1	failure
Case 2	successful
Case 3	successful
Case 4	successful
Case 5	successful
Case 6	successful
Case 7	successful
Case 8	successful
Case 9	successful

The ignition of Case 1 in Cases 1, 2, and 3 failed. Combined with the velocity field of the cold flow field and the concentration field of oil and gas components, it was found that the ignition position of Case1 was near the blunt body wall of the stabilizer, where the initial fire core was easily blown downstream due to the influence of the incoming flow from the air seam in front of it. As a result, the formed high-temperature area could not be spread to the interior of the return area. And finally, the core is blown out. Therefore, for the actual ramjet combustor, it is not recommended to set the igniter near the blunt body wall of the stabilizer. Although Case 2 is located outside the return zone, the high-temperature zone generated by the mixture around the initial ignition core can diffuse to the return zone, so it can be successfully ignited. Cases 4, 5, and 6 are all located inside the reflux zone. After the oil and gas mixture within the flammability limit is ignited around the initial ignition core, the high-temperature zone will spread upstream to the root of the reflux zone under the action of the reflux zone, where a new ignition source is formed to ignite the new mixture. Cases 3, 7, and 8 were all located at the boundary of the reflux area, and the oil and gas distribution in the blunt-body of the flame stabilizer was relatively uniform, and the three ignition locations were all successfully ignited. Case 9 can be regarded as the superposition of multiple ignition sources, so if one ignition source is in a successful ignition location, the entire jet ignition will be successful. This is also one of the reasons why the initial fire core formed by the hot jet igniter is large and ignition is easier to succeed. Based on the above analysis, when the ratio of oil and gas mixtures is within the flammability limit range, the selection of ignition position has an important influence on the success of ignition. When the high-temperature zone formed by the initial ignition core can be diffused into the reflux zone and maintain sufficient temperature, the ignition can be successful. On the contrary, the initial high-temperature zone of the initial fire core cannot diffuse and develop in the environment of continuous cooling of the incoming flow, and the ignition fails. Therefore, when setting the ignition position, it is not recommended to set the igniter near the blunt body wall of the stabilizer. The ignition position should be as close as possible to the low-speed reflux area for ignition based on the characteristics of the flow field in the combustion chamber.

4. Conclusions

In this paper, based on the specific structure and working principle of the ramjet, the ramjet combustion chamber is modeled, the flow field characteristics and combustion process of the ramjet combustion chamber are numerically simulated, and the flow field and flame development process are analyzed. The following conclusions are obtained:

(1) For the blunt-body of the flame stabilizer, the overall stable flame area is relatively large when the intake speed is low. However, it is also found that the strength of the return flow area at low inlet velocity is not as strong as that at high inlet velocity, and there is a phenomenon of oscillation and separation of the steady flame area. Therefore, the steady flame area formed at high inlet velocity is relatively small but more stable, with stronger resistance to external disturbance.

(2) Ignition location and flow field structure have an important influence on ignition success when the oil/gas ratio is appropriate. The ignition position should be as close to the reflux area as possible so that the flame can spread smoothly to the root of the reflux area to form a stable ignition source.

References

[1] HE Lining. (2006) Principle of aircraft propulsion system[M]. Beijing: National Defense Industry Press.
[2] Spalding D B. Combustion and mass transfer: a textbook with multiple-choice exercises for engineering students[M]. Oxford: Pergamon Press, 1979.
[3] Alvani R F, Fairweather M. Ignition characteristics of turbulent jet flows[J]. Chemical Engineering Research and Design, 2002, 80(8): 917-923.
[4] Veynante, D., Lucas, et al. Numerical simulation of the transient ignition regime of a turbulent diffusion flame[J]. AIAA journal, 1991, 29(5): 848-851.
[5] Bray K, Domingo P, Vervisch L. Role of the progress variable in models for partially premixed turbulent combustion[J]. Combustion and flame, 2005, 141(4): 431-437.
[6] Mastorakos E, Baritaud T A, Poinsot T J. Numerical simulations of autoignition in turbulent mixing flows[J]. Combustion and Flame, 1997, 109(1): 198-223.
[7] Kang T, KYRITSIS D C. Methane flame propagation in compositionally stratified gases[J]. Combustion science and technology, 2005, 177(11): 2191-2210.
[8] Greenberg J B. Finite-rate evaporation and droplet drag effects in spherical flame front propagation through a liquid fuel mist[J]. Combustion and flame, 2007, 148(4): 187-197.
[9] Hélie J, Trouvé A. A modified coherent flame model to describe turbulent flame propagation in mixtures with variable composition[J]. Proceedings of the Combustion Institute, 2000, 28(1): 193-201.
[10] Zette Strom. An Experimental Study of Side Dump Ram-jet Combustor [R]. ISABEL 85-7024.
[11] ZHANG Bin. (2014) Research on ignition simulation and characteristics of aero-engine combustor [D]. Nanjing University of Aeronautics and Astronautics.
[12] JIANG An-lin, TIAN Ye. (2020) Experimental Investigation on Combustion Characteristics in a Kerosene-Fueled Scramjet Engine [J]. Journal of Propulsion Technology.
[13] WANG Xiaodong. (2004) Effects of Fuel Injection Scheme on the Mixing and Combustion Efficiency of a Scramjet Combustor [J]. Acta Aeronautical et Astronautical Clinical, 25 (6): 556-559.

Journal of Physics: Conference Series **2489** (2023) 012011 doi:10.1088/1742-6596/2489/1/012011

Feasibility of autonomous orbit keeping with optimal fuel consumption using drag-free CubeSat

Rui LIU *, Donghua WEI, Yingying LIU, Yiheng LIU

School of Astronautics, Northwestern Polytechnical University

liurui@nwpu.edu.cn

Abstract. In this paper, the composition of the drag-free CubeSat platform was introduced first. Then the control strategy of drag-free CubeSat orbit keeping was proposed. Finally, the simulation with the control scheme of the Homan transfer orbit was compared. By comparing and analyzing the fuel consumption of the drag-free track and Homan transfer track, it is demonstrated that the continuous drag compensation of the drag-free control scheme can significantly reduce the consumption of thruster mass, and the drag-free technology has good advantages in track stability.

1. Introduction

These CubeSat, a low-cost nanosatellite designed using a modular, standardized concept, was proposed and defined as a design and manufacturing standard by Caltech and Stanford University in 1999. Science magazine announced CubeSat technology as one of the top 10 scientific breakthroughs of the year in 2014. CubeSats are widely used with commercial components, which have the characteristics of low cost, short cycle time, and fast response, and have broad application prospects. At present, the operating environment of the CubeSat network is mostly in the low orbit region, and the atmospheric drag is the most important interference factor. Due to the limited fuel carried by micro-satellites, the on-orbit life and practicability of micro-satellites are greatly limited. It is of great significance to study the use of towing-free technology to maintain the orbit of micro-satellites to prolong their life and the practicability of micro-satellites.

The principle of drag-free technology is to wrap an inertial detection component inside the spacecraft and make it operate in a purely gravitational environment through the shielding effect of the external spacecraft against non-conserved forces such as the atmosphere and solar pressure. The external spacecraft then tracks the orbit of the inertial detection unit and uses the micro-thrusters to exert sustained thrust to counteract the effects of non-conservative forces such as the atmosphere, the solar pressure, and the disturbance of the spacecraft itself, the whole spacecraft can achieve almost pure gravity under the flight state. The schematic of a drag-free spacecraft was shown in Figure 1.

There are four main applications of drag-free technology: navigation, earth science, fundamental physics, and astrophysics. Different application fields have different requirements for the performance of drag-free spacecraft [1]. Drag-free technology can eliminate non-conservative forces, resulting in the removal of large uncertain factors such as atmospheric drag and solar pressure perturbation in the spacecraft dynamics equation, which is convenient for a given orbit. NASA has demonstrated that using drag-free technology can reduce navigation errors and fuel consumption for orbit maintenance [2]. Therefore, the drag-free technology can be used to maintain the satellite orbit under the condition of low orbit and strong interference force with less fuel consumption.

In this paper, based on the research of the "Soaring Series" CubeSat [3] of Northwestern Polytechnical University, we investigate the feasibility of autonomous orbit keeping with optimal fuel consumption using a drag-free CubeSat. A drag-free CubeSat platform capable of providing 1/2U load space has been designed.

2. Composition of the drag-free CubeSat

The design of the overall structure of the drag-free CubeSat must take into account the system complexity, device cost, and performance index requirements. In this paper, the 3U CubeSat platform is used as an example to carry out the research on the drag-free orbit-keeping technology, and the mass of the whole satellite is 4 kg. The solar array is attached to the surface of the satellite [4], the attitude control system uses an integrated attitude sensor, and the attitude control actuators used mainly include a three-axis magnetic torque device [5] and a three-axis reaction flywheel [6]. The micro thrusters for drag-free control and orbit keeping are mounted on the central axis of the flight direction, at the rear of the satellite, and the thrusters are selected from commercial modular products. The available thrusters are solid cold gas thrusters and ion electric thrusters.

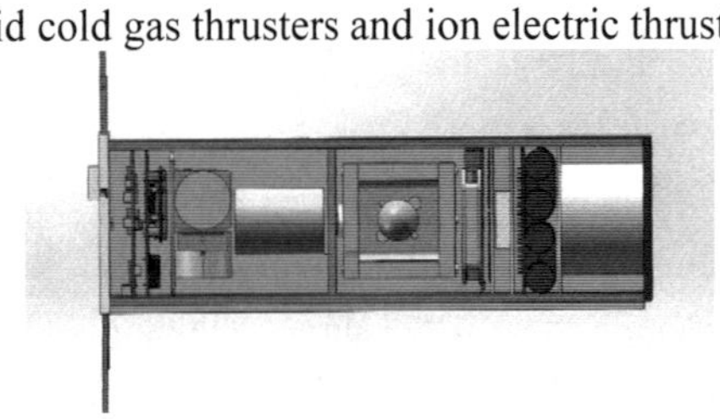

Figure 1. 3U Platform configuration of drag-free CubeSat

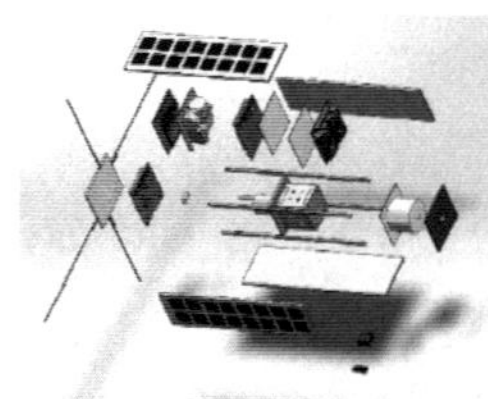

Figure 2. 3U Exploded view of drag-free CubeSat

The spherical detection mass block displacement mode drag-free technique is chosen to achieve orbit keeping of the CubeSat. The spherical detection mass block is located at the center of the satellite platform and uses relative position sensitizers to detect the relative position between the external spacecraft and the inspection mass, feeding back to the drag-free control system. As shown in figure 2, from left to right are the communication antenna, communication transceiver, GNSS receiver, three-axis reaction wheel, locking mechanism, drag-free system, three-axis magnetic torque device, on-board computer, gyroscope, power supply system, and solid cold gas thruster.

3. Strategies for maintaining the orbit of drag-free CubeSat

3.1. Dynamics of drag-free CubeSat

The motion of a drag-free CubeSat is a combination of two parts: the orbital motion following the detection mass and the body attitude motion. The purpose of the drag-free control of the CubeSat is to eliminate the effect of the non-gravitational force and to make the cube body follow the detection mass in a drag-free motion. The drag-free motion can be described by the following equation.

$$\mathbf{a}_l(t) = \frac{1}{m}(\boldsymbol{f}_d(\mathbf{r}, \mathbf{v}, \mathbf{q}, t) + \boldsymbol{f}(t)) \tag{1}$$

Here, $\boldsymbol{a}_l$ represents the acceleration of all forces other than gravity, which is zeroed in the drag-free control, $\boldsymbol{f}_d$ represents the environmental moment other than the gravitational force, and $\boldsymbol{f}$ represents the actuator moment.

In the spherical detection mass block displacement mode drag-free control problem, the CubeSat can be viewed as the relative motion process of two CubeSats, the drag-free CubeSat, and the detection mass block. By simplifying the theory of relative motion of CubeSat, the equations of relative motion of the detected mass block and the drag-free CubeSat in the dynamic coordinate system are:

$$\begin{cases} \ddot{x} - 2n\dot{y} - 3n^2 x = f_x \\ \ddot{y} + 2n\dot{x} = f_y \\ \ddot{z} + n^2 z = f_z \end{cases} \tag{2}$$

Where, x, y, and z denote the relative distance between the detected mass block and the drag-free cubic star, respectively. $\dot{x}$, $\dot{y}$, and $\dot{z}$ denote the relative motion velocity of the detected mass block and the drag-free cubic star, respectively. $\ddot{x}$, $\ddot{y}$, and $\ddot{z}$ denote the acceleration of the relative motion of the detected mass block and the non-trailing cubic star, respectively. And $n = \sqrt{\mu / a^3}$.

$$\Delta \boldsymbol{f} = \boldsymbol{f}_c - \boldsymbol{f}_s \tag{3}$$

where $\boldsymbol{f}_s$ and $\boldsymbol{f}_c$ are the acceleration vectors of the forces acting on the detected mass block and the drag-free cubic star. f_x, f_y, and f_z denote the components of $\Delta \boldsymbol{f}$ in each axis, respectively.

The fate state variables are: $x_1 = x$, $x_2 = y$, $x_3 = z$, $x_4 = \dot{x}$, $x_5 = \dot{y}$, $x_6 = \dot{z}$

$$\begin{cases} \dot{x}_1 = x_4 \\ \dot{x}_2 = x_5 \\ \dot{x}_3 = x_6 \\ \dot{x}_4 = 3n^2 x_1 + 2n x_5 + f_x \\ \dot{x}_5 = -2n x_4 + f_y \\ \dot{x}_6 = -n^2 x_3 + f_z \end{cases} \tag{4}$$

The amount of life control is $\boldsymbol{u} = [u_1 \quad u_2 \quad u_3]^{\mathrm{T}}$, then the state space of the system is described as:

$$\dot{\boldsymbol{x}} = \boldsymbol{A}\boldsymbol{x} + \boldsymbol{B}\boldsymbol{u} \tag{5}$$

$$\boldsymbol{y} = \boldsymbol{C}\boldsymbol{x} \tag{6}$$

Where

$$\boldsymbol{A} = \begin{bmatrix} 0 & 0 & 0 & 1 & 0 & 0 \\ 0 & 0 & 0 & 0 & 1 & 0 \\ 0 & 0 & 0 & 0 & 0 & 1 \\ 3n^2 & 0 & 0 & 0 & 2n & 0 \\ 0 & 0 & 0 & -2n & 0 & 0 \\ 0 & 0 & -n^2 & 0 & 0 & 0 \end{bmatrix}, \quad \boldsymbol{B} = \begin{bmatrix} 0 & 0 & 0 \\ 0 & 0 & 0 \\ 0 & 0 & 0 \\ 1 & 0 & 0 \\ 0 & 1 & 0 \\ 0 & 0 & 1 \end{bmatrix}, \quad \boldsymbol{C} = \begin{bmatrix} 1 & 0 & 0 & 0 & 0 & 0 \\ 0 & 1 & 0 & 0 & 0 & 0 \\ 0 & 0 & 1 & 0 & 0 & 0 \\ 0 & 0 & 0 & 1 & 0 & 0 \\ 0 & 0 & 0 & 0 & 1 & 0 \\ 0 & 0 & 0 & 0 & 0 & 1 \end{bmatrix} \tag{7}$$

According to the linear constant system energy-controlled rank criterion and energy-viewing rank criterion, $\mathrm{rank}\ \boldsymbol{Q}_c = \mathrm{rank}\ \boldsymbol{Q}_o = 6$, this system is energy-controlled energy-viewing.

3.2. Drag-free control

The CubeSat drag-free control law strives to be simple and easy to implement to meet the in-orbit real-time requirements, and when using state feedback, its information is large and complete, allowing the response characteristics to be freely governed without increasing the system dimensionality. Therefore, the state feedback control law is designed as:

$$\boldsymbol{u} = -\boldsymbol{K}\boldsymbol{x} + \boldsymbol{v} \tag{8}$$

$\boldsymbol{v}$ is the reference input and $\boldsymbol{K}$ is the matrix such that the closed-loop system formed by the feedback

$$\dot{x} = \left(\boldsymbol{A} - \boldsymbol{B}\boldsymbol{K}\right)\boldsymbol{x} + \boldsymbol{B}\boldsymbol{v} \tag{9}$$

Asymptotically stable, indicating that all its eigenvalues have negative real parts, the system achieves state feedback calming. As long as the system is fully controllable, the poles of the system can be realized in any configuration.

4. Simulation Analysis

The advantages of the drag-free spacecraft in terms of fuel consumption and orbital status are investigated by comparing the drag-free orbit-keeping scheme with the periodic lifting orbit-keeping scheme.

4.1. Drag-free track simulation

The initial circular orbit is 350 km. The atmospheric drag of the low-orbiting satellite is the largest disturbance factor, considering the atmospheric drag uptake, the simulated drag coefficient is 2.2, the head-on area is the cross-sectional area, the 3 U CubeSat cross-sectional area is 0.01 m^2, and the spacecraft mass is 4 kg. The initial conditions of the drag-free track simulation are shown in Table 1.

Table 1 Table of initial conditions for simulation without dragging track

Initial conditions	Parameter	Initial conditions	Parameter
Orbital coordinate system type	Classical	Right ascension at ascending intersection	90 °
Semi-major axis of the orbit	6721 km	Eccentricity of orbit	0.00016
Orbit inclination	60 °	Orbital coordinate system	ICRF
Argument of perigee	180 °	True anomaly	0 °
orbital altitude	350 km	Orbital period	5491.46 s
Atmospheric density model	Jacchia-Roberts	Face area	0.01 m^2
Atmospheric drag coefficient	2.2	Reflection coefficient	1.0
CubeSat scale	3U	CubeSat quality	4 kg

With the above orbit-keeping strategy, the orbital inclination and orbital altitude of the drag-free CubeSat vary as shown in figure 3 and figure 4 respectively.

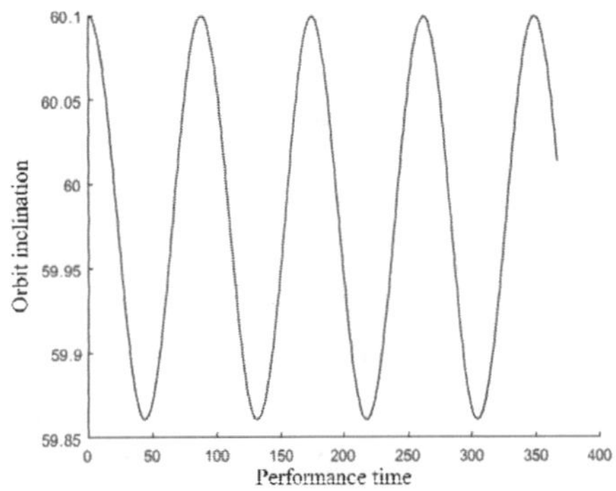

Figure 3. Variation of inclination angle

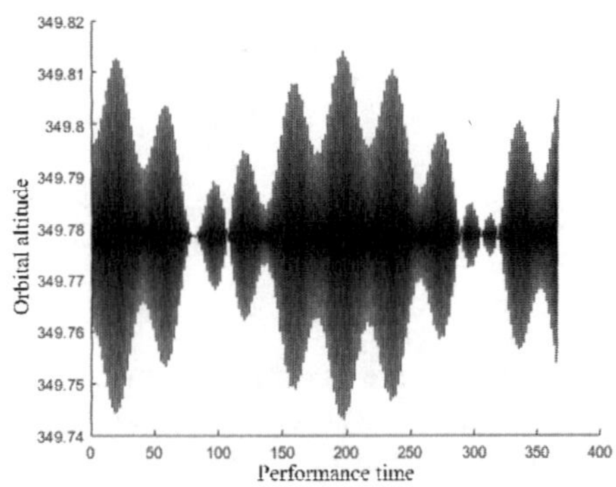

Figure 4. Variation curve of orbital height

From the simulation results, it can be seen that in a drag-free orbit under the effect of gravity and non-spherical regenerative forces only, the drag-free CubeSats with almost no height decay due to the absence of atmospheric drag interference.

The impulse provided by the thruster over time is shown in Figure 5.

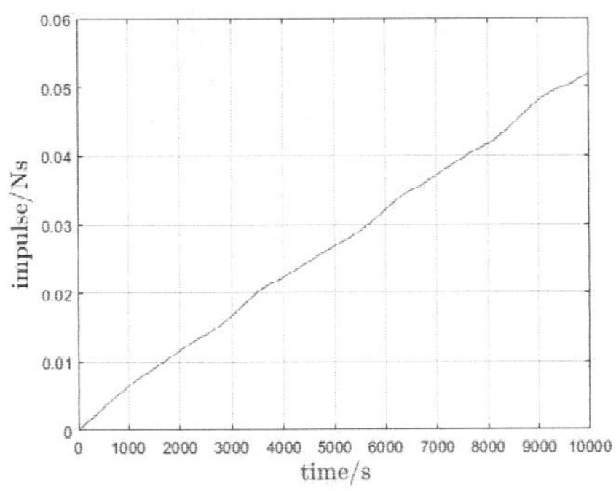

Figure 5. The impulse provided by the thruster

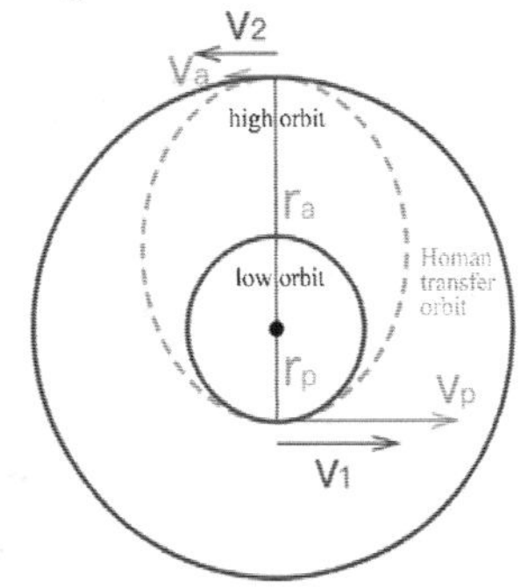

Figure 6. Schematic diagram of Homan transfer orbit

4.2. Comparative Simulation of Periodic Lifting of Homan Transfer

Ideally, since only two engine thrusts are required for the Hohmann transfer orbit, the satellite lift orbit will be a Hohmann transfer orbit with relative fuel saving. Schematica diagram of Homan transfer orbit is showed in figure 6. The satellite decayed from 350 km altitude orbit by atmospheric drag and decays to 345 km in 340 orbit9al cycles (21 days) on average. Altitude change curve of 350 km orbit is showed in figure 7. Thus, the thrusters are corrected for orbital altitude with a correction period of 5 km per decay. Find the Homan transfer orbit and Transfer Time T.

$$\Delta v = v_r \ln \frac{m_0}{m_1} \tag{10}$$

where m_0 and m_1 represent the initial mass of the satellite before the Homan transfer and the termination mass after the Homan transfer is completed, respectively, and is v_r the jet velocity. The fuel consumption equation is derived as:

$$\Delta m = m_0 - m_1 = m_0 \left(1 - e^{-\frac{\Delta v}{v_r}}\right) \tag{11}$$

4.3. Analysis of the conditions and advantages of using drag-free technology

The Homan transfer orbit scheme is simulated under the same conditions as the drag-free control scheme, the thruster-specific impulse used in the simulation is 65 s, and the gravitational acceleration at 350 km altitude is about 8.8 m/s^2.The calculated speed difference of the Homan transfer orbit is 0.0029 km/s, and the transfer time T is 0.7624 h, which means the fuel consumption of a single orbit transfer is about 0.0202 kg. Meanwhile, based on the total impulse provided by the drag-free control thrusters, it can be calculated that the drag-free CubeSat orbits for 21 days and the speed difference is 2.25825 m/s, fuel consumption 0.0158 kg. In the same case, the drag-free control method saves 21.8% of the thruster mass compared to the Homan transfer orbit control. A comparison of the fuel consumed by the two orbit-keeping methods for different time spans is shown in Figure 8.

It can be concluded that the drag-free technique is more beneficial to maintain the spacecraft's orbital altitude. The drag-free technique has an advantage in orbital stability during long spacecraft operations.

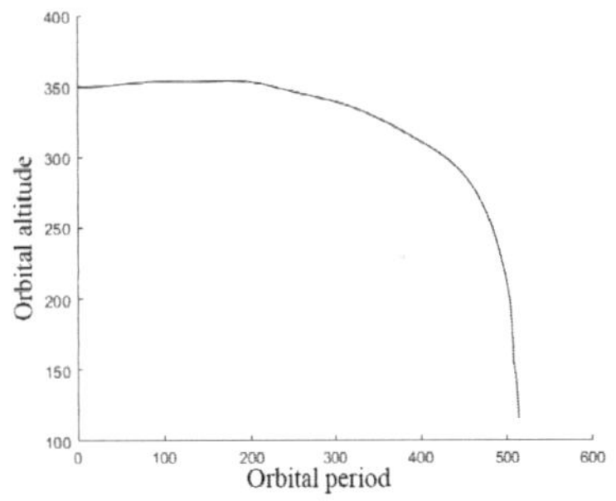

Figure 7. Altitude change curve of 350 km orbit

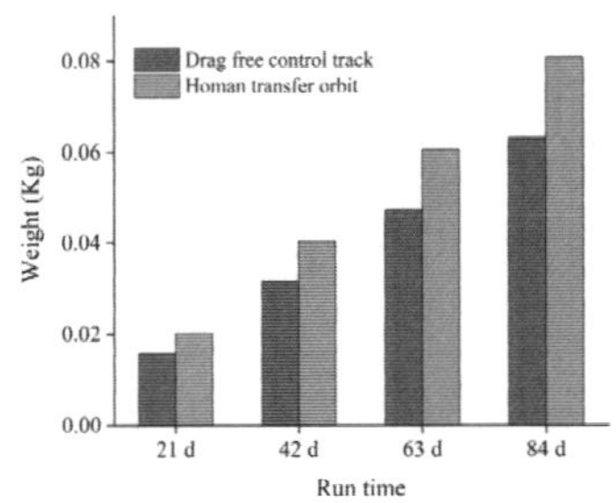

Figure 8. Fuel consumption comparison

5. Conclusions

This paper combined the drag-free technology with the CubeSat platform, proposes a drag-free CubeSat orbit maintenance control strategy, and finally conducts a comparison simulation with the Homan transfer orbit control scheme. It is verified that under the same conditions, the continuous drag compensation of the drag-free control scheme can significantly reduce the thruster mass consumption compared with the Homan transfer orbit control using intervals. Thus, it is confirmed that the drag-free CubeSat can significantly extend the service life for a given thruster mass. The fuel-saving advantage of drag-free CubeSat in orbit maintenance makes them more promising for operating CubeSat constellations and CubeSat networks in large-drag near-Earth orbits.

Acknowledgments

The research was supported by the Shaanxi Provincial Key R&D Program (2020KWZ-16).

References

[1] Deng J, Cai Z, Chen K, Shi X, Yu J and Li H 2019 Drag-free control and its application in China's space gravitational wave detection *Chin. Opt.* **12(3)** 503-14

[2] Nguyen A N and Conklin J W 2015 Three-axis drag-free control and drag force recovery of a single-thruster small satellite *J. Spacecraft. Rockets.* **52(6)** 1640-49

[3] Liu Y, Zhou J, Liu G, Zhang J, Bai B, Li P, Huang H and Liu R 2019 Development and prospect of "Aoxiang" series CubeSats *J. Astronautics.* **40(10)** 122-31

[4] Li P, Zhou J, Yu X and Liu G 2018 Design and on-orbit validation of electric power system for Aoxiang-Sat *ACTA. Energiae. Solaris. Sinica.* **39(04)** 1002-7

[5] Liu G and Zhou J 2017 Speed estimation and control strategy for micro-reaction wheel *J. Zhejiang Univ.-SCI A* **51 (12)** 2436-43

[6] Bai B, Zhou J and Wang S 2017 Multi-objective optimization design and implementation of a magnetorquer with magnetic core for CubeSat *J. Astronautics* **38(7)** 766- 71

The Impact on Absolute Along-Track Station-Keeping of Coordinate Additional Perturbation

Qi Liu[1], Shuge Zhao[1,*], Xiaoyu Li[1], Xuelei Deng[1] and Zhen Zhang[2]

[1]CASIC Space Engineering Development Co., Ltd., Beijing 100854, China

[2]Beijing Institute of Electronic System Engineering, Beijing 100854, China

shugezhao@163.com

Abstract. The purpose of the current paper is to investigate how the coordinated additional perturbation influences the along-track station-keeping. The absolute along-track station-keeping method has been developed and can achieve high-precision maintenance when the orbits are near-pole in previous studies. The accuracy of station-keeping deteriorates when the orbits are not near-pole. The reason is that the additional perturbation of the Mean Equator Mean Equinox of the J2000.0 (J2000) coordinate system makes a great difference in the absolute station-keeping between different inclinations, which has not been studied before. First, the impact factor model is constructed to measure the influence of the coordinate additional perturbation. Second, the impact factors of different orbits are analyzed and it is found the impact factor increases gradually with the decrease of inclination. Third, how impact factor indicating the accuracy of station-keeping is investigated and coordinate systems are recommended based on the impact factors. Finally, the numerical simulations of station-keeping in the J2000 coordinate system and True Equator Mean Equinox (TEME) coordinate system are carried out. Results show that the TEME coordinate system is more advantageous in the station-keeping mission than J2000 when the orbits are not near-pole.

1. Introduction

Mega satellite constellations in low earth orbit (LEO) can provide complete global coverage and improve the quality of service with lower signal propagation delays [1]. Thus, the construction of the LEO constellation has been attracting increasing attention all over the world. Programs represented by Starlink, OneWeb, Telesat, et al. have already started their deployment [2]. The constellation station-keeping is defined as keeping each satellite of the constellation in its desired “position” with periodic orbit-adjustment maneuvers [3]. With consideration of both constellation coverage and space safety, it is imperative to implement station-keeping control [4]-[6].

From the perspective of the control reference, the constellation station-keeping is divided into absolute control and relative control. The reference position of each satellite is predefined and not affected by other satellites in the absolute station-keeping. In contrast, the absolute positions are not concerned but the relative positions of satellites are controlled in the relative station-keeping. The absolute station-keeping method is relatively reliable due to the application of single satellite control techniques [7]. Thus, the absolute station-keeping for the LEO constellation has been a research highlight.

The high-precision absolute station-keeping methods for LEO constellation based on limit cycle were proposed in [7]-[9]. The semimajor axis change is determined by fitting the mean phase difference. It is assumed that the semimajor axis changes linearly due to atmospheric drag in [7] and quadratically in [8]. To realize high-precision station-keeping, the semimajor axis change is calculated iteratively according to the high-precision orbital prediction onboard [9].

Although the effects of the Earth's non-spherical harmonics, atmospheric drag, lunisolar attractions, and solar radiation pressure were taken into account in previous studies, the impact on the absolute station-keeping of the coordinate additional perturbation has not been studied. In orbital mechanics, the Mean Equator Mean Equinox of the J2000.0 (J2000) coordinate system is widely used. While the gravitational potential is constructed in an Earth-fixed coordinate system whose basic plane is the true equator. The deviation between the true equator of the Earth-fixed coordinate system and the mean equator of the J2000 coordinate system, which is caused by the precession and nutation of the Earth, results in the additional perturbation of the J2000 coordinate system [10]-[14].

The purpose of this paper is to investigate the impact of the coordinated additional perturbation on the along-track station-keeping. The paper is organized as follows. First, the impact factor model for measuring the effectiveness of the additional perturbation on the along-track station-keeping is established. Second, the impact factors of different orbits are analyzed. Then, how impact factors indicating the influences of the additional perturbation on station-keeping are investigated, and coordinate systems are recommended based on the impact factors. Finally, numerical simulations are performed to verify the station-keeping accuracy in the recommended coordinate system.

2. Model

2.1. The relationship between station-keeping period and orbital height

This paper is built based on [7]. The station-keeping period is

$$\Delta T_s = 2\sqrt{\frac{\Delta\lambda_{max} - \Delta\lambda_{min}}{\Delta\lambda_2}} \tag{1}$$

where $\Delta\lambda_{max}$ and $\Delta\lambda_{min}$ are the upper limit and lower limit of the desired mean phase difference interval; $\Delta\lambda_2$ is the second-order coefficient determined by the semimajor axis decay rate due to atmospheric drag

$$\Delta\lambda_2 = -\frac{3}{4}\sqrt{\frac{\mu}{a^5}}\dot{a}_d \tag{2}$$

where a is the semimajor axis, $\dot{a}_d$ is the semimajor axis decay rate, μ is the gravitational constant.

For near-circular orbit, the semimajor axis decay rate under the effect of atmospheric drag can be approximated in [15].

$$\dot{a}_d = -na^2\rho B \tag{3}$$

where $n = \sqrt{\mu/a^3}$ is the mean motion; ρ is the atmospheric density; $B = C_d S/m$ is the ballistic coefficient; C_d is the drag coefficient; S/m is the area-mass ratio of the satellite.

Substituting Equation (3) into Equation (2) results in

$$\Delta\lambda_2 = \frac{3}{4}\frac{\mu\rho B}{a^2} \tag{4}$$

The expression of the station-keeping period is obtained by substituting Equation (4) with Equation (1)

$$\Delta T_s = 4a\sqrt{\frac{\Delta\lambda_{max} - \Delta\lambda_{min}}{3\mu\rho B}} \tag{5}$$

Defining $\Delta\lambda_{\min} = -\Delta\lambda_{\max}$, the expression of ΔT_s is rewritten as

$$\Delta T_s = 4a\sqrt{\frac{2\Delta\lambda_{\max}}{3\mu\rho B}} \tag{6}$$

2.2. The long-periodic term of phase due to coordinate additional perturbation

The phase is defined as the sum of the argument of perigee and the mean anomaly. At time t, the long-periodic term of the phase caused by the additional perturbation is

$$\lambda_l(t) = (-\csc i)Q \tag{7}$$

where i is the orbit inclination. Q is determined by

$$Q = (2004''.3T)\cos\Omega + 8''.0\sin(\Omega - \Omega') + 1''.2\sin(\Omega + \Omega') \tag{8}$$

where Ω is the right ascension of ascending node (RAAN) of the satellite; Ω' is the mean ecliptic longitude of ascending node of the Moon; and T is the number of Julian centuries since the J2000.0 epoch

$$T = \frac{\mathrm{JD}(t) - \mathrm{JD}(\mathrm{J2000.0})}{36525} \tag{9}$$

where JD is the Julian Date. Thus, the influence of the coordinate additional perturbation increases with the extension of the interval from mission time to the J2000.0 epoch according to Equations (7)-(9).

In a short time, t can be approximated as constant. With a variation period of 18.6 years [16], Ω' varies much slower than Ω. Hence, the period of the long-periodic phase due to the additional perturbation is approximately equal to the period of Ω for any inclination according to Equations (7) and (8).

The time rate of Ω mainly depends on the J_2 term

$$\dot{\Omega} = -\frac{3}{2}J_2\left(\frac{R_e}{p}\right)^2 n\cos i \tag{10}$$

where p is the semi-latus and R_e is the Earth's mean equatorial radius.

Substituting Equation (10) into Equation (11) yields the period of the long-periodic term of the phase due to the additional perturbation

$$\Delta T_p = \frac{2\pi}{\dot{\Omega}} \tag{11}$$

2.3. Impact factor of coordinate additional perturbation

To measure the effect of J2000 coordinate additional perturbation on station-keeping, the impact factor of coordinate additional perturbation is defined as

$$\beta = \frac{\Delta\lambda_l}{\Delta\lambda_{\max}} \cdot \frac{\Delta T_s}{\Delta T_p} \cdot 100\% \tag{12}$$

where $\Delta\lambda_l$ is the amplitude of the long-periodic term of the phase due to the additional perturbation, i.e.,

$$\Delta\lambda_l = \frac{1}{2}\left[\max\left(\lambda_l(t)\right) - \min\left(\lambda_l(t)\right)\right] \tag{13}$$

The upper limit $\Delta\lambda_{max}$ is set according to the mission requirement. ΔT_s and ΔT_p are obtained by Equation (6) and Equation (11), respectively.

3. Numerical simulation

3.1. Impact factor of coordinate additional perturbation

In this subsection, the impact factors of coordinate additional perturbation in different orbits are analyzed. The orbit epoch is 1 Jan 2023 00:00:00.000 UTCG. The mean phase difference for each satellite is kept in the interval $[-0.06°, 0.06°]$, i.e., $\Delta\lambda_{min} = -0.06°$ and $\Delta\lambda_{max} = 0.06°$.

Figure 1 depicts the impact factor varying with the orbital height and inclination. The impact factor becomes larger as the orbital height increases and the inclination decreases. But the impact factor is more sensitive to the variation of the inclination than the orbital height. The impact factors are quite large when the orbits are not near-pole.

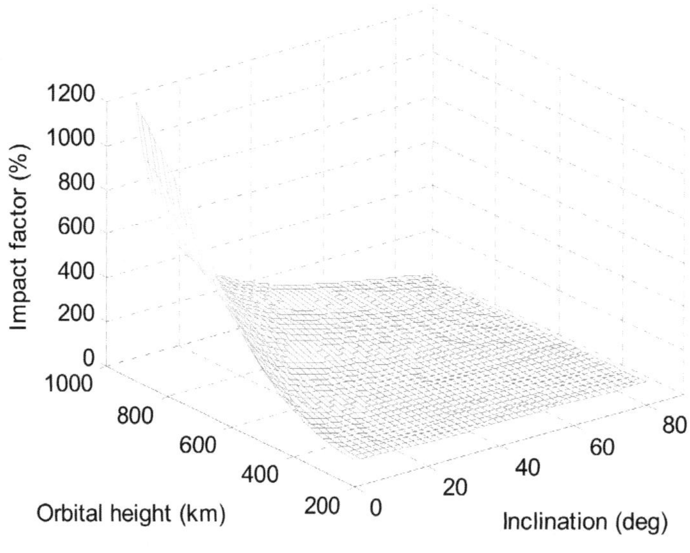

Figure 1. The impact factor on station keeping of the J2000 coordinate system.

Specifically, the impact factors of the coordinate additional perturbation at orbits with the same altitude of 800 km and different inclinations are presented in Table 1. At the inclination of 86°, the impact factor is 6.51%. While at the inclination of 50°, the impact factor increases to 82.35%. It indicates that there are great differences in the effect of the additional perturbation on the station-keeping concerning inclinations.

Table 1. Impact factors of coordinate additional perturbation for different orbits.

Orbital height and inclination	Station-keeping period (day)	Coordinate additional perturbation		
		Period (day)	Amplitude (deg)	Impact factor (%)
800 km, 50 deg	21	85	0.200	82.35
800 km, 62 deg	21	116	0.174	52.18
800 km, 74 deg	21	198	0.158	27.87
800 km, 86 deg	21	783	0.146	6.51

3.2. Simulation and analysis

This subsection investigates how the impact factor indicates the accuracy of the station-keeping mission by employing the absolute station-keeping method in the J2000 coordinate system.

The duration of the station-keeping mission is 120 days. The orbits are near-circular at the altitude of 800 km with inclinations as shown in Table 1. The area-mass ratio of the satellite is 0.0175 m^2/kg with a drag coefficient of 2.2. The sample period of the GPS observables is 2 hours. The one-sigma three-dimensional position and velocity errors of GPS observables are 5 m and 0.02 m/s, respectively. A model of Earth's gravitational field (4×4, World Geodetic System 1984 – Earth Gravitational Model 1996) and the atmospheric drag (US 1976 Standard Atmospheric Density Model) are included in the simulation. The orbit epoch and the mean phase difference limits are the same as those in Subsection 3.1.

The station-keeping results under different inclinations in the J2000 coordinate system are presented in Figure 2. The blue dashed lines represent the upper limit and the lower limit of the mean phase difference interval, and the red solid line represents the actual mean phase difference. It is intuitively clear that with the decrease of inclination i, i.e., with the increase of impact factor β, the accuracy of station-keeping deteriorates gradually.

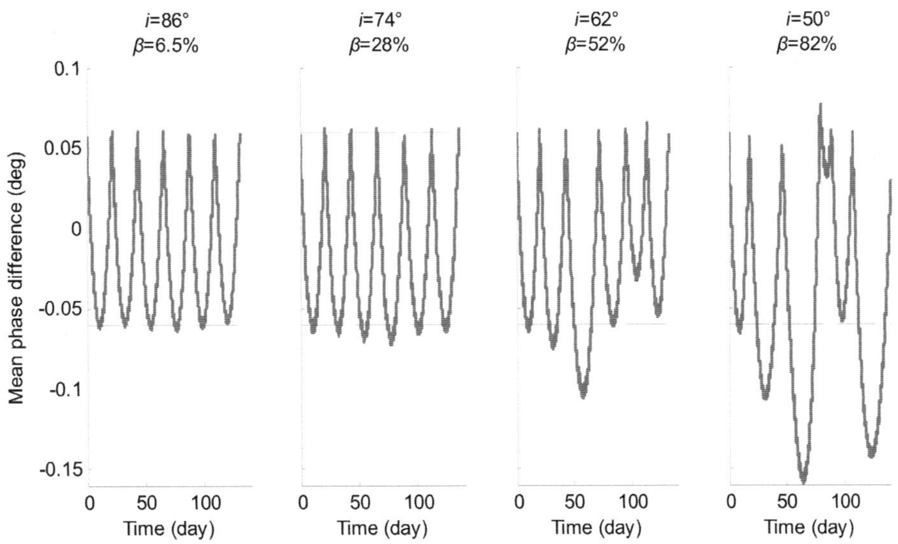

Figure 2. The comparison of station-keeping under different inclinations in the J2000 coordinate system.

When the impact factor β is 6.5%, the mean phase difference is controlled in the interval $[-0.06°, 0.06°]$ as expected. When the impact factor increases to 28%, the mean phase difference exceeds the lower limit a bit. When the impact factor increases to 52%, the mean phase difference exceeds more than -0.1°, and the accuracy of station-keeping deteriorates. When the impact factor increases to 82%, the mean phase difference exceeds more than -0.15°, and the accuracy of station-keeping deteriorates seriously. Hence, the accuracy of station-keeping can be indicated based on the impact factor.

Furthermore, the impact factor could be used to recommend the coordinate system for the station-keeping mission as shown in Table 2. The J2000 coordinate system could be selected when the impact factor is less than 50%, and the True Equator Mean Equinox (TEME) coordinate system could be selected when it is more than 50%.

Table 2. The coordinate system recommended based on the impact factors.

β	Effect	Coordinate system recommended
0~20%	Very weak	J2000
20%~50%	Weak	J2000/TEME
50%~80%	Strong	TEME/J2000
>80%	Very strong	TEME

As shown in Table 1, the amplitude of the long-periodic term of the phase due to the J2000 coordinate additional perturbation is an order of magnitude larger than the station-keeping limit. The reason is that the deviation between the true equator at mission time and the mean equator at the J2000.0 epoch increases with the extension of the interval from the mission time to the J2000.0 epoch. For the TEME coordinate system, the basic plane is the true equator and the basic axis points to the equinox of the J2000.0 epoch. The impact on along-track station-keeping of the coordinate additional perturbation should be mitigated.

Since the true equator oscillates in space due to the precession and nutation of the Earth, the TEME coordinate system is a quasi-inertial frame. The additional perturbation of the TEME coordinate system should be considered. However, the additional perturbation of the TEME coordinate system has a second-order effect, and it can be ignored in practice.

3.3. Simulation verification

To verify the station-keeping accuracy in the TEME coordinate system, a station-keeping simulation is performed. The orbit is near-circular at an altitude of 800 km with an inclination of 50°. The other initial conditions and the dynamic models are the same as those in Subsection 3.2.

Figure 3 presents the comparison of station-keeping results in the J2000 coordinate system and the TEME coordinate system. The mean phase difference appears large fluctuation and irregularity when the station-keeping is performed in the J2000 coordinate system. In contrast, the mean phase difference exhibits a specific periodicity pattern and is well controlled within the given interval in the TEME coordinate system. It demonstrates that the TEME coordinate system is very suitable for high-precision station-keeping when the orbits are not near-pole.

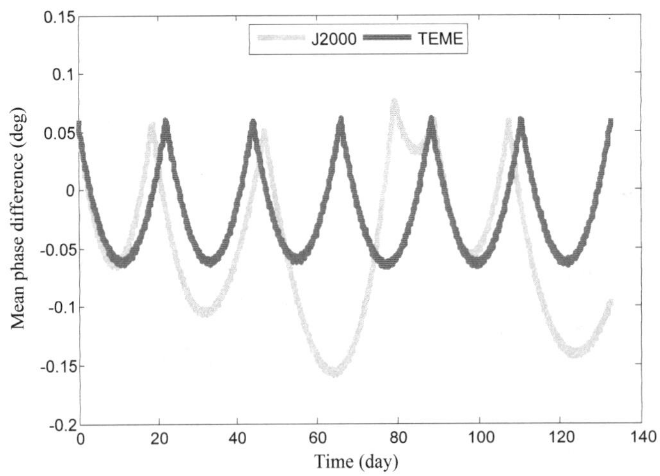

Figure 3. The comparison of station-keeping in the J2000 and TEME coordinate system.

4. Conclusion

In this paper, the impact factor model for measuring the effect of the J2000 coordinate additional perturbation on along-track station-keeping is established. The impact factor becomes larger as the orbital height increases and the inclination decreases. And the impact factor is more sensitive to the variation of the inclination than the orbital height. With the decrease of inclination, i.e., with the increase of impact factor, the accuracy of station-keeping in the J2000 coordinate system deteriorates gradually. When the impact factor is more than 50%, the TEME coordinate system is recommended. For orbits with an inclination of 50°, the absolute control method is capable of high-precision station-keeping in TEME coordinate system.

Besides, the True Equinox of Date (TOD) coordinate system, whose basic plane is also the true equator, is also applicable to station-keeping when the orbits are not near-pole.

References

[1] Ravishankar C, Gopal R, Ben Ammar N, Zakaria G and Huang X L. (2020) Next-generation global satellite system with mega-constellations. *Int. J. Satell. Comm. N*, **39**: 1–23.
[2] Juknaite K, Smith K and Crisp N H. 2021. Low latency broadband internet satellite constellations: technology, risks, and global impact. In: *ASCEND 2021*, Las Vegas, Nevada & Virtual.
[3] Chao C C. (2005) Applied Orbit Perturbation and Maintenance. *Aerospace Press,* California.
[4] Hernando-Ayuso J and Bombardelli C. (2021) Low-Thrust Collision Avoidance in Circular Orbits. *J. Guide. Control Dynam.* **44**: 983-95.
[5] May S L, Gehly S, Carter B A and Flegel S. (2018) Space debris collision probability analysis for proposed global broadband constellations. *Acta Astronaut.* **151**: 445-55.
[6] Raghunath K and Killick E. 2022. The Business and Economics of Space Sustainability. In: *ASCEND 2022*, Las Vegas, Nevada & Online.
[7] Liu Q, Xiang K H, Zhao S G, He Q, Li X Y and Zhang N. (2021) A high precision phase keeping method for LEO constellation. *J. Astronaut.* **42**: 1377-84.
[8] Qi Y, Li X G and Lin X X. (2022) Autonomous maintenance strategy for constellation configuration by satellite using GNSS data fitting. *Spacer. Eng.* **31**: 12-18.
[9] Sun Y, Cao J, Wu S G and He Y F. (2021). Adaptive absolute station-keeping method for LEO mega-constellation. *Mech. Eng.* **43**: 680-86.
[10] Kozai Y. (1960) Affect precession and nutation on the orbital elements of a close-earth satellite. *Astron. J.* **65**: 621-633.
[11] Lambeck K. (1973) Precession, nutation and the choice of reference system for close earth satellite orbits. *Celest. Mech.* **7**: 139-195.
[12] Liu L and Tang J S. (2008) On the additional perturbation due to the change of the coordinate system of Earth satellites and the choice of reference systems. *Chin. J. Space Sci.* **28:** 164–68.
[13] Liu L, Zhao Y H, Zhang W, Wang Y R and Wang J S. (2011). Coordinate additional perturbations to Mars orbiters and choice of corresponding coordinate system. *Chin. Astron. Astrophys.* **35**: 188–98.
[14] Du Y J, Zhang F Z, Xu T H, Gao F and Xu G C. (2021) Correction of precession-nutation and polar motion in analytical solutions of satellite equations of motion. *Adv. Space Res.* **68**: 4229-41.
[15] Chobotov V A. (2002) Orbital Mechanics. 3rd ed. *AIAA*, Reston, VA.
[16] Liu L. (2000) Orbit theory of Spacecraft. *National Defence Industry Press*, Beijing.

Buckling and Post-buckling Analysis of Composite Hat Stiffened Panels under Axial Compression Loads

Zhe Wang [1, 2, a*], Binwen Wang [1, 2, b], Rui Bao [1, c], Xiangming Chen [2, d], Qingxianglong Liang [2, e], and Fei Yu [2, f]

[1] School of Aeronautic Science and Engineering, Beihang University, Beijing, China

[2]Aeronautics Science and Technology Key Laboratory of Full-Scale Aircraft Structure and Fatigue, Aircraft Strength Research Institute of China, Xi'an, Shaanxi, China

[b]email: asriwbw@vip.163.com, [c]email: rbao@buaa.edu.cn, [d]email: asrichenxm@avic.com, [e]email:liangqxl@avic.com, [f]email: yuf078@avic.com,

*Corresponding author: [a]email: wangzhe_avic@163.com

Abstract: Owing to their unique material design-ability and excellent weight reduction properties of carbon fibre reinforced polymers(CFRP), commercial aircraft OEMs have been innovating design and manufacturing approaches to wider the application of advanced composites. In this paper, the bearing capacity of composite hat-stiffened panels is determined by an improved engineering calculation method and numerical analysis method. The effectiveness of both methods is verified by the test. The advantage of the engineering estimation method is simple and fast, which is suitable for the parametric analysis of the structure in the preliminary design phase. The finite element method can better simulate the test process and is suitable for the comprehensive evaluation of structural strength. The combination of the two methods has a high reference value for rapid design and selection.

1. Introduction

Owing to their unique material design-ability and excellent weight reduction properties of carbon fibre reinforced polymers(CFRP), commercial aircraft OEMs have been innovating design and manufacturing approaches to wider the application of advanced composites[1]. The failure modes are typically dominated by the buckling behaviour of stiffened composite panels. Therefore, the stability of composite stiffened panels has become a key issue studied by many scholars. Approaches that can accurately predict the buckling threshold and loading-bearing capacity of stiffened composite panels are therefore of great importance for industry uptake.

Many studies have been published on the failure behaviour of stiffened composite panels, both theoretically, experimentally, and numerically. Some researchers presented the effective width method and column analogy approaches to approximately determine the ultimate bearing capacity of stiffened panels [2-4]. Song et al. [5] studied the stability of thin-walled stiffened composite panels used for wing surface under compression and shear loads, in order to determine the safety margin of the structure. Yu et al. [6] investigated the applicability of the composite flat panel compression buckling method under idealized boundary conditions in skin local buckling calculation of hat-stiffened panels. By using a nonlinear finite element analysis, Liu et al. [7] investigated the compression-induced failure mode of stiffened composite panels in the post-bucking stage.

Zhao et al. [8] proposed a numerical model to predict the ultimate strength of stiffened composite panels by implementing a user-define damage model into commercial FE software Abaqus to progressively replicate the failure process of the structure under compression. Zhang et al. [9] explored the influence of the cladding layer to the failure loads by experiments and simulation.

In this paper, an analytical method was developed to rapidly determined the buckling results of the stiffened composite panel configuration studied. Analytical predations agreed with the FE simulation and experimental results, validating the effectiveness of the proposed method.

2. Specimen configuration and material properties

As shown in Figure 1, the flat composite panel are strengthened by five hat stringers and two frames. Both ends of the specimen were embedded into rigid blocks to ensure the uniformity of compression during the test. Figure 1 shows the in-plane dimensions of specimen and, the dimensions for the cross section of the stringer. The width of stiffened panels is 1002mm and the length is 992mm. The interval of stiffeners is 210mm across the width, as presented in Figure 2.

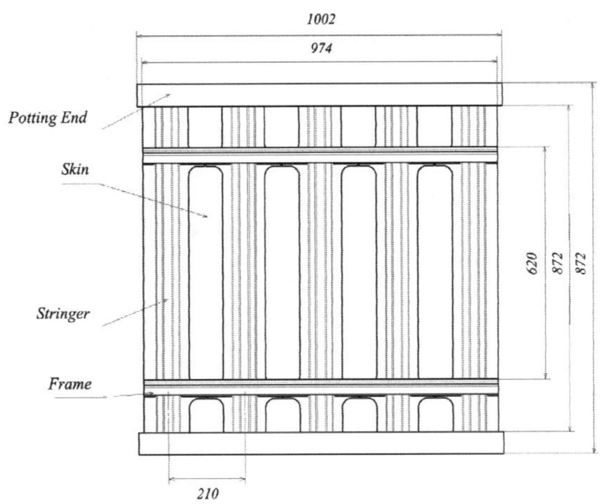

Figure 1. Dimensions of the specimen

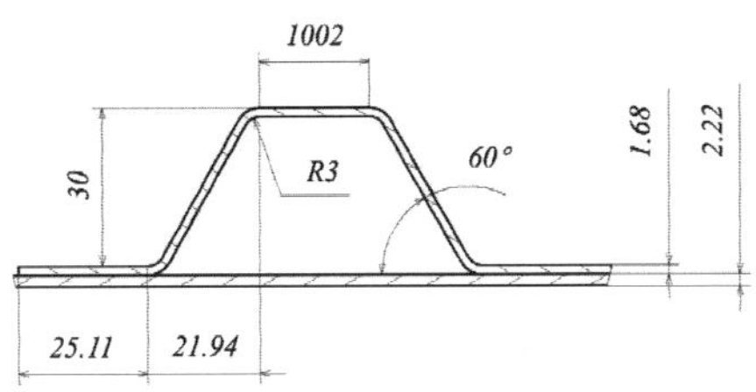

Figure 2. Cross-sectional view of the hat stiffener

The material of the skin and stiffener was X850 composite. The nominal ply thickness is 0.184mm. The material of the frame is 2024-T42 aluminum alloy. The material properties are presented in Table 1 and Table 2. The lay-up information is shown in Table 3.

Table 1. Mechanical properties of X850 composite

Parameter	Value	Parameter	Value
E_1/GPa	154	X_t/MPa	3259
E_2/GPa	8.5	X_c/MPa	1626
v_{12}	0.35	Y_t/MPa	54.6
G_{12}/GPa	4.2	Y_c/MPa	253
		S_{12}/MPa	147
		S_{23}/MPa	102.5

Table 2. Mechanical properties of 2024-T42 Al

Elastic modulus /MPa	*Poisson's ratio*
72345	0.33

Table 3. Lay-up information

Component	Layup information
Skin	[45/-45/-45/90/45/0] s
Stringer	[45/0/0/-45/90/-45/0/0/45]

3. Test procedure

The specimen was installed on the bottom platform of test machine driven by hydraulic pressure, as shown in Figure 3. The maximum loading capacity is 2000 kN. The axial compression is applied by progressively lowering the top platform in a displacement controlled manner. During the test, the specimen is placed vertically on the lower platform.

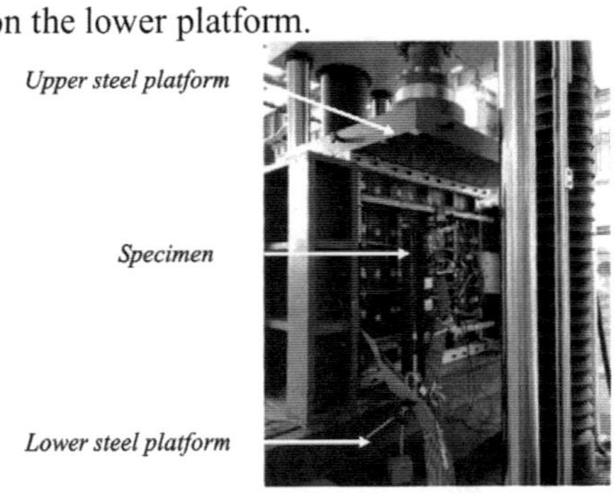

Figure 3. Compression test setup

The arrangement scheme of strain gauges is shown in Figure 4. There are five sections from the upper end to the lower end of the specimen. The first number of the strain gauge represents the section number.

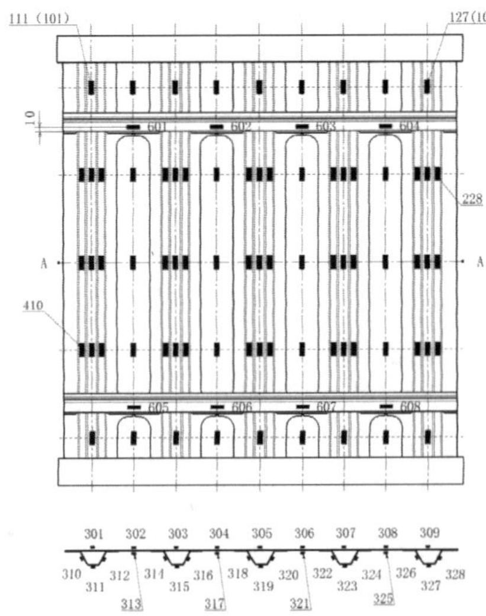

Figure 4. The arrangement scheme of strain gauges

4. Experimental results

When the load reaches 388.50kN, the strain gauges No. 104 and No. 502 buckle first. When the load increased to 441.48kN, the skin between the stringers of the middle three sections occurred to buckle.

The specimen is damaged when the load reaches 776.85kN. The failure mode of the specimen is that the fracture occurred between the upper end and the adjacent frame, as shown in Figure 5.

Figure 5. Failure modes of specimen

5. Engineering method calculation

5.1 Buckling load of orthotropic plates

The buckling load of orthotropic laminated plates under compression loads, which is simply supported on four sides can be calculated according to the classical theory, and the minimum value of the longitudinal half wave number *m* can be calculated as

$$N_{xcr} = \frac{\pi^2}{b^2}\left\{D_{11}\left(\frac{mb}{a}\right)^2 + D_{22}\left(\frac{a}{mb}\right)^2 + 2[D_{12}+2D_{66}]\right\} \quad (1)$$

For a long flat plate ($a/b \geq 4$) or when the length-width ratio of the plate is $a/b = m\sqrt[4]{D_{11}/D_{22}}$, by the following formula, the buckling load can be calculated

$$N_{xcr} = \frac{2\pi^2}{b^2}\left\{\sqrt{D_{11}D_{22}} + [D_{12}+2D_{66}]\right\} \quad (2)$$

For a thin-walled plate with stringers, it can generally be treated as simply supported on two long sides, and its local buckling load is calculated according to Eq. (2).

The orthotropic laminated plates under compression loads and simply supported at its two loading edges and two edges clamped, which buckling load can be calculated according to the classical theory. And the minimum value of the longitudinal half wave number *m* can be calculated as

$$N_{xcr} = \frac{\pi^2}{b^2}\left\{D_{11}\left(\frac{mb}{a}\right)^2 + 5.33D_{22}\left(\frac{a}{mb}\right)^2 + 2.665[D_{12}+2D_{66}]\right\} \quad (3)$$

For a long flat plate ($a/b \geq 4$), the buckling load can be calculated as

$$N_{xcr} = \frac{\pi^2}{b^2}\left\{4.6\sqrt{D_{11}D_{22}} + 2.665[D_{12}+2D_{66}]\right\} \quad (4)$$

The axial compression buckling threshold of the orthotropic laminated long plate ($a/b \geq 4$) with four edges clamped is the same as the orthotropic laminated long plate with two loading edges simply supported and the other two edges clamped, which can be calculated according to Eq. (4).

The buckling load of the long plate with two loading edges fixed and two edges simply supported is the same as the long plate with four edges simply supported, which can be calculated according to Eq. (2).

5.2 Skin width calculation

The skin width and the boundary conditions are taken into calculation to determine the local buckling threshold. Three different methods are discussed to calculate the equivalent width of the skin in Figure 6. Simply supported and clamped also need to be considered into analysis of boundary conditions.

As shown in Figure 6, the equivalent skin width is selected to be spacing between the centerlines of two adjacent flanges, where $b_2 = (b_1 + b_3) / 2$. The equivalent boundary conditions are assumed to be simply-supported for the four edges of the extracted sub-structure.

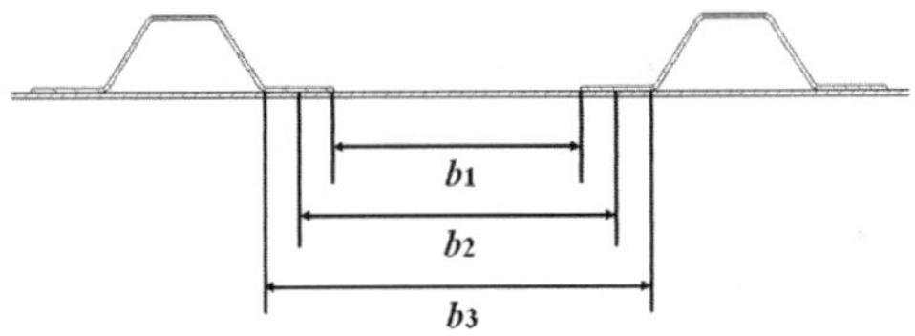

Figure 6. Different selections of the equivalent stiffener spacing.

5.3 Buckling loading analysis

The buckle of the skin under the hat stringer is not considered in the calculation, because the width of the skin under the hat stringer is less than the width of the skin between stringers. Meanwhile, it can be considered that the free edges on both sides will not buckle before the skin between stringers, because the stiffness of the free edges on both sides is strong enough.

For hat stiffened panels, because of the large torsional stiffness of the stiffeners, the calculation accuracy of considering the elastic support of the stiffeners is close to that of the clamped. The width of the plate element must be modified after buckling load calculation.

When the skin is initially buckled, the bearing capacity of stringers is calculated as

$$N_x = \frac{(A_{11} - A_{12} * A_{12} / A_{22})_{stringer}}{(A_{11} - A_{12} * A_{12} / A_{22})_{skin}} N_{xcr} \qquad (5)$$

5.4 Bearing capacity analysis

According to the column analogy method, the determination of compression failure load was conducted. Three different failure models were adopted to estimate, which are the long-column failure, the medium-long column failure, and the short-column failure. Figure 7 shows the discussion with three failure models based on the slenderness ratio $\lambda = L'/\rho$.

The stiffener crippling is typical failure mode in the short column. The common failure mode is global buckling when the slenderness ratio in the long column. Failure completely caused by local buckling, which is the typical mode in the medium-long column. The failure stress value can be calculated as

$$\bar{\sigma}_{co} = \left[1 - \left(1 - \frac{\sigma_{cr}}{\bar{\sigma}_{cc}}\right)\frac{\sigma_{cr}}{\sigma_r}\right]\bar{\sigma}_{cc} \qquad (6)$$

where $\bar{\sigma}_{co}$ represents the panel average failure stress; σ_{cr} represents the skin local buckling stress; $\bar{\sigma}_{cc}$ represents the panel results when the slenderness ratio in the short column; σ_r represents the global buckling load without considering stiffness degradation; σ_r represents the global buckling stress calculated without considering stiffness degradation.

An analytical approach was proposed to approximately determine the crippling stress value of the stiffened panel by calculating the minimum buckling stress value of local plate elements. This crippling stress can be regarded as the ultimate failure load when the effective slenderness ratio of the stiffener falls into the "medium-long column" regime, otherwise the failure load is primarily governed by the Euler's buckling load for larger slenderness ratio.

$\sigma_{cr} \approx \bar{\sigma}_{cc}/2$ is adopted in the preliminary estimation. According to the Euler formula, the global buckling stress can be calculated as

$$\sigma_r = \frac{\pi^2 \bar{E}_x}{(L'/\rho^2)} = \frac{c\pi^2 \overline{E_x I}}{A}\frac{1}{L^2} \qquad (7)$$

where ρ represents the least radius of gyration (see Eq.(8)); I represents the inertia moment to the centroidal axis of panels; A represents a section of the stiffened panel; c represents boundary coefficient of the stiffened panel; L represents the length of panels; L' represents the effective column length (see Eq.(9)); $\overline{E}_x$ represents the equivalent elastic modulus; E_{xi} represents the axial elastic modulus of the ith plate element(see Eq.(10)); A_i represents a sectional area of the ith plate element; t_i represents the thickness of the ith plate element.

$$\rho = \sqrt{I/A} \quad (8)$$

$$L' = L/\sqrt{c} \quad (9)$$

$$E_{xi} = \left(A_{11} - A_{11}^{2}/A_{22}\right)_i / t_i \quad (10)$$

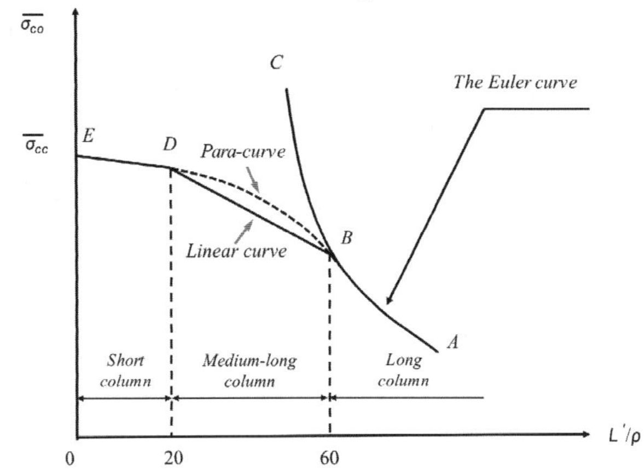

Figure 7. Schematic showing the variation of load bearing capacity versus the slenderness ratio λ

5.5 Engineering method calculation

When the width is chosen as b_2, the side is treated as simply supported and the buckling load is 402.79kN. When the width is selected as b_3, the side is treated as clamped and the buckling load is 423.65kN. The error compared with the test are 3.68% and 9.05% respectively. The skin local buckling stress is 90.4 MPa by the method in Section 5.1. The critical buckling stress of the stringer is 480.7 MPa, which is the minimum buckling stress value for the plate elements.

The stringer minimum buckling stress value is 480.7 MPa. According to the Tsai-Wu criterion, the stringer strength can be calculated. The ultimate strength is 911.8 MPa. The global buckling stress value is 225.0 MPa. Therefore the global buckling load can be calculated as 813.8 kN. The failure stress value of the stiffened panel is 323.9 MPa, and the corresponding failure load is 1171.5 kN. Owing to the failure load calculated by the column analogy method is larger than the global buckling load. The failure load should be corrected as 813.8 kN. The calculation error is 4.76%, which compared with failure load of the test.

6. Numerical analysis

6.1 Finite element model

A stiffened panel model (see Figure 8) was established. The shell element was used for skin and stringer. The solid element was used for potting end and metal components. The interface was simulated by the cohesive zone model. The boundary conditions of the finite element model are demonstrated in Figure 8. The bottom of the specimen is constrained to X-direction and Z-direction displacement. To simulate the restriction effect on both sides of the frame, the side of the frame can move only along the X-direction. The upper end of the specimen is constrained to the Z-direction displacement and applied uniaxial compression displacement $U1$=5mm.

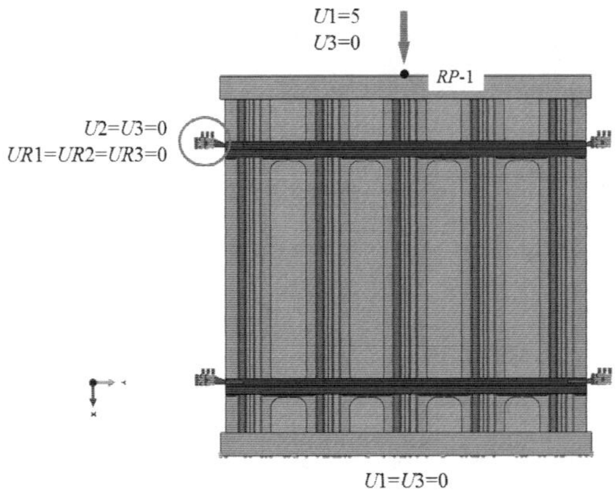

Figure 8. The numerical analysis model

6.2 Failure criterion

The failure criterion was the Hashin criterion. To simulate the interface behaviour between the stringer and skin accurately, the cohesive zone model is adopted. The interface failure criterion is based on the continuous damage mechanism of the traction-displacement relationship between the connected elastomers.

$$\begin{bmatrix} t_n \\ t_s \\ t_t \end{bmatrix} = \begin{bmatrix} K_{nn} & & \\ & K_{ss} & \\ & & K_{tt} \end{bmatrix} \begin{bmatrix} \delta_n \\ \delta_s \\ \delta_t \end{bmatrix} \qquad (11)$$

Where, t_n denotes the normal traction, t_s and t_t denote the tractions along two tangent directions. K_{nn} , K_{ss} and K_{tt} are the corresponding stiffness coefficients respectively. In this paper, it's recommended that $K_{nn} = K_{ss} = K_{tt} = 10^6 N / mm^3$. δ in Eq.(11) is relative displacement vector at the interface, where the subscript n denotes the normal direction, s and t denote the in-plane shear directions.

The quadratic nominal stress criterion is employed to determine the initial interface damage

$$d = \left\{ \frac{t_n}{t_n^0} \right\}^2 + \left\{ \frac{t_s}{t_s^0} \right\}^2 + \left\{ \frac{t_t}{t_t^0} \right\}^2, \begin{cases} d < 1 \\ d = 1, failure \end{cases} \qquad (12)$$

Where, d is the interface failure index, the damage initiates when d equals one. t_n^0 represents tensile strength of interface, t_s^0 and t_t^0 represent shear strength of interface.

The damage variable D was determined according to the function of energy release rates associated to three fracture modes, i.e., G_{IC}, G_{IIC}、G_{IIIC}.

$$D = \left\{ \frac{G_I}{G_{IC}} \right\}^\alpha + \left\{ \frac{G_{II}}{G_{IIC}} \right\}^\alpha + \left\{ \frac{G_{III}}{G_{IIIC}} \right\}^\alpha, \begin{cases} D < 1 \\ D = 1, failure \end{cases} \qquad (13)$$

Where, the coefficient α is chosen to be 1. the value of D ranges from 0 to 1, which represents the degree of damage. The interface is completely damaged when $D = 1$.

The cohesive zone model which characterizes three-dimensional constitutive relation is shown in Figure 9.

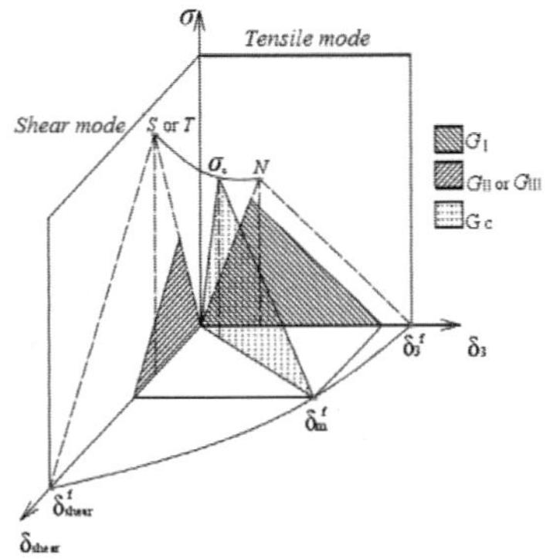

Figure 9. Bilinear mixed-mode softening law

6.3 Verification

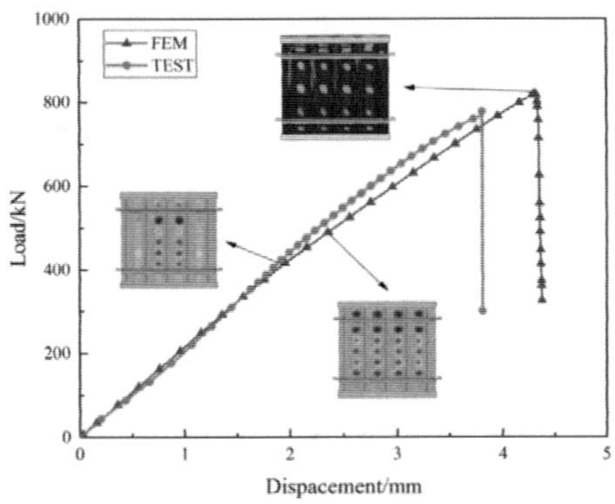

Figure 10. The load-displacement curve between FEM and the test

Figure 10 indicates the load-displacement curve between the test and FEM. Figure 11 indicates the comparison results of load-strain curves. Simulation results show that the buckling load is 415.81kN, the failure load is 820.09kN, and the error compared with the test is 7.03% and 5.57% respectively. Figure 12 shows the failure modes prediction by the FE analysis, showing a good agreement with the experimental data.

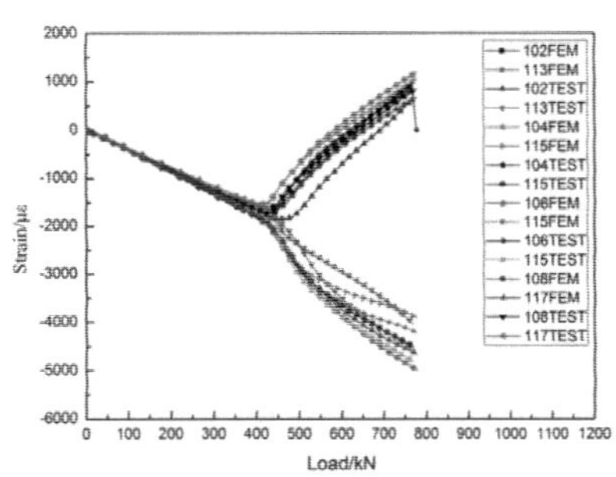

(a) Section No.1

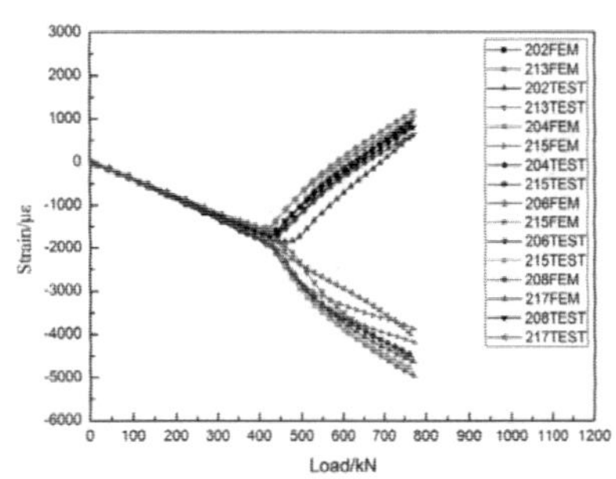

(b) Section No.2

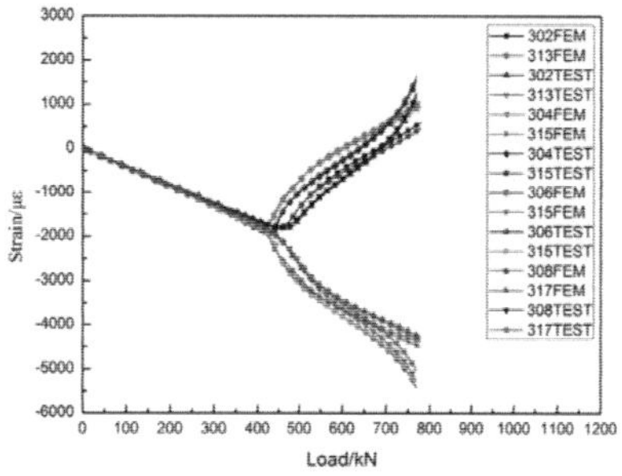

(c) Section No.3

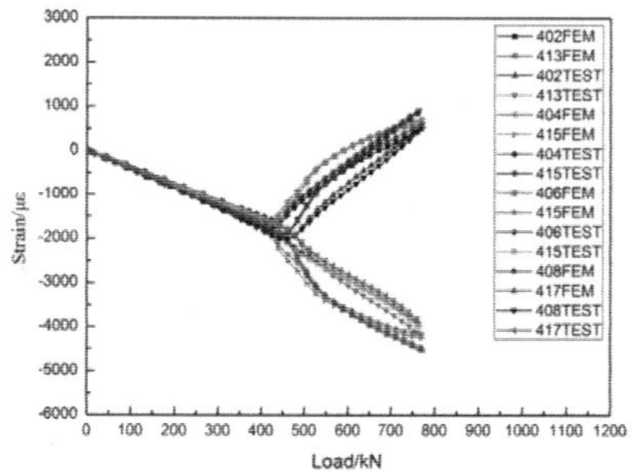

(d) Section No.4

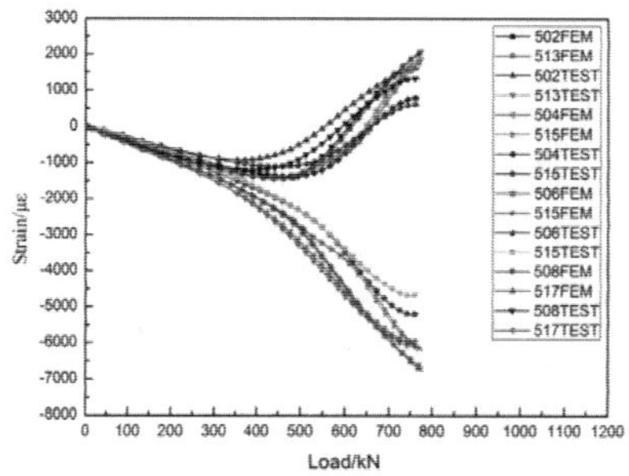

(e) Section No.5

Figure 11. Load-strain curve between FEM and test

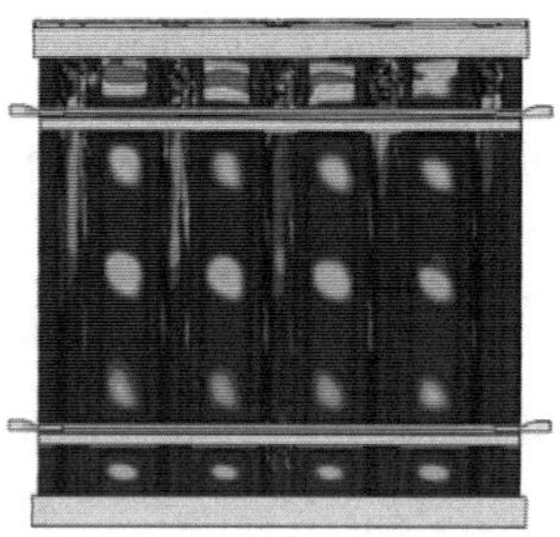

Figure 12. The failure mode of the specimen

7. Conclusions

In this paper, based on the engineering estimation method and FEM, the strength of stiffened panels under axial compression is predicted. Compared with the test, The following conclusions can be obtained:

1) Two methods for the effective width and boundary conditions of the skin between stringers for buckling load calculation are proposed. The error of buckling load and failure load between the prediction by the engineering method and the test value is 3.68% and 9.05% respectively, which is within the allowable range in engineering. It is recommended that b_2 be preferred as the effective width of the skin, and the side can be treated as simply supported;
2) An analytical approach was proposed to approximately determine the crippling stress of the stiffened panel, by calculating the minimum buckling stress of local plate elements. This crimpling stress can be regarded as the ultimate failure load of the stiffened panel when the effective slenderness ratio of the stiffener falls into the "medium-long column" regime, otherwise the failure

load is primarily governed by the Euler's buckling load for larger slenderness ratio. The error between the engineering estimation result and the test value is 4.76%

3) The error between the prediction of buckling load and failure load by the FEM based on the Hashin criterion and cohesive element model and test value are 7.03% and 5.57% respectively. Moreover, the failure mode can be predicted accurately.
4) The advantage of the engineering estimation method is simple and fast, which is suitable for the parametric analysis of the structure in the preliminary design phase. The finite element method can better simulate the test process and is suitable for comprehensively evaluating structural strength. The combination of the two methods has a high reference value for rapid design and selection.

Acknowledgments
The research was supported by the Aviation Science Foundation of China (2020Z055023002).

References

[1] Tang J M. Current status and trends of advanced composites in aerospace. Spacecraft Environment Engineering, 30(4):352-359.
[2] MICHAEL C Y. Composite airframe structures[M] Hong Kong: Hong Kong Conmilit Press Ltd,1992:397-422.
[3] VON KARMAN T, SECHLER E E, DONNELL L H, *et al.* The strength of thin plates in compression[J]. Transactions of the ASME, 1932,54(2):53-57.
[4] PEVZNER P, ABRAMOVICH H, WELLER T. Calculation of the collapse load of an axially compressed laminated composite stringer-stiffened curved panel—An engineering approach[J]. Composite Structures,2002:50-52.
[5] SONG G, CUI D G, DONG L J, *et al.* Application of buckling/post-buckling analysis of composite stiffened plates[J]. Acta Materiae Compositae Sinica, 2017,34(1):96-103.
[6] YU Z B. Research on compression buckling analysis method of flat stiffened panels made of carbon fiber composite material[J]. Jiangsu Science and Technology Information,2015(22):63-65.
[7] LIU C Y, XU X W, CHEN K. Numerical analysis of post-buckling and bearing capacity of composite stiffened plates considering debonding[J]. Acta Materiae Compositae Sinica,2010,27(6):158-166.
[8] ZHAO W T, FAN S B, LIU W H, *et al.* Reliability analysis of the ultimate compressive bearing capacity of composite stiffened plates[J]. Journal of Computational Mechanics,2016,33(6):856-862.
[9] ZHANG G F, SUN X S, WU C L, *et al.* Failure mechanism analysis and design of omega stiffened composite panel[J]. Acta Materiae Compositae Sinica,2017,34(11):2479-2486.

Dynamic CFD/CSD Coupling Research for Flexible Micro Aircraft

Lili CHEN[1]*, Wenqing YANG[2] and Wei WANG[1]

[1]AVIC The First Aircraft Institute, Xi'an, 710089, China

[2]School of Aeronautics, Northwestern Polytechnical University, Xi'an, 710072, China

lilichen8511@outlook.com

Abstract. Bio-inspired flapping wing aircraft have made progress in recent years. Flying birds can fly in various low Reynolds numbers (10^3-10^5) with high flight stability and maneuvering ability by use of the flexible deformation of skeletal muscle structures or thin-film structure wings. Early studies mainly focused on the aerodynamics and kinematics of the flapping wings, with relatively little research on the flexible structure, and the aerodynamic structure coupling characteristics of micro flappings are still unclear. At present, the focus of research is to evaluate the impact of structural deformation on the aerodynamic characteristics of flexible micro-flapping wing. A CFD/CSD coupling solver are developed and several technical challenges related to the CFD/CSD interface are resolved in this study (CFD/CSD data exchange method and moving mesh generation). The aerodynamic characteristics of five flexible wings with different structural parameters are calculated and compared. The results show that structural parameters play an important role in aerodynamic performance, which also indicates that the potential aerodynamic performance can be improved by the optimal structural design.

1. Introduction

Birds provide the best example of the development of micro flapping wing studies. Nearly all birds can fly through long periods of natural evolution. Insects especially have particularly flying capability that allow them to adapt their specific range to Reynolds number. They usually can operate at a wide range of low Reynolds numbers (10^3-10^5) with outstanding stability as well as mobility by use of flexible deformation of the wings which generated under aerodynamic as well as inertial forces.

At present, aeroelastic research of flapping wing mainly relies on wind tunnel experiment technology. Ho [1] and Wu Ping [2] et al. Investigated the aeroelasticity characteristics of mirco-flappings with different flexible structures by experimental study. They summarize that flexible deformation plays a crucial part in aerodynamics.The periodic structural distortion and deformation are generated, resulting in leading edge eddy current rupture and life loss. In flexible torsional motion, on the other hand, thrust increases significantly. Calculation studies are relatively few, but there are still some outstanding studies on the aeroelasticity calculation of flapping wing [3-10].

The focus of this study is to evaluate the effect of structural deformation on the aerodynamic characteristics of the flexible flapping wing.

A computational dynamic aeroelastic coupling framework is established to simulate complex flows around a micro-flexible flapping wing, and to study the effects of flexible wing. The deformation and flow structures of five groups of flexible wings with different structural parameters are compared, and the influence mechanism of flexible wing on aerodynamics is studied.

2. Methodology

2.1. CFD solver

The governing equation of fluid dynamics is the RANS equation:

$$\frac{D}{Dt}\iiint_{\Omega} W dV + \iint_{\partial\Omega} \overline{\overline{H'}} \bullet n dS = \iint_{\partial\Omega} \overline{\overline{H_v}} \bullet n dS \tag{1}$$

A moving mesh method is used to generate mesh automatically for flexible flapping wing. Detailed information about grid generation technologies has been presented in many of our previous studies. C-O grid of flapping wing structure is used for CFD simulation. Tangential, radial and spanwise mesh sizes are 176, 56 and 60, respectively. Fig. 1 shows the grid for simulating the aeroelastic characteristics, which is generated by inhouse program.

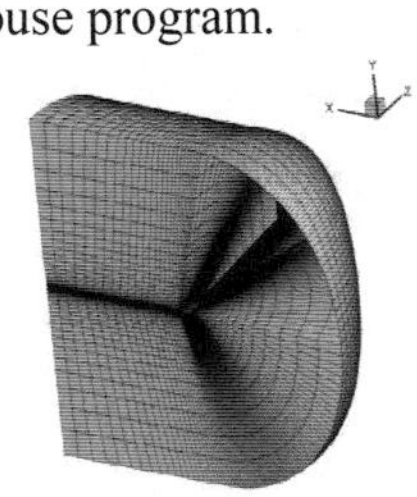

Fig 1. Flapping wing moving C-O grid

2.2. CSD solver

The governing equation for structural dynamics is given by

$$[M]\{\ddot{q}\}+[K]\{q\}=\{F(t)\} \tag{2}$$

In this paper, modal analysis solutions are used to solve equation (2). By introducing a modal matrix Φ, equation (2) reduces the size by preserving the former "m" mode. All results in this article were obtained while retaining the top five patterns. Node deformation can be written as

$$\{q\}=[\Phi_m]\{\overline{q}\} \tag{3}$$

where, each column of $[\Phi_m]$ contains the eigenvector corresponding to a pattern. By plugging in this equation from equation (3), equation (3) can be converted to

$$[\Phi_m]^T[M][\Phi_m]\{\ddot{\overline{q}}\}+[\Phi_m]^T[K][\Phi_m]\{\overline{q}\}=[\Phi_m]^T\{F(t)\} \tag{4}$$

$$[\overline{M}]\{\ddot{\overline{q}}\}+[\overline{K}]\{\overline{q}\}=\{\overline{F}(t)\} \tag{5}$$

2.3. Fluid-structure Coupling Method

A loose coupling method is chosen to solve the coupling problem between fluid and structure solver. The periodic aerodynamics are calculated first by a CFD solver, then by a CSD solver to calculate the periodic velocity,acceleration as well as structural displacement, and then again until both the structural deformation and aerodynamics converge.

3. Results

This paper aims to investigate the effect of structural parameters (distribution of front-axle beam, cant beam, and wing rib with different diameters) on the aero-elastic characteristics of a flexible flapping wing. Five groups of flexible wings with different distributions of beam diameter are discussed here, as seen in Table 1. The location distribution of the beam is chosen from the actual flapping wing aircraft, as seen in Fig. 2.

Fig 2. Flapping wing model

Table 1. Five groups with different distributions of beam diameter

	Front-axle beam	Cant beam	Wing rib
wing 1	1.5mm	1.0mm	1.0mm
wing 2	1.5mm	1.5mm	1.0mm
wing 3	2.0mm	1.5mm	1.0mm
wing 4	2.0mm	1.5mm	1.5mm
wing 5	2.0mm	2.0mm	2.0mm

A single-degree of freedom(DOF) motion is defined. The motion is described as: $\psi(t) = \psi_1 \cos(\omega t)$, where, $\psi(t)$ is the flapping angle in the t time step, ψ_1 is the flapping-amplitude, ω is the frequency. The calculated state is: wind-velocity is 8m /s, flapping-frequency is 6Hz, and flapping-amplitude is 350.

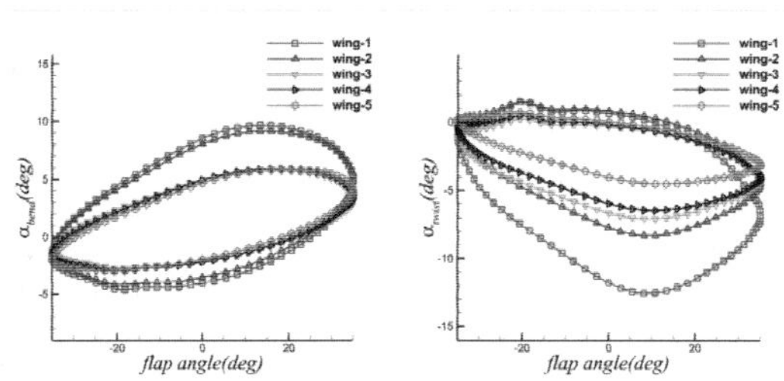

Fig 3. Periodic deformation phase plot of flexible flapping wing

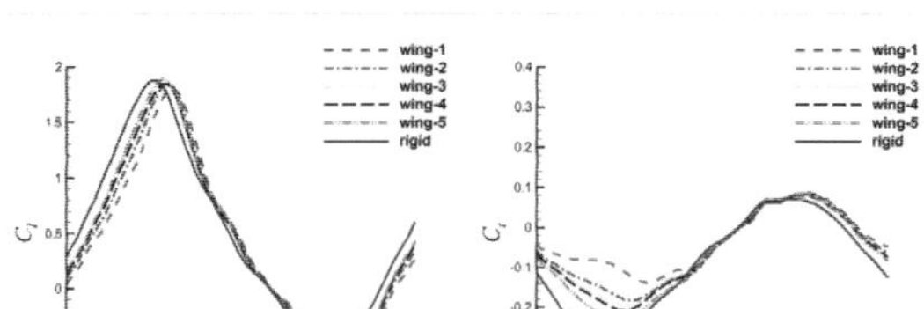

Fig 4. Lift and thrust curves

Fig. 3 shows the deformation at the tip. Figure 4 shows the time history of force. Significant structural deformation can be seen for all five wings selected. However, due to the different diameters of a beam, characteristics of deformation are looking different: bending and twist deformation amplitude are decreasing from Wing 1 to Wing 5. Force coefficients are also diverse under the influence of structural deformation. Through the integration of the periodic force coefficient, the average lift and thrust coefficient is got and compared. The results show that Wing 5 gets the best average lift characteristics and the worst thrust characteristics, and Wing 1 gets the best average thrust characteristics and the worst lift characteristics. However, Wing 2 gets the relatively optimal lift as well as thrust characteristics.

In order to better understand the influence of structural parameters on aerodynamics, the flow pressure contours are studied in detail.The results of selected wingspan positions and representative moments on wings 1, 2 and 5 are shown.

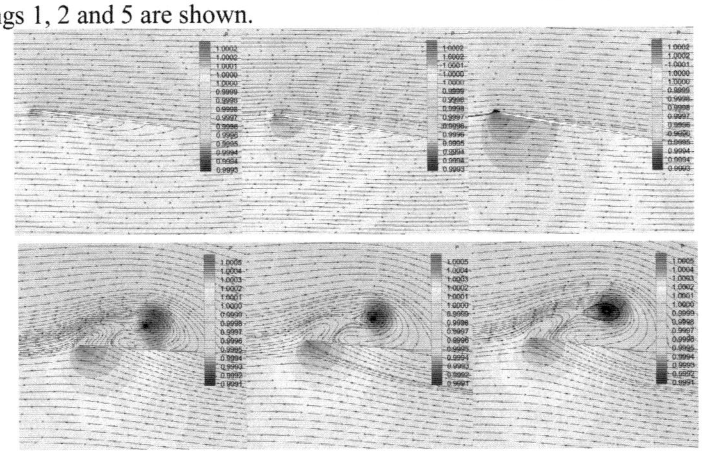

t/T=0

t/T=1/4

(a) wing 1 (b) wing 2 (c) wing 5

Fig 5. Pressure and streamlines in the stroke cycle around the 50% half wingspan section of the airfoil

t/T=0 t/T=1/4

Fig 6. Pressure distribution along the wing semi-span of rigid and flexible wings

It is seen in Figure 3 — Figure 6 that at time of 0T and T/4, the corresponding thrust is decreased from Wing 5 to Wing 1, and it can be explained that at that time, the negative twist deformation is decreased from Wing 1 to Wing 5, which cause the loss of instantaneous angle of attack, aerodynamic force vertical to airfoil chord line produce a component in the horizontal direction, resulting in a significant increase of the thrust coefficient. Whereas, from wing 1 to Wing 5, the leading edge suction is enhanced, resulting in an increase of the component of aerodynamic force in a vertical direction (lift).

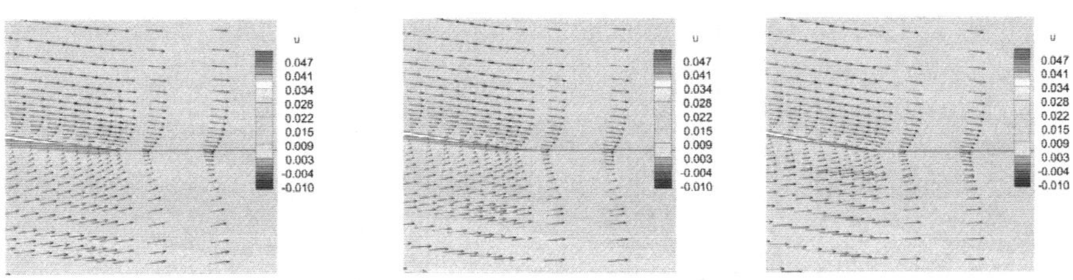

t/T=0

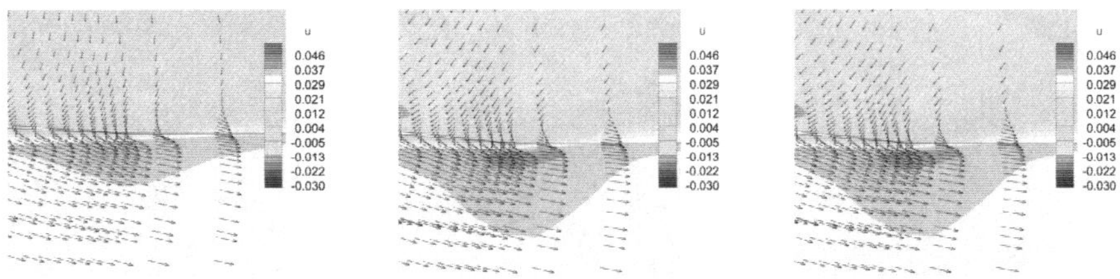

t/T=1/4

(a) wing 1 (b) wing 2 (c) wing 5

Fig 7. Speed types at four different moments in a 50% half span stroke cycle

Fig. 7 shows the chordwise velocity vector at 50% semi-span section. It can be seen that at 0T and T/4, the chordwise velocity of wing 5 is larger compared with Wing 1 and Wing 2, which is equivalent to 'jet' flow, increasing thrust.

4. Conclusion

A CFD/CSD coupling framework for calculating flexible flapping wing micro-aircraft is proposed and the results of different structural parameters are reported. Several important conclusions of this numerical study are as follows:

- in considering the parameters of the scope, structure was proved to have beneficial effects on aerodynamic parameters.
- proved to be the leading edge suction air power generation is very important.
- trailing edge near the string to the velocity vector is proved that it is important for the generation of thrust.
- Future work will address with complex plane geometry shape and material distribution of aerodynamics of flapping wing - structure coupling characteristics.

It can be seen from this study that the flexible structure of flapping-wing has a significant impact on the aerodynamic characteristics. Therefore, by optimizing the flexible structural of flapping-wing to improve flight performance in the future research on aerodynamic characteristics .

References

[1] Ho, S., Nassef, H., Pornsinsirirak. (2003). Unsteady aerodynamics and flow control for flapping wing flyers. Progress in Aerospace Sciences, 39(8), 635–681.

[2] Chen, S., Li, H., Guo, S.(2018). Unsteady aerodynamic model of a flexible flapping wing. Aerospace Science and Technology, 80, 354–367.

[3] Rodriguez, D. L., Aftosmis, M. J. (2016). Optimization of Flexible Wings with Distributed Flaps at Off-Design Conditions. Journal of Aircraft, 53(6), 1731–1745.

[4] Kim, Y. , Cho, H. (2018). Advanced structural analysis based on reduced-order modeling for gas turbine blade. Aiaa Journal, 56(8), 1-5.

[5] Hong, G., Kaneko, S., Mitsume, N. (2021). Robust fluid–structure interaction analysis for parametric study of flapping motion. Finite Elements in Analysis and Design, 183–184.

[6] Nagai, H. , Nakamura, K. , Fujita, K. (2021). Development of tailless two-winged flapping drone with gravity center position control. Sensors and Materials, 33(3), 859-872.

[7] Su, X. , Wang, N. (2018). Numerical study on effects of mult-control surface position on aeroelastic behavior of forward-swept wing. 27-37.

[8] Tsatsas, I. , Pontillo, A. (2022). Aeroelastic damping estimation for a flexible high-aspect-ratio wing. Journal of Aerospace Engineering(2), 35-40.

[9] Soneda, K. , Yokozeki, T. (2021). Multi-fidelity Aeroelastic Simulation of a Morphing Wing Trailing Edge. In: AIAA Scitech 2021 Forum.

[10] Gogulapati, A., & Friedmann, P. P. (2014). Approximate Aerodynamic and Aeroelastic Modeling of Flapping Wings in Forward Flight. AIAA Journal, 52(1), 212–218.

PID control and simulation of moving mass quadcopter UAV

Tianfang Xu

School of Mechanical and Aerospace Engineering, Jilin University, Changchun, Jilin, 130022, China

Afang916486326@163.com

Abstract. Based on the quadrotor drones we see today, in this paper, a mass center control mechanism installed on the body is proposed to control the flying attitude of the UAV. The moving mass control technology changes the position of the center of mass by moving the position of the mass block of the system. The technology is used in a wide range of aircraft, from drones to gravity satellites. The moving mass center mechanism mentioned in this paper is composed of three sliding blocks that can slide along the spatial axes and are controlled by a linear motor. The linear motor controls the slider movement to generate the torque along different channel directions, to realize the attitude change of the UAV. In this paper, through the kinematics analysis of the quadrotor UAV, combined with PID control and simulation, finally realized the modification of the UAV's flight attitude with moving mass technology. The simulation results show that the quadrotor UAV can complete the attitude conversion stably with the help of the mass center control mechanism.

1. Introduction

1.1 Technical background

UAVs have been widely used in many walks of life in recent years because they have the advantages of vertical take-off and landing, single flight or companion flight, and low manufacturing costs. These advantages allow the UAV to be widely used in military surveillance, traffic information collection, large-scale multi-agent planning, power system detection, disaster first aid kits, and other fields. Since it is ably applied in so many fields, it also benefits from the high coupling, strong nonlinearity, many controlled variables, and sensitivity to external interference, which make the quadcopter an ideal platform for verifying various advanced control technologies and real-time navigation control strategies [1].

On this basis, this paper will use the moving mass control center to achieve control simulation of drones. Moving mass control is a classic control method that achieves flight control by changing the centroid of the aircraft, also known as mass moment control, and its basic principle is to change the position of the aircraft centroid by moving the sliding mass block installed in the fuselage to achieve attitude adjustment. [2]

1.2 Literature review

Chen Baihui et al. proposed a new dual-slider control strategy for the rotation channel moving mass center control of fixed-wing aircraft and designed a slider motion control device that can integrate the inside of the wing sir. [3] Kefan Wang et al. proposed a solar UAV moving mass center control scheme with the battery as the mass slider and the slide rail arranged along the extension direction for the normal layout of the solar UAV for the cross-directional control problem. [4] Xiaoyi Qiu et al. proposed a single-slider moving mass center UAV layout scheme with a smaller time lag and simpler structure, analyzed the influence of slider parameters on the dynamic characteristics of moving mass centroid UAV, on this basis, gave the ideal installation position of the slider, and studied the change of control efficiency of moving mass center UAV layout scheme with speed. [5]

1.3 Main research content

First of all, the three attitude channels of the UAV will be installed with a centroid adjustment mechanism, which is different from the previous method of adjusting the rotor motor speed to form a speed difference to form a torque to adjust the attitude, and the slider mass offset will be directly used to control.

Secondly, this paper will debug the attitude control of the UAV under the condition of moving mass heart and give the debugging situation by Pid fuzzy control.

2. Method

Common quadcopter UAVs are H-type and X-type, and this article selects X-type UAVs as research objects.

2.1 Total coordinate system's establish

First of all, for the research of drones, we need to establish a coordinate system, usually using a coordinate system that is relatively fixed to the earth's surface. Taking any point o on the earth's surface as the coordinate origin, the ox-axis starts from that point and points due north, the oy-axis starts from this point and points due east, and the oz-axis starts from this point and is perpendicular to the oxy plane upwards, and is generally considered to be inertial coordinate systems.

2.2 UAV coordinate system's establish

In this coordinate system, it is also necessary to establish a particle coordinate system of the UAV, as shown in Figure 1. The origin is at the mass of the UAV o_b, from the center of mass to the nose of the ox_b axis, in the plane parallel to the rotor plane vertical ox_b axis, pointing to the right side of the oy_b axis, oz_b axis and oxy_b constitute the right-hand coordinate system. Our force analysis of drones is in this coordinate system.

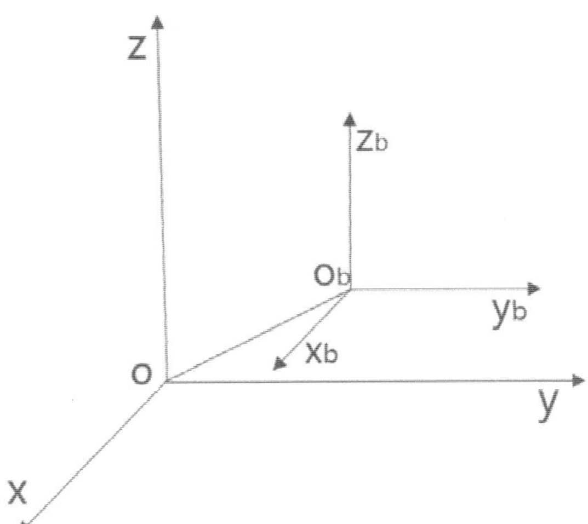

Figure 1. Ground coordinate system and UAV coordinate system

2.3 Conversion of the UAV's attitude angle

There is a certain conversion relationship between the coordinate systems, and the three attitude control channels of the UAV are pitch, yaw, and roll. Through the analysis, it can be known that the angles on these three channels, namely the pitch angle θ, the yaw angle ψ, and the rolling angle φ the matrix formed.

$$R_v^b = \begin{pmatrix} cos\theta cos\psi & cos\theta sin\psi & -sin\theta \\ sin\varphi sin\theta cos\psi - cos\varphi sin\psi & sin\varphi sin\theta sin\psi + cos\varphi cos\psi & \sin\varphi cos\theta \\ cos\varphi sin\theta cos\psi + sin\varphi sin\psi & cos\varphi sin\theta sin\psi - sin\varphi cos\psi & cos\varphi cos\theta \end{pmatrix}$$

2.4 Drone motion analysis

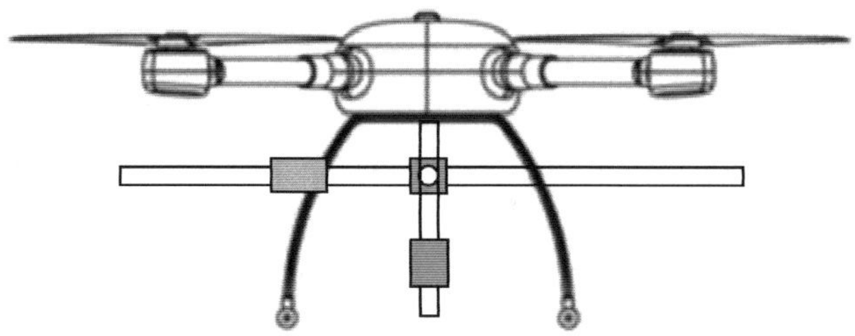

Figure 2. Front view

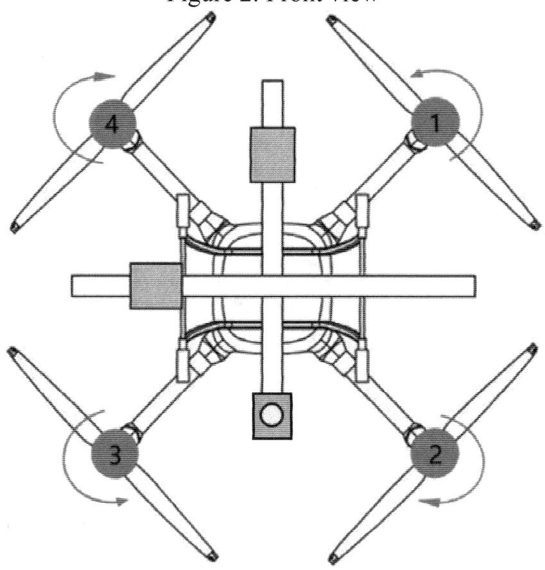

Figure 3. Top-up view

The external forces affected by the selected UAV(shown in Figure 4) in the work are gravity, the lift generated by the rotor, air resistance, and lift, when the flight speed is not high, and in windless conditions, we ignore the air resistance and lift.

When the general UAV realizes pitch movement, the electric machine synchronously increases (or decreases) the speed of the No. 1 and No. 4 rotors, and synchronizes (increases or) decreases the speed of the No. 2 and No. 3 propellers as shown in Figure 3, which will form a torque at the center of mass so that the UAV makes a pitch movement, and the moving, mass center UAV is in the case of not changing the rotor speed, changing the slider position to produce torque to achieve the same effect, that is, pitch motion.

The same is true when rolling motion, the general UAV through the motor synchronization increases (or decreases) the speed of the 1 and 2 rotors, and synchronizes (increasess or) reduces the speed of the 3 and 4 propellers, which will form a torque at the center of mass so that the UAV makes a pitch movement, and the moving mass center UAV is in the case of not changing the rotor speed, by changing the slider position, to produce torque to achieve the same effect, that is, rolling motion.

For yaw movements, drones are achieved by reverse torsional force. In the process of rotor rotation, due to the air resistance effect, a reverse torque force opposite the direction of rotation will be formed to overcome the influence of reverse torsion, four rotors, two forward rotations, two reversals, and diagonally from 4 each rotor rotation direction is the same; The size of the reverse torque force is related to the rotor speed when the four motor speeds are the same, the reverse torque force generated by the four rotors is balanced with each other, and the quadcopter aircraft does not rotate; When the speed of the four motors is not the same, the unbalanced reverse torque force will cause the quadcopter to rotate horizontally, to achieve yaw movement, such as the speed of the No. 1. No. 3 motor is increased, the No. 2 and No. 4 motor speed is reduced, the four axes will rotate horizontally, because the overall lift is unchanged, so it will not lead to rise/fall, and the slider does not play a major role in controlling yaw movement.

2.5 Control system analysis

For the control system design of the moving mass quadcopter UAV, there are high requirements for the accuracy of the position, and this requirement can be well met by attitude control, so this paper uses PID fuzzy control as the basic system. At present, the most commonly used and easiest method in rotor control is PID control, which has the characteristics of simple structure and easy realization for linear control systems, and is widely used, or PID control algorithm is the mainstream algorithm of rotorcraft at present. Compared with traditional PID control, the PID fuzzy algorithm is different in that fuzzy control is a control method that uses the basic ideas and theories of fuzzy mathematics, and in the traditional PID control process, the accuracy of control is a factor that directly affects the control results. Therefore, repeated debugging of parameters is a key part of PID control simulation. However, for more complex systems, there will be more parameters that need to be adjusted, and the greater the error. Therefore, the choice of PID fuzzy control to simplify the dynamic process of the system fuzzy control has been widely used in the field of intelligent control due to the mature theoretical research, relatively simple implementation, and wide adaptability.

As shown in Figure 4, the system has three sliders and four rotors to control separately, so there are seven control variables, of which the slider is the main control variable. We enter the fuzzy processor by inputting three attitude angles of the drone and then input it to the attitude controller, and after noise processing, it becomes the displacement signal δ of the slider to control the slider, and finally control the motor of the rotor, and by controlling the rotor speed, it together forms the attitude control closed loop of the drone.

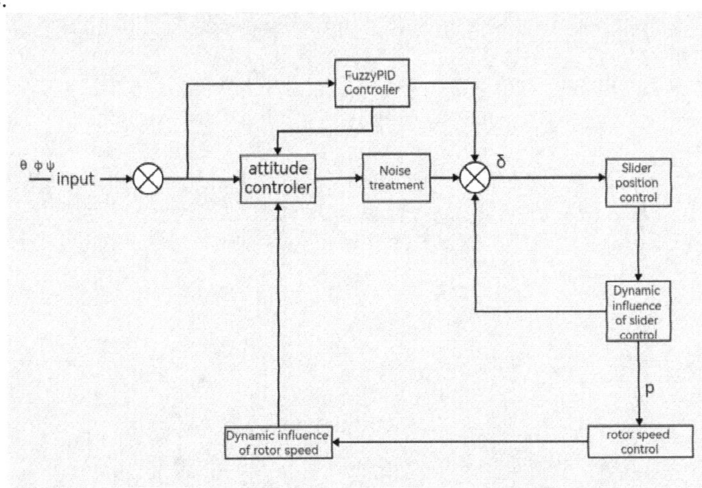

Figure 4. Control system schematic

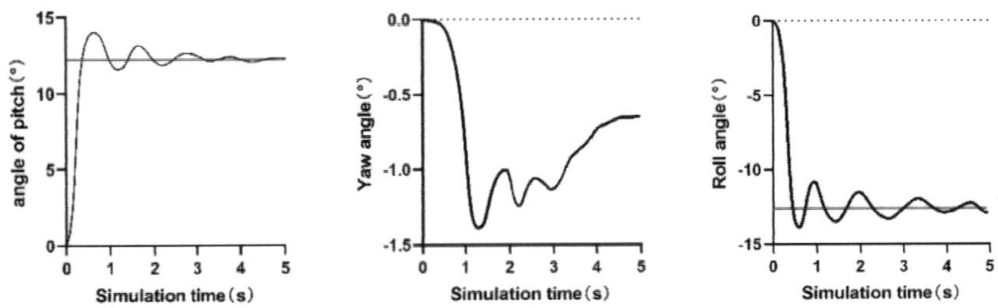

Figure 5. Attitude control simulation of drones

As shown in Figure 5, the slider control system is not a uniform curve, there is a certain large fluctuation at the beginning of the movement stage, and the subsequent movement will also maintain a certain fluctuation under the action of the moment of inertia, but it will gradually decrease, and it will tend to be stable after a period, indicating that the system can complete the control function.

3 Conclusion

In this paper, a system is designed for the PID control of a moving mass quadrotor UAV, which first models the UAV, then analyzes its kinematics, and designs a control system to prove its feasibility by simulation. Through simulation, we verified that the designed system can meet the requirements.

Since the movement of the quadrotor UAV is nonlinear, the linear system designed by PID can only represent the movement of the UAV to a certain extent, so we do not consider the influence of environmental factors and the delay error caused by the reaction speed of the motor, in reality, to ensure that the simulation results are closer to reality. But these problems are still something that needs to be considered and solved in the future.

Reference

[1] WANG Cheng, YANG Jie, YAO Hui & XI Jianxiang. (2018). Review of quadcopter UAV flight control algorithms. Electro-Optics and Control (12), 53-58.

[2]LI Jianqing, GAO Changsheng, JING Wuxing & DU Huajun. (2016). Control problem of down-controlled single-slide moving mass aircraft. Science in China: Technical Science (10),1048-1056.

[3] CHEN Baihui, HAN Lulu, WANG Xiaoming & FENG Xiaogang. (2022). Design, analysis, and flight test of moving mass vehicle based on in wing double slider. Flight Mechanics (03),81-87.doi:10.13645/j.cnki.f.d.20220411.007.

[4] WANG Kefan, QIU Xiaoqi, GAO Changsheng & JING Wuxing. (2021). Control technology of transverse heading moving mass center of solar UAV in near space. Air and Space Defense (04), 29-36.

[5]QIU Xiaoqi, GAO Changsheng & JING Wuxing. (2022). Dynamic analysis and immunity control of fixed-wing UAV with moving mass center. Journal of Beijing University of Aeronautics and Astronautics (03),430-437.doi:10.13700/j.bh.1001-5965.2020.0573.

Journal of Physics: Conference Series **2489** (2023) 012016 doi:10.1088/1742-6596/2489/1/012016

Fatigue life assessment of fuselage panel modification based on flight measurement

Fu wen Liu[1*], Bo Huang[1], Ying Chun Xiao[1], and Yu Yang[1]

[1] Structural Health Monitoring and Intelligent Structure Laboratory, AVIC Aircraft Strength Research Institute of China, Xi'an 710065, China.

[*]Corresponding author's e-mail:liufuwen623@126.com

Abstract. During the modification of the aircraft fuselage panel, the fatigue performance of the structure changes. To ensure the safety and reliability of commercial aircraft, the fatigue life assessment of the modified structure is required. In this paper, the detailed fatigue rating method was used for fatigue assessment. The strain information of the modified structure during the flight was obtained by installing fiber Bragg grating sensors. Through the finite element analysis, the fatigue critical parts were determined and the measured data before modification were converted into the data after modification. The detailed fatigue ratings of the fatigue critical parts were calculated. The fatigue life of the critical parts was evaluated based on the linear fatigue damage cumulative rule. According to the results, the most critical part was determined, which provides a basis for the determination of the modification plan of the fuselage panel.

1. Introduction

Airworthiness regulations require that structural integrity must be maintained throughout the life cycle of commercial aircraft. Currently, the airworthiness regulations and certification for commercial aircraft structures cover the life cycle from design, manufacturing, and testing, to continuous airworthiness, which ensures the safety of commercial aircraft. However, with long-term flights, repairs, and modifications, the actual load deviates from the design load significantly. The rapid accumulation of aircraft fatigue and corrosion damage has become the main factor affecting flight safety and maintenance costs. After the Aloha air crash in 1988, the structural integrity of aging commercial aircraft has become the focus of research in aviation. At the same time, the methods of damage tolerance assessment, analysis, and verification of Widespread Fatigue Damage (WFD) and determination of Limit of Validity (LOV) are constantly updated in the industry[1]. In 2019, cracks were detected in more than 50 B737-NG fuselage frame pickle forks around the world, the component that connects the wing structure, landing gear, and fuselage[2]. The theoretical life of the pickle fork is 90,000 cycles, while the aircraft in which the cracks were found flew less than 30,000 cycles. Seriously, the cracks were not found during the routine inspection and maintenance process but were discovered accidentally during the comprehensive inspection after a series of B737-MAX air crashes.

A common feature leading to the above air crashes is the lack of monitoring of the use of individual aircraft structures, which makes it impossible to assess the difference between individual aircraft relative to the full-scale fatigue testing. Therefore, Structural Health Monitoring (SHM) technology for commercial aircraft has been rapidly developed in recent years[3-7]. Fiber Bragg Grating[8] (FBG) has a good application prospect because of its excellent anti-electromagnetic interference and Wavelength Division Multiplexing performance. In 2016, the Fraunhofer Institute for Structural Durability and System Reliability in Germany replaced the ATR 72-600 upper fuselage panel with a composite panel

Journal of Physics: Conference Series **2489** (2023) 012016 doi:10.1088/1742-6596/2489/1/012016

and then used 40 FBG sensors (16 for longitudinal stain, 16 for transversal strain, and 8 for temperature compensation) to monitor strain and 54 piezoelectric sensors to monitor damage to the skin and stringer[9].

In this paper, FBG sensors are used to measure the load of the aircraft during flight and the life of the fatigue critical parts in the modified structure is calculated, which supports the determination of the modification scheme.

2. Flight measurement of fatigue load of panel modification

2.1. Introduction of panel modification

The fuselage panel is modified to add an antenna, which is located at the top of the fuselage, specifically at the top BS790 station, between stringer S-1 and S-2R, as shown in Figure 1 below. As the antenna is installed to make a hole in the skin, it needs to use the external doubler, and the satellite phone antenna is fixed to the doubler by 4 screws, see Figure 2 for details.

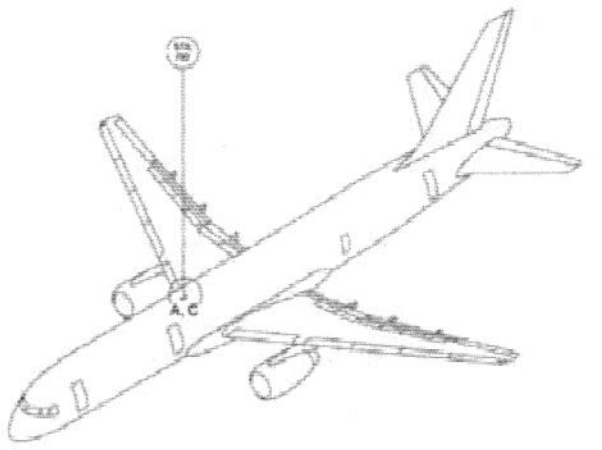

Figure 1. Panel modification installation location.

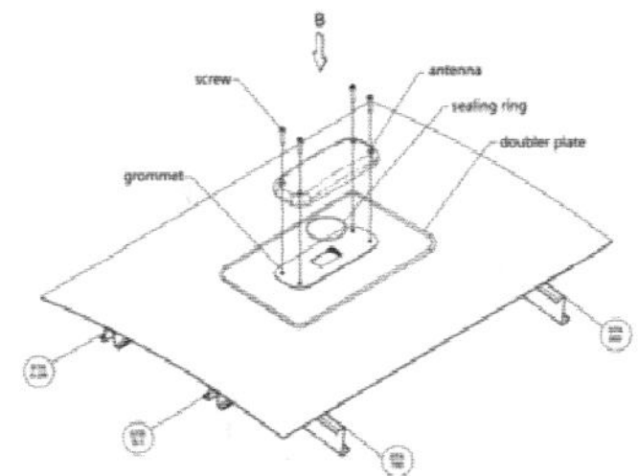

Figure 2. Satellite phone antenna installation program.

2.2. Flight load acquisition and processing

The strain of the panel modification was obtained by installing FBG sensors, the specific installation scheme before modification was shown in Figure 3 below, and the measured data of the sensors with larger loads were selected for analysis, i.e., the data of sensors No. 1 and No. 2. Sensors No. 1 and No. 2 correspond to the circumferential and longitudinal directions, respectively. The flight with the longest range and the largest commercial load was selected as the severe working condition for analysis, and the measured data of the flight are shown in Table 1.

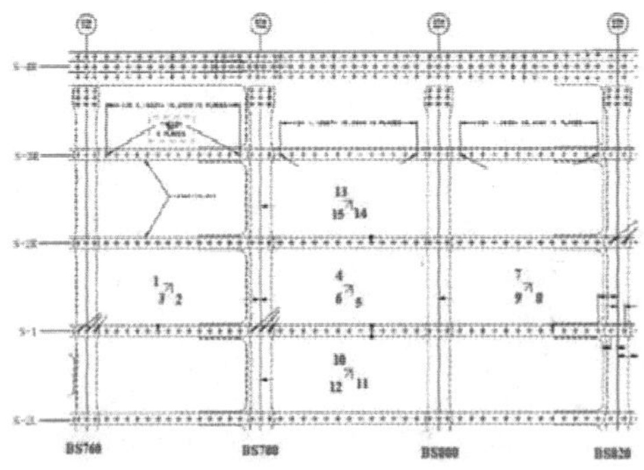

Figure 3. Sensor installation solution.

Table 1. Measured flight information.

Date	Aircraft status	Departure time	Arrival time	Flight hours	Loading rate	Maximum altitude (m)
2021-6-26	flight mission	5:23:00	9:09:00	3:46:00	92%	10706

Typically measured data during the flight, such as flight altitude, and longitudinal and circumferential strain are shown in Figure 4, Figure 5, and Figure 6.

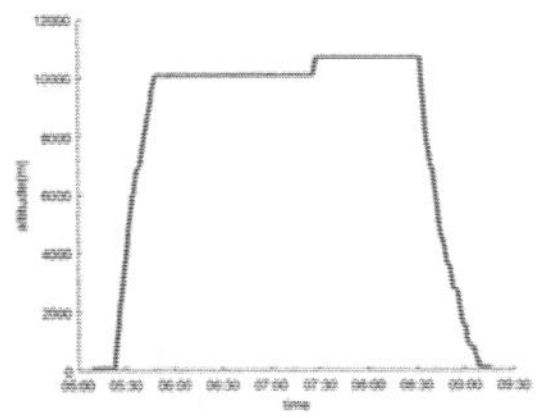

Figure 4. Time-altitude curve.

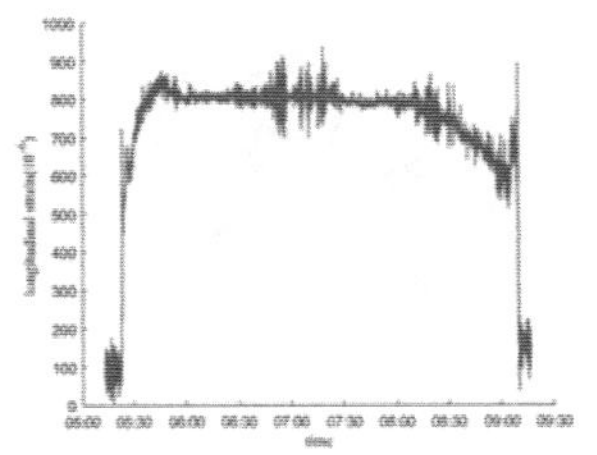

Figure 5. Time- longitudinal strain curve.

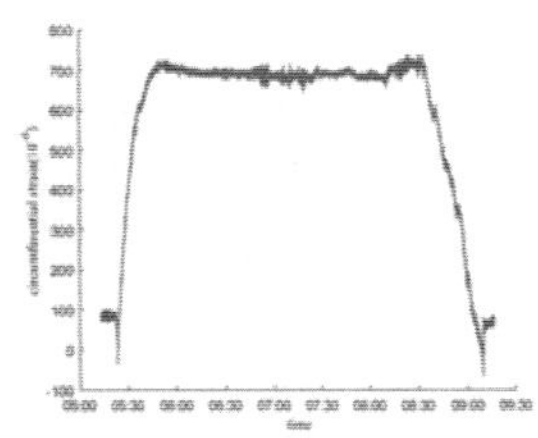

Figure 6. Time- circumferential strain curve.

It is assumed that the fuselage skin is plane stress. The strain data can be calculated according to the plane stress to obtain the longitudinal and circumferential stresses, and the time-stress curves are shown in Figure 7 and Figure 8.

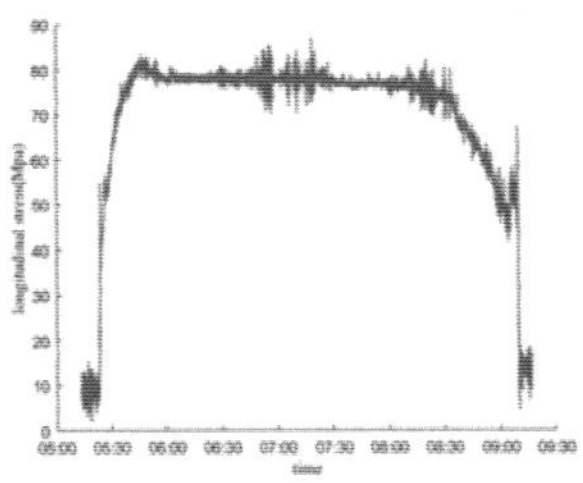

Figure 7. Time- longitudinal stress curve.

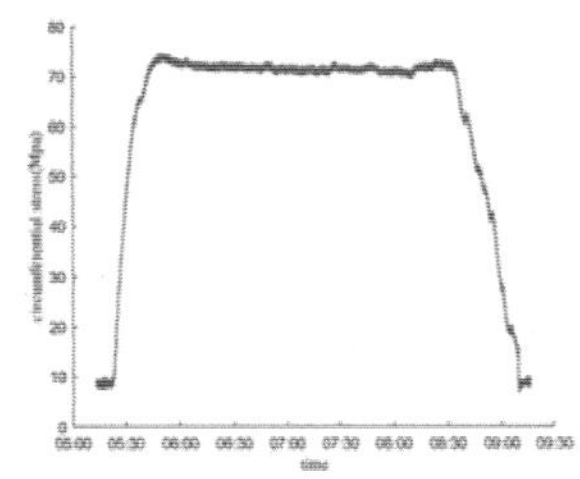

Figure 8. Time- circumferential stress curve.

The stress data are processed by removing the continuous equal data points, extracting the peak and valley values, and filtering small-variation. The purpose is to reduce the data volume and extract the data contributing to fatigue damage. Rain flow counting is performed on the reduced data to obtain the mean stress, stress amplitude, and cycle number of each cyclic stress of a flight, which provides input for the calculation of fatigue life.

2.3. Load analysis

The hole edge stress is solved by establishing a detailed sub-model and interpolating the load of the global model to the boundary of the sub-model, as shown in Figure 9 below. While the rivet elements are established to analyze the hole stress in the connection, and the calculation process is shown in Figure 10 and Figure 11.

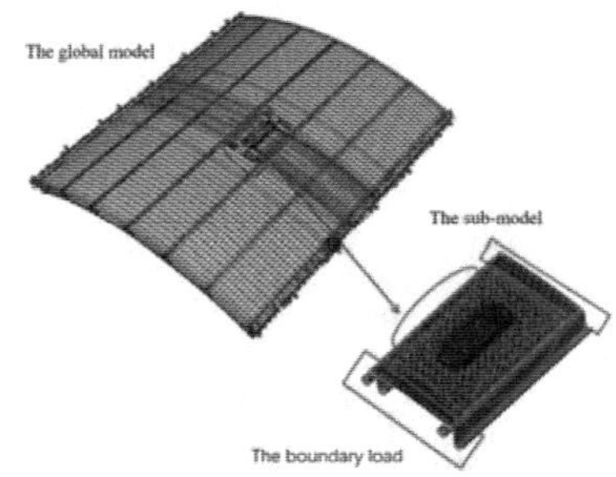

Figure 9. The global model and sub-model.

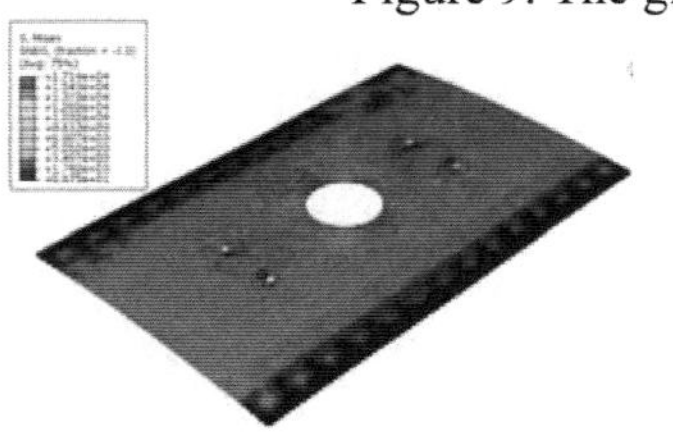

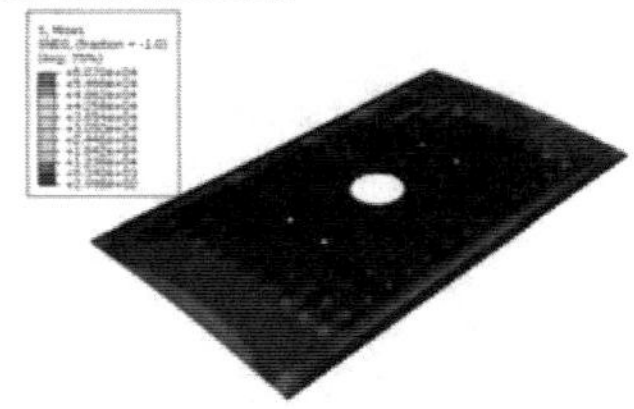

Figure 10. Finite element analysis results of the doubler.

Figure 11. Finite element analysis results of the skin.

Through the finite element analysis, it can be found that the fatigue critical parts of the panel modification are the satellite antenna hole and the edge part of the skin-doubler connection, as shown in Figure 12 below.

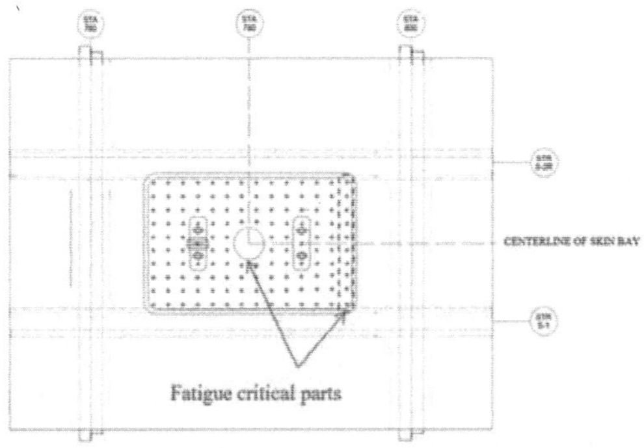

Figure 12. Fatigue critical part distribution.

Through the analysis of the finite element and the measured data, the ratio of the load at the fatigue critical parts to the load at the sensor position can be obtained, thus converting the measured load spectrum before the modification into the measured load spectrum after the modification, as shown in Table 2.

Table 2. Results of finite element analysis of fatigue critical parts and sensor installation locations.

Serial number	Stress analysis site	Reference load direction	Location	Stress (Mpa)	load at critical parts /load at FBG location
1	skin-doubler connection	longitudinal direction	the first row of rivets	139	1.8
2		circumferential direction	the first row of rivets	47	0.7

3	satellite antenna hole	longitudinal direction	hole edge	92	1.2
4	FBG1 location	longitudinal direction	between stringer S-1 and S-2R, BS790 station	77	/
5	FBG2 location	circumferential direction	between stringer S-1 and S-2R, BS790 station	67	/

3. Fatigue life assessment of panel modification based on measured load

3.1. Detail fatigue rating calculation

For the satellite antenna hole, the detail fatigue rating (DFR) is calculated by the following equation:

$$DFR = DFR_{base} \cdot U \cdot R_c \quad (1)$$

According to Table 3 in reference [10], the information about the 2024-T3 thin plate can be obtained, the surface roughness of the flat plate $R_a = 3.2$, drilled and not chamfered. According to the finite element analysis, the stress concentration is $K_{tg} = 2.0$, $DFR_{base} = 59 \times 3.1/2.0 = 91.4MPa$.

Table 3. DFR_{base} for the aluminum hole

Alloy & Status	Flat surface roughness R_a	DFR_{base}			
		Drilled hole		Reamed hole	Hole peening（BAC5730）maximum aperture =10 mm
		Unchamfered	Chamfered < 0.25 mm	Chamfered > 0.25 mm	
2024-T3 thin plate	3.2	59	69	76	90
	1.6	59	76	83	92
	0.8	59	83	90	92

Remark: DFR_{base} in the table is based on the stress concentration factor of the central hole K_{tg}= 3.1.

For the satellite antenna hole, $R_c = 1.44$, $U = 1, DFR = 131.6MPa$.

The detail fatigue rating of the skin-doubler connection is calculated by taking the longitudinal and circumferential directions as reference loads, respectively. For the skin-doubler connection, the DFR value is calculated according to the allowable value of the single shear hard point joint:

$$DFR = DFR_{base} \cdot A \cdot B \cdot C \cdot D \cdot E \cdot U \cdot R_c \quad (2)$$

As with the satellite antenna hole described above, the process is not described. The calculated DFR values for the longitudinal and circumferential directions are 114.7 MPa and 110.3 MPa, respectively.

Based on the above analysis, the detailed fatigue ratings of the satellite antenna hole and the edge part of the skin-doubler connection are shown in Table 4.

Table 4. Detail fatigue rating results.

Serial number	Fatigue critical parts	Reference load direction	Location	DFR (MPa)
1	skin-doubler connection	longitudinal direction	the first row of rivets	114.7
2		circumferential direction	the first row of rivets	110.3
3	satellite antenna hole	longitudinal direction	hole edge	131.6

3.2. Fatigue life assessment of panel modification

According to the detailed fatigue rating method, the number of cycles to failure at the i-th stress is

$$N_i = 10^5 \left[\frac{\sigma_{m0} - 0.53DFR}{0.47\left(\sigma_{m0} - \sigma_{mi}\right)DFR} \sigma_{ai} \right]^B \tag{2}$$

Where σ_{mi} and σ_{ai} are the mean stress and stress amplitude at the i-th stress, σ_{m0} is the ultimate strength of the material, and B is the slope of the S-N curve.

The structural damage at the i-th stress is

$$D_i = \frac{1}{N_i} \tag{3}$$

Based on the linear fatigue damage cumulative rule, the cumulative damage during a flight is calculated as

$$D = \sum_{i=1}^{k} D_i = \sum_{i=1}^{k} \frac{1}{N_i} \tag{4}$$

Finally, the fatigue life can be calculated as

$$\lambda = \frac{1}{D} = \frac{1}{\sum_{i=1}^{k} \frac{1}{N_i}} \tag{5}$$

Using the flight data on the severe working condition selected in section 2.2 as the basis for the life calculation of the panel modification, considering the load transfer between the monitoring parts and the fatigue critical parts before and after the modification, the load of the fatigue critical parts is obtained. The fatigue life of the satellite antenna hole and the first row of rivets in the skin-doubler connection is calculated according to the detailed fatigue rating method, and the calculation results are shown in Table 5. From the calculation results, the first row of rivets in the longitudinal direction of the skin-doubler connection is the most critical part, and the life is evaluated as 384,893 cycles.

Table 5. Fatigue analysis results based on typical flight data.

	The first row of rivets		Satellite antenna hole
	Longitudinal direction	Circumferential direction	Longitudinal direction
DFR(MPa)	114.7	110.3	131.6
Life assessment $N_{95/95}$	384893	1967698	1093439
Most critical part determination	√		

4. Conclusion

In this paper, the detailed fatigue rating method was adopted to assess the fatigue life of the fuselage panel modification. The FBG sensors were installed to measure the strain of the modified structure during flight. Through the finite element analysis, the fatigue critical parts were determined and the measured data before modification were converted into the data after modification. The measured stress data of typical flights were selected, and the fatigue life based on the single take-off and landing was calculated. According to the calculation results, the most critical part is determined to be the first row of the rivets in the longitudinal direction of the skin-doubler connection, and the fatigue life is evaluated to be 384,893 cycles, which shows that the fatigue life of the modified structure was much higher than the design life, and the modification solution of the fuselage panel has a high safety margin.

Journal of Physics: Conference Series **2489** (2023) 012016 doi:10.1088/1742-6596/2489/1/012016

References

[1] Lin, J. H., Wang, B. W. (2022) Evolution history of airworthiness regulations for aircraft structural fatigue designs. Aeronautical Science and Technology, 33: 39-51.

[2] Matthew, G., 2019. Cracks found on Boeing's 737 NG pickle forks. https://www.engineering.com/story/cracks-found-on-boeings-737-ng-pickle-forks.

[3] Arjen, K., 2019. Fiber optic sensing for structural health monitoring. In: 30th ICAF Symposium. Krakow.

[4] Wada, D., Igawa, H., Tmayama, M., et al. (2019) Flight demonstration of aircraft wing monitoring using optical fiber distributed sensing system. Smart Materials and Structures, 28:055007.

[5] Kim, J. H., Park, Y., Kim, Y. Y., et al. (2015) Aircraft health and usage monitoring system for in-flight strain measurement of a wing structure. Smart and Structures, 24:105003.

[6] Wei, Y. L., Gao, F. P. (2022) Architecture design method for structural health monitoring system (SHM) of commercial aircraft. Computer Measurement &Control, 30: 38-43.

[7] Bockenheimer, C., Speckmann, H., Team, I. 2013. Validation, verification, and implementation of SHM at Airbus. In: Proceedings of the 9th International Workshop on Structural Health Monitoring (IWSHM 2013). Stanford.10-12.

[8] Hill, K. O., Meltz, G. (1997) Fiber Bragg grating technology fundamentals and overview. Journal of Lightwave Technology, 15:1263-1276.

[9] Lilov, M., Siebel, T., Asins, C.C., et al, 2016. Development of a Structural Health and Load Monitoring System and its Implementation into an Aircraft Flying Demonstrator. In: 8th European Workshop on Structural Health Monitoring (EWSHM 2016). Bilbao.

[10] Zheng, X. L. (2003) Durability and damage tolerance design of commercial aircraft metal structures. Aviation Industry Press, Beijing.

Numerical Analysis of Pressure Load on Hypersonic Wing

Zheng Chao[1], Shubo Jiang[*] and Guangming Zhang

Nanjing Tech University, College of Electrical Engineering and Control Science, Nanjing, China

Email: 1219653431@qq.com,;jiangshubo@njtech.edu.cn

Abstract. The flight vibration environment of hypersonic vehicles has always been a hot topic in the field of aerospace at home and abroad. In this paper, the simple wing model and the outer flow field grid of the wing are established by using fluent software. A series of environmental parameter conditions are set in the hypersonic flow field, and the steady and unsteady flow fields are analyzed. The distribution characteristics of the pressure load on the wing surface in the hypersonic flow field are obtained. The results show that the pressure load measured at the head position of the wing model is larger, and the pressure load at the fuselage position is smaller and changes more gently. With the increase of Mach number, the pressure load on the wing model will also increase. Then MatLab software is used to convert the time domain curve of wing surface pressure load into a power spectral density curve, which provides load input for subsequent random vibration response simulation on finite element software.

1. Introduction

The vibration environment encountered by hypersonic aircraft during flight is very severe [1]. When the aircraft wall is subjected to strong pressure loads, it is easy to cause sustained vibration of the aircraft structure, and the resulting dynamic response may lead to fatigue damage to missile equipment and airframe structure [2]. Therefore, it is necessary to predict the pressure load on hypersonic missiles.

At present, many scholars at home and abroad mainly focus on cavity tests, flight tests, theory, and empirical estimation of hypersonic vehicle pressure environments [3]. With the development and application of computational fluid dynamics (CFD), CFD numerical calculation method has become an important means to study the aerodynamic characteristics of aircraft. In this paper, we use fluent software to establish the grid of the wing's external flow field and analyze its fixed length and unsteady flow field [4]. This analysis fully considers the influence of various factors on the pressure environment, avoids the error caused by the theory and estimation method, and considers the influence of Mach number, angle of attack, and other parameters on the pressure environment.

The numerical analysis method is based on the N-S equation, and given the parameter conditions under different conditions, the equations applicable to different flow fields are obtained. Turbulence numerical simulation methods are divided into three types: direct numerical simulation (DNS), Reynolds average method (RANS), and large eddy simulation (LES) [5].

The direct numerical simulation method (DNS) does not use any turbulence model. By directly solving the complete three-bit unsteady N-S equations, the time evolution of all instantaneous turbulent motion in the three-dimensional flow field can be calculated. This method has high accuracy, but it takes a long time to calculate and requires a high computer. The large eddy simulation method (LES) uses a direct numerical simulation method to simulate large-scale turbulent movement and uses a sub-grid scale model to simulate the impact of small-scale turbulent movement on large-scale turbulent movement.

This method is more suitable for flow separation areas with large-scale eddy currents, but it is not accurate for near-wall flow simulation. Reynolds average method (RANS) applies the turbulence statistical theory to average the time of the unsteady N-S equation and solve the amount of time required in the project[6].

Now, because the calculation model in this paper is relatively simple, and there is no large-scale separated vortex structure in the flow field, to save resources, this paper selects Reynolds average method (RANS) and selects an equation S-A turbulence model suitable for the wall boundary layer flow.

2. Finite element analysis

2.1. Computational model

The two-dimensional wing model is selected as the calculation model. The wing model is symmetrical from top to bottom, which can reduce the calculation amount. The average aerodynamic chord length of the wing is 0.64m, the wingspan is 1.19m, the aspect ratio is 3.8, and the root tip ratio is 0.56.

In this paper, the flow field calculation module in the widely used fluent finite element software is used to analyze the flow field of the hypersonic wing. The model in this paper is relatively simple and suitable for the use of structural grids. The model needs to pay attention to the subtle changes in the internal flow of the boundary layer and can capture small vortices in the flow field. Therefore, the grid is divided more intensively. The symmetry plane grid of the wing fluid domain is established as shown in the Figure 1and Figure 2, with a grid number of 264571 units, and 281126 nodes.

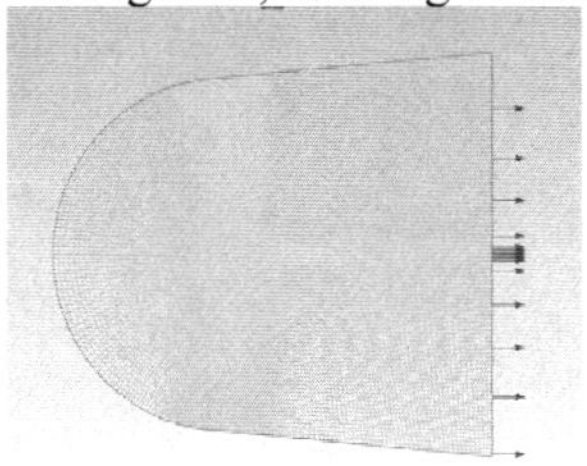

Figure 1. Grid division of overall flow field area

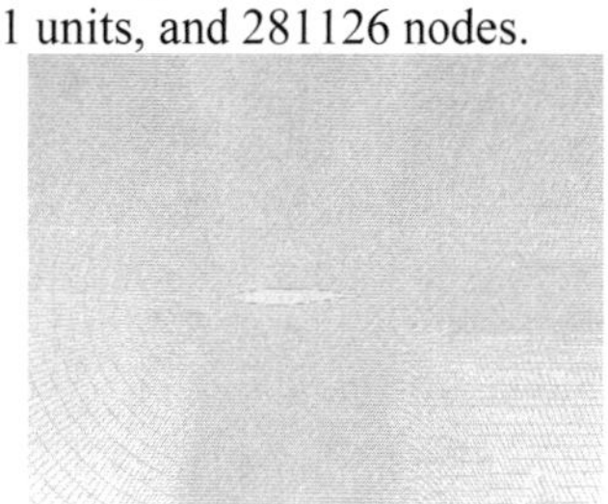

Figure 2. Mesh generation around the wing

2.2. Parameter condition setting

The parameter conditions in fluent software in turn are set, the air parameters at an altitude of 10km above sea level (Table 1) for the incoming flow parameters of the calculation model are selected, the Mach number of the incoming flow as 6 is selected, and the angle of attack of the incoming flow is selected as 0 °. The boundary condition sets the wall as an adiabatic condition, and the parameters of the velocity inlet and free inflow and outflow boundary are defined by the inflow parameters.

Table 1. Incoming gas parameter setting

Parameter conditions	Parameter value
Altitude	10
Air density	0.41351
Static pressure	26500
Static temperature	223.25
Sound velocity	299
Dynamic viscosity	1.46e-5
Kinematic viscosity	3.5e-5

In this model, the finite volume method based on density is used to simulate the hypersonic flow field. The gas of the model is selected as the ideal gas, and the turbulence model selected is the Spalart Allmaras model. The pressure far fields of the inlet and outlet in the boundary conditions are set according to the parameter conditions in Table 1.

Then the model starts to perform stable calculations, and the free flow parameter is selected as the initialization parameter of the calculation domain. Considering the number of grids and the required calculation time, the number of iteration steps is set to 8000, and the transient calculation starts after the calculation convergence.

2.3. Analysis of aerodynamic results

It can be seen from the pressure nephogram and the wall pressure curve that a strong oblique shock wave is generated at the leading edge of the wing, causing the air to be compressed violently. The pressure at the leading edge of the wing increases sharply to $3.6 * 10^4$ Pa, the temperature reaches 1500K, and the gas compression density is 2.38kg/m3. At the same time, the airflow speed here is close to 0. Then the speed increases from zero, the pressure starts to decrease rapidly, and the lowest pressure reaches $2.25 * 10^4$ Pa.

Wall pressure is a broadband load, which contains large turbulent vortices with low frequency and high amplitude and small turbulent vortices with high frequency and low amplitude. In the process of finite element simulation, if you want to obtain the pulsating pressure caused by shock wave/boundary layer interaction, turbulent boundary layer, separated flow, and other factors through calculation, you need to select the appropriate physical time in the unsteady calculation of the flow field, and its time step should be less than the period of the turbulent vortex. Therefore, the total time is 0.004s, and the physical time step is $2 * 10^{-6}$s. Taking a point near the wall to get the curve of the pressure change of the sampling point with time is shown in Figure 5.

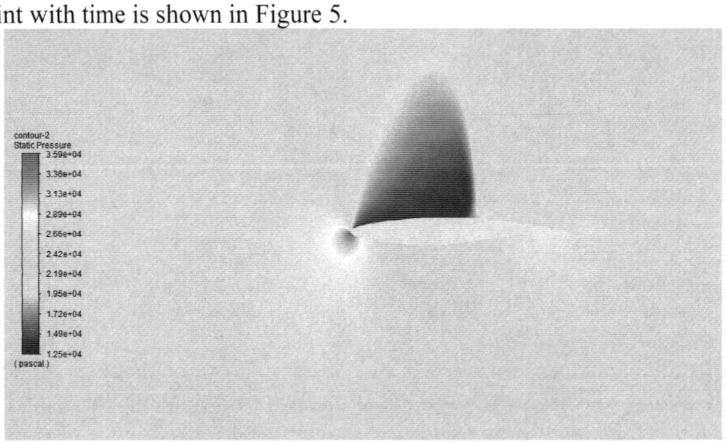

Figure 3. Airfoil pressure nephogram

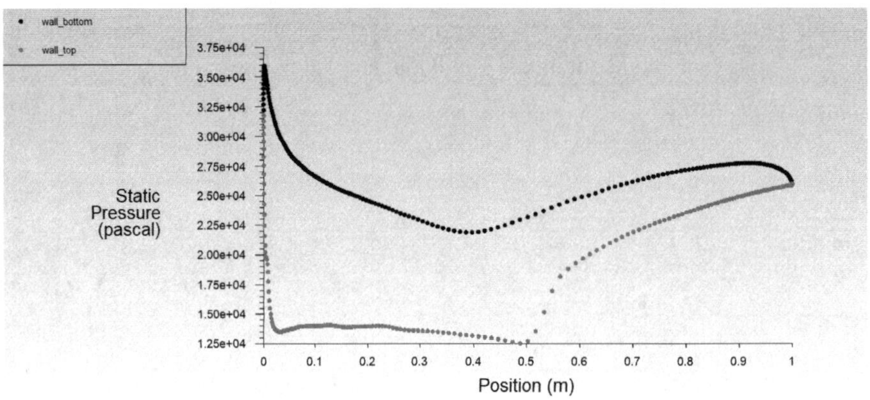

Figure 4. Pressure distribution curve on upper and lower surfaces of the airfoil

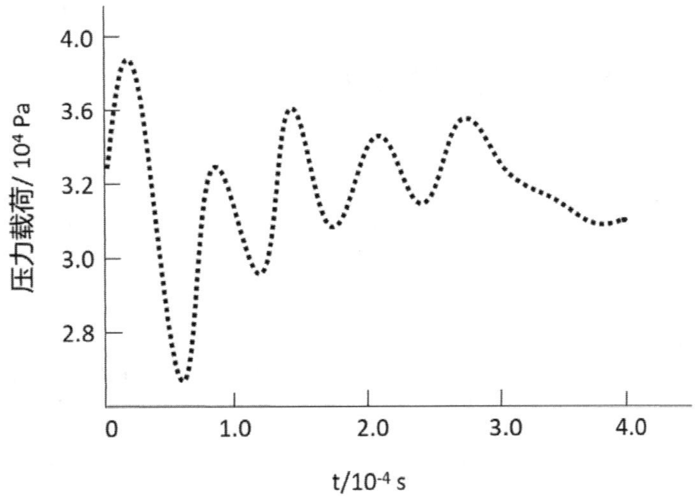

Figure 5. Time domain variation curve of airfoil surface pressure

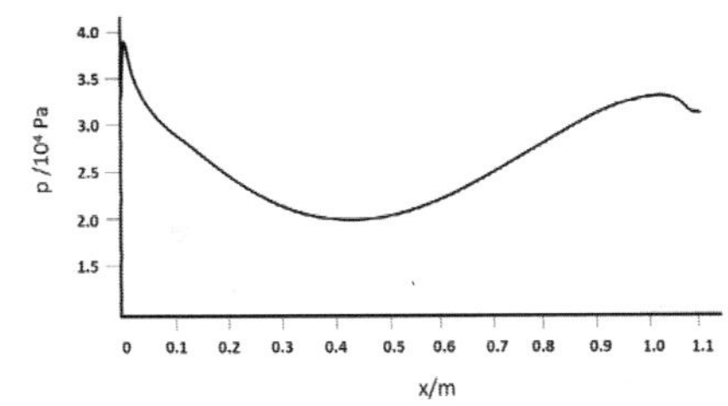

Pressure distribution at Ma=5

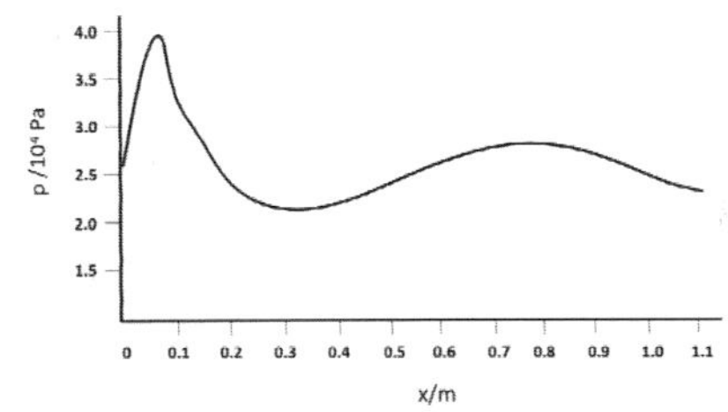

Pressure distribution at Ma=6

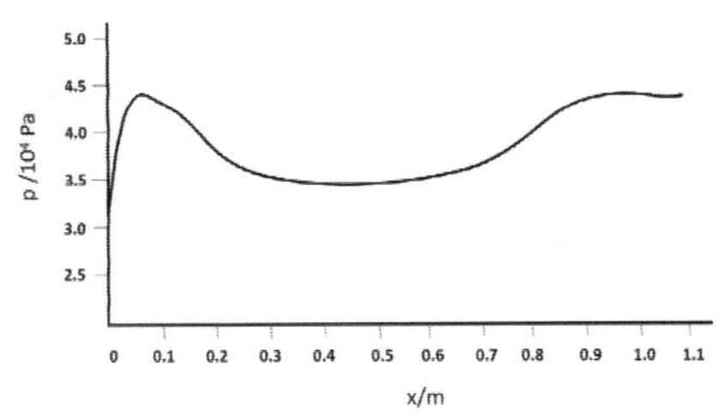

Pressure distribution at Ma=7

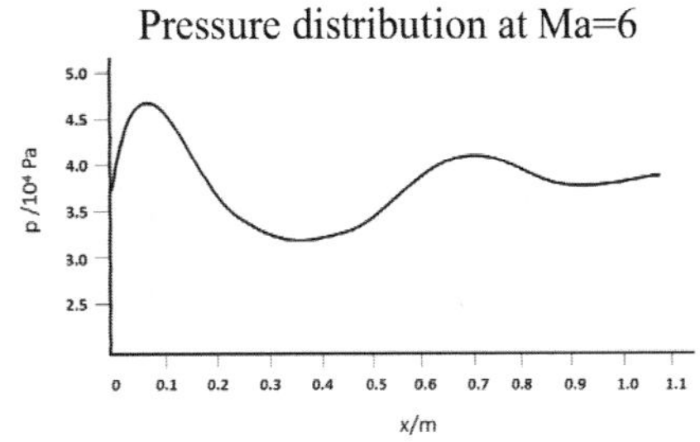

Pressure distribution at Ma=8

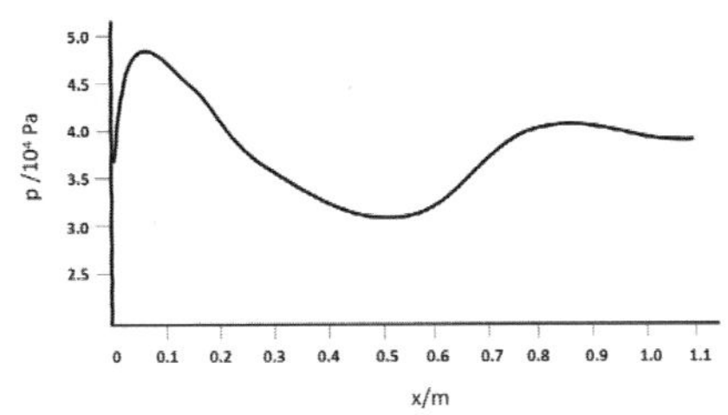

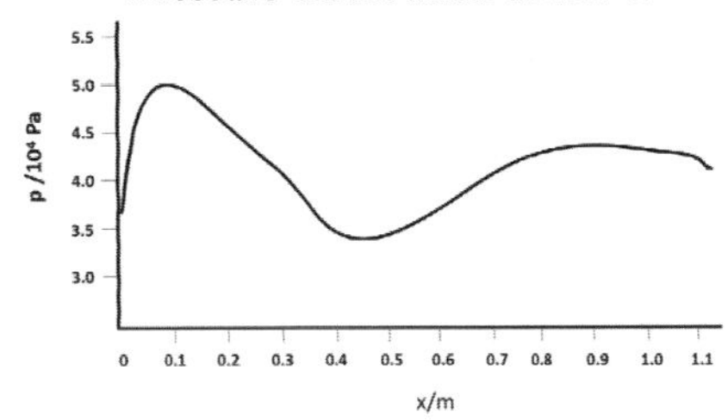

Pressure distribution at Ma=9 Pressure distribution at Ma=10

Figure 6. Variation curve of wall pressure value along x direction at different Mach numbers

It can be concluded from Figures 3 and 4 that in the hypersonic boundary layer, the pressure at the shock wave position at the leading edge of the wing is very high, the pressure behind the shock wave decreases, and the aerodynamic pressure after the shock wave changes little. It can be seen from Figure 6 that with the increase of incoming Mach number, the magnitude of pressure load in the boundary layer gradually increases.

As the pressure load belongs to a random load, this random process can be considered an ergodic process of various states. According to the time domain change of the fluctuating pressure value in a small period, the power spectral density analysis program can be compiled using the engineering software MATLAB. The power spectral density curve of the wing pressure load can be obtained, which provides a reference for the subsequent random vibration response experiment input load.

3. Conclusion

This paper uses fluent software and the CFD (computational fluid dynamics) method to establish the flow field grid of the wing model in the software and analyzes the distribution characteristics of the pressure load on the wing surface in the hypersonic flow field by inputting the relevant incoming gas parameters. At the same time, the influence of each number and other factors on the pressure environment on the wing surface is studied and analyzed, and the corresponding conclusions are drawn. The obtained airfoil surface pressure time domain change curve is converted into a power spectral density curve through MatLab, which provides load input for subsequent random vibration response simulation on finite element software. This paper presents a set of engineering methods to predict the fluctuating pressure environment, which is a new progress in the research topic of the prediction of the fluctuating pressure environment of hypersonic vehicles and can also provide some basis and reference for hypersonic ground environment simulation tests. However, because the wing model and flow field grid established in this paper are relatively simple, and the parameter conditions are also set in the ideal state, the real situation will be much more complex than in the simulation. In the subsequent research work, a more detailed aircraft model will be established, and the relevant pressure load prediction simulation will be further improved after considering more complex flow field parameter conditions.

References

[1] Dolling D S, Or C T 1985 J. Unsteadiness of the shock wave structure in attached and separated compression ramp flows. *Experiments in Fluids* 3(1): 24-32

[2] Pirozzoli S, Grasso F 2006 *J.* Direct numerical simulation of impinging shock wave/turbulent boundary layer interaction at M= 2.25. *Physics of fluids* 18(6): 065113

[3] Dressage J P, Dupont P, Debiève J F 2006 *J.* Unsteadiness in shock wave boundary layer interactions with separation. *Aerospace Science and Technology* 10(2): 85-91

[4] Houbolt J C 2006 *J.* On the estimation of pressure fluctuations in boundary layers and wakes. *General Electric Report TIS 66SD296*

[5] Yang Zepeng, Wang Suozhu, Gao Zhenxun 2022 *J.* Studies on effects of wall temperature variation on heat transfer in the hypersonic laminar boundary layer. *International Journal of Heat and Mass Transfer* 190

[6] Bai Tian, Duan Zhiwei, Xiao Zhixiang 2022 *J.* Depth effects on the cavity-induced transition at hypersonic speed by DNS. *International Journal of Heat and Fluid Flow* 97

Journal of Physics: Conference Series **2489** (2023) 012018 doi:10.1088/1742-6596/2489/1/012018

Helicopter Load Factor Limit Protection Method Based on A Predictive Control Algorithm for Vertical Maneuvering

Xiaoxiao Liu, Pan Li, and Zhenhua Zhu

National Key Laboratory of Rotorcraft Aeromechanics, NUAA, Nanjing 210016, China

Email: lipan@nuaa.edu.cn

Abstract. The purpose of this paper is to solve the problem of the helicopter load factor limit exceeding caused by inappropriate pilot control in vertical maneuvering flight. Through the collective stick control power restriction, an load factor limit protection method based on a load factor predictive algorithm is proposed. First, the load factor prediction is realized by the change rate of the collective stick position given by the actuator and the control derivative of the collective stick position to vertical acceleration, and then the vertical control power is restricted according to the predicted load factor, so as to achieve the purpose of helicopter load factor protection.

1. Introduction

According to statistics, more than 60% of flight accidents in the world are caused by human factors annually, of which about 48.6% are caused by improper pilot control [1]. A large proportion of flight accidents caused by improper pilot control is due to the pilot's control that makes the flight state of the aircraft exceed its flight envelope boundary. For helicopters, the limitations of residual power, structural strength, rotor stall speed, and critical Mach number make it possible to fly normally only in a certain range of parameters such as flight speed, flight altitude, load factor, engine torque, and rotor speed parameters. The flight envelope protection of the aircraft is of great significance to use its full flight capability, reduce the pilot control workloads and improve the flight safety of the aircraft. For example, when the pilot is close to the edge of the flight envelope, he needs to monitor and judge whether the helicopter exceeds its flight envelope boundary, which will undoubtedly distract the pilot's attention, greatly increase the pilot's psychological burden, and affect the flight performance of the helicopter. If the pilot does not control the helicopter properly and the helicopter exceeds the flight envelope boundary, it will easily cause an irreversible accident. For example, if a helicopter exceeds its load factor envelope during a maneuvering flight, it will easily cause helicopter parts' fatigue damage and structural damage, and then endanger flight safety [2]. Therefore, it is important to study the load factor protection method for helicopters.

Load factor protection is a common function in fixed-wing aircraft. Generally, at low speed, the angle of attack is protected; at high speed, the load factor protection is performed, and the load factor of the fixed wing is related to its airspeed and angle of attack. Different from the fixed wing, the load factor of the helicopter is related to the airspeed and fuselage angle of attack, but also to the collective pitch in the forward flight state, and mainly to the collective pitch in hover and low-speed flight conditions. The load factor protection of helicopters has been researched both at home and abroad, but it is less applied in actual models. Abroad, Horn[3]-[5] et al adopted the dynamic trim concept of maneuvering flight state, that is, without considering the dynamic response process of fast variables such as angular rate in the transient maneuvering process, they established the quasi-steady dynamic equations of boundary parameters such as angle of attack, load factor, and rotor torque, and established the full-envelope nonlinear functional relationship between the boundary parameters of the flight envelope and the slowly-varying flight state quantities and control quantities (which can be obtained directly by sensors) using a large number of flight test and flight simulation data by the neural network method. Because of the dynamic trim concept, the boundary parameters predicted by the model are naturally ahead of the actual transient boundary parameters by a certain time, so the control margin of each axis related to the boundary of the flight envelope can be estimated by using the boundary parameters predicted in advance and the linear inverse model of the neural network model. And changing the stick force gradient and stick force gradient inflection point of the joystick according to the size of the control margin restricts the pilot's manipulation through tactile reminders to avoid exceeding the flight boundary. Bateman [6] et al used the flight test data and flight simulation data to directly establish the nonlinear function relationship between the current sensor state parameters and the flight envelope boundary parameters after a short period of time through the polynomial neural network model. The neural network model and its inverse are used to predict the control margin, and the pilot's control is prompted and restricted by the joystick force sensing system, so as to realize the protection of the boundary parameters of the flight envelope. Thaiss[7] et al realized load factor protection by limiting the output of the command by the command model in the command model tracking control structure. Kimball [8] et al realized the protection of rotor shaft torque and rotation speed by designing feedback control law based on sensor data on V-22 Osprey.

The above flight envelope protection methods are mainly divided into two types: one is based on sensor data through the design of feedback control law to achieve the limitation of the flight envelope parameters, and the other through the flight envelope boundary parameter prediction technology predicts the flight state distance from the flight envelope boundary, through the cockpit interactive information to prompt the pilot to avoid the flight envelope boundary. These two methods are not the simplest and most effective methods for simply limiting the load factor. For the feedback control method, on one hand, an accurate load factor estimation algorithm is required, and on the other hand, it will have some impact on the maneuvering quality. The load factor protection method based on the load factor ahead prediction technology is not mature and there is no relevant information to show its application in practice. On the one hand, it is difficult to achieve relative accuracy in the whole load factor envelope. On the other hand, this load factor prediction technique currently commonly uses a neural network model, which is trained by relying on a large amount of flight test data and simulation test data, and this neural network model has time complexity and space complexity, and it is difficult to test and verify the control software of the neural network. In addition, because this type of method mostly adopts the way of actively changing the force gradient of the joystick to achieve load factor protection, it is more dependent on the active side stick, and it is difficult to apply to the conventional central joystick and passive side stick.

These problems lead to the poor engineering implementation ability of this load factor protection method based on neural network model prediction, while this paper aims to establish a load factor predictive algorithm with simple and reliable control structure and high engineering implementability. The predictive algorithm used in this paper is reliable because the magnitude of load factor generated by the pilot lifting a certain amount of the collective stick is directly related to the control derivative of collective stick position to vertical acceleration in the helicopter's control derivative matrix, and this control derivative can be obtained in engineering by the helicopter test flight or wind tunnel test.

The load factor protection of helicopters includes load factor protection in hover and low-speed flight conditions and load factor protection at large forward speeds, the former can be achieved by limiting the collective pitch, and the latter can be achieved by limiting both the collective pitch and the longitudinal cyclic pitch. As a preliminary study, this paper focuses on the load factor protection of helicopters in hover and low-speed flight conditions and studies the design method of the load factor protection control law for helicopter vertical maneuvering flight state.

2. Design and Simulation of Vertical Maneuvering Load Factor Protection Control Law for Helicopters

2.1. Load factor Characteristic Analysis in Helicopter Vertical Maneuvering Flight

The load factor of a helicopter refers to the ratio of all external forces except gravity to gravity on the helicopter (usually limiting the vertical load factor) [9]. The load factor in the vertical maneuvering flight of the helicopter is caused by the aerodynamic force of the helicopter. The aerodynamic force of the helicopter in conventional configuration includes the aerodynamic force of the rotor, fuselage, horizontal tail, vertical tail, and tail rotor. Among these forces, the vertical component of rotor aerodynamic forces (the thrust of the rotor) is the main part of the vertical load factor; in addition, there is also the vertical component of the fuselage and horizontal tail aerodynamic forces, mainly its lift; the vertical tail and tail rotor basically provide no vertical aerodynamic force.

When the helicopter is hovering at low a speed, due to the small dynamic pressure, the lift of the horizontal tail and the fuselage can be basically ignored compared with the rotor thrust, and the rotor thrust is mainly determined by the collective pitch. Therefore, in this state, the helicopter's load factor can only be limited by the collective pitch of the rotor. That is, it is only necessary to design the helicopter load factor protection control law for the vertical rate control (RC) response type of the vertical channel and the vertical collective pitch direct control (DC) response type in the helicopter full-authority fly-by-wire flight control system.

2.2. Design and simulation analysis of control power restriction method based on load factor predictive algorithm for helicopter vertical maneuvering

Before proceeding to the design of the load factor protection method, the helicopter flight dynamics model and the basic flight control law used in this paper are introduced. The helicopter flight dynamics model is a nonlinear model from the literature [10], which considers the airfoil unsteady/dynamic stall, dynamic wake distortion caused by maneuvering flight, blade elastic deformation effect, and engine dynamic characteristics. The basic flight control law is adopted as the command model tracking control structure, which includes the command model, inverse model, and stability loop, and its schematic diagram is shown in Figure 1.

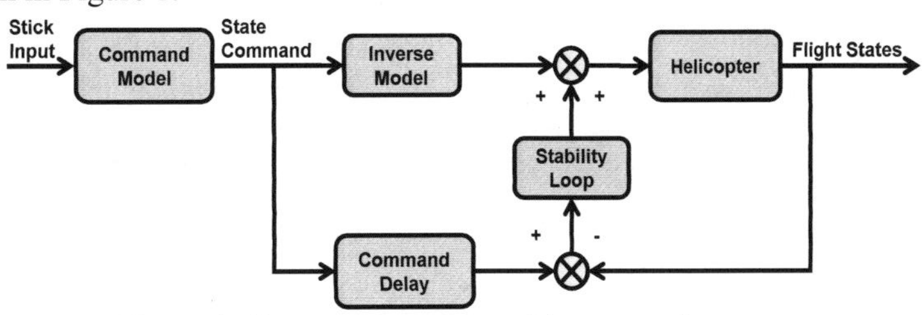

Figure 1. Command mode tracking control structure.

The direct restriction method of the command model given in the literature [7] is only applicable to the command model tracking control structure or dynamic inverse control structure and only to the vertical RC response type in it. In contrast, the control power restriction method based on the load factor predictive algorithm can be adapted to any control structure because the essence of the method is to restrict the control power of the collective stick, and therefore it is applicable to both the vertical RC response type and the vertical DC response type. The following is an example of the vertical DC response type to introduce the control power restriction method based on the load factor predictive algorithm, which is shown schematically in Figure 2.

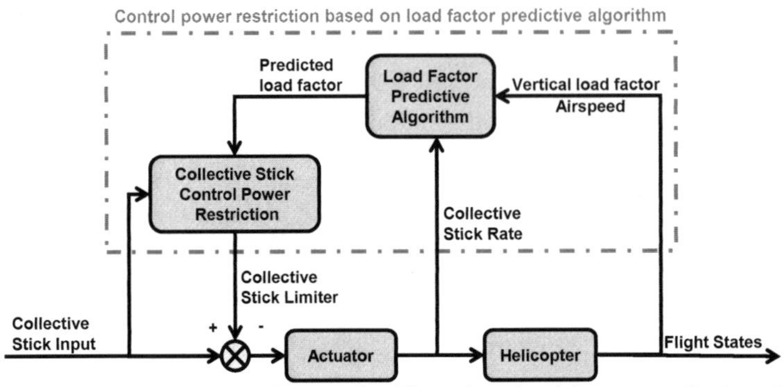

Figure 2. Control power restriction method based on load factor predictive algorithm.

There are two core modules in the control structure of control power restriction based on the load factor predictive algorithm, one is the load factor prediction module and the other is the control power restriction module, and the method used in the load factor prediction module is introduced first.

2.2.1. Vertical maneuvering load factor predictive algorithm.

The design idea of the load factor predictive algorithm is to achieve the load factor prediction based on the rate of change of the collective stick given by the actuator and the control derivative of the collective stick position to vertical force in the helicopter's control derivative matrix, which is implemented as follows:

First, the helicopter nonlinear flight dynamics model is linearized with small perturbations to obtain the helicopter's control derivative matrix in the current state, and then the collective stick position increment in a small period of time in the future can be extrapolated based on the rate of change of the collective stick position given by the actuator at this time, and the collective stick position increment is shown as follows:

$$\Delta\delta_{col} \approx v_{col} \cdot \Delta t \tag{1}$$

$\Delta\delta_{col}$ is the collective stick position increment, v_{col} is the rate of change of the collective stick position given by the actuator at this time, Δt is the predicted time interval. Then, using this increment and the control derivative of collective stick position to vertical acceleration, the vertical acceleration after that time can be obtained and then transformed into load factor, and this incremental load factor can be approximated as:

$$\Delta n_z \approx \frac{\Delta \dot{w}}{g} \approx \frac{\frac{\partial a_z}{\partial \delta_{col}} \cdot \Delta\delta_{col}}{g} = \frac{\frac{\partial F_z}{\partial \delta_{col} \cdot m} \cdot \Delta\delta_{col}}{g} \tag{2}$$

$\frac{\partial F_z}{\partial \delta_{col}}$ is the control derivative of collective stick position to vertical force, $\frac{\partial a_z}{\partial \delta_{col}}$ is the control derivative of collective stick position to vertical acceleration, $\Delta \dot{w}$ is vertical acceleration increment, Δn_z is load factor increment. This load factor increment plus the load factor at this time is the predicted load factor, and the predicted load factor is shown below:

$$n_{z\,pre} = \Delta n_z + n_{zcur} \tag{3}$$

$n_{z\,pre}$ is the predicted load factor, n_{zcur} is the load factor at the current moment. Once this predicted load factor exceeds the limited load factor, it is time to start restricting the control power of the collective stick, so as to achieve the purpose of load factor limit protection.

2.2.2. Control Power Restriction Method.

The control power restriction of the collective stick means: once the predicted load factor exceeds the limited load factor, the control effect of the collective stick's increment shall be restricted, and it is not directly restricted, but restricted in a compensatory way. A schematic diagram of the control power restriction of the collective stick is shown in Figure 3.

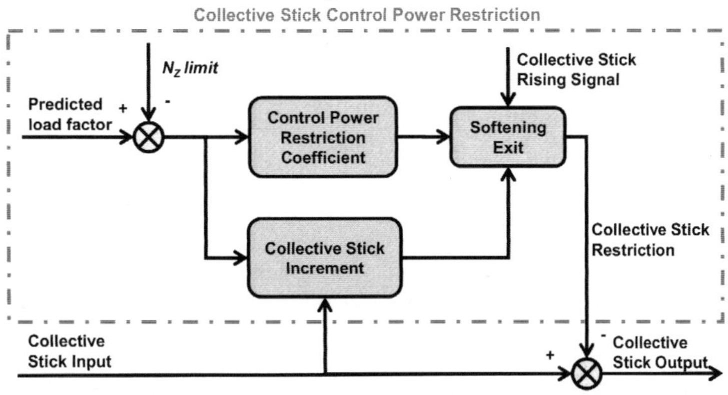

Figure 3. Structure diagram of collective stick control power restriction.

Its working principle is: when the predicted load factor is less than the limited load factor, that is, when the load factor deviation is less than 0, the control power restriction coefficient of the collective stick is 0, and the collective stick command's restriction is not performed. When the load factor deviation is greater than 0, the control power restriction coefficient is a number between 0 and 1, and the control power of the collective stick is restricted. In addition, if the control power of the collective stick is only restricted in the process of lifting the collective stick, it will only have a protective effect on the load factor in the lifting process. Once the collective stick position remains constant, in other words, the increment of the collective stick position is 0, the control power restriction module will not work, and the load factor will immediately rise. Therefore, a softening exit model is added to the restricted structure, and the collective stick's lifting signal is used as a switch. When the signal changes from 1 to 0, the load factor restriction doesn't exit immediately but delays the exit for achieving a better load factor protection effect.

2.2.3. Simulation analysis of the accuracy of the load factor predictive algorithm.

In hover, when the vertical response type is DC, the collective stick is operated with a fast and large square wave to test the accuracy of the load factor predictive algorithm. The collective stick's position command is shown in Figure 4 and the predictive simulation results are shown in Figure 5.

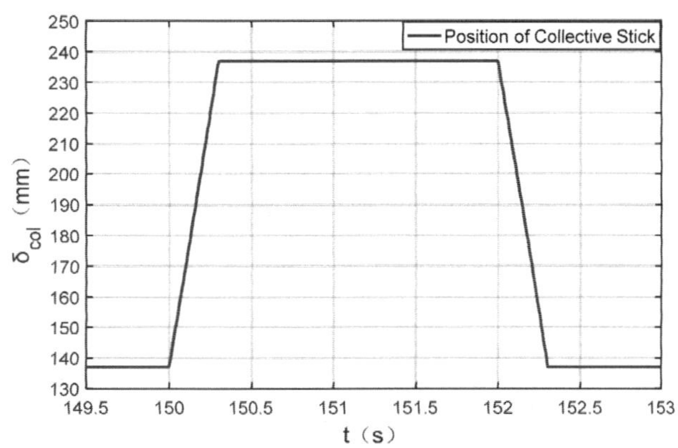

Figure 4. Collective stick's position.

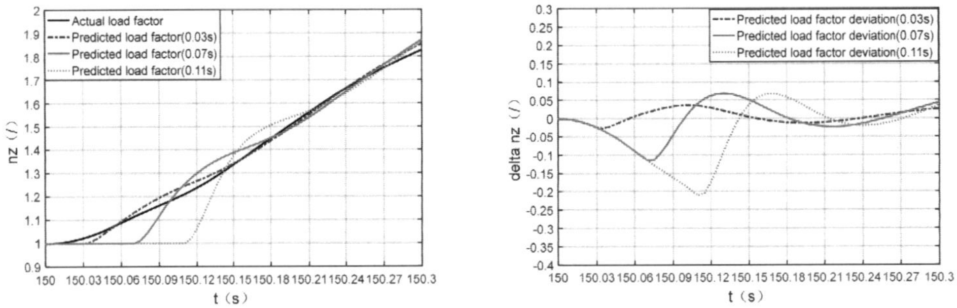

(a) Load factor comparison in lifting collective stick phase

(b) Predicted load factor deviation for different predictive time intervals

Figure 5. Response diagram of predicted overloads and actual overloads.

As can be seen in Figure 5, with different predictive time intervals, the predicted load factor will be different. The predicted load factor stays at 1 for a period of time just after the maneuver starts, which is caused by the fact that the rate of change of the collective stick position used for load factor prediction is still zero before the maneuver starts, and the length of this period is equal to the predictive interval used. Subsequently, the predicted load factor increases from 1, gradually approaching the actual load factor, and the smaller the predictive time interval, the earlier this approach starts. During the hold time of the collective stick, the predicted load factor with different predictive times is close to the actual load factor, and the predicted load factor keeps fluctuating slightly around the actual load factor, and the smaller the predictive time interval is, the smaller the deviation of the predicted load factor from the actual load factor is. However, it should be noted that the predictive time interval should not be chosen too small. If the predictive time interval is too small, it means that the reaction time left for the control system is also very small, and the protection effect will be very poor, even if the predictive effect of the load factor is very accurate. Considering this problem, the predictive time interval is chosen as the middle value, that is, 0.07 s.

2.2.4. Effectiveness analysis of the load factor protection based on the control power restriction method. We use the above-mentioned collective stick's position command to validate the effectiveness of the load factor protection method.

The main parameters in the design of the control power restriction method based on the load factor predictive algorithm are: load factor limit value is set to 1.7, the predictive time interval is set to 0.07 s, collective stick control power restriction coefficient is set to 0.9, exit softening time constant is set to 0.6 s, and exit softening limit speed is 40 mm/s. Figure 6 shows the response results of the rapid and large square wave manipulation of the collective stick in the vertical DC response type in hover.

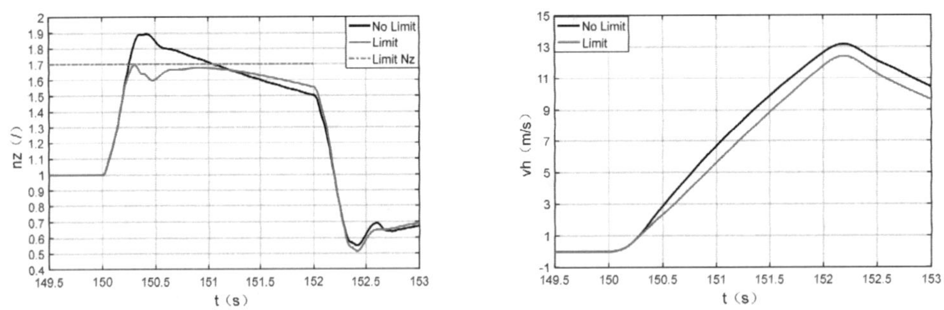

(a) Load factor response

(b) Vertical speed response

Figure 6. Collective stick square wave manipulation's response diagram.

It can be seen from Figure 6 that the vertical load factor is limited to the vicinity of the desired load factor of 1.7, indicating that the control power restriction method based on the load factor predictive algorithm can effectively protect the helicopter load factor from exceeding the limit in vertical maneuvering flight. And the restriction method has little effect on the vertical speed response.

2.3. Simulation analysis of the method adaptation under different flight states

The key link in the control power restriction method based on the load factor predictive algorithm is to use the control derivative of collective stick position to vertical acceleration to achieve the load factor prediction, and the control derivative of the helicopter is different in its different flight states. Therefore, to achieve a good load factor protection effect in different flight conditions of a helicopter, it is necessary to analyze the effect of different flight conditions of a helicopter on the control derivative.

The different flight states of the helicopter include its weight, altitude, climb rate, and airspeed. Since this paper mainly focuses on the load factor protection method for the vertical maneuvering flight state, the effect of airspeed on the control derivative is not considered first. In addition, the influence of helicopter weight on the control derivative is also easy to analyze. It is easy to see from Equation (2) that the relationship between helicopter weight and the control derivative is inversely proportional. And the other flight states are kept constant, the greater the helicopter weight is, the smaller the control derivative is. Therefore, the next step is mainly to analyze the influence of helicopter altitude and climb rate on the control derivative, and the helicopter at different altitudes and climb rates are trimmed and linearized to obtain the control derivative of collective stick position to vertical acceleration in different states, and the results are shown in Figure 7.

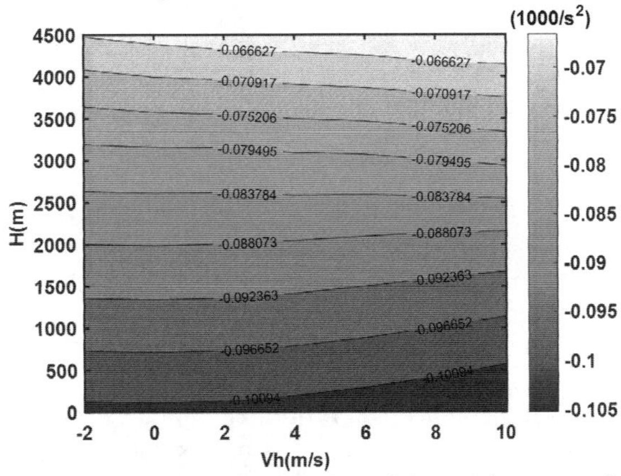

Figure 7. The control derivative of collective stick position to vertical acceleration.

As can be seen in Figure 7, the control derivative of collective stick position to vertical acceleration varies greatly with altitude, being larger at lower altitudes and smaller at higher altitudes. As the climb rate of the helicopter increases, the control derivative of collective stick position to vertical acceleration basically does not change much. At low altitudes, the control derivative increases slightly with increasing climb rate, and at high altitudes, the control derivative decreases slightly with increasing climb rate. In summary, to use this restricting method for load factor protection, the effect of helicopter altitude on the control derivative of collective stick position to vertical acceleration should be considered first, and the second one is the climb rate. The effect of the climb rate on the control derivative can be disregarded if the restricting method is further simplified.

Based on the above simulation analysis, the effect of different altitudes and climb rates on the control derivative of collective stick position to vertical acceleration is taken into account based on the original method. The control derivative of the collective stick to vertical acceleration in the current state is obtained by two-dimensional interpolation of the control derivatives at different heights and climb rates. Then carry out the load factor prediction and the control power restriction of the collective stick position.

Under the same maneuvering and control parameters, the load factor protection effect of the improved method is tested for different helicopter flight states, including the load factor protection effects at different altitudes in hover and at different climb rates. The simulation results are shown in Figure 8.

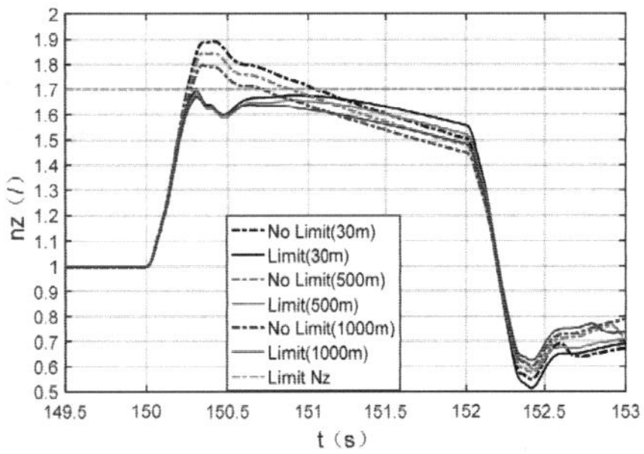

(a) Load factor response at different heights

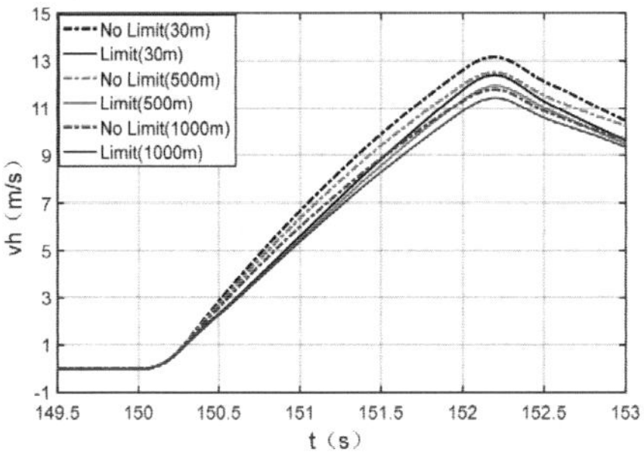

(b) Vertical speed response at different heights

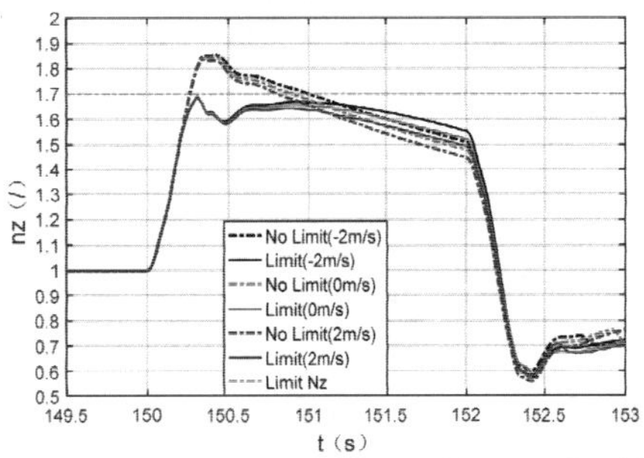

(c) Load factor response at different climbing rates

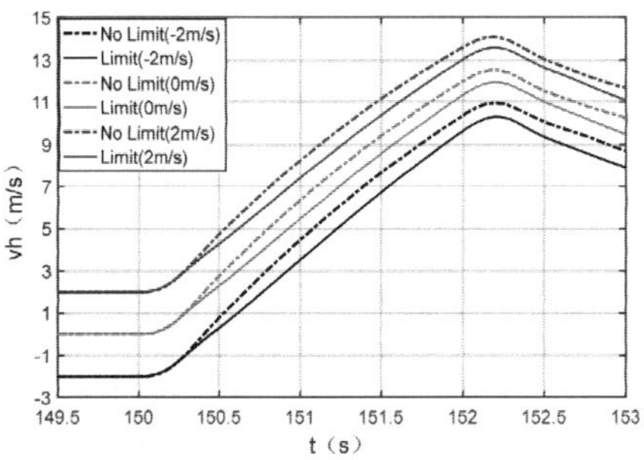

(d) Vertical speed response at different climbing rates

Figure 8. Comparison of maneuvering response in helicopter different states.

From Figure 8, it can be seen that the same maneuver will produce different load factors when the helicopter is in hover at different altitudes, and the lower the altitude, the larger the load factor. The improved control power restriction method based on the load factor predictive algorithm can solve the load factor exceeding problem of the helicopter at different altitudes. The same maneuvers produced similar load factors for different climb rates at the same altitude, which was caused by the small difference in the control derivatives for the three different climb rates used in the simulations. The improved method can also solve the load factor exceeding problem for different climb rates. In addition, the improved method still does not affect the normal vertical speed response.

2.4. Analysis and simulation of the effect of control derivative error on the applicability

The control derivative of collective stick position to vertical acceleration is usually obtained from flight tests or wind tunnel tests, and there is definitely an error between the test data and the real control derivative of the helicopter. Therefore, the next simulation is to analyze the effect of the error of the control derivative on the effects of load factor protection.

The control derivative of collective stick position to vertical acceleration is obtained by trimming and linearizing the helicopter nonlinear dynamics model, the control derivative is regarded as the true value of the helicopter's control derivative. Artificially adjust the value of the control derivative used in the restricting method, after a large number of simulation tests, it is possible to obtain the error between the control derivative and its true value in a range that still meets the good load factor protection effects, and the results of the simulation test are shown in Figure 9.

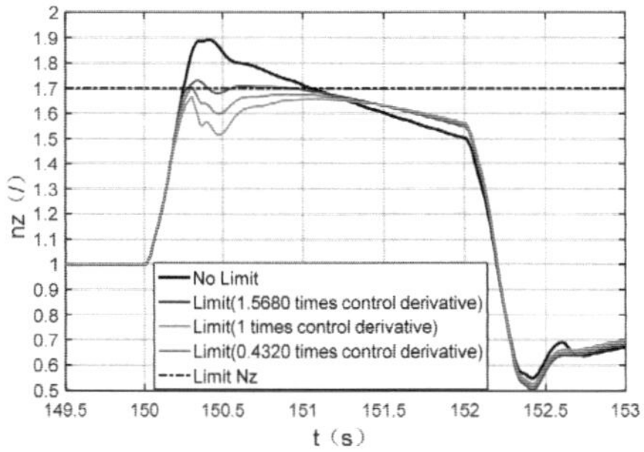

(a) Overall diagram of load factor response

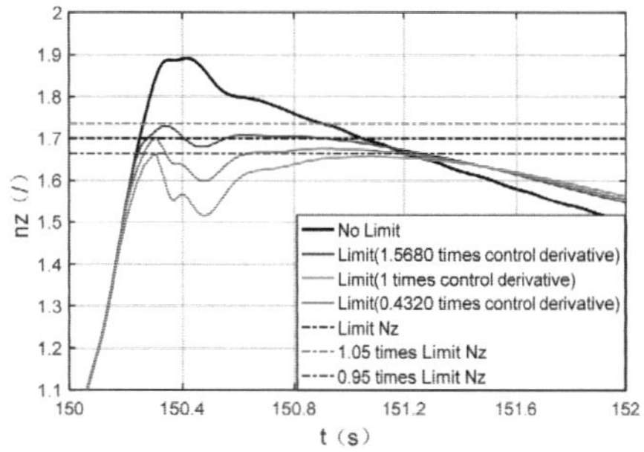

(b) Partial diagram of load factor response

Figure 9. The effect of the error of the control derivative on the effect of load factor protection.

According to the simulation test results in Figure 9, it can be concluded that if we want to apply the control power restriction method to limit the load factor to 1.7 and its restrictive error does not exceed 5%, then the allowable error of the adopted control derivative cannot exceed 56.8%. The results show that the proposed method has good robustness and engineering practicability.

3. Conclusion

The proposed method of control power restriction based on a load factor predictive algorithm can solve the problem of helicopter load factor exceeding caused by inappropriate pilot control in vertical maneuvering flight, and this method has little effect on the vertical speed response and can achieve the purpose of load factor protection without excessively affecting the play of the helicopter's flight performance. And the method can effectively solve the load factor protection problem at different altitudes and different climb rates and has good robustness. In addition, this paper also gives the error tolerance of the control derivatives used in the proposed restriction method for practical engineering applications.

References

[1] Zhao A, Meng Z, Gao H and Gao Y. (2017) The Research on the Flight Envelope of Aircraft Effect on Safety of Flight.AIRCRAFT DESIGN,37: 11-16;

[2] Chen Y, Wang Y, Zhang L and Chen Y. (2015) Helicopter Flying Speed-overload Envelope Expansion Test Method Research, Science Technology and Engineering,12: 4;

[3] Horn J, Calise A and Prasad J. (2001) Flight Envelope Cueing on a Tilt-Rotor Aircraft Using Neural Network Limit Prediction, Journal of the American Helicopter Society,46(1): 23-31;

[4] Horn J, Calise A and Prasad J. (2002) Flight envelope limit detection and avoidance for rotorcraft, Journal of the American Helicopter Society, 47: 253-262;

[5] Horn J, Calise A and Prasad J. 1998. Flight envelope limiting systems using neural networks.In: 23rd Atmospheric Flight Mechanics Conference. Portland, OR, U.S.A. 4459;

[6] Bateman A, Ward D and Barron R. 1998. Piloted simulation evaluation of a neural network limit avoidance system for rotorcraft. In:24th Atmospheric Flight Mechanics Conference. Portland, OR, U.S.A. 4252;

[7] Thaiss C, McColl C and Horn J. 2014. Rotorcraft real time damage alleviation through load limiting control. In:55th AIAA/ASMe/ASCE/AHS/SC Structures, Structural Dynamics, and Materials Conference. National Harbor, Maryland. 1528;

[8] Kimball and Dwayne F. (1987) Recent Tilt Rotor Flight Control Law Innovations, Journal of the American Helicopter Society, 32: 33-42;

[9] Li Y.(2021) Flight test verification of maximum overload coefficient for helicopter autorotation flight. Science Technology and Engineering, 21: 2085-2092;

[10] Li P and Chen R. (2010) Formulation and Validation of a Helicopter Model for Pull-up Maneuver Simulation. Acta Aeronautica et Astronautica Sinica, 31(12): 2315-2323.

Health Assessment of Mission-oriented TT&C Antenna

Lei Guo[1], Jihui Xu[1], Weihui He[2]

[1.] Air Force Engineering University, Shaanxi, 710043, China

[2.] Xi'an Satellite Control Center, Shaanxi, 710043, China

116302384@QQ.COM

Abstract: To solve the problem that it is difficult to quantitatively evaluate the health status of TT&C antennas and that the health status cannot be effectively associated with the ability to perform tasks, this paper proposes a systematic evaluation method for the health management of aerospace TT&C antennas based on the functional mission of TT&C antennas, which closely associates the equipment health status with the ability to perform tasks. According to the functional mission of equipment, the overall health level is evaluated through the health status of the underlying indicators of the ability to perform tasks and equipment performance. Firstly, the assessment index system of the mission-oriented TT&C antenna is analyzed, and then the assessment of the target is decomposed into the assessment of multiple underlying indicators using the analytic hierarchy process (AHP), which is widely used in health assessment. The quantitative assessment is made according to the specific indicators of the lowest level, and their importance to the target assessment is determined layer by layer, so as to achieve the purpose of assessing the target health status. On this basis, the health level of the TT&C antenna is determined according to the maximum membership principle. The evaluation index system and method have important reference significance for the follow-up test and identification work.

1 Introduction

At present, China's aerospace TT&C antennas are characterized by a large volume, complex structure, and high degree of modularization. According to the different tasks they perform, their models are also diverse and expensive. In the past, the assessment of the health status of aerospace TT&C antennas was completely dependent on the observation of the equipment's operating status by managerial staff in routine usage, and no quantitative data provided support in the entire process. The final results were not only subjective and unconvincing but also insufficient to make an objective and accurate assessment of the health status of the aerospace TT&C antennas. Another common method was to collect historical health data of aerospace TT&C antennas, analyze and process these historical data purposefully, build an assessment system of the health status of aerospace TT&C antennas[1] by comprehensively considering historical and current operating data of TT&C antennas, and evaluate the health status of aerospace TT&C antennas. However, historical failures were not inextricably linked to the capacity of the equipment to conduct tasks, which made it tough to evaluate the relationship between equipment health and mission performance which managers were more concerned about. By now, research on the security capability[2][3] of mission-oriented equipment and collaboration efficiency[4][5] of the unmanned aerial vehicle (UAV) has been carried out. The main evaluation methods are QFD, AHP, etc.,[6][7] while they mainly focus on the effectiveness evaluation with no research on health assessment of mission-oriented TT&C antennas. In the field of TT&C antenna assessment, research work[8] has been actualized on the health evaluation of the TT&C network from three aspects, namely, the control center, TT&C node, and communication link. What's more, research on aspects including single installation status[9], performance[10], and effectiveness evaluation[11]has been performed from the perspective of subsystems and performance indicators such as monitoring, base-band, channel, antenna servo feeder, time-frequency, etc. Those evaluated objects in the research are not different from the health assessment of a single installation in this paper, and the indicator system is not distinctly related to the task, but the establishment method and evaluation model of the indicator system is of reference value for this study.

2 Health Assessment Method for TT&C Antenna

2.1 Assessment Procedure

The traditional evaluation index system of TT&C antennas is mainly established by dividing primary subsystems such as the monitoring, base-band, channel, and antenna servo feeder and determining the indicators of each subsystem. Based on the health evaluated by the difference between the performance test value and the reference value, the correlation between the indicators and capabilities, and the relationships of serial-parallel links within the system and among the systems cannot be grasped. The mission-oriented health assessment of TT&C antennas in this paper is to evaluate the health status of the equipment from the perspectives of mission capability, results of synthetic indicator test, and historical fault data of the equipment. Because the health status of the components in the TT&C antennas in the actual operation is uncertain and ambiguous at the index level, the ambiguous synthesis method based on health can be adopted to evaluate the health of the TT&C antennas. When the health degree of the aerospace TT&C antennas can be defined between zero and one based on a certain underlying index, the health state of the aerospace TT&C equipment in a certain functional mission can be manifested mathematically through results of specific performance index test and link status of equipment, which is beneficial to the mission-oriented assessment of the overall health state of the aerospace TT&C equipment. Considering characteristics of the hierarchical structure and ambiguity of the health status in the evaluation index system for aerospace TT&C antennas, the assessment is illustrated in Figure 1, including six steps, namely, establishing the state set and evaluation set, extracting and normalizing data, stipulating the importance of each level, creating an ambiguous judgment matrix of the health index, evaluating health membership, and determining the overall health status.

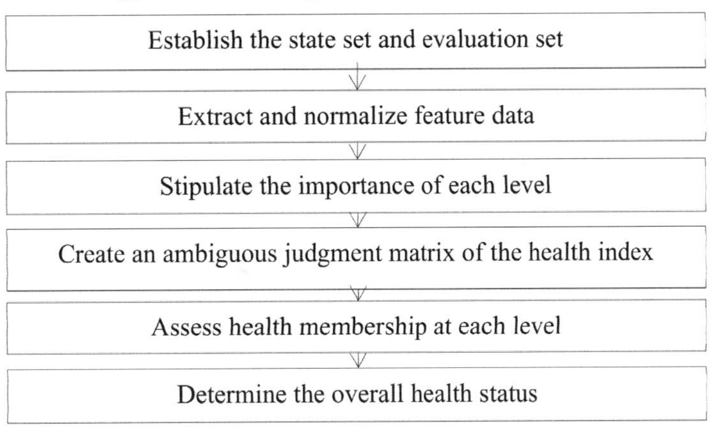

Figure 1. Assessment Procedure.

2.2 Mission-oriented Assessment System

This paper puts emphasis on the construction of a performance index system targeting mission capability and feeds back information containing the failure rate and reliability of the equipment linked to its health assessment. Five respects are primarily involved in the mission-oriented status assessment: Ka-band data transmission, X-band data transmission, S-band telemetry, S-band remote control, and S-band status assessment of the external test. The Ka-band data is shown in Figure 2.The health assessment of the equipment can be associated with each functional mission from two aspects of key performance and link status.

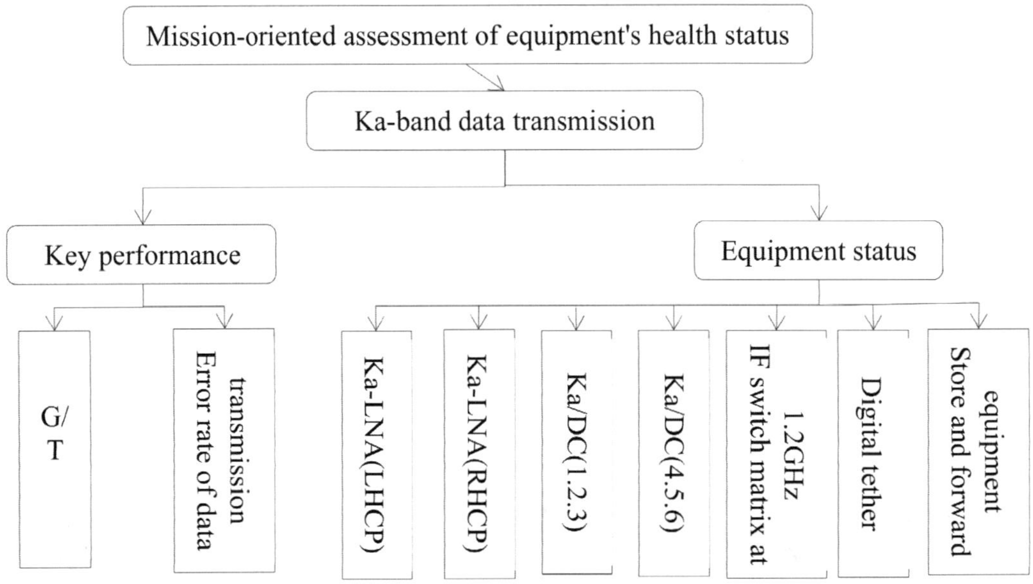

Figure 2. Mission-oriented Assessment of Health Status.

Taking the health assessment of S-band telemetry as an example, its relevant key indicators include G/T and the error rate of telemetry. The corresponding relationship between the indicators and the health of link tasks is displayed in Table 1, where X is calculated by virtue of the specific parameters of the equipment.

Table 1. Key Performance Indicators of Telemetry.

Indicators	Indicator value	Description of Health	Score
G/T	≥ XdB/K	Normal	
	≥ XdB/K ≤ XdB/K	Decrease	
	≤ XdB/K	Badly deteriorate	
Error rate of telemetry	Meet and exceed the requirements	Normal	
	E/N0 is worse than the requirements within XdB	Decrease	
	E/N0 is worse than the requirements over XdB	Badly deteriorate	

Analysis of the main and standby links: based on the equipment, fault, and health status description, the corresponding relationship between the fault status of equipment and mission health is sorted out. Specific examples are displayed in Table 2.

Table 2. Assessment of Equipment's Link Status.

Equipment	Fault	Health description
Field amplifier LNA (RHCP)	LNA3 and LNA4 are normal	System is normal
	LNA3 or LNA4 fails	System runs with the unidirectional link
	LNA3 and LNA4 fail	System fails

Field amplifier LNA (LHCP)	LNA1 and LNA2 are normal	System id normal
	LNA1 or LNA2 fail	System runs with the unidirectional link
	LNA1 and LNA2 fail	System fails
Down converter S-D/C (1, 2, 3)	All equipment is normal, and the corresponding matrix ports are normal	System is normal
	Only one converter fails	The operation of the system has no backup
	Only two converters fail	The function of the system has defects
	Three converters fail	System fails
Down converter S-D/C (4, 5, 6)	All equipment is normal, and the corresponding matrix ports are normal	System is normal
	Only one converter fails	The operation of the system has no backup
	Only two converters fail	The function of the system has defects
	Two or more converters fail	System fails
70MHz IF switch	The equipment is normal	System is normal
	The equipment fails	System fails
TT&C base-band	All equipment is normal, and the corresponding matrix ports are normal	System is normal
	Only one TT&C base-band	System runs with the unidirectional link
	Two TT&C base-bands fail	System fails

3. Assessment Calculation Based on AHP

3.1 Stipulation of the Importance of Each Level

Different performance indicators at different levels will have different effects on the health status of TT&C antennas. Stipulating the importance of each indicator at each level is the key to the assessment of the equipment's health status. AHP is applied to determine W = {W1, W2, L, Wi, L, Wn}, the importance weight vector of the first-level indicators, Wi = {Wi1, Wi2, L, Wij, L, Wim}, that of the second-level indicators, and Wij = {Wij1, Wij2, L, Wijk, L, Wijo}, that of the third-level indicators. The relationship between each level is agreed with the iterative inclusion. The influencing factors of each level and each index can be regarded as the judgment information about the failure status of the aerospace TT&C antennas, which can be corrected by stipulating the importance of each level in combination with historical usage data. That mainly depends on whether the equipment is a single point in the link. The more backups, the lower the weight of the corresponding health assessment value.

3.2 Creation of An Ambiguous Judgment Matrix of the Health Indicator

The health data of indicators at all levels of aerospace TT&C antennas can be summarized to determine the health level of target-layer equipment performance. Many functions can solve the membership, and different ones have their applicability. Indicators and health objectives at all levels can be utilized as the basis for selection. The ridge distribution membership function is harnessed to solve the health status level membership of each indicator, as shown in Formulas (1) - (4). This function is featured by the wide main value interval and gentle transition zone.

$$r_{\mathrm{I}}\left(h_i\right)=\begin{cases}1 & \left(h_{\mathrm{ij}}=1\right) \\ 0.5+0.5\sin\left[N\left(h_{\mathrm{ij}}-0.9\right)/0.2\right] & \left(0.8<h_{\mathrm{ij}}<1\right) \\ 0 & h_{\mathrm{ij}}\leq 0.8\end{cases} \tag{1}$$

$$r_{\mathrm{II}}\left(h_{\mathrm{ij}}\right)=\begin{cases}0 & \left(h_{ij}=1\right) \\ 0.5-0.5\sin\left[N\left(h_{ij}-0.9\right)/0.2\right] & \left(0.8<h_{\mathrm{ij}}<1\right) \\ 0.5+0.5\sin\left[N\left(h_{ij}-0.65\right)/0.3\right] & \left(0.5<h_{ij}\leq 0.8\right) \\ 0 & h_{ij}\leq 0.5\end{cases} \tag{2}$$

$$r_{\mathrm{III}}\left(h_{\mathrm{ij}}\right)=\begin{cases}0 & \left(h_{\mathrm{ij}}>0.8\right) \\ 0.5-0.5\sin\left[N\left(h_{\mathrm{ij}}-0.65\right)/0.3\right] & \left(0.5<h_{\mathrm{ij}}\leq 0.8\right) \\ 0 & h_{\mathrm{ij}}\leq 0.5\end{cases} \tag{3}$$

$$r_{\mathrm{IV}}\left(h_{\mathrm{ij}}\right)=\begin{cases}0 & \left(h_{\mathrm{ij}}>0.5\right) \\ 0.5-0.5\sin\left[\aleph\left(h_{\mathrm{ij}}-0.25\right)/0.5\right] & \left(0<h_{\mathrm{ij}}\leq 0.5\right) \\ 1 & h_{\mathrm{ij}}=0\end{cases} \tag{4}$$

Thereby, the ambiguous judgment matrix of the jth second-level indicator of the ith first-level one with the health data as the assessment standard can be obtained:

$$R_{\mathrm{i}}=\begin{bmatrix} r_{\mathrm{I}}\left(h_{\mathrm{i1}}\right) & \cdots & r_{\mathrm{IV}}\left(h_{\mathrm{i1}}\right) \\ \vdots & \ddots & \vdots \\ r_{\mathrm{I}}\left(h_{\mathrm{im}}\right) & \cdots & r_{\mathrm{IV}}\left(h_{\mathrm{im}}\right)\end{bmatrix} \tag{5}$$

3.3 Assessment of Health Membership of Each Level

On the basis of the above, by integrating the membership and importance of the health data of the aerospace TT&C antennas, the membership evaluation vector of the health status of the jth indicator under the ith one can be acquired as follows:

$$H_{\mathrm{i}}=W_{\mathrm{i}}\bullet R_{\mathrm{i}}=\left[W_{\mathrm{i1}}\cdots W_{\mathrm{im}}\right]\bullet\begin{bmatrix} r_{\mathrm{I}}\left(h_{\mathrm{i1}}\right) & \cdots & r_{\mathrm{rV}}\left(h_{\mathrm{i1}}\right) \\ \vdots & \ddots & \vdots \\ r_{\mathrm{i}}\left(h_{\mathrm{im}}\right) & \cdots & r_{\mathrm{iV}}\left(h_{\mathrm{im}}\right)\end{bmatrix} \tag{6}$$

$$H_{\mathrm{i}}=\left\lfloor h_{\mathrm{iI}}h_{\mathrm{iII}}h_{\mathrm{iIII}}h_{\mathrm{iIV}}\right\rfloor,\quad \mathrm{i}=1,2,\cdots,\mathrm{n}; \tag{7}$$

Furthermore, the membership evaluation vector of the health status of the upper level can be obtained by using Formula (6), up to the highest level, and the final assessment result can be represented by

$H = \lfloor h_{I} h_{II} h_{III} h_{IV} \rfloor$. According to the maximum membership principle, the health level of the aerospace TT&C antennas can be determined.

4 Application Cases

Taking the mission-oriented health assessment of equipment as an example, the importance of the five missions in its specific tasks. Scores of Ka-band data transmission, X-band data transmission, S-band telemetry, S-band remote control, and S-band external test are shown in Table 3.

Table 3. Scores of Five Missions' Importance.

Mission	Ka-band data transmission	X-band data transmission	S-band telemetry	S-band remote control	S-band external test
Score	85	95	80	90	70
Weight	0.20	0.23	0.19	0.21	0.17

By normalization, M = [0.20 0.23 0.19 0.21 0.17], which belongs to the second-level indicator weight.

Concerning the stipulation of the third-level indicator importance, take Ka-band data transmission as an instance to determine the weights of key performance indicators and the health of links. Shown in Table 4:

Table 4. Scores of Weights of Key Performance Indicators and Health of Links.

Factor	Key performance	Link status
Score	90	85
Weight	0.51	0.49

By normalization, M1 = [0.51 0.49], which belongs to the third-level indicator weight.

In terms of the fourth-level indicator assessment, take the external test as an example. Scores of the key performance importance and health status in the mission of the measurement of S-band distance, velocity, and angle are listed in Table 5.

Table 5. Indicator Assessment of S External Test.

Indicator	Accuracy of distance (A)	Accuracy of velocity (A)	Accuracy of angle (A)	Accuracy of distance (B)	Accuracy of velocity (B)	Accuracy of angle (B)	Accuracy of distance (C)	Accuracy of velocity (C)	Accuracy of angle (C)
Score	75	85	65	85	95	75	80	90	70
Weight	0.10	0.12	0.09	0.12	0.13	0.10	0.11	0.13	0.10
Health status	1	1	0.80	0.95	0.90	0.85	1	0.95	0.95

In Table 5, A refers to standard TT&C; B non-coherent spread spectrum; and C integration of TT&C data transmission. M51 = [0.10, 0.12, 0.09, 0.12, 0.13, 0.10, 0.11, 0.13, 0.10], h51=[1, 1, 1, 1, 1, 0.85, 1, 1, 1].

Table 6. Link Assessment of S External Test.

Type	LN (LHCP)	LN (RHCP)	S-D/C (1, 2, 3)	S-D/C (4, 5, 6)	70MHz IF switching matrix	Track base-band
Score	75	75	85	85	90	95
Weight	0.14	0.14	0.17	0.17	0.18	0.20
Health status	1	0.95	0.8	0.95	1	0.95

In Table 6, Link status of measurement of S-band distance, velocity, and angle: M52 = [0.14, 0.14, 0.17, 0.17, 0.18, 0.20], h52 = [1, 1, 0.8, 1, 1, 1].

By Substituting figures into the calculation, the calculated results of each piece of equipment can be divided into four cases, excellent, good, fault, and failure. In the calculation, 0.95 represents excellent; 0.80 good; 0.65 fault; 0.55 failure. Calculate the health status of the third layer (the health status of the fourth layer needs to be used, that is, set h11, h12, etc.).

Taking the H11 calculation as an example, the value in R is calculated by the ambiguous judgment matrix. According to matrices M11 and h11, the value of H11 can be obtained by using Formula (6):

$$R11 = \begin{bmatrix} 0.5 & 0.5 & 0 & 0 \\ 0.85 & 0.15 & 0 & 0 \end{bmatrix}$$

$$H11 = M11*R11 = [0.54, 0.46]\begin{bmatrix} 0.5 & 0.5 & 0 & 0 \\ 0.85 & 0.15 & 0 & 0 \end{bmatrix} = [0.661, 0.339,0,0]$$

According to the maximum membership principle, this part should be regarded as excellent, and thus, H11 = 0.95 is represented by 0.95.

Similarly, H12 = [0.75, 0.145, 0.105, 0.0] = 0.95, H21 = [0.46,0.54, 0,0] = 0.80, H22 = [0.665, 0.26, 0.075, 0.0] = 0.95, H31 = [0.542, 0.458, 0.0, 0.0] = 0.95, H32 = [0.647, 0.3437, 0.0, 0.0] = 0.95, H41 = [0.556, 0.444, 0.0, 0.0] = 0.95, H42 = [0.646, 0.283, 0.07, 0.0] = 0.95, H51 = [0.707, 0.293, 0.0, 0.0] = 0.95, H52 = [0.753, 0.247, 0.0, 0.0] = 0.95.

$$R = \begin{vmatrix} 0.85 & 0.15 & 0 & 0 \\ 0 & 1 & 0 & 0 \\ 0.85 & 0.15 & 0 & 0 \\ 0.85 & 0.15 & 0 & 0 \\ 0.85 & 0.15 & 0 & 0 \end{vmatrix}$$

$$M = \begin{vmatrix} 0.20 & 0.23 & 0.19 & 0.21 & 0.17 \end{vmatrix}$$

By using Formula (6), it is calculated that H = [0.6545, 0.3455, 0.0, 0.0] and that its maximum position is 1. Therefore, it can be considered that the equipment is in excellent health.

5. Conclusion

According to the functional missions of the equipment, this paper assesses the health status from five aspects: Ka-band data transmission, X-band data transmission, S-band telemetry, S-band remote control, and S-band external test. Each mission is divided into two dimensions: key performance and link health status. A mission-oriented assessment indicator system is established. Through the AHP, the close connection between the health assessment and the mission performance of TT&C antennas is innovatively realized. In the meantime, the maximum membership principle is applied to combine the quantitative calculation results with the qualitative evaluation. Based on a large number of tests and statistical data of certain equipment, the rationality and feasibility of the model algorithm in this paper are fully verified. The assessment indicator system and methods provide support for equipment health assessment, operation, and maintenance management, and ensure the capability of mission performance of the equipment. What's more, it is a significant reference for the in-service assessment in test verification.

References

[1] Kong F. Q., Ni X. D. A Study on Health Assessment Technology of Data-driven Airborne Radar[J]. Scientific and Technological Innovation,2021,33(11):105-108.
[2] Zhao L. H., Ruan Y. J., Zhu Y. L., et al. Analysis Method for Requirement Satisfaction of Mission-oriented Equipment's Security Capability[J]. Command Control & Simulation,2022, 44(4):6.
[3] Cao Y. H., Liao X. H., Hu X. F., et al. A Study on the Comprehensive Capability Evaluation of Aerospace TT&C Equipment's Security System[J]. Journal of Equipment Academy,2009, 20(5):5.
[4] Yin H., Hou T. T., Li D. G. Evaluation of Manned/Unmanned Aerial Vehicle Cooperative Effectiveness for Typical Tasks[J]. Unmanned Systems Technology,2021, 4(1):9.
[5] Li J., Qin G. L., Zhang K. C., et al. Operational Effectiveness Evaluation of TT&C System Based on Game Theory[J]. Fire Control & Command Control,2019, 44(1):5.
[6] Du Z. L., Ren X. Evaluation Method for Mission-oriented Requirement Satisfaction of Equipment System Based on QFD[J]. Measurement & Control Technology,2022, 41(3):6.

[7] Liu Z. J., Qiao Y., Liu B. Design and Implementation of A Mission-oriented Effectiveness Evaluation System[J]. Computer and Information Technology,2022(003):030.
[8] Wu T., Gao H. L., Chen J. C., et al. Health Evaluation Method for Ground TT&C Network[J]. Journal of Spacecraft TT&C Technology,2017, 36(1):7.
[9] Zhou L. X., Ding Z. H., Song Y. Z., et al. A State Evaluation Method for TT&C Equipment Based on Ambiguous Comprehensive Evaluation[J]. Measurement & Control Technology,2018, 37(B11):6.
[10] Ren M., Zhou W. J., Guo J. H., et al. Comprehensive Performance Evaluation of Range TT&C System Based on AHP[J]. Telecommunication Engineering,2012, 52(12):5.
[11] Mao L., Zhou C. G., Zhang Y. X., et al. Application of Ambiguous AHP in Effectiveness Evaluation of Ship-borne TT&C System[J]. Electronic Design Engineering,2014, 22(8):4.

Civil Aircraft Airworthiness Noise Measurement

Ling Chen[1,a], Qiang Li[1,b], Yizhi Zhang[1,c], Liang Xu[1,d]

[1]Shanghai Aircraft Design & Research Institute

[a]chenling1@comac.cc, [b]liqiang3@comac.cc, [c]zhangyizhi1@comac.cc, [d]xuliang3@comac.cc

Abstract. Civil aircraft noise is the main source of the airport and its surrounding areas, in recent years, with the continuous development of the air transport industry, aircraft noise has attracted more and more public attention, according to the International Civil Aviation Convention Annex 16 Volume I [1] and China Civil Aviation Regulations CCAR36 [2] Aircraft Type and Airworthiness Certification Noise Regulations. The noise limit requirements for transport aircraft are put forward. According to the requirements of the regulations, it is the key to determine that the aircraft meets the requirements of the regulations to obtain the noise data through flight tests and to obtain the equivalent perceived noise level through data processing. In the process of noise certification flight test, the noise measurement system should be arranged accurately in strict accordance with the requirements of airworthiness clauses, to accurately measure and record the noise and obtain the required data. In this paper, the noise airworthiness certification flight test of a certain type of civil aircraft is taken as an example, and the noise measurement system of civil aircraft is designed.

1. Introduction

Airworthiness noise is used to assess the noise impact of aircraft take-off and landing on the ground, which is generally shown by flight tests. Airworthiness noise measurement of civil aircraft is formulated according to the specific requirements of China Civil Aviation Regulations CCAR36 [2] Noise Regulations for Aircraft Model and Airworthiness Certification, which mainly includes the requirements of aircraft noise measurement equipment received on the ground, the aircraft noise measurement system received on the ground, and the framework of certification noise determination system and post-processing process.

2. Aircraft noise measurements as received on the ground

Equipment requirements for the measurement of aircraft noise as received on the ground typically consist of a microphone system, a recording and playback system, and an analysis system.

2.1. Microphone System

The sensor system consists of a windshield, microphone, preamplifier, extension cable, pseudo pure tone ground reflecting disc, and bracket accessories. According to the function classification, it includes a noise sound pressure level microphone system and a pseudo pure tone noise sound pressure level microphone system, and the two systems use the same type of microphone and supporting equipment [3].

- In the reference direction, the free-field sensitivity level of the microphone and its preamplifier is within $\pm$ 1.0 dB of the calibration check frequency sensitivity level at one-third octave center frequencies in the range from 50 Hz to 5 kHz inclusive and does not exceed ± 2.0 dB at center frequencies of 6.3 kHz, 8 kHz, and 10 kHz.

- For sinusoidal sound waves, the free-field sensitivity levels of the microphone system at sound incidence angles of 30°, 60°, 90°, 120°, and 150° do not differ from 0° (normal incidence) by more than the values given in Table 1 at the center frequencies of each one-third octave band in the range from 50 Hz to 10 kHz. The difference in free-field sensitivity between two adjacent acoustic incident angles in Table 1 must not exceed the tolerance for the larger of the two angles.

Table 1. The maximum difference of microphone incidence angle from normal incidence (dB).

Central frequency(kHz)	Angle of incidence 30 °	60 °	90 °	120 °	150 °
0.05 ~ 1.6	0.5	0.5	1.0	1.0	1.0
2.0	0.5	0.5	1.0	1.0	1.0
2.5	0.5	0.5	1.0	1.5	1.5
3.15	0.5	1.0	1.5	2.0	2.0
4.0	0.5	1.0	2.0	2.5	2.5
5.0	0.5	1.5	2.5	3.0	3.0
6.3	1.0	2.0	3.0	4.0	4.0
8.0	1.5	2.5	4.0	5.5	5.5
10.0	2.0	3.5	5.5	6.5	7.5

2.2. Recording and reproducing system

- For frequencies below 12.5 kHz, the level of aliasing that may occur during conversion shall be at least 50 dB below the upper limit of the linear operating range of the device, with a sampling rate of not less than 28 kHz. Before the signal is digitized, it must be processed by an anti-aliasing filter.
- A stable sinusoidal electrical signal for input to the entire measurement system (including all microphone system equipment except the microphone), the signal level of which shall not exceed 5 dB equivalent to the calibrated sound pressure level on the reference level range. The time-averaged signal level at any one-third octave band center frequency from 50 Hz to 10 kHz, as indicated by the readout equipment, shall not exceed ± 1.5 dB over the calibration check frequency. The frequency response of the measurement system (including the analog-to-digital conversion equipment) from 10 kHz to 11.2 kHz must be within ± 0.3 dB of the 10 kHz frequency response.
- A linear operating range of at least 50 dB below the upper limit of the stage for all suitable stages and stable sinusoidal electrical signals input to the measurement system (including all microphone system equipment except microphones) at the one-third octave band center frequencies of 50 Hz, 1 kHz, 10 kHz and, if not, the calibration check frequency. Stage non-linearity not exceeding ± 0.5 dB.
- On the reference range, the sound level relative to the calibrated sound pressure level must be at least 5 dB below the upper limit of the range but shall not exceed 30 dB.
- An overload indicator must be included in the recording and playback system, which can give an overload indication in the event of an overload on any of the relevant stages.

2.3. Analytic System
The analysis system is used for real-time data processing of noise data, ground meteorological data monitoring, and judging the validity of test points by combining track data and air meteorological data. The basic requirements for the frequency bandwidth, channel configuration, and gain setting of the analysis are as follows:

- The output of the analytical system must consist of one-third octave band sound pressure levels as a function of time obtained by processing the noise signal (previously recorded) by an analytical system having the following characteristics: a set of 24 one-third octave filters or equivalent, centered from 50 Hz to 10 kHz inclusive. The response and average characteristics of the system, in principle, the output of an arbitrary one-third octave filter are squared, averaged, and displayed or stored as a time-averaged sound pressure level. The continuous sound pressure level sampling interval for spectral analysis must be 500 $\pm$ 5 ms. The analysis system does not process the sound pressure signal and the duration of data loss does not exceed 5 ms during data readout or analyzer reset. The analysis system must operate in real-time from 50 Hz to at least 12 kHz (inclusive). This requirement applies to all operating channels of a multichannel spectrum analysis system.
- In the range of center frequency from 50 Hz to 10 kHz (inclusive), the minimum standard of a one-third octave band analysis system is the electrical performance requirement of Class 2 in the revised IEC61260.
- The resolution of the sound pressure level, whether displayed or stored, must be 0.1 dB or better.

3. Aircraft noise measurement system
Please follow these instructions as carefully as possible so all articles within a conference have the same style on the title page. This paragraph follows a section title so it should not be indented.

3.1. Test Equipment Installation
- Overflight and approach noise measurement points are located on the extension line of the runway center line, not more than 1000 meters away from the end of the runway entrance. The lateral noise measurement point is located at the sideline position which is 450 meters away from the flyover/approach noise measurement point and the connecting line is perpendicular to the runway center line.
- There shall be no weeds and any objects irrelevant to the noise measurement within the circle with the microphone position as the center and 7.62 m as the radius at each measuring point position, and the area shall be filled with sand and compacted, and there shall be no snow within the circle with the microphone position as the center and 15.24 m as the radius.
- The pseudo pure tone sound pressure level microphone system shall not affect the noise measurement of the noise sound pressure level microphone system and shall be placed on the compacted ground at a position not less than 3.5 m away from the microphone.
- The microphone is arranged on the preamplifier in a thread-fixing mode, and the preamplifier is fixed on the top end of the tripod by a special clamp.
- The arrangement of the noise sound pressure level microphone system requires that the center of the sensitive element of the noise sound pressure level microphone is 1.2 m higher than the local ground by adjusting the height of the tripod. Adjust the direction of the microphone according to the pr-calculated angle, so that the sensitive element is basically located in the plane determined by the pr-established aircraft reference flight path and measurement points (as shown in Figure 1), ensure that the incidence angle of aircraft noise is within 30 °, and report the direction of the sensor to the authority. The microphone bracket must be placed to minimize interference to the sound measurement.

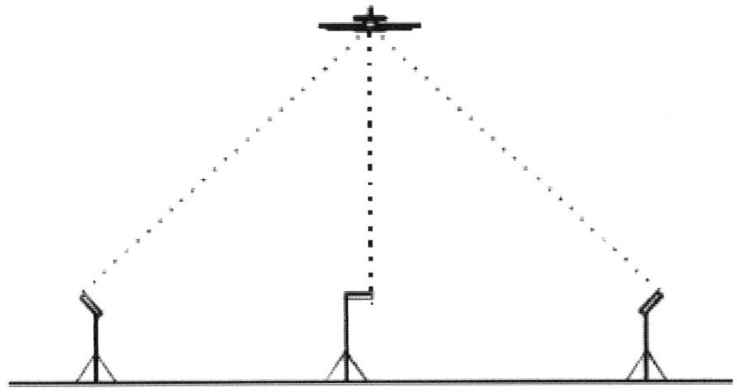

Fig. 1 The plane determined by the flight path and the measuring point of the sensing element

- Pseudo-pure tone noise sound pressure level microphone system layout requirements: the microphone is installed in an inverted position, and its microphone-sensitive element is located 7 mm above the ground reflection disc and parallel to the ground reflection disc [3]. The ground reflection disc should be flush with the ground and buried on the ground, and there should be no gap between the disc and the ground and below it. The microphones were mounted on a radius perpendicular to the direction of flight of the test airplane, from the center of the disk to 3/4 of a radius from the outer edge, as shown in Figure 2.

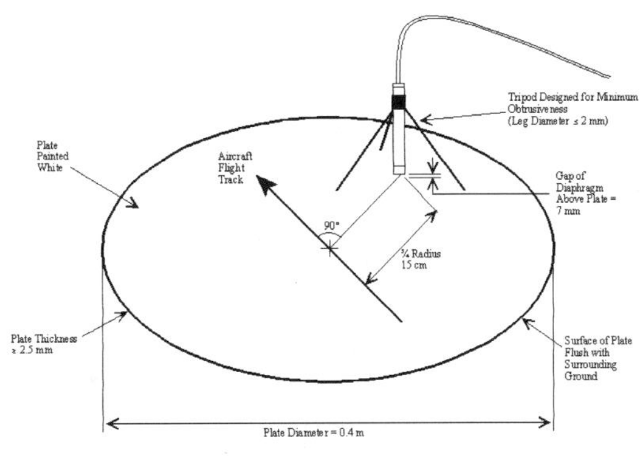

Figure 13 **Configuration for ½ in. Inverted Microphone Over Plate**

Fig. 2 Arrangement of the microphone inverted on the disc

- After the arrangement of the positions of the measurement points is finished, connecting a microphone of each measurement point to a preamplifier, connecting the preamplifier to a multi-channel analysis system by using a low-noise cable for data acquisition, recording and analysis, and analyzing the time history of noise signals in real-time, wherein the multi-channel analysis system is connected to network bridge equipment.
- GPS antenna is connected to the analysis system to realize the time synchronization of the system. The network camera is connected to that analysis system, and the output data and the noise data are transmitted to a central measurement station for the monitor.
- The microphone, preamplifier, and webcam are powered by the analysis system, which is powered by the battery.

3.2. Ground Received Aircraft Noise Measurement System

Ground Received Aircraft Noise Measurement System refers to the equipment combination used to measure the sound pressure level of aircraft at the ground measurement point, which consists of a microphone system, recording and playback system, analysis system, and calibration system [2]. The aircraft noise test system architecture is shown in Figure 3.

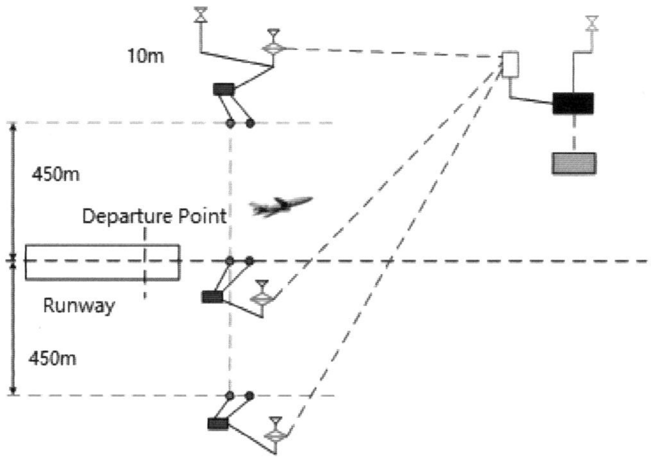

Fig. 3 Architecture diagram of aircraft noise test system received on the ground

The noise data measurement system comprises a windshield, a microphone system, a pseudo pure tone identification system, a recording and replaying system, an analysis system, and a calibration system and specifically comprises two lateral noise measurement points, one approach/flyover noise measurement point, and one central measurement point.

This system uses wireless network transmission architecture for system data management and exchange. The microphone system in the noise data measuring system is used for sensing lateral, flyover, and approach sound pressure levels, converting the lateral, flyover, and approach sound pressure levels into electrical signals, and transmitting the electrical signals to data acquisition equipment at each measuring point, packing and recording data by the data acquisition equipment at each measuring point, and transmitting the data to a central measuring point in a wireless transmission mode. The noise data of each sound pressure measurement point, the flight path data, and the air weather data are combined with the central measurement point to carry out data correction so as to judge the validity of the test point. The system uses a GPS signal as the clock source to meet the system time synchronization requirements.

4. Airworthiness Noise Measurement and Post-processing

To get the airworthiness noise level, in addition to the noise received on the ground, it is necessary to consider the meteorological data measurement system, flight path data measurement system, telemetry transmission system, airborne real-time monitoring data processing, and monitoring system.

4.1. Airworthiness noise measurement system

The airworthiness noise test system uses network architecture for system data management and exchange. The microphone system in the noise data measuring system is used for sensing lateral, flyover, and approach sound pressure levels, converting the lateral, flyover, and approach sound pressure levels into electric signals, and transmitting the electric signals to data acquisition equipment at each measuring point, packing and recording data by the data acquisition equipment at each measuring point, and transmitting the data to a central measuring point. Aircraft GPS data telemetry is transmitted to the ground, converted into engineering quantities by real-time processing, and transmitted to the central measurement point [4]. The air weather data is imported into the central measurement point through the storage medium. The noise data of each sound pressure measurement point, the flight path data, and the air weather data are combined with the central measurement point to carry out data correction so as to judge the validity of the test point.

In addition, the data from 10 meters high ground meteorological measurement point is also transmitted to the central measurement point through the lateral measurement point for monitoring, to determine whether the meteorological data meet the test requirements. Each subsystem uses a GPS signal as the clock source to meet the demand of time synchronization of the whole system. The principal block diagram of the certification noise test system is shown in Figure 4.

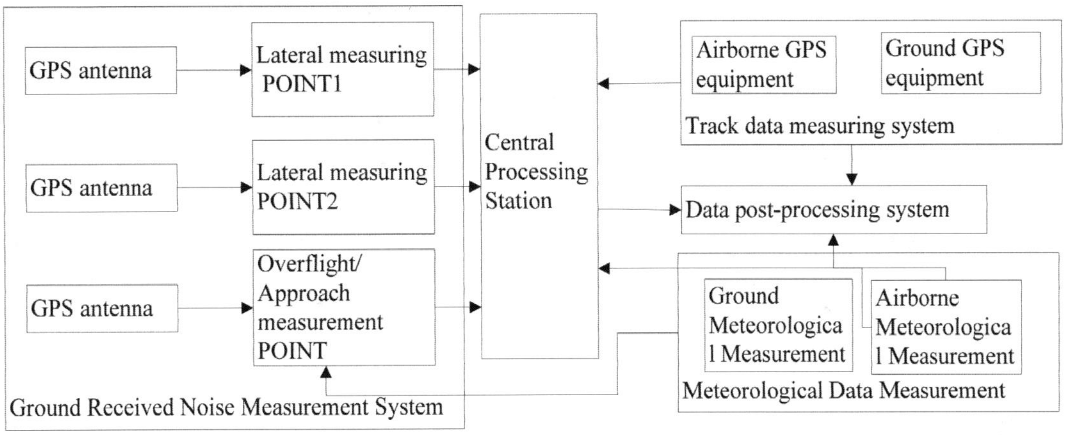

Fig. 4 Principle block diagram of airworthiness noise measurement system

4.2. Data post-processing

Regulations require that a uniform metric, effective perceived noise level (EPNdB), be used when processing data [5]. The noise certification data processing flow is shown in Figure 5.

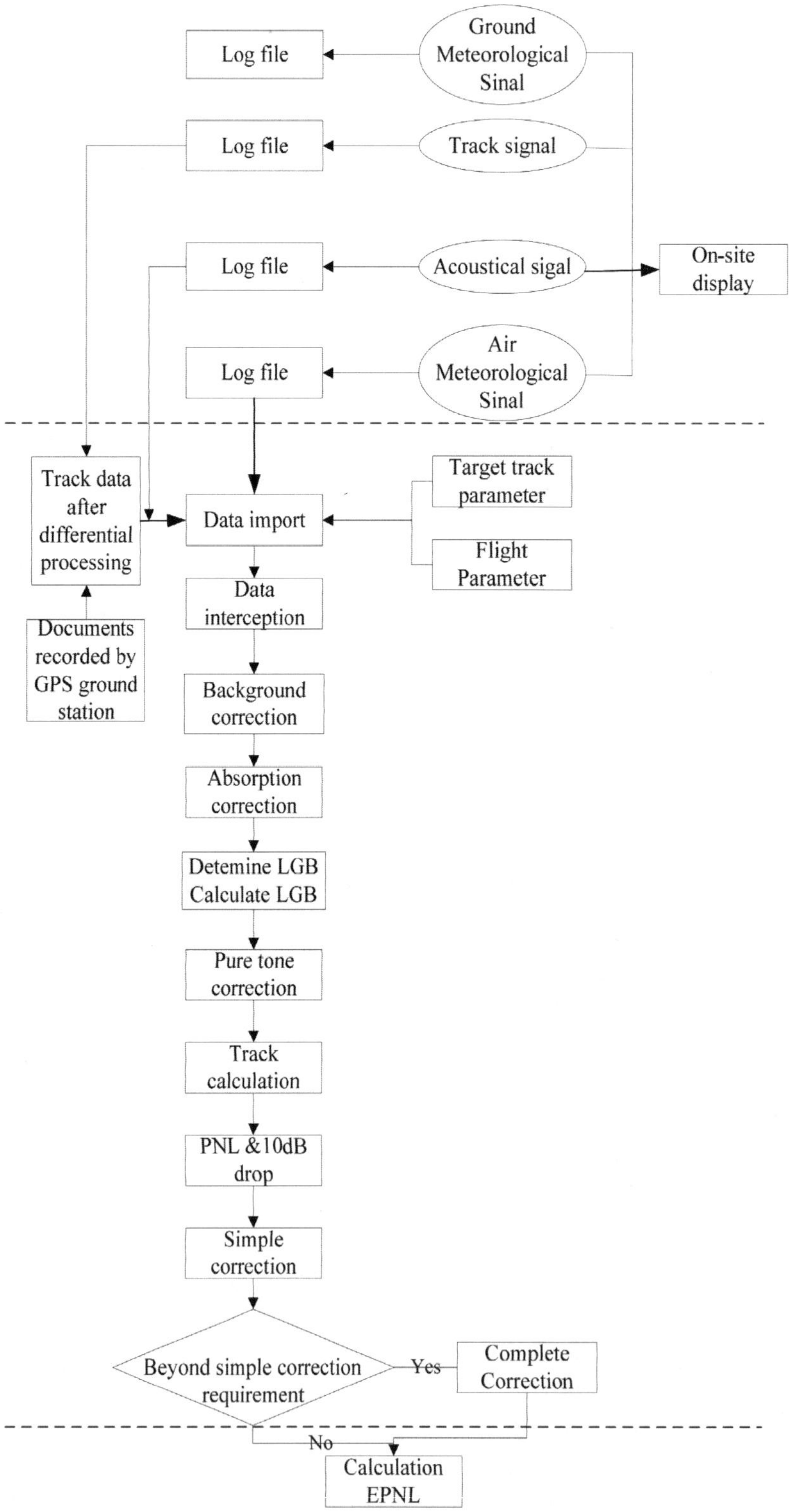

Figure 5 Data Processing Flowchart

5. Conclusion

This paper summarizes the noise measurement equipment and measurement system received on the ground, gives the airworthiness noise measurement and analysis system, and gives the post-processing flow chart, using this system and post-processing flow, we can get the airworthiness noise simply and effectively, which has reference significance for the subsequent civil aircraft airworthiness noise determination.

References

[1] ICAO International Standards and Recommended Practices, Environmental Protection, Annex 16, Volume Ⅰ, Aircraft Noise, Third Edition, Amendment 7, 2002
[2] CCAR-36-R1, Regulations on noise for type and airworthiness certification of aircraft, 2017
[3] Environmental Technical Manual, Volume Ⅰ, Procedures for the Noise Certification of Aircraft, Third Edition, 2018, ICAO-DOC-9501
[4] AC-36-AA-2008-04, Noise Regulations for Aircraft Type and Airworthiness Certification. 2008.
[5] Yan Guohua, Wang Xingke, 2009, Conf. of the 22nd National Symposium on Application of Vibration and Noise

Research on WBS-RBS Oriented Security Risk Assessment Index System for Space Domain Private Network

Rui Wang

Air Force Engineering University

317275959@qq.com

Abstract: To solve the problems such as unclear overall situations, incomplete risk analysis, and lack of targeted evaluation indicators of the security of the special network in the aerospace field, this paper takes the special network in the aerospace field as the research object, uses the WBS-RBS method to analyze the risk of network security, and uses the OSM model to establish an evaluation indicator system for the identified risks, to provide an indicator model for the subsequent security detection and evaluation of the special network in the aerospace field.

1. Introduction

With the continuous development of China's space industry, space launch missions have increased significantly year by year, and the amount of China's orbit spacecraft has exceeded 500. The special network in the space field is the core backbone network that carries the space launch and long-term spacecraft management tasks and is the key infrastructure of the space system. It is facing unprecedented challenges at the level of technical development and strategic security. As the underlying support network system for safeguarding the use of national space assets management, it is very likely to be selected as the potential target of network attacks. In September this year, the national official media and Beijing Qihu Technology Co., Ltd. released an investigation report on the network attacks on Western Polytechnic University by foreign countries, revealing that the "Specific Intrusion Action Office" under the US National Security Agency continued to attack and steal secrets from North-western Polytechnic University, stealing more than 140GB of its core data. This cyber security incident has profoundly revealed that cyberspace has become a key area of military, economic, and technological development, and competition among countries, and has also sounded an alarm bell for the security of special networks in the space field.

Given the current severe and complex network security situation, it is urgent to research and design a scientific and effective network security detection and evaluation index model, comprehensively detect, explore, and evaluate the security vulnerabilities and weaknesses of the special network in the aerospace field, and purposefully improve the comprehensive network security protection level to ensure the safe, stable, and reliable operation of the aerospace system.

At present, the research of network security evaluation indicators in China mainly focuses on the establishment of an organizational framework and business system, and the relevant technical system and standard system are still in the research stage. Among them, Zhang Xugao [1] proposed an information system security situation assessment model based on the interval matrix correction method given the problems existing in the current information system security situation assessment model. Zhou Chao [2] explored the application of a quantum neural network algorithm in information security risk assessment and proposed an information security risk assessment method based on a quantum gate circuit neural network. The information security evaluation index model established by Cai Yue [3] uses the analytic hierarchy process and the information entropy theory to evaluate the dual model fuzzy evaluation method. Xin Qian [4] adopted the unit evaluation model of entropy weight neural network and the overall evaluation model based on the fuzzy comprehensive evaluation method to achieve the goal of hierarchical evaluation of information systems. Chen Zheyu and Lin Mingwei [5] proposed a hesitation fuzzy language envelope analysis model based on an analytic hierarchy process to evaluate the network security of edge nodes. Gao Yunyun [6] explored the method of using the K-means algorithm to conduct mining analysis for clustering analysis, and finally obtained clustering results by clustering scores, and proposed corresponding security strategies according to the characteristics of each category. Zhang Chunjie [7] proposed a security risk assessment method based on incomplete information static attack and defense game model for the uncertainty between attack and defense in the control system. Zhao Chen [8] built the information system security evaluation model by using the analytic hierarchy process and fuzzy evaluation method. Wang Yuhe [9] proposed an evidential reasoning algorithm and BRB security assessment method.

The author comprehensively analyzed several mainstream security assessment methods and model algorithms in the industry, combined with the characteristics of the special network security system in the aerospace field and the actual security protection requirements, studied and designed the method of using WBS (Work Breakdown Structure) and RBS (Risk Breakdown Structure) to visualize multi-dimensional network security risk indicators, and analyzed and sorted them out, The hierarchical structure of "Object Strategy Measure" (OSM) is introduced, and finally, the idea of constructing a WBS-RBS based security risk analysis and evaluation index system for special networks in the aerospace field is proposed.

2. WBS-RBS-oriented security risk indicators for special networks in the aerospace field

2.1. Build WBS structure for special network security work in the aerospace field

WBS is mainly used to decompose the overall project task into specific work elements. It is generally used to refine and decompose the work with complex structures, to facilitate the implementation, inspection, and evaluation of specific work. According to the special network security in-depth protection system and security protection objects in the aerospace field, the WBS work breakdown structure is used to divide the network security protection tasks into five elements, including information transmission security, carrier network security, information terminal security, application software security, and security management system, according to the top-down creation method. Information transmission involves ground circuits and satellite communication circuits. The bearer network involves network equipment and network protocols, and the application software involves data and software. The specific security WBS work breakdown structure is shown in Figure 1.

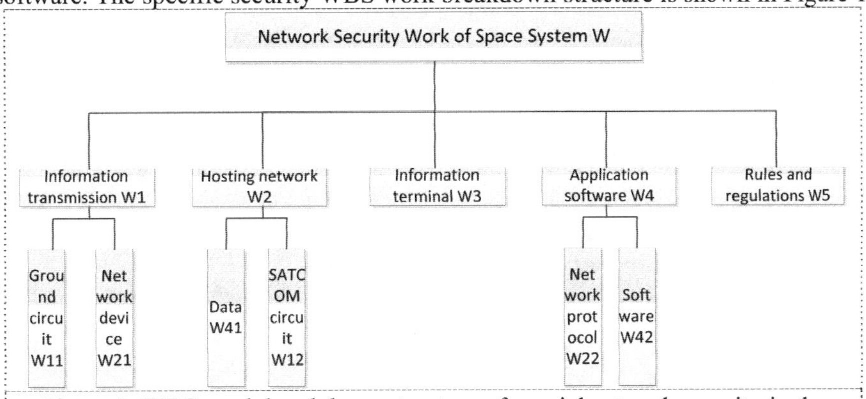

Figure 1. WBS work breakdown structure of special network security in the aerospace field

2.2. Build RBS structure of special network security risk in the aerospace field

RBS mainly decomposes the hierarchical structure of identified risks according to the risk category. Using the RBS risk decomposition structure, it decomposes the special network security risks in the aerospace field into five elements: physical environment security risks, communication network security risks, regional boundary security risks, computing environment security risks, and network security management risks. The security risks of the physical environment mainly come from the security of the channel and the computer room environment. Communication network security risk mainly comes from the security of network interaction and communication protocol, regional boundary security risk comes from the risk of intranet data interaction and international site access, computing environment security risk comes from the risk of the terminal, software, data, network, and security equipment, and network security management risk mainly comes from the risk of security operation and maintenance and the efficiency of security equipment. The specific security RBS risk breakdown structure is shown in Figure 2.

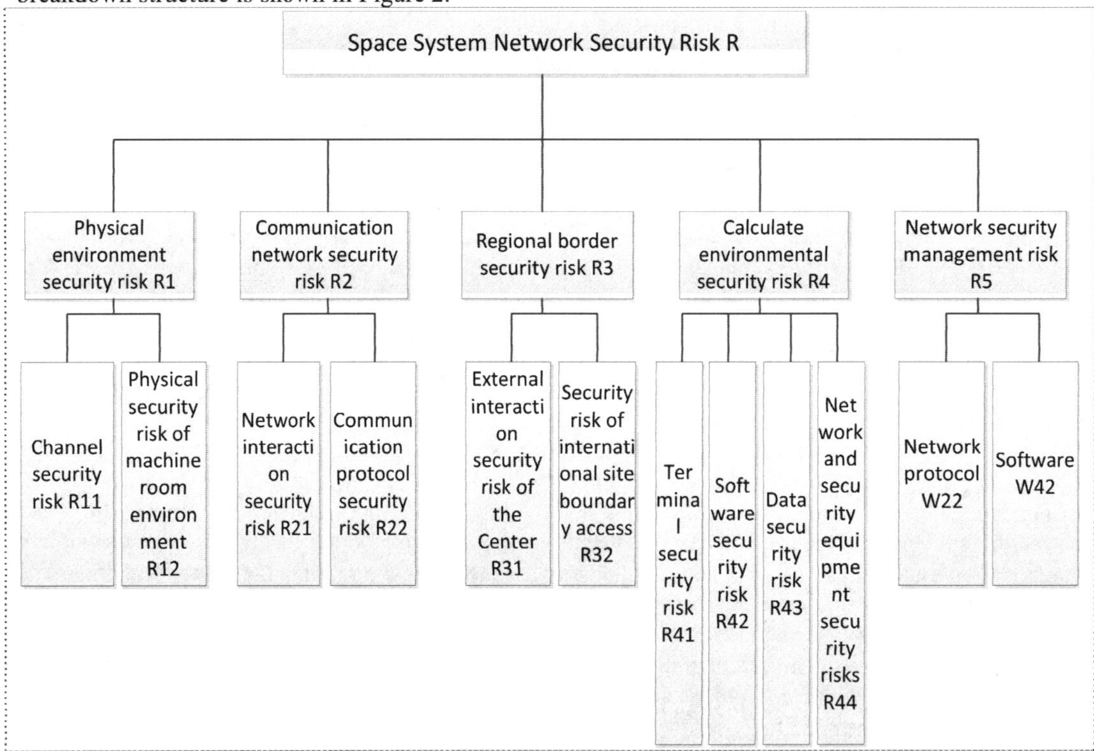

Figure 2 Risk decomposition structure of private network security RBS in the aerospace field

2.3. Build WBS-RBS safety risk index coupling matrix

The WBS work breakdown structure and RBS risk breakdown structure of the special network security are coupled in the aerospace field, whether the coupled basic element elements have security risks is judged, the risk sources and representative risk points are determined through analysis, and the WBS-RBS security risk index coupling matrix of the special network is constructed in the aerospace field, as shown in Table 1.

Table 1 WBS-RBS security risk index coupling matrix of the special network in the aerospace field

Elements of work / Risk Factor		W1		W2		W3	W4		W5
		W11	W12	W21	W22	W3	W41	W42	W5
R1	R11	1	1	0	0	0	1	0	0
	R12	1	1	1	0	1	0	0	0
R2	R21	0	0	1	0	1	0	0	0
	R22	0	1	0	1	0	0	0	0
R3	R31	0	0	0	0	0	0	0	0
	R32	0	0	1	0	0	0	0	0
R4	R41	0	0	0	0	1	0	1	0
	R42	0	0	0	0	0	0	1	0
	R43	0	0	0	0	1	0	1	0
	R44	0	0	0	0	0	0	0	0
R5	R51	0	0	0	0	0	0	0	1
	R52	0	0	0	0	0	0	0	1

By summarizing and sorting out the identified security risk indicators, since the risks faced by various computer room environments are the same, W11/R12, W12/R12, W21/R12, and W3/R12 can be unified into the security risk indicators of the computer room environment; The security risk indicators of W3/R41 computer terminals and W42/R41 software operation terminals can be unified into the security risk indicators of computer terminals; The security risk indicators of W5/R51 network security operation and maintenance management and W5/R52 security protection equipment management can be consolidated into the security risk indicators of operation and maintenance management. After consolidation, there are 14 safety risk indicator elements, as shown in Table 2.

Table 2. Summary of Network Security Risk Indicator Elements

NO.	Elements of safety risk indicators
one	Safety risk indicators of ground circuit intrusion
two	Security risk indicators of satellite communication link being intercepted and counterfeited by electromagnetism
three	Safety risk indicators of various computer room environments
four	Security risk indicators without internal network isolation
five	Security risk indicators of network boundary not isolated
six	Security risk indicators of network protocols
seven	Security risk indicators of wireless channel protocols
eight	Security risk indicators of uncontrolled network access
nine	Security risk indicators of computer terminal security
ten	Safety risk indicators of software
eleven	Safety risk indicators for data transmission over the ground
twelve	Security risk indicators of data transmission through SATCOM
thirteen	Safety risk indicators of software data
fourteen	Security risk indicators of operation and maintenance management

3. The idea of building the security evaluation index system of the special network in the aerospace field

3.1. The idea of index system construction

To make the evaluation results objectively and truly reflect the security situation of the information system, the selection of evaluation indicators should, on the one hand, be based on the relevant national and industrial standards, and be subdivided into relevant inspection and evaluation items according to the standards, so that the evaluation indicators can effectively measure the compliance of the information system and judge whether it meets the basic standards. On the other hand, the selection of evaluation indicators should also consider the actual situation of the information system, which can objectively and truly reflect the key points and risk points in the network security work, so that the indicators can accurately judge whether the special network security system in the aerospace field can control the threats and hidden dangers and whether the business system, risk hidden dangers, and vulnerabilities can be protected.

1. Scientific principle. On the construction of the indicator system, it should be able to reflect the objective laws of the assessment object, scientifically reflect the level of network security of the assessed object and the potential risks, and objectively and comprehensively reflect the true relationship between the indicators.

2. Principle of typicality. The selection of evaluation indicators should reflect the comprehensive situation of the evaluation object and be representative. Especially in some cases, if the number of indicators needs to be reduced, it can also be easy to carry out data calculation, so that the credibility of the evaluation results will not be reduced [10].

3. Principle of practicality. In the design of evaluation indicators, special attention should be paid to the operability of the follow-up in practice. In particular, the indicator design needs to be able to quantify and compare, and unify the measurement unit and calculation method of indicators, so that it is very practical in the specific evaluation process.

In the systematic analysis and carding of this paper, there are many network-security factors and corresponding evaluation indicators. Therefore, to improve the efficiency of detection and evaluation and highlight key factors, this paper has summarized and combed the indicator content in combination with the OSM model, and referred to relevant theories and actual situations, to form a more objective and scientific detection and evaluation indicator system.

3.2. OSM model

The OSM model is an abbreviation for Object, Strategy, and Measure. By defining the research goal, understanding the target behavior path and indicator hierarchy, establishing the corresponding strategy, and disassembling the strategy tasks, we can obtain the required evaluation indicators. This model is usually used for the evaluation of enterprise management.

The traditional indicator system construction method is not targeted and is mainly oriented to compliance indicators. The construction of the special network security evaluation indicator system in the aerospace field can use the OSM model to map the risk points and indicator measures one by one, making the construction of the indicator system more targeted and the evaluation indicators more objective.

Based on the OSM model, the Object is the network security protection capability, and then the strategy and measure are layered for the research objectives. The bottom-up approach is adopted. First, the identified risk problems are used to determine the evaluation indicators from the risk perspective, and then the evaluation indicators are summarized to form strategies, and finally, the network security evaluation indicator system is formed [11].

3.3. Design of security evaluation index system for the special network in the aerospace field

Based on the above analysis results of special network security risk indicators in the aerospace field, combined with the OSM model, the bottom-up approach is adopted to start with the risk points and formulate targeted evaluation indicators for different risk points. Then, according to the categories of evaluation indicators, the evaluation strategy is formed and the evaluation objectives are finally formed.

According to the security characteristics of the space TT&C system and the potential risks it faces, the security assessment system indicators and specific assessment items are sorted out. The content of the security indicators of the special network in the space field is preliminarily selected as shown in Table 3.

Table 3. Summary of network security indicators

NO.	Risk point	Corresponding evaluation index	Corresponding indicator category
1	Risk of ground circuit intrusion	Physical Access Control	Physical security
2		Ground circuit encryption	
3	Risk of electromagnetic interception and counterfeiting of SATCOM link	Electromagnetic protection of SATCOM circuit	
4		SATCOM link encryption	
5	Risks of various computer room environments	Machine room environment security	
6	Risk of no internal network isolation	Internal network isolation (VLAN division, ACL setting)	network security
7	Risk that the network boundary is not isolated	Boundary isolation (firewall, gateway)	
8	Risks of network protocols	Intrusion detection	
9	Risk of wireless channel protocol	Flow monitoring	
10	Risk of uncontrolled terminal network access	Terminal Access Control	Host security
11	Risk of computer terminal security	Localization rate of terminal	
12		Bug fix	
13		Virus killing	
14		Host control	
15	Software risks	Software code audit	Application security
16		Software Access Control	
17		Software vulnerability repair	
18	Risk of data transmission over the ground	Apply Encryption	
19	Risk of data transmission through SATCOM		
20	Risk of software data	Application Access Control	data security
21		Static data encryption	
22		Data operation audit	
23	Risks of network security operation and maintenance management	Personnel capacity	Mocha ITOM
24		Rules and regulations	

4. Instance application

The research results in this paper are applied to the security detection and evaluation of aerospace private networks. Based on the identified risk points and the corresponding evaluation indexes, the fuzzy comprehensive evaluation method and analytic hierarchy process are used to construct the detection and evaluation model, and the network security detection and evaluation work is carried out. According to the analysis of the situation of detection and evaluation, the risk identified by the WBS-RBS method and the index system built on this basis can better evaluate the existing problems of network security, for the subsequent development of targeted reinforcement to provide a certain follow-up.

5. Conclusion

In this paper, the WBS-RBS method is used to decompose and identify the security risks existing in the special network in the aerospace field, and the identified multi-dimensional network security risk indicators are coupled into concrete indicator elements. Using the OSM hierarchical structure model, corresponding evaluation indicators are obtained by comparison one by one. Finally, a WBS-RBS-oriented approach to building the security risk analysis and evaluation indicator system for the special network in the aerospace field is proposed, It provides a reference for the subsequent construction of a complete network security detection and evaluation model and the development of risk assessment.

Acknowledgments:

First, I would like to thank my tutor Professor Xu Jihui for his painstaking efforts and frequent use of personal rest time on holidays to guide my thesis so that I can complete the thesis writing. Secondly, I would like to thank Yuan Shuai for his academic help. It is through the discussion with him that the paper is closer to the reality of the unit.

References:

[1] Zhang Xugao Research on information system security situation assessment based on interval matrix correction method [D]. Civil Aviation University of China, 2020. DOI: 10.27627/d.cnki.gzmhy.2020.000383

[2] Zhou Chao, Pan Ping, Huang Liang. Information Security Risk Assessment Based on Quantum Gate Circuit Neural Network [J]. Computer Engineering, 2018, 44 (12): 39-45. DOI: 10.19678/j.issn.1000-3428.0051383

[3] Cai Yue Research on the Construction and Application of Maritime Information System Security Assessment Model [D]. Beijing Jiaotong University, 2020. DOI: 10.26944/d.cnki.gbfju.2020.002771

[4] Xin Qian Research on Security Assessment of Civil Aviation Information System for Classification Protection [D]. Civil Aviation University of China, 2018

[5] Chen Zheyu, Lin Mingwei Hesitant fuzzy language envelope analysis model based on analytic hierarchy process and its application in edge node network security assessment [J]. Computer Application Research, 2021,38 (01): 209-214. DOI: 10.19734/j.issn.1001-3695.2019.09.0589

[6] Gao Yunyun Research on the Application of Data Mining Technology in the Evaluation and Decision of Network Security Classification Protection [D]. Beijing Jiaotong University, 2021. DOI: 10.26944/d.cnki.gbfju.2021.003167

[7] Zhang Chunjie Research on information security risk assessment technology of industrial control system based on game theory [D]. Changchun University of Technology, 2021. DOI: 10.27805/d. CNKI. GCC.2021.000610

[8] Zhao Chen Evaluation on Information Security System Construction of China Southern Airlines Dalian Branch Based on Fuzzy Comprehensive Evaluation [D]. East China Jiaotong University, 2020. DOI: 10.27147/d.cnki.ghdju.2020.000106

[9] Wang Yuhe Research on Industrial Control Network Security Assessment Method [D]. Harbin University of Technology, 2019

[10] Yao Yongzhang Research on the Evaluation Method and Application of the Standardized Construction of the Communist Youth League of an Enterprise [D]. Hunan University, 2014

[11] Li Ning, Wang Xiao, Yang Yu. Research on the Evaluation Method of Technical Standards for Electric Power Enterprises Based on Analytic Hierarchy Process [J]. China Standardization, 2017 (20): 32-34+38

Noise Attenuation by Half Flat tip Serrated Trailing Edge in Rotating Blades

Jieyan Chen, Jinlong Xie, Hsiao Mun Lee*

School of Mechanical and Electrical Engineering, Guangzhou University, Guangzhou, Guangdong Province, China

* Corresponding author: hmlee@gzhu.edu.cn

Abstract-Two rotating blade models with different half-flat tip trailing edge serrations were proposed. The flow characteristics and aerodynamic noise were simulated using the large-eddy simulation model and Ffowcs Williams and Hawking's acoustic analogy, respectively. The effect of half-flat tip trailing edge serrations in reducing noise was investigated. Results showed that the proposed half-flat tip trailing edge serration designs can effectively reduce broadband noise, and the effectiveness of noise reduction was related to the wavelength-to-amplitude ratio of the half-flat tip trailing edge serrations. The Wide-serrated blade had better noise reduction performance, and the local noise reduction according to the simulations can be as high as 3.3 dBA. Besides, it was found that the aerodynamic performance of the rotating blades was not detrimentally affected by the introduction of half-flat tip serration.

1. INTRODUCTION

Rotating machinery is one of the most common mechanical equipment in production and life, such as fans, wind machines, drones, and wind turbines. Generally, the noise of rotating machinery can be mainly divided into two categories. The first one is the noise caused by the vibration of mechanical transmission processes such as gear bites. The second one is the aerodynamic noise caused by the unstable airflow on the blade. The mechanism of aerodynamic noise is complicated since it is heavily related to the dynamic flow field of the rotating machinery. Recently, the mitigation of aerodynamic noise has received more and more attention from researchers and engineers in the application of low-noise rotating machinery [1]. The implementation of the serrated trailing edge (TE) in blades is found to be an effective way to reduce aerodynamic noise.

Howe [2-3] studied the noise reduction effectiveness of serrated TE based on aerodynamic acoustics theory. Results showed that the noise attenuation occurred in a specific frequency range when $\omega \geq U_c/\delta$, where, ω was the radial frequencies, δ was the thickness of the TE boundary layer and U_c was the velocity of the mainstream. It was also found that when the serrated angle was less than 45° and $\lambda/h \leq 4$, the serrated TE achieved a noise reduction effectiveness of about $10\times\log[1+(4h/\lambda)^2]$ dB, where λ is the wavelength and h is the amplitude. Gruber et al. [4] studied the noise reduction mechanism of NACA6512 airfoil with a serrated TE through noise experiments. The results showed that the design criteria of serrated TE were: $St_\delta = f\delta/U_c < 1$, $h/\delta > 0.5$, and λ were sufficiently small, where St_δ was the Strouhal number, ω was the radial frequency of the noise, δ was the thickness of the TE boundary layer, U_c was the convective velocity of the turbulent eddy passing the TE, λ was the spanwise wavelength, and h was the amplitude. Chong et al. [5-6] conducted a preliminary study on the reduction of airfoil broadband self-noise by a non-flat plate serrated TE structure. It was noticed that the non-flat plate serration not only reduced the broadband noise by about 2 dB to 8 dB but also suppressed the high-frequency noise. Zhe et al. [7] studied the aerodynamic and aero-acoustic properties of TE-serrated propellers under different serrated parameters and flight conditions through Particle Image Velocimetry (PIV) and acoustical experiments. They found that the increase of TE serrated width-to-height ratio led to an increase in noise attenuation effectiveness without significant changes in the aerodynamic performance. Yang et al. [8] used wind tunnel measurement to study the aerodynamic and noise reduction performance of the TE serrated propeller. The results showed that the TE serrated propeller can achieve noticeable broadband noise reduction, and the amount of noise reduction was dependent on the Strouhal number. Lee et al. [9-10] studied the noise reduction effect of ceiling fans and propellers through a series of experiments and numerical simulations. It was revealed that the applications of serrated TE in ceiling fans and propellers can effectively reduce the aerodynamic noise by about 13.9 dBA and 5.8 dBA, respectively.

As abovementioned, TE serration has been successfully applied to reduce the aerodynamic noises of airfoils and rotating blades over the past decades. Most of these studies were mainly using experimental approaches to characterize the TE serrated effect on rotating blades and hence could incur high costs. Therefore, the main objective of the current effort is to use numerical simulation methods to investigate the effectiveness of flat-tip TE serrations in reducing the aerodynamic noises of rotating blades. The effects of different serrated parameters were investigated to shed some light on the design of low-noise rotating blades.

2. NUMERICAL SIMULATION

2.1. Geometric model

As shown in Figure 1, the in-house designed rotating blade is demonstrated and the e864-thick airfoil is applied for the cross-section of the blades. In the present study, the rotating blade has a radius (R) of 241.3 mm and is divided into 18 e864 thick airfoil sections. It is proposed due to ensure adequate serrated structure. The chord (c) and the twist angle (β) of the rotating blade evolve with the relative section radius (r).

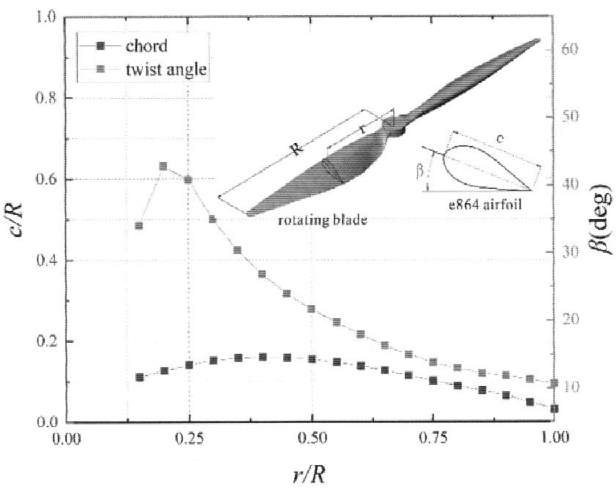

Figure 1. Geometric details – rotating blade

The flat tip TE serrations were applied on TE starting from $0.5R$ until the blade's tip, as shown in Figure 2. According to the finding by Howe [2], a good noise reduction effect can be achieved for a TE airfoil when the serrated angle (ε) is less than 45°, and the wavelength to amplitude ratio (λ/h) is less than 4.0. Hence, the current design of the flat tip TE serrated structure followed the guidelines proposed by Howe, as shown in TABLE I. The average wavelengths (λ) for two different flat-tip TE serrations were fixed at $0.015R$. The λ/h of two half flat tip TE serrated blades (Narrow-serrated blade and

Wide-serrated blade) were 0.008*R* and 0.011*R*, respectively. The noise reduction of these two designs was predicted to be 3 dB to 5 dB according to the theory proposed by Howe [3].

TABLE I. DESIGN PARAMETERS OF HALF FLAT TIP THE SERRATIONS.

Design parameters		ε (°)	λ(mm)	λ/h
Baseline blade		0	0	0
Half flat tip TE serrated blades	Narrow-serrated blade	25	0.015R	0.008R
	Wide-serrated blade	34	0.015R	0.011R

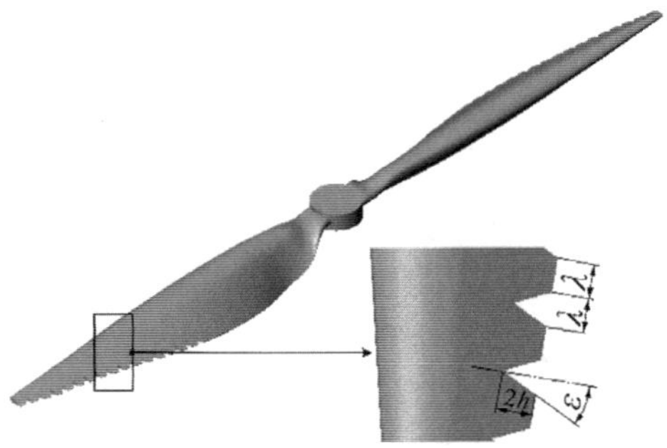

Figure 2. Half flat tip TE serrated blade and design parameters

2.2.Numerical model

Figure 3 shows the computational domain and boundary conditions. In the present study, ANSYS FLUENT was used as the simulation platform. The geometry of the computational domain was a cylinder coaxial with a rotating blade, which can be divided into two parts: stationary domain and rotational domain. The rotational domain was set in the center of the stationary domain, and the rotating blade was placed at the center of the rotational domain. According to Stahuda et al. [11], to ensure the minimum effect of the boundary conditions on the flow at the blade disc, the stationary domain radius was five times the blade radius, and the stationary domain height was ten times the diameter of the blade. The two domains were connected using domain interfaces that were used to transfer data between the stationary and rotational frames.

The rotation speed of the blade was set at 3000 rpm. The inlet boundary was configured to be a pressure inlet where the total gauge pressure was set to 0 Pa, and a pressure outlet with a gauge pressure of 0 Pa was designated as the outlet boundary. The steady-state findings were first obtained using the RNG k-ε turbulence model, which was then applied as the initial conditions for the unsteady-state Large-Eddy simulation (LES). The Coupled scheme was deployed for the pressure-velocity coupling with PRESTO utilized for pressure discretization, bounded central differencing for momentum term, and bounded second-order implicit discretization for time term. The unsteady-flow simulation time step size was chosen to $\Delta t = 2.5\times10^{-4}$ s, by which the noise prediction had a maximum frequency of 2000 Hz. Meanwhile, the Ffowcs Williams and Hawkings (FW-H) acoustic model was enabled to capture the change of pressure with time at the sound source surface, and hence the acoustic variables of interest can be calculated from the postprocessing. In particular, the corresponding spectral characteristics were obtained by the Fast Fourier Transform (FFT) transformation. Four noise monitoring points that were positioned 600 mm away from the center of the rotating blade were proposed, as shown in TABLE II.

TABLE II. LOCATION COORDINATES OF NOISE MONITORING POINTS.

Noise monitoring points	P1	P2	P3	P4
X(mm)	0	0	0	0
Y(mm)	600	424.26	0	-424.26
Z(mm)	0	-424.26	-600	-424.26

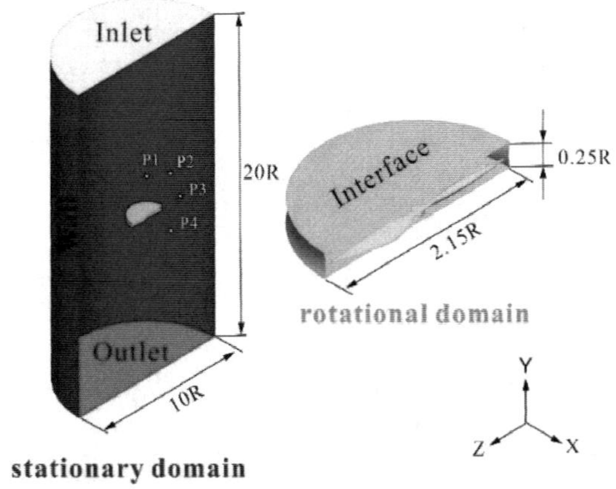

Figure 3. Computational domain

Vortices in the LES turbulence model are classified into two types: large-scale eddies and small-scale eddies. The large-scale eddies are obtained by solving the modified Navier-Stokes equations directly, whereas the small-scale eddies were modeled with an appropriate sub-grid-scale (SGS) model. Following are expressions for the continuity and modified Navier-Stokes equations:

$$\frac{\partial \overline{u}_i}{\partial \overline{x}_i} = 0 \tag{1}$$

$$\frac{\partial \rho \overline{u}_i}{\partial t} + \frac{\partial \rho \overline{u}_i \overline{u}_j}{\partial x_j} = \frac{\partial}{\partial x_j}\left[v \left(\frac{\partial \overline{u}_i}{\partial x_j} + \frac{\partial \overline{u}_j}{\partial x_i} \right) \right] + \frac{\partial \tau_{ij}}{\partial x_j} - \frac{\partial \overline{p}}{\partial x_i} \tag{2}$$

where, x_i and x_j represent the physical space. $\overline{u}_i$ and $\overline{u}_j$ are the fluid velocity components in the physical space. t stands for time. $\overline{p}$, ρ and v represent the static pressure, the density, and the kinematic viscosity, respectively.

As the SGS model (τ_{ij}), the Smagorinsky-Lilly model was chosen and calculated as follows:

$$\tau_{ij} - \frac{1}{3}\tau_{kk}\delta_{ij} = -2\mu_t \overline{S}_{ij} \tag{3}$$

where τ_{kk} denotes the isotropic part of SGS stresses. $\mu_t = \rho L_s^2 \sqrt{2\overline{S}_{ij}\overline{S}_{ij}}$ stands for the SGS turbulent viscosity. $\overline{S}_{ij} \equiv \left(\partial \overline{u}_i / \partial x_j + \partial \overline{u}_j / \partial x_i \right)/2$ denotes the rate-of-strain tensor. $L_s = \min\left(\kappa d, C_s \Delta\right)$ is the mixing length for SGS. d denotes the distance to the nearest wall. $\kappa = 0.41$ and $C_s = 0.1$ are the von Kármán constant and the Smagorinsky constant, respectively. $\Delta = V^{1/3}$ represents the local grid scale, where V is an element's volume.

The acoustic model integrated into FLUENT is based on the FW-H equation as indicated in Equation (4).

$$\frac{1}{c^2}\frac{\partial^2 p'}{\partial t^2} - \nabla^2 p' = \frac{\partial}{\partial t}\left[\rho_0 v_n \delta(f)\right] - \frac{\partial}{\partial x_i}\left[P_{ij} n_j \delta(f)\right] + \frac{\partial^2}{\partial x_i \partial x_j}\left[T_{ij} H(f)\right] \tag{4}$$

where, c is the far-field sound and p' is the sound pressure at the far field. ρ_0 and v_n are the ambient fluid density and the normal velocity of the control surface, respectively. P_{ij} is the fluid stress tensor. $T_{ij} = \rho u_i u_j + P_{ij} - \delta_{ij} c^2 \rho'$ is the Light hill's stress tensor. $\delta(f)$ and $H(f)$ represent the Dirac Delta function and the Heaviside function $\delta(f) = H'(f)$. In Equation (4), the first term on the right is the monopole source, known as thickness noise. It is the noise mainly generated by the displacement of fluid in the process of blade rotation. The second term is the dipole source, known as "loading noise", which comes from the interaction between the fluid flow and the rotating blade. The third term is the quadrupolar source. which is the noise mainly generated by fluid stress. This term is only considered when the blade tip is in the supersonic and transonic states, and hence is ignored in the present study.

2.3. Mesh independence

To achieve good accuracy of the simulation results with reasonable computation cost, a mesh independence study was conducted. The validation was performed for the thrust of the baseline blade and three sets of mesh were employed. Finally, when the elements were up to 8 million, the thrust varied by less than 0.5%, as shown in TABLE III.

TABLE III. THE NUMBER OF MESH ELEMENTS AND BASELINE BLADE THRUST FOR EACH COMPUTATION.

Grid number	Mesh 1	Mesh 2	Mesh 3
Elements (in million)	7.25	8.97	13.03
Thrust (N)	5.50	5.53	5.53

2.4. Evaluated parameters

The SPL represents an abbreviation for sound pressure level. To compute the equivalent A-weighted SPL (LA_{eq}) for each piece of data, noise contributions from each frequency are taken into consideration. LA_{eq} is defined as:

$$LA_{eq} = 10\log\sum_{i=1}^{n} 10^{0.1SPL_i} \tag{5}$$

To observe the effectiveness of blade noise reduction, the reduction of the LA_{eq} (ΔLA_{eq}) is obtained by (6), where LA_{eqB} and LA_{eqS} are the LA_{eq} produced by the baseline blade and half flat tip TE serrated blades, respectively.

$$\Delta LA_{eq} = LA_{eqB} - LA_{eqs} \tag{6}$$

The Blade Passing Frequency (BPF) denotes the sweep frequency of the rotating blade, which can be expressed as:

$$\Delta f = \frac{k \cdot n \cdot Z}{60} \tag{7}$$

where n denotes rotating speed. Z represents the number of blades. f indicates the fundamental frequency when k=1. f is the harmonic of the blade passing frequency when k=2. In this study, the fundamental frequency is 100 Hz when k=1.

The change rate of the thrust force (ΔT) is defined as (8), where $T_{baseline}$ and $T_{serrated}$ represent the thrust forces generated by the baseline blade and half flat tip TE serrated blades, respectively.

$$\Delta T = \frac{T_{serrated} - T_{baseline}}{T_{baseline}} \tag{8}$$

3. Results and Discussion

3.1. Acoustic performance study

TABLE IV shows ΔLA_{eq} obtained by the Narrow-serrated and Wide-serrated blades under 3000 rpm rotating speed for the four positions. It can be seen that the Wide-serrated blade has a noise attenuation effect in all four positions. The best noise attenuation effect is at Position 1 with a noise reduction of 3.3 dBA. The least noise reduction effect is noticed at Position 4 with only about 0.2 dBA. Compared with the Wide-serrated blade, the noise reduction from the Narrow-serrated blade is not satisfactory, and only a marginal noise reduction effect is obtained at Position 1 and Position 3. While, at Positions 2 and 4, the obtained noise is even worse. These results are in line with the experimental findings reported by Chong [5] and Zhe [7] that a Wide-serrated blade with a larger wavelength-to-amplitude ratio outperformed the narrower one. Nonetheless, they are in contradict Howe's theory [3] that the narrow-serrated serration with a smaller wavelength-to-amplitude ratio is more advantageous in achieving better noise reduction. The reason should be that Howe's theory was based on flat TE serration. The flat tip TE serrated structure in this study is a non-flat serrated structure, and the bluntness at the serration roots is unavoidable. When the flow field passes through the roots of the non-flat serration, vortex shedding will be induced, which leads to the vortex shedding noise. The smaller the wavelength-to-amplitude ratio, the blunter the roots of serration. As a result, the vortex shedding noise is more significant in the Narrow-serrated blade designs, contributing to an increase in the overall sound pressure level. Hence, it can be suggested that a Wide-serrated TE structure could be a potential noise mitigation measure to reduce the aerodynamic noise from rotating blades.

TABLE IV. ΔLA_{eq} (DBA) OBTAINED BY THE NARROW-SERRATED AND WIDE-SERRATED BLADES UNDER 3000 RPM ROTATING SPEEDS AT FOUR POSITIONS.

Position	P1	P2	P3	P4
Narrow-serrated blade	0.3	-1.0	0.6	-0.5
Wide-serrated blade	3.3	0.7	1.0	0.2

To further understand the influence of the flat tip TE serrated structure on the aerodynamic noise of the rotating blade at different positions, the noise spectrum is analyzed as shown in Figure 4. The noise spectrum of the baseline blade, Narrow-serrated blade, and Wide-serrated blade under 3000 rpm rotating speed for the four monitored positions are presented. In principle, the aerodynamic noise of rotating blades has two different components: the first one is the rotational noise by which the frequency spectrum is composed of all-order harmonics of BPF; the second one is the broadband noise which has a continuous frequency behavior. When the SPL of rotating noise is significantly higher than that of broadband noise, the noise energy of rotating blades is supposed to be more concentrated on rotating noise.

Figure 4 (a) shows that at Position 1, the trends of SPLs for the baseline, Narrow-serrated, and Wide-serrated blades decrease initially from 0 Hz to about 160 Hz, and then increase gradually to about 300 Hz and thereafter decrease until 600 Hz. For the frequencies ranging from 600 Hz to 2000 Hz, the evolutions of SPLs of the baseline blade, Narrow-serrated blade, and Wide-serrated blade are different. In this frequency range, the SPLs of the baseline blade generally increase, while the SPLs of the Wide-serrated blade decrease consistently. Nevertheless, the SPLs for the Narrow-serrated blade decrease slightly. It is speculated that the flat tip TE serrated structure has reduced the turbulence intensity near the TE of the rotating blade, thereby attenuating the broadband noise in this frequency range. Similar results were reported in Yang's study [8] that the presence of TE serrated structure tended to reduce the broadband noise. The TE serrated blade with wider serrations achieved better aerodynamic noise reduction could be due to the reduced turbulence as compared to the design with narrow serrations.

Figures 4 (b), (c), and (d) show that at Positions 2, 3, and 4, the evolutions of SPLs of the baseline, Narrow-serrated, and Wide-serrated blades are like that at Position 1. It manifests the similar phenomenon that broadband noise reduction is the main cause of the overall noise reduction for the rotating blades.

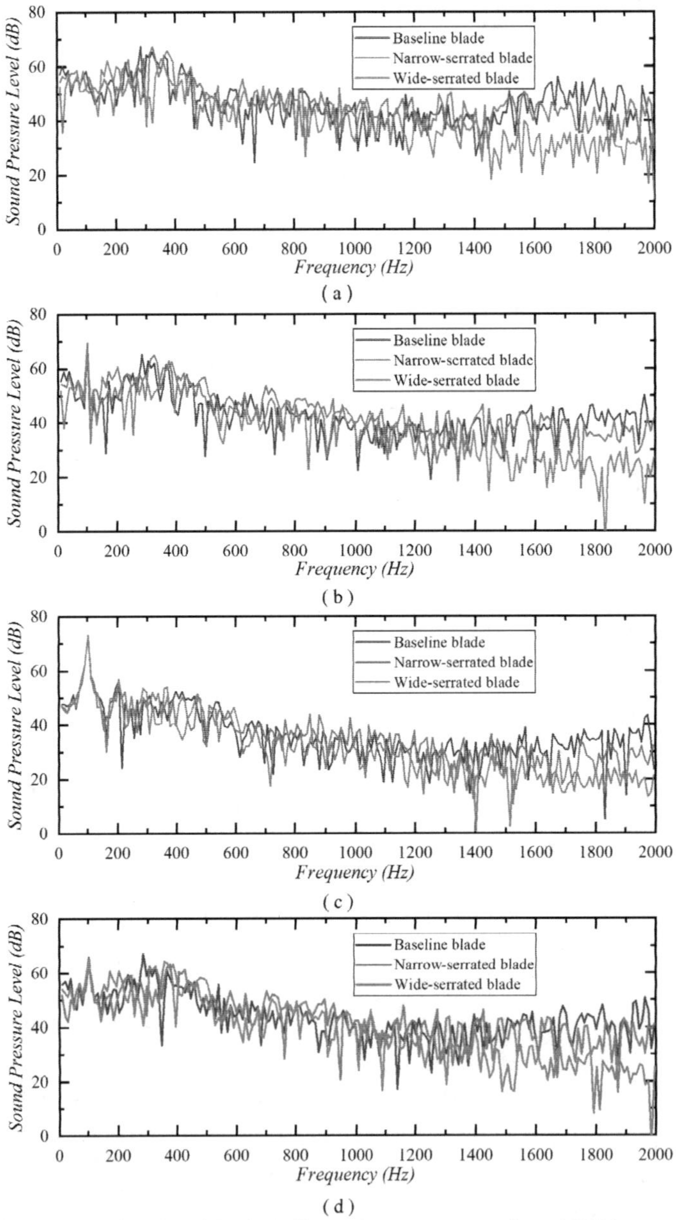

Figure 4. Sound pressure levels of the baseline, Narrow-serrated, and Wide-serrated blades with 3000 rpm at different positions. (a) Position 1 (b) Position 2 (c) Position 3 (d) Position 4

3.2. Aerodynamic performance study

The data in TABLE V summarizes the change rates of thrust force (ΔT) obtained by the half flat tip TE serrated blades in the present study. It can be found that both half-flat tip TE serrated blades produced slightly higher thrust than the baseline blade. Differing from the results of noise reduction effectiveness, the Narrow-serrated blade with a larger bluntness achieved a better thrust force than the wider one. The slightly improved aerodynamic performance may be attributed to the fact that the selected rotating blades in the present study are considered thick airfoil rotating blades. According to Relevant [12], the bluntness of flat tip serrated structure for thick airfoil rotating blades is beneficial to improving the comprehensive aerodynamic performance.

TABLE V. ΔT (%) OBTAINED BY THE HALF FLAT TIP TE SERRATED BLADES UNDER 3000 RPM ROTATING SPEEDS.

	Narrow-serrated blade	Wide-serrated blade
ΔT (%)	1.81	0.94

4. CONCLUSION

In the present study, two blade models with different tip TE serrations were proposed and simulated using the LES and the FW-H acoustic analogy methods to investigate the effects of serrations on the aerodynamic and noise-reduction performances of rotating blades. Simulations were conducted at a rotating speed of 3000 rpm, and four monitoring points surrounding the rotating circumference were characterized to study the parametric effects of serration structures. Results demonstrated that the Wide-serrated blade generally had better noise reduction performance than the Narrow-serrated blade. The best noise reduction of the Wide-serrated half-flat tip TE blade can be as high as 3.3 dBA. According to the frequency spectrum of noise, the flat tip TE serrated structure was more effective in attenuating broadband noise in the frequency region above 600 Hz. Interestingly, the aerodynamic performance was not deteriorated by fabricating half flat tip TE serrations on the blades according to the simulation. It should be noted that this study has only examined the performances of two half-flat tip serrations on the rotating blades. A substantial work with the consideration of more serration geometries could be necessary to draw a full conclusion in the future.

ACKNOWLEDGMENT

This research was funded by the Natural Science Foundation of Guangdong Province [2019A1515012223, 2021A1515012269] and the Guang-zhou Basic Research Program - City School (College) Joint Funding Project [202102010384, 202102010410].

REFERENCES

[1] Lee, S., Ayton, L., Bertagnolio, F., Moreau, S., Chong, T. P., Joseph, P. (2021) Turbulent boundary layer trailing-edge noise: Theory, computation, experiment, and application. Progress in Aerospace Sciences. 126: 1-56.

[2] Howe, M. S. (1991) Aerodynamic noise of a serrated trailing edge. J. Journal of Fluids and Structure. 5: 33-45.

[3] Howe, M. S. (1998) Noise produced by a sawtooth trailing edge. J. Acoustical Society of America Journal, 90: 482-487.

[4] Gruber, M., Joseph, P. F., Chong, T. P. (2011) On the mechanisms of serrated airfoil trailing edge noise reduction. In 17th AIAA/CEAS Aeroacoustics Conference (32nd AIAA Aeroacoustics Conference). Portland. pp. 1-23.

[5] Chong, T. P., Joseph, P. F., Gruber, M. (2013) Airfoil self-noise reduction by non-flat plate type trailing edge serrations. J. Applied Acoustics, 74: 607–613.

[6] Chong, T. P., Vathylakis, A., Joseph, P. F., Gruber, M. (2013) Self-Noise Produced by an Airfoil with Nonflat Plate Trailing-Edge Serrations. J. Aiaa Journal, 51: 2665-2677.

[7] Zhe, N., Hui, H. (2017) An Experimental Study on the Aerodynamics and Aero-acoustic Characteristics of Small Propellers. In: Aiaa Science & Technology Forum & Exposition. California. pp. 1-17.

[8] Yang, Y., Li, Y., Liu, Y., Arcondoulis, E.J.G., Wang, Y., Huang, B. et al. (2019) Aero-acoustic and aerodynamic investigation of multicopper rotors with serrated trailing edges. In: 25th AIAA/CEAS Aeroacoustics Conference. Delft. pp. 1-12.

[9] Lee, H. M., Lim, K. M., Xie, J.L., Lee, H.P. (2018) Reduction of Ceiling Fan Noise by Serrated Trailing Edge. J. Fluctuation and Noise Letters. 17: 1850026(1-12)
[10] Lee, H. M., Lu, Z., Lim, K. M., Xie, J.L. (2019) Quieter propeller with the serrated trailing edge. J. Applied Acoustics. 146: 227-236.
[11] Stahuda, M., Karczewski, M., Obidowski, D., Jozwik, K. (2016) Development of a CFD model for propeller simulation. J. Mechanics and Mechanical Engineering. 20: 579-593.
[12] Yang, R., Li, R., Zhang, S., Li, D. (2010) Computational Analyses on Aerodynamic Characteristics of Flatback Wind Turbine Airfoils. J. Journal of mechanical engineering. 46: 106-110

The High-speed Aircraft Trajectory Prediction Method Based on Deep Learning Network

Yu Chuan Hu, Jing Dai, Juan Yi and Zhong Hang Fan

[1]China Academy of Launch Vehicle Technology, Beijing, China

huyuchuan1998@163.com

Abstract. This paper applies deep learning technology to the trajectory prediction research of high-speed aircraft in order to predict the medium- and long-term trajectory. Hence, a hybrid model combining the time series convolution neural network (TSCNN) and the long short-term memory (LSTM) is put forward to predict position and velocity of high-speed aircraft. In the TSCNN, convolution operations are used to extract trajectory features of time and space to predict the future trajectory. However, the error will be generated during prediction the process, so the LSTM network can be tapped to correct it by analyzing why it is produced in a bid to address error accumulation. According to simulations conducted, the model can lead to high-precision prediction.

1. Introduction

The high-speed aircraft in this paper is a lift-body vehicle whose flight trajectory is controlled by aerodynamic forces, with its velocity over Mach 5. The traditional inertial trajectory prediction method is not applicable to the long maneuvering flight of high-speed aircraft, so it is essential to study new trajectory prediction techniques [1].

Existing flight prediction algorithms extrapolate results according to the equations of motion, on the basis of the idea of the Kalman filtering [2,3]. The accuracy of the results is mainly dependent on that of target motion model. Among those methods, the interactive multiple model (IMM) is regarded as the most efficient one to track low-speed maneuvering targets, while it is not accurate enough to track high-speed maneuvering ones [4]. Besides, geometric methods, such as shallow neural networks and polynomial fitting, are only applicable to short-term prediction [5,6]. With the development of artificial intelligence (AI), deep learning technology with strong non-linear mapping ability makes it possible to predict the medium- and long-term trajectory [7].

Based on the deep learning network, this paper builds the TSCNN to predict the medium- and long-term trajectory. In a bid to avoid error accumulation, the LSTM is used to correct the error generated during the prediction process. This model finally leads to satisfying results.

2. Correlation theory

2.1. Convolution neural network

This part briefly introduces some layers related to the CNN, including the convolution layer, the batch normalization (BN) layer, the ReLU activation function, and the regression layer.

The convolution layer encompasses various feature maps extracted from the inputs by convolution operations. Compared with other networks, the CNN features the local connection and weights sharing, both of which can decrease complexity of models and accelerate convergence.

In the BN layer, data from the last layer will be standardized, and then two parameters, γ and β , are used to map the standardized data. This layer can prevent vanishing gradient and over-fitting and also accelerate convergence. Mapping operation is expressed as:

$$\hat{x} = x \cdot \gamma + \beta \tag{1}$$

where $\hat{x}$ denotes the standardized data; γ is used to realize scaling, and β is used to realize shifting.

The ReLU function is often used as the nonlinear activation function of the convolution layer, expressed as:

$$f(x) = \max\{0, x\} \tag{2}$$

where x denotes the inputs, $f(x)$ outputs. It can better back-propagate gradients while ensuring non-linearity.

The regression layer is generally used to calculate the half mean square error (MSE) loss, which is expressed as:

$$Loss = \frac{1}{2}\sum_{i=1}^{R}(t_i - y_i)^2 \tag{3}$$

where R is the number of responses, t_i the labels, y_i the predicted results. The problem of high-speed aircraft trajectory prediction discussed in this paper is a typical regression one, so the last layer of the prediction network is the regression layer [8].

2.2. LSTM network

The LSTM network can control the information transmission by opening and closing the input gate, the output gate and the forget gate to solve problems that may occur in other recurrent neural network (RNN). Those problems include vanishing gradients and exploding gradients [9].

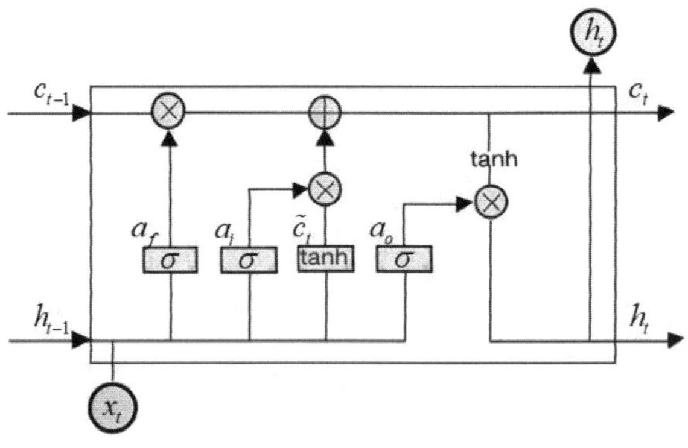

Figure 1. LSTM network structure

The structure of the LSTM network is shown in Figure 1, where three inputs and two outputs are included. x_t denotes the inputs of the current moment, and c_{t-1} and h_{t-1} denote the state vectors of the last moment. Two outputs are the state vectors of the current moment, respectively c_t and h_t .

The forget gate used to control the influence of the memory of the last moment on the current memory acts on the state $\tilde{c}_t$. The input gate that controls how much new memory is received to the long-term memory acts on the inputs x_t and the state h_{t-1} . Then a new state c_t is acquired. The output gate acts on the state c_t . The LSTM network finally outputs h_t . The equations are expressed as [10]:

$$\begin{cases} a_f = \sigma\left(W_f\left[h_{t-1}, x_t\right] + b_f\right) \\ \tilde{c}_t = \tanh\left(W_c\left[h_{t-1}, x_t\right] + b_c\right) \\ a_i = \sigma\left(W_i\left[h_{t-1}, x_t\right] + b_i\right) \\ c_t = a_f \tilde{c}_t + a_i c_{t-1} \\ a_o = \sigma\left(W_o\left[h_{t-1}, x_t\right] + b_o\right) \\ h_t = a_o \cdot \tanh\left(c_t\right) \end{cases} \tag{4}$$

where W_f , W_c , W_i and W_o are the weights; b_f , b_c , b_i and b_o are the gated unit bias; σ is generally the sigmoid activation function; a_f , a_i and a_o are the three control variables.

3. Trajectory prediction methods

3.1. The dynamic model of aircraft

To study the flight characteristics of high-speed aircraft and provide training samples for machine learning, it is necessary to establish a high-speed aircraft dynamic model and conduct simulations to obtain a standard trajectory. Assuming that the earth is a rotating sphere, the variables describing the motion include radial distance r, longitude λ, latitude ϕ, the Earth-relative velocity V, flight-path angle and θ azimuth angle ψ . The motion equations are expressed as [11]:

$$\begin{cases} \dfrac{\mathrm{d}r}{\mathrm{d}t} = V\sin\theta \\ \dfrac{\mathrm{d}\lambda}{\mathrm{d}t} = \dfrac{V\cos\theta\sin\psi}{r\cos\phi} \\ \dfrac{\mathrm{d}\phi}{\mathrm{d}t} = \dfrac{V\cos\theta\cos\psi}{r} \\ \dfrac{\mathrm{d}\psi}{\mathrm{d}t} = \dfrac{L\sin\sigma}{mV\cos\theta} + \dfrac{V\tan\phi\cos\theta\sin\psi}{r} + 2\omega_{\mathrm{e}}(\sin\phi - \cos\psi\tan\theta\cos\phi) + \dfrac{\omega_{\mathrm{e}}^2 r\cos\phi\sin\phi\sin\psi}{V\cos\theta} \\ \dfrac{\mathrm{d}\theta}{\mathrm{d}t} = \dfrac{L\cos\sigma}{mV} - \dfrac{g\cos\theta}{V} + \dfrac{V\cos\theta}{r} + 2\omega_{\mathrm{e}}\sin\psi\cos\phi + \dfrac{\omega_{\mathrm{e}}^2 r}{V}\cos\phi(\sin\phi\cos\psi\sin\theta + \cos\phi\cos\theta) \\ \dfrac{\mathrm{d}V}{\mathrm{d}t} = -\dfrac{D}{m} - g\sin\theta + \omega_{\mathrm{e}}^2 r(\cos^2\phi\sin\theta - \cos\phi\sin\phi\cos\psi\cos\theta) \end{cases} \tag{5}$$

where m is the mass, ω_{e} the Earth angular velocity, g the gravitational acceleration and σ the bank angle; L and D are respectively the aerodynamic lift and drag.

3.2. Trajectory prediction methods based on deep learning

3.2.1. TSCNN trajectory prediction network

The trajectory prediction of high-speed aircraft is a time series prediction problem, and its future flight states are both related to the current and historical states [12]. The traditional CNN can not solve this problem. Thus, given the time series properties of trajectory, a novel TSCNN is proposed to predict the trajectory of high-speed aircraft.

As shown in Figure 2, the structure of the TSCNN includes a input layer, convolution blocks, a fully connected (FC) layer and a output layer. First, the input is a 2D matrix formed by expanding the state column vectors at different moments along the time axis. Then, after five convolution operations, two depth concat layers, one transposition layer and one FC layer, the outputs are the prediction results at the next moment. Each convolution block is composed of three layers, including a convolution layer, a BN layer and a ReLU layer. The convolution kernels designed for five convolution operations are shown in table 1. Ker1 and Ker2 are used to extract features of the inputs to obtain feature maps (Conv1, Conv2). Ker4, Ker5 are used to extract features along the time axis and the state axis respectively to obtain feature maps (Conc4, Conv5).

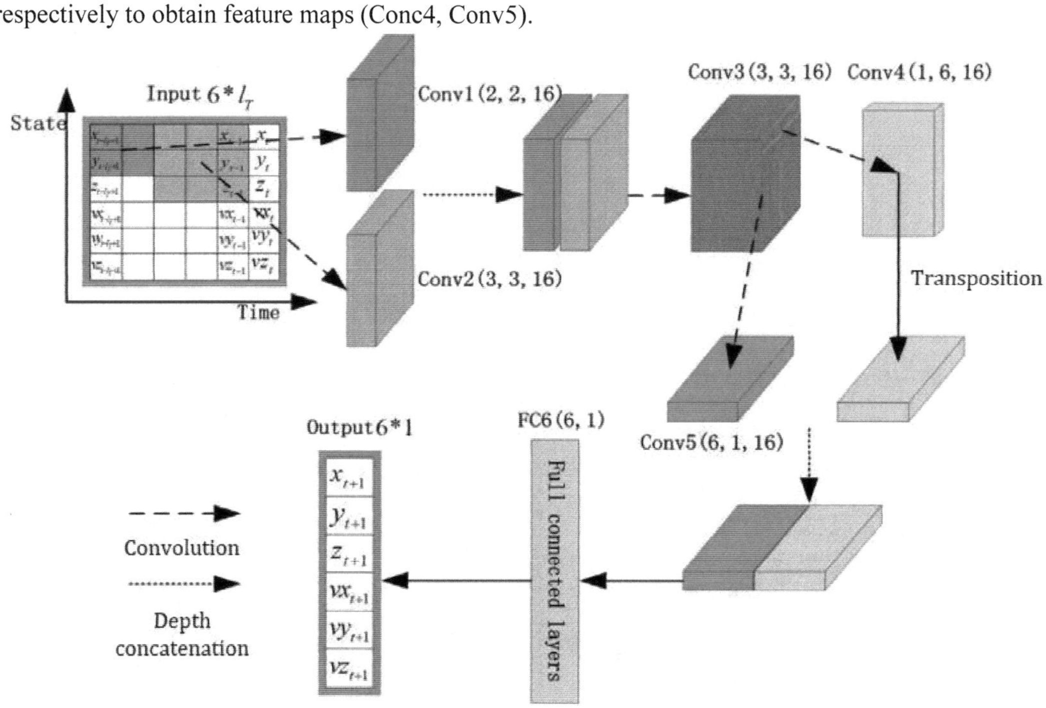

Figure 2. The structure of TSCNN

Table 1. Convolution kernels

	Scales	Number
Ker1	2×2	16
Ker2	3×3	16
Ker3	3×3	16
Ker4	1×6	16
Ker5	6×1	16

The inputs of TSCNN are expressed as:

$$s(t)=\left[x_t,y_t,z_t,vx_t,vy_t,vz_t\right]^T \tag{6}$$

$$S(t)=s(t)=\left[s_{t-l_T+1} \ \cdots \ s_{t-1} \ s_t\right] \tag{7}$$

where $s(t)$ denotes current aircraft states, including position and velocity information; $S(t)$ denotes inputs of the TSCNN, including historical and current information; l_T represents how much the states at the next moment are dependent on historical information.

The TSCNN outputs are the states at the next moment, and its equation expressed as:

$$s_{pred}(t+1)=\left[\hat{x}_{t+1},\hat{y}_{t+1},\hat{z}_{t+1},\hat{v}x_{t+1},\hat{v}y_{t+1},\hat{v}z_{t+1}\right]^T \tag{8}$$

The principle of trajectory predictor based on the established TSCNN is shown in Figure 3.

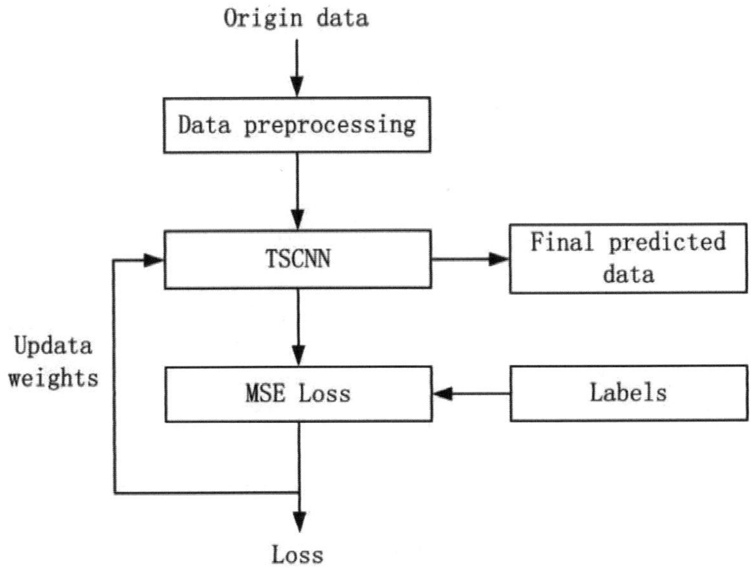

Figure 3. Trajectory predictor

The original data is observed during the tracking process and used as training samples for the TSCNN model. The total trajectory data set is:

$$Set = \{Traj^1, Traj^2, \cdots, Traj^N\} \tag{9}$$

where $Traj^N$, a flight trajectory of high-speed aircraft, is a sequence of time, including training inputs and labels.

$$Traj = \{Inputs : S(t), Labels : s(t+1)\} \tag{10}$$

The pre-processing normalizes the possible value range of the data to reduce the amplitude range. The loss function, which uses the half MSE function, calculates the error between the outputs and labels of the trajectory predictor that will be improved by an optimizer.

3.2.2. LSTM-TSCNN trajectory prediction network

In order to predict the medium- and long-term trajectory of high-speed aircraft, we replace the oldest data in the inputs with the predicted data and therefore realize the multi-step prediction of trajectory in a sliding manner. In this way, the previous error is inevitably introduced into subsequent prediction so that the prediction error gradually accumulates. Therefore, it is very necessary to correct the error by introducing a correction value in the prediction process so as to improve the accuracy of the medium- and long-term trajectory prediction.

The LSTM-TSCNN model is obtained on the basis of the TSCNN model. The LSTM network predicts the difference between the states and the TSCNN prediction results. The final prediction results are the sum of the LSTM network outputs and the TSCNN outputs. The prediction process is shown in Figure 4.

Table 2. Initial conditions

$h_0(km)$	$v_0(m/s)$	$\theta_0(^\circ)$	$\lambda_0(^\circ)$	$\phi_0(^\circ)$	$\psi_0(^\circ)$
69	6000	-1	0	0	110

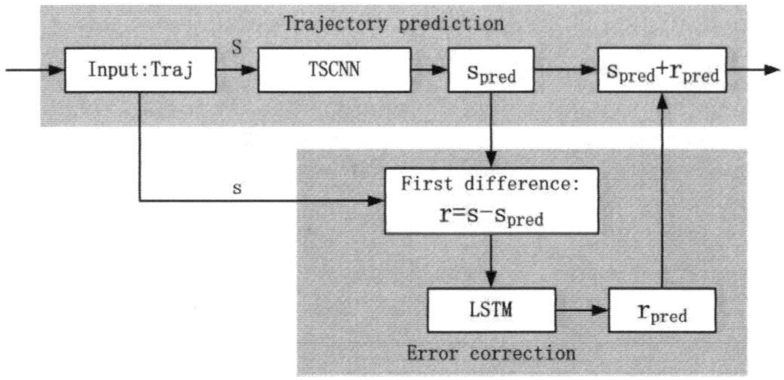

Figure 4. The structure of LSTM-TSCNN

The model is composed of TSCNN and LSTM network, and the accuracy of the prediction results is improved by the method of differential prediction. The final output is:

$$\hat{s} = s_{pred} + r_{pred} \tag{11}$$

here, r_{pred} is the corrected value, s_{pred} the predicted result of the TSCNN, $\hat{s}$ the final result. Compared with the TSCNN, the predicted trajectory based on LSTM-TSCNN is much closer to the actual one.

4. Experimental results

To verify the effectiveness of the proposed method, a large number of simulations are carried out. First, the observation data is generated by the 3DOF motion model. Then, the trajectory prediction simulations are carried out using the proposed method.

4.1. Sample generation

This paper uses the CAV-H vehicle as the object to verify the trajectory prediction algorithm. Its main characteristic parameters include its mass of 709 kg and the reference area of 0.48 m^2 [13]. The initial conditions are shown in table 2.

Assuming that the observation period is 0.1s, the skip-glide trajectory is drawn in different coordinate systems, and the height feature in the Earth-Centred Fixed Coordinated System (ECF) is shown in Figure 5. The training samples are the observation data of positions and velocity under the launch coordinated system. The data of 400-500s is used as samples to predict the trajectory of 500-650s, as shown in Figure 6.

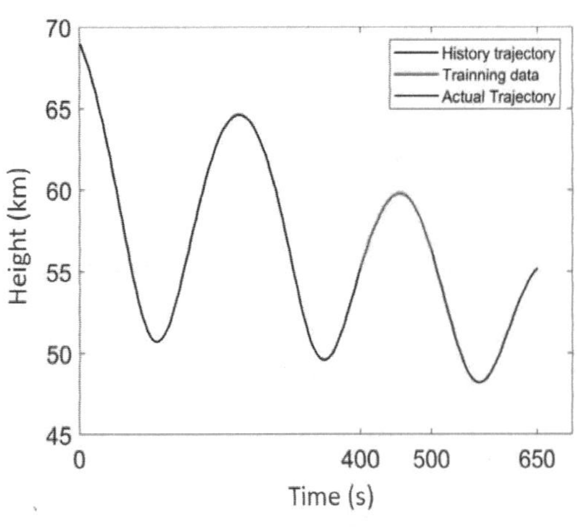

Figure 5. Height characteristic

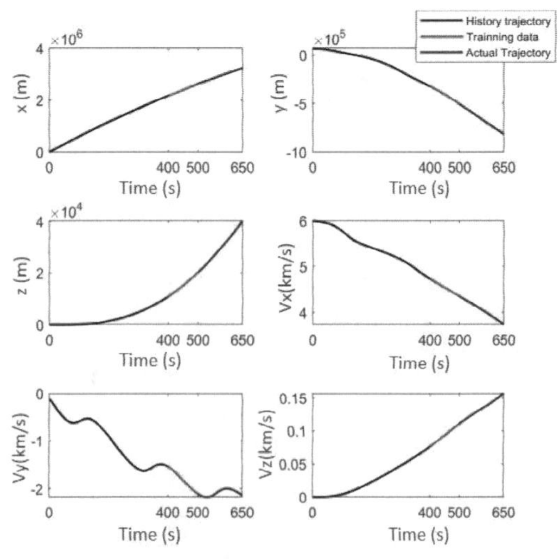

Figure 6. Position and velocity characteristics

4.2. Trajectory prediction

To verify the performance of the proposed hybrid model, we compare four methods, including the TSCNN, the LSTM-TSCNN (the LSTM network includes 50 hidden nodes), the FC network and the Singer-Kalman model. The FC network consists of five FC layers with 50 nodes. The sample data consists of 80% of training set, 10% of validation set and 10% of test set. The related hyper-parameters of these three deep learning models are shown in table 3.

Table 3. Initial configuration

Hyper-parameter	Input (2D matrix)	Output (2D matrix)	Optimizer	Learn rate	Epochs	Mini-batch Size
Value	6×6	6×1	SGDM	0.01	60	56

After training, four models are used to predict the flight trajectory in the next 150s. The absolute error of the position and velocity at the 150th second is shown in table 4. The concrete prediction results are shown in Figure 7 and Figure 8. The total position error is mainly related to that in the X direction, while the total velocity error is mainly related to that in the X and Y directions. In the medium- and long-term trajectory prediction, compared with the traditional Singer-Kalman model, whose accuracy is dependent on that of the target motion model, the strong non-linear mapping ability of the deep neural network model is better. Hence, the results of the former are less accurate than those of the latter. The accuracy of the TSCNN is higher than the FC network, with an increase of 10.96% at the 150th second. This is because CNN can extract features by sharing weights and leverage historical information by adding time series properties. Compared with the TSCNN model, the prediction accuracy of the hybrid model has been greatly improved due to the introduction of error correction. The total error of position and velocity at the 150th second are respectively 7.29 km and 9.84 m/s. The results demonstrate that the accuracy of prediction increases by 50.9%.

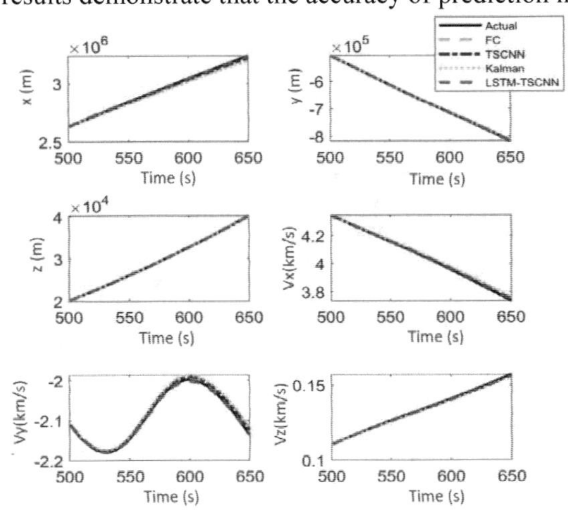

Figure 7. Predicted trajectory

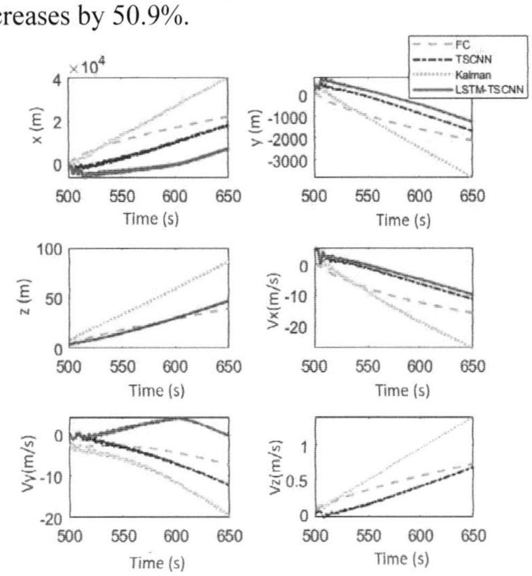

Figure 8. Prediction error

Table 4. Comparison of performance

Absolute error	Kalman	FC	TSCNN	LSTM-TSCNN
Velocity(m/s)	33.61	17.51	16.78	9.84
Position(km)	40.39	22.41	18.43	7.29

5. Conclusion

This paper discusses the deep learning method for trajectory prediction of high-speed aircraft, using the LSTM-TSCNN to solve the problem of medium- and long-term trajectory prediction. To verify the effectiveness of the proposed method, the prediction accuracy of the Kalman filtering method and various deep learning methods are compared. The results show that the total prediction error of position and velocity of the method proposed is respectively 7.29 km and 9.84 m/s at the 150th second. Compared with other methods, the LSTM-TSCNN has higher prediction accuracy, which can effectively solve the problem of the medium- and long-term trajectory prediction of high-speed aircraft.

References

[1] Pollack J H. Boost-glide weapons and us-china strategic stability[J]. The Nonproliferation Review, 2015, 24(7):155-164;
[2] Fan Y, Zhu W, Bai G. A cost-effective tracking algorithm for hypersonic glide vehicle maneuver based on modified aerodynamic model[J]. Applied Sciences, 2016, 6(10):312-329;
[3] Fan Y, Lu F, Zhu W. et al. A hybrid model algorithm for hypersonic glide vehicle maneuver tracking based on the aerodynamic model[J]. Applied Sciences, 2017, 7(2): 159-168;
[4] Qin L, Li J L, Zhou D. Tracking for near space target based on IMM algorithm[J]. System Engineering and Electronics, 2014, 35(7):1243-1249;
[5] Ritz S G, Hartfield R J, Dahlen J A. et al. Near real-time characterization of unknown missiles in flight using computational intelligence[C]//2015 IEEE Aerospace Conference, 2015;
[6] Barratt S T, Kochender M J, Boyd S P. Learning Probabilistic Trajectory Models of Aircraft in Terminal Airspace From Position Data[J]. Transactions on Intelligent Transportation Systems, 2018, 20(9):3536-3545;
[7] Hochreiter S, Bengio Y, Frasconi P. Gradient flow in recurrent nets:the difficulty of learning long-term dependencies[J]. Field Guide to Dynamical Recurrent Networks, IEEE Press, 2009;
[8] Zhou M, Shi Z W, Ding H P. Aircraft classification in remote-sensing images using convolution neural networks[J]. Journal of Image and Graphic, 2017, 32(5): 702-708;
[9] Cai Y L, Deng Y L, Su Y H. LSTM Ballistic Classification and Prediction Method for Hypersonic Vehicles[C]//CCSSTA21st 2020, 2020;
[10] Ma L, Tian S. A Hybrid CNN-LSTM Model for Aircraft 4D Trajectory Prediction[J]. IEEE Access, 2020, 8(7):134668-134680;
[11] Wang T, Zhang H B, Li Y Y. Landing footprint generation of entry vehicle based on Gauss pseudospectral method[J]. Journal of National University of Defense Technology, 2016, 38(3):75-80;
[12] Liu Y H, Wang H G. Control-oriented UAV highly feasible trajectory planning: A deep learning method[J]. Aerospace Science and Technology, 2021, 11(9):1-11;
[13] PHILLIPS T H. A common aero vehicles(CAV) model description and employment guide[R]. Corporation for AFRL and AFSPC, 2003;

Key Failure Mode Identification Method of LED System Based on Fuzzy TOPSIS

Bing Zhang[1], Zhaojun Hao[1], Lifang Wang[1] and Lexiao Li[2]

[1.] Systems Engineering Research Institute, Beijing, China

[2.] NO.722 Research Institute, CSSC, Wuhan, China

Zhangliqi857@126.com

Abstract. The premise of the system reliability analysis is to effectively discern the patterns of product failures at all levels. To deal with the problem that the diversity of failure modes of complex products represented by LED systems cannot be analyzed comprehensively, in this paper, a key failure mode identification method based on Fuzzy TOPSIS is proposed. First, the construction method of the failure relevance tree is given to support the comprehensive identification of potential failure modes. Then, the decision rule of node failure weight is mined by the rough set method, the systematic evaluation of the node failure weight is realized with the TOPSIS method, and the key failure modes are identified. Finally, a case validation of the proposed method in this paper is performed.

1. Introduction

Currently, LED has been widely used in many engineering fields such as national defense, aerospace, and automobile industry. Due to the characteristics of high power, high heat, and long working time, many researchers focus on the life and reliability analysis of high-poser LED [1]. The LED system is composed of an LED matrix module, heat dissipation, power drive, electrode, etc.

In terms of reliability and life analysis of LED systems, researchers have carried out a lot of basic research work around its failure mode and mechanism [2-4]. M. H. Chang of the University of Maryland systematically collated and analyzed the failure modes, causes, effects, and mechanisms of LED [4]. Due to the complexity and multi-level characteristics of the LED lighting system, the failure modes of the LED system are complex and diverse, and each module has multiple failure modes. From the perspective of failure physics, to analyze the reliability of the entire system, it is necessary to analyze each failure mode. However, considering the feasibility and reality, it is impossible to consider all failure modes of each component. Therefore, a few critical failure modes should be selected as the basis for system reliability research according to the Pareto principle. The report of the US Department of Energy also pointed out that the LED drive power supply is a weak component in the LED system [5], which leads to up to 52% of total system failures [6].

At present, it is difficult to collect enough data to accurately identify the key failure modes of the LED system. Therefore, given the complex structure and various failure modes of LED system, from the design point of view, based on the idea of Axiomatic Design [7], this paper constructs a fault relevance tree structure from the function to the physical structure to the potential failure mode and systematically identifies the failure mode set of LED system based on the fault relevance tree. Further, the node fault weights are systematically and comprehensively evaluated and sequenced, and the key failure mode sets are finally identified, providing support for reliability analysis and other work.

2. Construction of fault relevance tree for LED system

To systematically explore the potential failure modes of LED systems at all levels, this paper uses the mapping idea in Axiomatic Design to determine the "functional domain", "physical domain", and "fault domain", forming the "function structure fault" correlation analysis idea.

As shown in Figure 1, the function and structure decomposition is carried out vertically from top to bottom, and the function is subdivided horizontally. The hierarchical relationship of failure modes can be obtained based on the results of the functional structure analysis. Through the hierarchical relationship of failure modes, each failure mode of the system can be found layer by layer, and then the identification of key failure modes can be carried out.

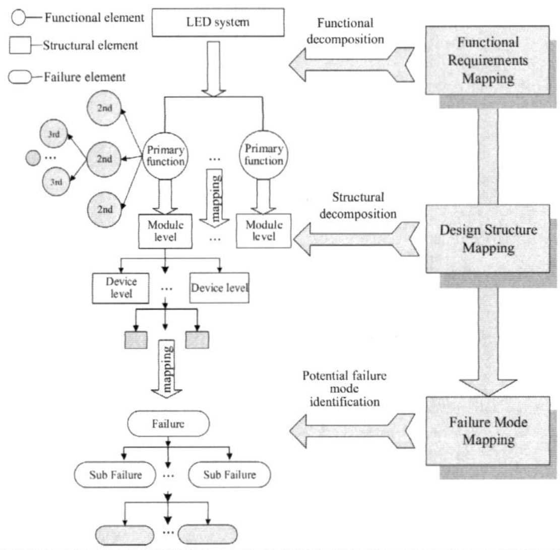

Figure 1. Schematic diagram of fault association tree model.

3. Key failure mode identification based on fault relevance tree

The failure modes of LED systems are complex and diverse, and the occurrence frequency and impact severity of each failure mode are different. The identification of key failure modes is the basis of reliability research. Based on the fault relevance tree, this paper adopts the decision rules of rough set [8] mining node fault weights and then uses the Fuzzy TOPOS method [9] to give a systematic and comprehensive evaluation and ranking of node fault weights.

The influence of failure on product reliability is analyzed from three aspects: probability, detectability, and consequence severity. To further clarify the specific meaning of each indicator in a specific external environment, it is necessary to determine the corresponding feature attribute set of each evaluation indicator in advance according to the statistical reliability information of the LED system. The corresponding feature attribute set of each indicator is shown in Table 1.

Table 1. Characteristic attribute set of evaluation indicator system.

Attribute / Indicator	Feature attribute C1	Feature attribute C2	Feature attribute C3	Feature attribute C4	Decision attribute D
Probability	Failure frequency	Failure type			Influence rate
Detectability	Design	Production	Storage	Use	Influence rate
Consequence severity	Remedial costs	Impact on the environment	Impact on the system	Impact on oneself	Influence rate

3.1. Determining weights and decision rules of node indicators based on rough sets

The weight (W_{IF}) of the above three indicators reflects the contribution of each indicator to the node weight, and $\sum_{j=1}^{3} W_{IF\,j} = 1$. First, feature attributes are collected as sample data, and all the characteristic attribute values contained in the sample data are analyzed through discrete normalization processing to form a decision table. According to the data in the decision table, the data with the same decision attribute are identified and the positive domain associated with each characteristic attribute value is solved. Finally, the cardinal ratio of each positive domain set is used to solve the weight of the index influence factor.

$$M_{C-c_i(D)} = 1 - \frac{card(POS_{C-c_i}(D))}{card(POS_C(D))} \tag{1}$$

Similarly, the decision rules of indicators at each node can also be determined by using the rough set method. First, based on the established decision table, the coordination of the decision table is maintained, and the same example rows and attribute columns in the decision table are eliminated to obtain a simplified decision table. Then, it is calculated through the corresponding conditional attribute core value table of the decision table, and the minimum decision rule set that meets the decision requirements is output.

3.2. Determining the evaluation fuzzy value of each indicator of the node according to the decision rules

In this paper, the fuzzy evaluation is quantified in the form of a triangular fuzzy number [10]. Generally, the general form of a triangular fuzzy number can be expressed as $M = (l, m, u)$, a feature membership function $\mu_M(x): R \rightarrow [0,1]$ is equivalent to the following formula:

$$\mu_M(x) = \begin{cases} 0, & x < l \ or \ x > u \\ (x-l)/(m-l), & l \le x \le m \\ (x-u)/(m-u), & m \le x \le u \end{cases} \tag{2}$$

where $l \le m \le u$, l, u, and m respectively represent the left value, right value, and median value of the fuzzy number. Assume that the fuzzy importance of variables is divided into six levels from low to high, and the transformation rule of triangular fuzzy number is shown in Table 2.

Table 2. Triangular fuzzy number transformation rule.

Fuzzy importance	Natural language	Triangular fuzzy number
w_1 (VLI)	Very weak importance	(0.05, 0.15, 0.25)
w_2 (LI)	Weakly important	(0.2, 0.3, 0.4)
w_3 (WI)	Secondary importance	(0.35, 0.45, 0.55)
w_4 (EI)	Equally important	(0.5, 0.6, 0.7)
w_5 (MI)	More important	(0.65, 0.75, 0.85)
w_6 (VI)	Very important	(0.8,0.9,1)

3.3. Ranking the Importance of Failure Modes Based on Fuzzy TOPSIS

The specific steps are as follows:

- Taking $r_{ij} = \frac{x_{ij}}{\sqrt{\sum_{i=1}^{n} x_{ij}^2}}$ as the statistic, a standardized decision matrix is established;
- Taking w_j as the weight coefficient, analyzing the statistics r_{ij}, and then the weighted statistics $v_{ij} = r_{ij} w_j$ is obtained.
- Determining positive ideal node A^+ and negative ideal node A^-.

$$A^+ = \left\{ \left(\max v_{ij} \middle| j \in J \right), \left(\min v_{ij} \middle| j \in J' \right) \right\} = \left\{ v_1^+, v_1^+, \cdots, v_n^+ \right\} \tag{3}$$

$$A^- = \left\{ \left(\min v_{ij} \middle| j \in J \right), \left(\max v_{ij} \middle| j \in J' \right) \right\} = \left\{ v_1^-, v_1^-, \cdots, v_n^- \right\} \tag{4}$$

where J represents the benefit indicator and J' represents the cost indicator.

- The distance between each failure mode node and the ideal point is analyzed and calculated according to the distance formula $S_i^+ = \sqrt{\sum \left(v_{ij} - v_j^+ \right)^2}, i = 1, 2, \cdots, n$ or $S_i^- = \sqrt{\sum \left(v_{ij} - v_j^- \right)^2}, i == 1, 2, \cdots, n$, and the expression of the relative closeness of each failure mode node can be obtained.

$$C_i^+ = \frac{S_i^-}{S_i^+ + S_i^-} \tag{5}$$

- The neutral decision is selected, that is, the optimistic and pessimistic coefficient is 0.5, and the expected value of node closeness can be calculated by the following formula.

$$I(C_i^+) = \frac{(l_i + 2m_i + u_i)}{4}, \tag{6}$$

whose relative weight value can be expressed as.

$$w_i = \frac{I(C_i^+)}{\sum_{i=1}^{n} I(C_i^+)}. \tag{7}$$

4. Case study

The fault relevance tree of the LED system is built, as shown in Figure 2. To simplify the analysis process, the main fault modes of the substrate in the fault domain are selected, namely FM1.2.1, FM1.2.2, and FM1.2.3, and the calculation process of node fault weight is demonstrated.

Step 1: Determine the weight of each indicator.

Based on statistical data, the impact factors of each evaluation indicator are quantified according to the above rough set model. Then, the sample data-oriented to decision attributes is shown in Table 3. The decision table system can be expressed as $T = (X, C \cup D, V, f)$, where the finite sample field is $X = \{x_1, x_2, x_3, x_4, x_5, x_6, x_7, x_8\}$, the index attribute set is $I = \{Index1, Index2, Index3\}$, and the decision attribute is $D = \{d\}$, $V = \{0,1,2\}$.

Table 3. Decision table of impact weight of evaluation indicator system.

Indicator / Sample No	Indicator attribute			Decision attribute
	Index 1	Index 2	Index 3	D
X1	1	1	0	0
X2	1	0	1	1
X3	1	0	2	1
X4	0	1	0	0
…				

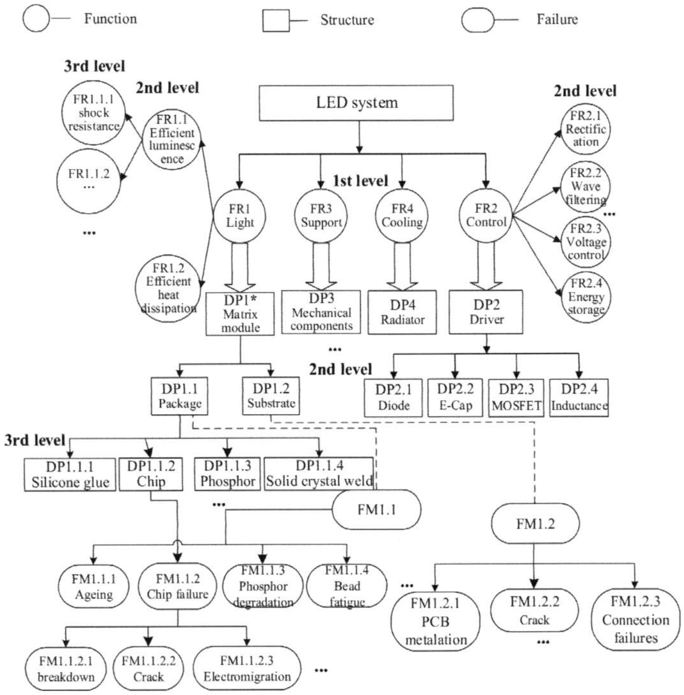

Figure 2. The fault relevance tree of LED system.

The specific solution process is as follows:

- First, we calculate the objective weight of each indicator.

$$X/I=\{\{X_1\},\{X_2\},\{X_3\},\{X_4\},\{X_5,X_7\},\{X_6,X_8\}\}$$

$$X/D=\{\{x_1,x_4,x_5,x_8\},\{x_2,x_3,x_6,x_7\}\}$$

$$X/(I-\{Index1\})=\{\{x_1,x_4\},\{x_2,x_5,x_7\},\{x_3\},\{x_6,x_8\}\}$$

$$X/(I-\{Index2\})=\{\{x_1\},\{x_2\},\{x_3\},\{x_4\},\{x_5,x_7\},\{x_6,x_8\}\}$$

$$X/(I-\{Index3\})=\{\{x_1\},\{x_2,x_3\},\{x_4,x_6,x_8\},\{x_5,x_7\}\}$$

$POS_I(D)=\{x_1,x_2,x_3,x_4\}$, $POS_{\{I-\{Index1\}\}}(D)=\{x_1,x_4\}$, $POS_{\{I-\{Index2\}\}}(D)=\{x_1,x_2,x_3,x_4\}$, $POS_{\{I-\{Index3\}\}}(D)=\{x_1,x_4\}$

Based on the positive domain value obtained, the influence factors can be obtained according to the formula $M_{\{I-\{Index(i)\}\}}=1-\dfrac{card(POS_{\{I-\{Index(i)\}\}}(D))}{card(POS_{\{I-\{Index(i)\}\}}(D))}$. The results are shown in Table 4.

Table 4. Table of influence factors of fault weight evaluation index system.

Indicator No	Index 1	Index 2	Index 3
Indicator influence factors	Probability 0.4	Detectability 0.1	Consequence severity 0.5

Step 2: Determine decision rules based on rough sets.

Index 3 (Consequence severity) of Node FM1.2 is selected as the analysis object. The score of decision attribute D is divided into six levels according to the triangular fuzzy number transformation in Table 2. The weight decision table can be obtained. After simplification, the simplest rule is solved, and Table 5 is obtained.

Table5. The simplest rule of weight decision.

Indicator / Sample No	Indicator attribute			Decision attribute
	Index 3-C1	Index 3-C2	Index 3-C3	D
X1''	1	1	—	1
X2''	1	2	—	2
X3''	2	1	2	3
X4''	2	2	2	4
X5''	2	—	3	5
X6''	3	1	3	5
X7''	3	3	2	6

Step 3: Determine the fuzzy value of each indicator based on the decision rules.

According to the decision rules determined in the above process and the analysis results of each node, the fuzzy value of its importance evaluation is preliminary determined as shown in Table 6.

Table 6. Evaluation value of index weight.

Node	Indicators		
	Index 1	Index 2	Index 3
FM1.2.1	VI	LI	MI
FM1.2.2	LI	VI	WI
FM1.2.3	VLI	WI	VI

Step 4: According to the fuzzy TOPSIS decision theory and the method of fuzzy approximation to the ideal, the relative weight of each node is calculated and the priority is determined.

The natural language evaluation value of each evaluation index is transformed corresponding to each node in Table 6, and a weighted standardized decision-making matrix is established according to the influence factors of each index in Table 4. The results are shown in Table 7.

Table 7. Weighted normalized decision matrix.

Node	Indicators		
	Index 1	Index 2	Index 3
FM1.2.1	(0.32, 0,.36, 0.4)	(0.02, 0.03, 0.04)	(0.325,0,375,0.425)
FM1.2.2	(0.08, 0.12, 0.16)	(0.08, 0.09, 0.1)	(0.25, 0.3, 0.35)
FM1.2.3	(0.02, 0.06, 0.1)	(0.035, 0.045, 0.055)	(0.4, 0.45, 0.5)

The ideal node and negative ideal node are found, and the results are shown in Table 8.

Table 8. Ideal and negative ideal nodes.

Node	Indicators		
	Index 1	Index 2	Index 3
A^+	(0.32, 0.36, 0.4)	(0.08, 0.09, 0.1)	(0.4, 0.45, 0.5)
A^-	(0.02, 0.06, 0.01)	(0.02, 0.03, 0.04)	(0.25, 0.3, 0.35)

According to the method of fuzzy approximation to the ideal, the distance between the evaluation value of each node and the ideal node and the negative ideal node is calculated according to the data in the table above. Index 1 in FM1.2.1 is taken as an example:

$$S_1^+ = \{\left[(0.32,\ 0,.36,\ 0.4)-(0.32,\ 0.36,\ 0.4)\right]^2 + \left[(0.02,\ 0.03,\ 0.04)-(0.035,\ 0.045,\ 0.055)\right]^2 + \left[(0.325,\ 0,375,\ 0.425)-(0.325,\ 0.375,\ 0.425)\right]^2\}^{1/2} = (0.015, 0.015, 0.015)$$

Similarly, the calculation results shown in Table 9 can be obtained.

Table 9. Distance between failure relevance weight and ideal nodes.

Ideal node distance	Weight	Negative ideal node distance	Weight
S_1^+	(0.15,0.15,0.15)	S_1^-	(0.117,0.117,0.397)

S_2^+	(0.322,0.322,0.364)	S_2^-	(0.028,0.028,0.036)
S_3^+	(0.309,0.309,0.397)	S_3^-	(0.015,0.015,0.015)

The calculation results of the distance between each node and the ideal node can be given as follows.

$C_1^+ = \frac{S_1^-}{S_1^+ + S_1^-} = (0.438,0.438,0.725)$, $C_2^+ = \frac{S_2^-}{S_2^+ + S_2^-} = (0.082,0.082,0.096)$, $C_3^+ = \frac{S_3^-}{S_3^+ + S_3^-} = (0.046,0.046,0.036)$.

Finally, the method of fuzzy approximation to the ideal is used, and the triangular fuzzy number model is used to calculate the weight value of each node:

$I(C_1^+) = \frac{0.438 + 2 \cdot 0.438 + 0.725}{4} = 0.510$, $I(C_2^+)=0.085$, $I(C_3^+)=0.044$.

According to the weight value of each node, the relative weight of each node is calculated: $w_1 = \frac{0.510}{0.510 + 0.085 + 0.044} = 0.798$, $w_2 = 0.133$, $w_3 = 0.069$. The weight sorting of nodes is obtained: FM1.2.1, FM1.2.2, and FM1.2.3. The weight of FM1.2.1 is much higher than the other two nodes.

5. Conclusions

Aiming at the problem that the failure modes of LED systems are various and difficult to analyze one by one, according to the relevance analysis idea of "function-structure-fault", this paper establishes a fault relevance tree from the function layer to the structure layer, and then to the failure mode layer, which supports the comprehensive identification of the failure mode set of LED system. Then, the decision rules of node fault weight mining based on rough sets and the method of TOPSIS are used to calculate and sort the fault mode weights of each node, and the key failures are determined. Finally, the above methods are verified by examples, and the key failure modes of the LED lighting system are identified.

References

[1] Lasance C J M, Poppe A. Thermal Management for LED Applications. Solid State Lighting Technology & Application, 2013, 2.

[2] Meneghesso G, Meneghini M, Zanoni E. Recent results on the degradation of white LEDs for lighting. Journal of Physics D Applied Physics, 2010, 43(35):354007

[3] Pecht M G, Chang M H. Failure Mechanisms and Reliability Issues in LEDs. Solid State Lighting Technology & Application, 2013, 1:43-110

[4] Fan J, Yung KC, Pecht M. Physics-of-Failure-Based Prognostics and Health Management for High-Power White Light-Emitting Diode Lighting. IEEE Transactions on Device & Materials Reliability, 2011, 11(3):407-416.

[5] Anderson J., Bhatkal R., and Bradley D., "LED Luminaire Lifetime: Recommendations for Testing and Reporting," Next Generation Lighting Industry Alliance Report, Washington, DC, USA, 2011

[6] Almeida P S, Camponogara D, Dalla Costa M, et al. Matching LED and driver life spans a review of different techniques. IEEE Industrial Electronics Magazine, 2015, 9(2): 36-47

[7] Nam Pyo Suh, Xie Youbai, Axiomatic design, China Machine Press, 2004, Beijing. (In Chinese)

[8] Zhang Wenxiu, Rough Set Theory, and Methods, Science Press, 2001. (In Chinese)

[9] Bottani E, Rizzi A. A fuzzy TOPSIS methodology to support outsourcing of logistics services. Supply Chain Management, 2006, 11(4):294-308.

[10] Shui X Z. Study on Method for Triangular Fuzzy Number-Based Multi-Attribute Decision Making with Preference Information on Alternatives. Systems Engineering and Electronics, 2002, 24(8):9-3.

Research on automatic driving control strategy based on improved model predictive control

Jianzhen Chu[1,2], Qingyu Zhang[1,2], Zengxi Wang[1,2] and Tong Jia[1,2]

[1] Automotive Data of China (Tianjin) Co., Ltd., Tianjin, 300300, China

[2] China Automotive Technology & Research Center Co., Ltd., Tianjin, 300300, China

Corresponding author's e-mail: chujianzhen@catarc.ac.cn

Abstract. As a new technology direction in the automotive industry in recent years, automatic driving technology has played an important role in reducing traffic accidents, promoting energy conservation and emission reduction, and improving the efficiency of high-speed and urban roads. As one of the key technologies of autonomous vehicles, the performance of control algorithm greatly affects the smoothness of vehicle driving and the accuracy of track tracking. In order to consider the lateral control and longitudinal control accuracy at the same time, and ensure that the vehicle travels according to the given trajectory, this paper first establishes the vehicle kinematics model, and improves the traditional model predictive control algorithm, which realizes the high dynamic control of the desired trajectory while ensuring the vehicle stability and comfort. Finally, based on the built real vehicle test platform for automatic driving, the acceleration and deceleration test and the high-speed serpentine test are designed. By analyzing the test data, it can be seen that the control algorithm proposed in this paper can effectively achieve the vehicle trajectory tracking, and the control accuracy is improved.

1. Introduction

The advantages of autonomous vehicles in terms of intelligence and automation are more obvious than those of traditional vehicles. Automatic driving can be used in many scenarios, such as highways, urban roads, smart ports, and smart mines. The design of lateral controller and longitudinal controller is the core technology of the automatic driving vehicle control module, which determines whether the vehicle can cope with complex traffic scenarios. At present, the main trajectory tracking algorithms include PID control,linear quadratic regulator (LQR) tracking control, and model predictive control [1-5]. The improved model predictive control model proposed in this paper has the advantages of high control accuracy, fast response speed, and good robustness, and is suitable for a variety of vehicle types and traffic scenarios. To effectively verify the effect of the control strategy, different tests are designed based on the automatic driving real vehicle test platform. Through the comparative analysis of test results, the test vehicle performs well in control accuracy, response time, tracking effect, etc., which can meet the use requirements in different scenarios.

2. Automatic driving system

Autopilot software system usually adopts multi-layer architecture, in which the perception module, planning module and control module are important components. Among them, the perception layer perceives the driving environment through lidar, camera, and other sensors; The decision layer realizes the decision-making of vehicles based on the perception results, and constantly carries out path planning during the driving process, such as overtaking lane changing, stopping, obstacle avoidance, lane keeping, etc. The control layer controls the vehicle through the designed control strategy according to the information of the decision-making layer and uses the bottom controller to control the vehicle braking, acceleration, steering, etc. The overall architecture of the auto drive system is shown in Figure 1.

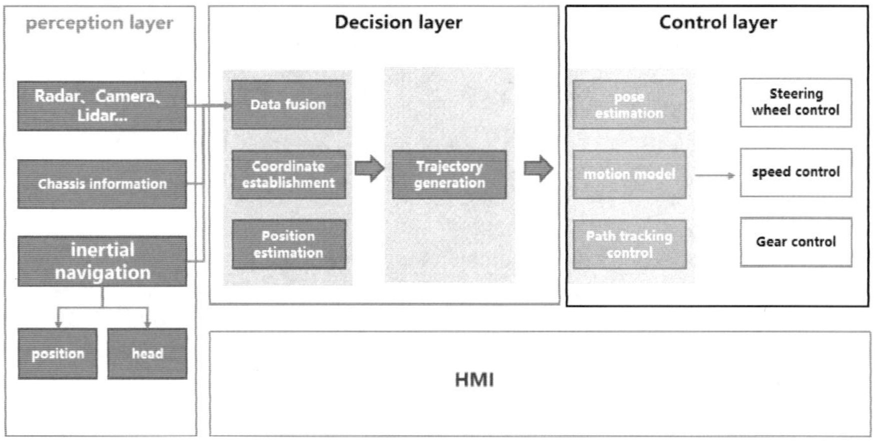

Figure 1. Auto drive system architecture

The bottom controller mainly collects the steering wheel angle information, gear position information, and accelerator/brake pedal travel information, and records the communication data in the form of an upper computer. At the same time, the instructions sent by the decision-making layer are combined with the current vehicle status information to be converted into vehicle control instructions, which are transmitted through the CAN bus to finally achieve vehicle control.

The automatic driving hardware system is modified based on the Geely XingYue-L model, and the vehicle supports drive-by-wire. By adding sensors and chassis wire control, the requirements of automatic driving for environmental information and vehicle control can be met, providing a basis for subsequent real vehicle tests. The hardware configuration layout and real vehicle schematic diagram of the automatic driving vehicle are shown in Figure 2 and Figure 3 respectively.

Figure 2. The layout of autopilot hardware

Figure 3. Schematic Diagram of Automatic Driving Real Vehicle Modification

Journal of Physics: Conference Series **2489** (2023) 012025 doi:10.1088/1742-6596/2489/1/012025

3. Control scheme

This paper adopts hierarchical control mode to adapt to the complexity and accuracy of automatic driving vehicle control. The control layer receives the trajectory information sent by the decision-making layer, determines the control mode according to the designed control strategy, calculates the transverse and longitudinal control results in combination with the vehicle status and body parameters, and sends the calculation results to the executive layer to realize the vehicle steering, driving, and braking. The decision layer, control layer, and execution layer communicate with each other in real-time and share data.

3.1. Kinematic and dynamic models

To achieve automatic vehicle tracking and ensure the tracking effect, it is necessary to carry out a kinematics and dynamics analysis of the vehicle.

Vehicle models are generally divided into kinematic models and dynamic models. Both models can be used for trajectory tracking control of autonomous vehicles. The kinematics model is built based on the geometric relationship of vehicles. The dynamic model is constructed based on the force analysis because the tire friction coefficient is considered. By modeling the kinematics and dynamics of an autonomous vehicle, it can play an important role in vehicle motion control performance and trajectory tracking effect [6-7].

Considering the complexity and practical application of vehicle motion on the ground, The vehicle model can be simplified according to the bicycle model, which has two degrees of freedom, as shown in the following figure 4:

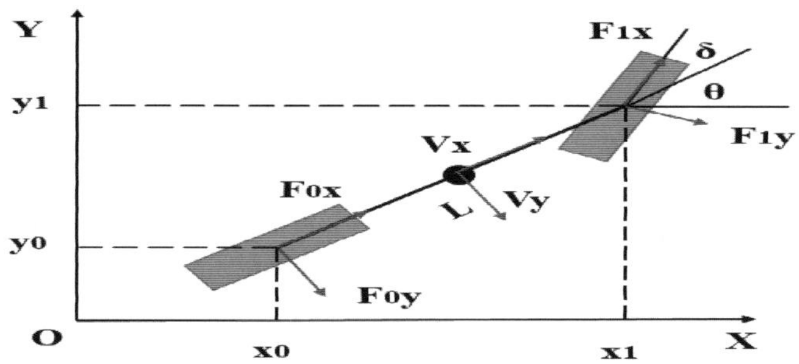

Figure 4. Kinematic and dynamic models

The kinematics model and dynamics model of the self-driving vehicle in this paper are constructed with reference to the world coordinate system. (x0, y0)/(x1, y1) are respectively used to represent the midpoint position of the front/rear axle of the vehicle, and Vx/Vy are respectively used to represent the lateral/lateral velocity at the vehicle centroid position; F0x/F1x represents the longitudinal force along the body direction borne by the front/rear axles of the vehicle, F0y/F1y represents the transverse force perpendicular to the body direction borne by the front and rear axles of the vehicle, and I represents the rotational inertia of the vehicle, θ Is the vehicle yaw angle, δ Is the front wheel angle of the vehicle with two degrees of freedom, m is the curb weight of the vehicle; L is the longitudinal distance from the front axle to the rear axle of the vehicle, where the distance from the centroid to the front/rear axle is expressed in a/b.

Ignoring the influence factors such as sideslip and roll during vehicle driving, Equation (1) is the simplified vehicle kinematics model.

$$\begin{bmatrix} x \\ y \\ \theta \end{bmatrix} = \begin{bmatrix} \cos\theta \\ \sin\theta \\ \tan\delta / L \end{bmatrix} v_x \tag{1}$$

Equation 2 is a simplified dynamic model. The transverse and longitudinal forces of the front and rear wheels under the stress of the car body are listed. Then, according to the small angle assumption, the linear model is transformed, and I, m, and δ bring in the force expression and finally obtain the dynamic model.

$$m\ddot{x} = m\dot{y}\dot{\delta} + 2F0x + 2F1x$$
$$m\ddot{y} = -m\dot{x}\dot{\delta} + 2F0y + 2F1y \quad (2)$$
$$I\ddot{\delta} = 2aF0y - 2bF1y$$

3.2. Control system strategy

The vehicle control strategy includes the lateral control strategy and the longitudinal control strategy. The longitudinal control realizes the control of the vehicle driving system and braking system, and the lateral control realizes the control of the vehicle steering system. In trajectory tracking control, it is usually necessary to introduce a preview control algorithm in consideration of factors such as system delay, that is, by setting preview distance, it is input into the control system for calculation.

Before testing the vehicle control system, it is usually necessary for the tester to record the map information in advance, and the decision-making system will calculate the expected trajectory. The input of lateral control system includes vehicle chassis information, map information, positioning information, etc. The system calculates the lateral error between the vehicle preview point and the desired track point based on the preview theory, and the final output is the desired front wheel angle of the vehicle. The drive controller converts the calculated expected angle into a control command and sends it to the vehicle chassis to realize the lateral control of the vehicle [8]. Longitudinal control calculates the expected acceleration of the vehicle according to the longitudinal position deviation or speed deviation of the desired track point and the current position and then converts it into control commands to control the accelerator and brake pedal, to achieve the purpose of controlling the vehicle.

In this paper, an automatic driving vehicle control system with lateral and longitudinal control coupling is established. Compared with the method of independently considering the lateral and longitudinal control systems, the controller is more in line with the vehicle kinematics characteristics. The control system design is shown in the figure 5.

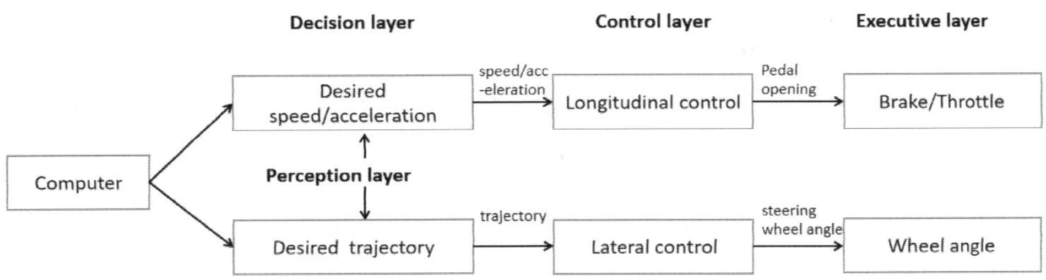

Figure 5 Lateral and longitudinal joint control system

3.3. Model predictive control

In this paper, the controller of automatic driving vehicle is designed based on the model predictive control(MPC) algorithm. MPC control algorithm design mainly includes prediction model building, rolling optimization function design and feedback correction. The idea of solving the optimal control output of the controller is to convert the control problem into solving the optimal control sequence in the finite prediction time domain, and finally output the first control quantity in the sequence as the optimal control quantity at the current time. The following figure 6 shows the control flow of the controller. The input is the desired tracking track, and the output is the control amount at the current time.[9].

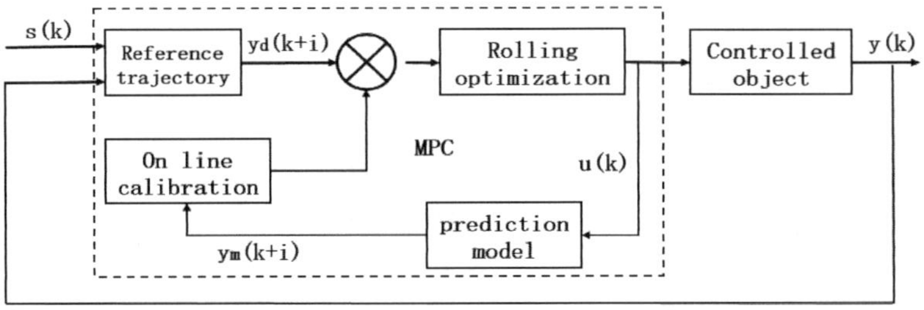

Figure 6 Structure Diagram of MPC System

The principle of MPC controller is shown in the figure 7. K time represents the current control time, the left axis represents the past time of the system, and the right axis represents the future time of the system. By segmenting the future time, the MPC controller only calculates the optimal control sequence in a finite time domain per cycle.

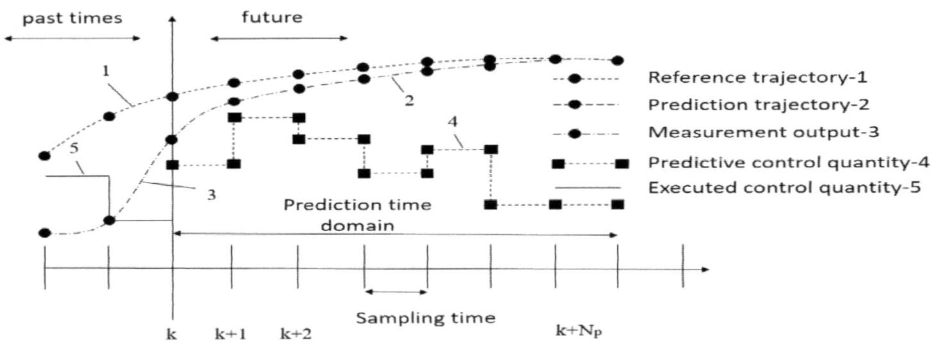

Figure 7 Model predictive control principle

The MPC controller predicts the output of a future sampling time at time k; At the k+1 moment, the real-time control quantity is optimized based on the improved MPC and referring to the optimization result at the k moment to find the optimal solution. The initial control time has s0 minimum sampling periods. At the same time, the maximum fluctuation amplitude of the predicted value of the uncontrollable input quantity within the control time [10]. A control interval contains MK minimum sampling times, and s0 decreases L2， the principle is shown in Figure 8.

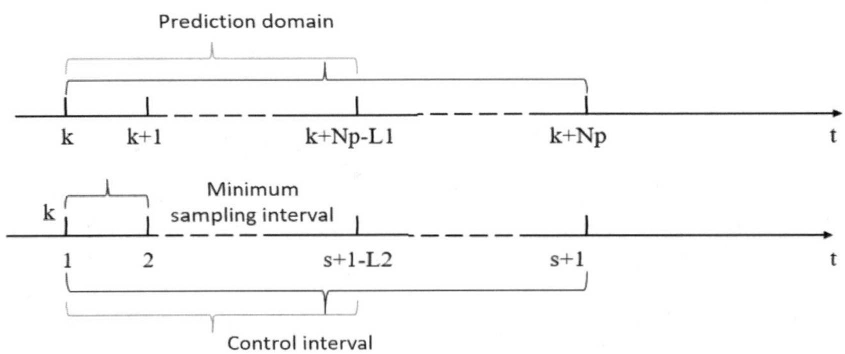

Figure 8 MPC prediction domain

Based on the above improved MPC principle, the self-vehicle position, yaw angle and vehicle speed are selected as the system state variables, and the vehicle front wheel angle and acceleration are selected as the control variables. At the same time, considering the self-vehicle position and expected position deviation, vehicle kinematics constraints, transverse and longitudinal parameter anti-interference and other factors, a model predictive controller is established.

4. Test

4.1. Automobile parameters

Automobile parameters are shown in Table 1.

Table 1 Automobile parameters

Automobile parameters	value
size	4770*1895*1689mm
weight（m）	1730kg
Vehicle front wheelbase（a）	1232mm
Vehicle rear wheelbase（b）	1468mm
Wheelbase（L）	2700mm
Maximum power(P)	110kW
Front wheel lateral stiffness(k1)	66900
Rear wheel lateral stiffness(k2)	62700

4.2. Test

4.2.1. Human Machine Interface (HMI) design

HMI is developed based on QT.As shown in Figure 9, including vehicle status information, such as real-time speed, vehicle position, vehicle attitude, steering wheel angle, accelerator, brake, lights, and other information; It includes path planning information, obstacle information around the vehicle, and real-time video information of the road in front of the vehicle (and its surroundings, for parking) shall be displayed in a visual two-dimensional or three-dimensional image; The automatic driving mode of the vehicle is turned on, the initialization function of each sensor device, the device status information, etc.

Figure 9 HMI interface

4.2.2. Acceleration and deceleration test

The acceleration and deceleration test mainly verifies the longitudinal control effect of the vehicle. The vehicle speed, expected speed and speed deviation obtained from the test are shown in the following figure 10.

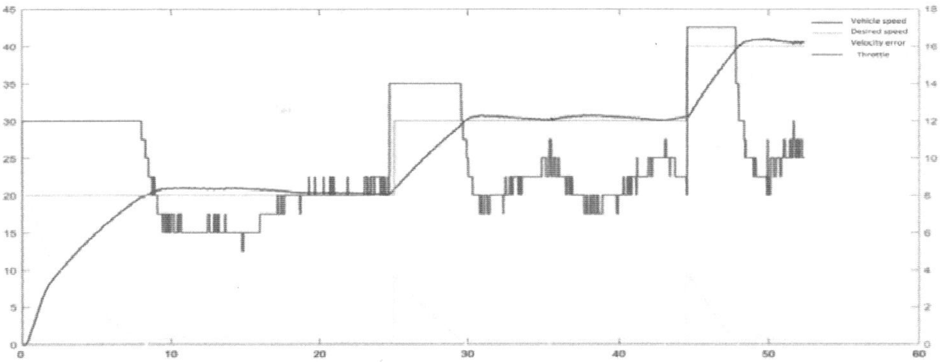

Figure 10. Vehicle speed, expected speed and speed deviation data

4.2.3. High-speed serpentine test

The high-speed serpentine test is mainly used to verify the lateral control effect of the automatic driving vehicle. The lateral acceleration and vehicle speed obtained by this test are shown in the following figure 11.

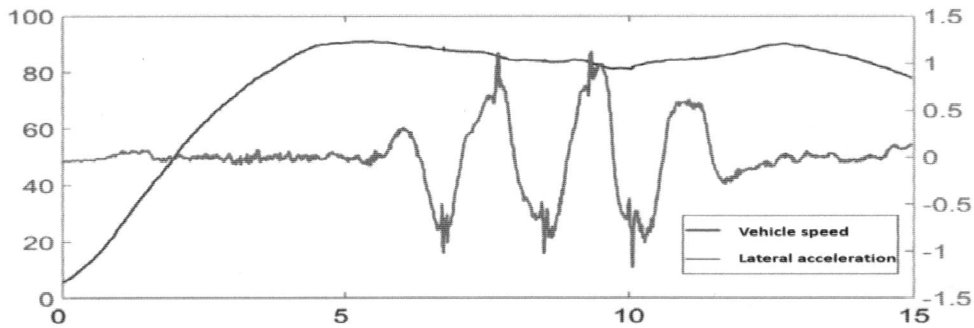

Figure 11. Vehicle speed and lateral acceleration data

Through the analysis of the test data, in the acceleration and deceleration test, the tracking effect of the vehicle to different speed conditions is good, the error is within the allowable range, and the control strategy is robust to the speed. In the high-speed serpentine test, the vehicle lateral acceleration control meets the test requirements, and the maximum lateral acceleration can reach 1m/s2. At the same time, for high-speed operation, selecting appropriate MPC parameters is very important for the actual tracking effect, and adjusting parameters can achieve better control effect.

5. Conclusion

Based on the improved MPC control strategy, this paper designs the transverse and longitudinal control system of the automatic driving vehicle, and proves the feasibility of the system through the actual vehicle test. In this paper, firstly, the kinematics and dynamics models of the autonomous vehicle are established, and the vehicle longitudinal and lateral control strategies are designed by proposing an improved model predictive control algorithm, and the preview tracking control algorithm is introduced to further ensure control accuracy. The control algorithm is developed based on ubuntu environment, written in C++language, and verified by the actual vehicle test.

The results of acceleration and deceleration test and high-speed serpentine test show that the control strategy can not only meet the accurate tracking of the desired speed in longitudinal control, but also meet the requirements of lateral acceleration in lateral control, and has a good control effect on both lateral control and longitudinal control.

References

[1] Jie L, Hai L. An algorithm using predictive control theory for an automatic drive[C]// Machine Learning and Cybernetics, 2002. Proceedings. 2002 International Conference on. IEEE Xplore, 2002.
[2] Wang Z, Zhang Q, Jia T, et al. Research on the control algorithm of an automatic driving robot based on improved model predictive control[J]. Journal of Physics: Conference Series, 2021, 1920(1):012116 (6pp).
[3] Tashiro T. Vehicle steering control with MPC for target trajectory tracking of autonomous reverse parking[C]// Control Applications (CCA), 2013 IEEE International Conference on. IEEE, 2013.
[4] Manzoor T, Xia Y, Zhai D H, et al. Trajectory tracking control of a VTOL unmanned aerial vehicle using offset-free tracking MPC[J]. Chinese Journal of Aeronautics, 2020, 33(7):2024-2042.
[5] Wang Z, Tan D, Ge G, et al. Optimal Trajectory Planning and Control for Automatic Lane Change of in-Wheel Motor Driving Vehicles on Snow and Ice Roads[J]. Automatic Control and Computer Sciences, 2020, 54(5):432-445.
[6] Qi L, Jiao X, Wang Z. Trajectory Tracking Control of Intelligent Driving Vehicle Based on Model Predictive Control[C]// 2020 Chinese Automation Congress (CAC). 2020.
[7] Lukman Yudokusumo, Lukman Yudokusumo, IG. Prasetya Dwi Wibawa, et al. Design and implementation program identification of traffic form in self-driving car robot[J]. Journal of Physics: Conference Series,2019,1367(1):012067 (6pp).
[8] Hoffmann G M, Tomlin C J, Montemerlo M, et al. Autonomous automobile trajectory tracking for off-road driving: Controller design, experimental validation, and racing[C]// American Control Conference, 2007. ACC '07. IEEE, 2007.
[9] Zhang C, Ji X, Zhang B, et al. Trajectory Tracking Control of Intelligent Car Based on Fuzzy Control[J]. Automobile Applied Technology, 2019.
[10] Kang D H, Bong J H, Park J, et al. Reinforcement Learning Strategy for Automatic Control of Real-time Obstacle Avoidance based on Vehicle Dynamics[J]. The Journal of Korea Robotics Society, 2017, 12(3):297-305.

Effectiveness Evaluation of Medium-sized Carrier Rockets Based on ADC Method

Jianjun Wang[1], Zhensen Xu[2], Tao Zhou[1], Jun Zhang[2,*] and Xiangqing Wu[2]

[1] China Academy of Launch Vehicle Technology (CALT), Beijing, 100076, China;

[2] College of Aerospace Engineering, Nanjing University of Aeronautics and Astronautics, Nanjing, Jiangsu, 210016, China

E-mail: zhangjunrdf@nuaa.edu.cn

Abstract. Using scientific evaluation methods to evaluate the mission effectiveness and cost-efficiency of rockets is one of the important topics in the development of rocket models. Aiming at the problem that the effectiveness evaluation of carrier rockets cannot be quantitatively analyzed, this paper takes the medium-sized carrier rockets, covering both new and old generation ones in service, as the object of establishing a simplified inherent capability index system. The ADC method is utilized to quantitatively evaluate the availabilities, dependabilities, and capabilities of various types of rockets and calculates the corresponding effectiveness and cost efficiencies when performing launch missions in LEO, GTO, and SSO orbits. The results show that Model II has excellent mission effectiveness, which is particularly suitable for medium and high orbit missions; the cost-efficiency of Model III is as good as that of the former, and it can complete many tasks; but Model I is outdated.

1. Introduction

Nowadays, the significance and application value of the systematic development of weapons and equipment have become increasingly prominent, which has become an extremely important and challenging frontier topic. Effectiveness evaluation plays an important role in the process of equipment system construction, which greatly boosts the improvement and development of equipment systems.

The so-called effectiveness refers to the completion level of a given task in a typical scenario for a system and the corresponding research covers two basic issues, i.e., determining the effectiveness criteria and establishing the mathematical model of effectiveness [1]. There are many methods of effectiveness evaluation, such as the Support Vector Machines (SVM) method [2], the Technique for Order Preference by Similarity to an Ideal Solution (TOPSIS) method [3,4], the Multiple Criteria Decision Making (MCDM) method [5], the Analytic Network Process (ANP) method [6], Availability, Dependability, and Capability (ADC) method [7], etc. ADC is one of the most widely applied methods for evaluating system effectiveness [8]. A large number of research works have scientifically modeled and evaluated different application scenarios by employing the ADC method [9,10]. However, there is still a lack of comparative analysis of similar equipment or factors, which cannot accurately reflect the real situation or effectiveness improvement of the system.

Taking Alliance-2 (Model I), Falcon-9 (Model II, without recycling), and Ariane-5 (Model III) medium-sized carrier rockets as the research object, we set up three typical orbital launch scenarios, such as Low Earth Orbit (LEO), Geostationary Transfer Orbit (GTO), and Sun-synchronous orbit (SSO), and then use the ADC method to evaluate the effectiveness of rockets in different orbits, and finally calculate and analyze the evaluated contributions of each type of the medium-sized carrier rocket system.

2. Evaluation model of the conventional launch mission

The ADC method is proposed by the Weapons Systems Effectiveness Industry Advisory Committee (WSEIAC) [8]. Two important factors of the ADC method, including the scientific nature of the algorithm and the application prospect, are comprehensively considered. It uses the availability of rockets before launch **A**, the dependability of rockets during flying **D**, and the inherent capability of rockets **C** to conduct comprehensive effectiveness evaluation **E** of the medium-sized carrier rocket system in carrying out conventional launch missions [11], namely:

$$\mathbf{E}=\mathbf{A}\cdot\mathbf{D}\cdot\mathbf{C} \tag{1}$$

As for the conventional launch mission in medium and high orbit, the evaluation of the system capability of the carrier rockets covers three subsystems, including the system effectiveness of the measurement and control system, the load system, and the rocket system [12].

2.1. Availability

The initial working state of the carrier rocket system is reflected by an availability vector, and different elements in the vector represent the probability that each subsystem is under different working situations. Then the availability vector of the equipment system composed of three subsystems can be expressed as:

$$\mathbf{A}=[a_1,a_2,\cdots,a_8] \tag{2}$$

where a_i is the probability that the system is when the task starts under the ith situation.

Assume that each subsystem has two working states when it starts to perform tasks, i.e., normal state N and fault state F. A parameter b_j^s is denoted as the probability of the system being in the normal working state when the system starts to execute the task. According to the availability theory, b_j^s can be expressed as:

$$b_j^s=\begin{cases}\frac{\text{MTBF}}{\text{MTBF+MTTR}}, & s=N\\ \frac{\text{MTTR}}{\text{MTBF+MTTR}}, & s=F\end{cases};\ j=\{c,\ z,\ h\} \tag{3}$$

where MTBF is the mean time interval between failures of subsystems, and MTTR is the mean time interval taken to repair the failure of the subsystem. If the availabilities of the three aforementioned subsystems are b_c^s, b_z^s, and b_h^s respectively, then the probability that the system a_i is given as:

$$a_i = b_c^s \times b_z^s \times b_h^s,\ i=\{1,2,\ldots,8\};\ s=\{N,\ F\} \tag{4}$$

Therefore, the carrier rocket system has eight states before launch as shown in Table 1:

Table 1. Statistics of initial statuses before launch.

States	b_c^s	b_z^s	b_h^s
a1	N	N	N
a2	N	N	F
a3	N	F	N
a4	N	F	F
a5	F	N	N
a6	F	N	F
a7	F	F	N
a8	F	F	F

2.2. Dependability

The dependability matrix **D** represents the state transition of the system during the task execution. This matrix takes the initial working state of the system as the benchmark and other states that may occur in the task of the system as the transfer object. Assume that there are n possible states of the system when working, and the system is in the ith ($i \leq n$) state when it starts. Then the system ends with the original i^{th} state or other n-1 possible states. Therefore, the dependability matrix **D** is an n-order matrix.

There are three stages during the flight of a carrier rocket. Stage I is from the take-off through ignition to the separation of the first part of the rocket. Stage II is from the separation of the first part to the separation of the second part of the rocket. Stage III is the separation of the second part to the separation of the satellite from the rocket [13]. Then the failure rate λ of the measurement and control system obtained from the availability model is written as:

$$\lambda=\begin{cases}\frac{1}{MTBF} & \text{for the measurement and control system}\\ 1-a_k & \text{for the load system and the rocket system}\end{cases} \tag{5}$$

where a_k is the reliability of the load system and rocket system, and T is the working duration of the subsystem in each flight stage.

The failure rate follows an exponential distribution. Denote p_c^s, p_z^s, p_h^s as the failure probabilities of the measurement and control system, the load system, and the rocket system during execution missions. Then it is achieved that:

$$p_i^s = 1-e^{-\lambda_i T_i} \text{ , } i = \{c, z, h\} \tag{6}$$

If d_{ij} is assigned as the probability that the system is in the a_i state at the beginning and in the a_j state at the end, then it can be reasonably given as:

$$d_{ij} = p_c^s \times p_z^s \times p_h^s \text{ , } i = \{1,2,\ldots,8\}; s = \{T, F\} \tag{7}$$

Since the carrier rocket cannot be repaired during flight, the elements below the diagonal of the matrix are all set as 0. The state of the subsystem may change during work. The change from the initial state to the final state of the subsystem is generally discussed. The dependability matrix $\mathbf{D_i}$ of the system during the ith flight stage of the carrier rocket can be expressed as:

$$\mathbf{D_i}=\begin{bmatrix} d_{11} & d_{12} & \cdots & d_{18}\\ 0 & d_{22} & & d_{28}\\ \vdots & & \ddots & \vdots\\ 0 & 0 & \cdots & d_{88}\end{bmatrix} \text{ , } i = \{1,2,3\} \tag{8}$$

Then the dependability of the carrier rocket during the full flight is given as:

$$\mathbf{D} = \mathbf{D_1} \times \mathbf{D_2} \times \mathbf{D_3} \tag{9}$$

2.3. Capability

The capability of the system to complete tasks is characterized by the inherent capability **C**. It is assumed that the inherent capabilities of the measurement and control system and the load system fully meet the requirements of conventional launch tasks in this paper. Then the inherent capabilities of the equipment are expressed by the capabilities of the rocket system (only when all subsystems work normally can the task be completed):

$$\mathbf{C^T}=[C_{TX}, 0, 0, 0, 0, 0, 0, 0] \tag{10}$$

Based on the task scenario during launch, the satisfaction degree of the current carrier rocket system capability is evaluated according to the launch requirements of the carrier rocket system in LEO, GTO, and SSO orbits. Select key indicators such as effective load, orbit entry accuracy, and fairing envelope to calculate the inherent capability C_{TX} of the rocket system and simplify the evaluation index system [11] as shown in Figure 1.

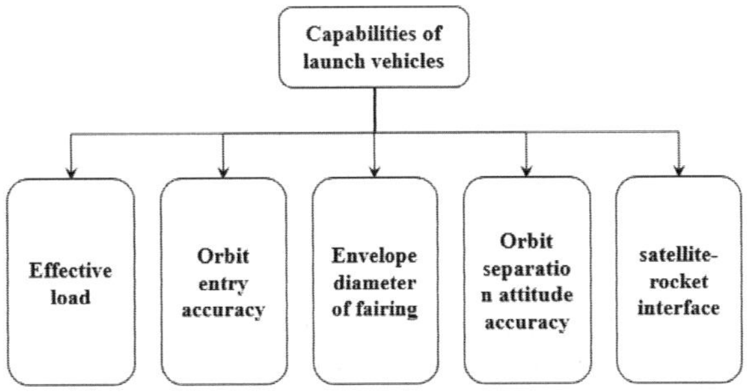

Figure 1. The figure of simplified capability evaluation index system.

The calculation equation is given as:

$$C_{TX} = C_y \cdot \left(C_{pr}+C_{fa}+C_{fl}+C_{xj}\right) \tag{11}$$

where C_y, C_{pr}, C_{fa}, C_{fl}, and C_{xj} are satisfaction indexes of the effective load, the orbital accuracy, the fairing envelope, the separation accuracy, and the satellite-rocket interface, respectively.

2.3.1. Satisfaction index of effective load.

The calculation equation for the effective load demand of the carrier rocket is given as:

$$C_y=\frac{\overline{C}_i}{\overline{C}} \tag{12}$$

where $\overline{C}$ and $\overline{C}_i$, respectively, are the effective load required for a launch mission and the effective load of rockets in the unit of the ton.

2.3.2. Satisfaction index of orbit entry accuracy.

Different evaluation standards of satisfaction index of orbit entry accuracy are built up according to different types of orbits. Therefore, calculation models of the satisfaction index of orbit entry accuracy in the LEO, GTO, and SSO are also different.

For the LEO, the calculation model is given as:

$$C_{pr}=\frac{\frac{53}{|\Delta a|}+\frac{35}{|\Delta h_p|}+\frac{0.17}{|\Delta i|}}{3} \tag{13}$$

For the GTO, the calculation model is given as:

$$C_{pr}=\frac{\frac{75}{|\Delta a|}+\frac{30}{|\Delta h_p|}+\frac{0.21}{|\Delta i|}}{3} \tag{14}$$

For the SSO, the calculation model is given as:

$$C_{pr}=\frac{\frac{5}{|\Delta a|}+\frac{0.003}{|\Delta e|}+\frac{0.08}{|\Delta i|}}{3} \tag{15}$$

where $|\Delta a|$ is the deviation of the semi-major axis in the unit of km, $|\Delta h_p|$ is the deviation of the perigee altitude in the unit of km, $|\Delta i|$ is the deviation of the track inclination in the unit of °, and $|\Delta e|$ is the eccentricity ratio.

2.3.3. Satisfaction index of fairing envelope.

Envelope evaluation of fairing varies with different orbits and different effective loads. The satisfaction indexes of fairing envelope under different orbit launch missions follow the exponential relationship as shown in Equations (16-18).

For rockets with a load capacity of 2000kg and below, the satisfaction index is given as:

$$C_{fa} = -0.089\cdot(C_{zl}-3.35)^2+1 \tag{16}$$

For rockets with a load capacity between 2000kg and 4000kg, the satisfaction index is given as:

$$C_{fa} = -0.057\cdot(C_{zl}-4.20)^2+1 \tag{17}$$

For rockets with a load capacity of 4000kg and above, the satisfaction index is given as:

$$C_{fa} = -0.037\cdot(C_{zl}-5.20)^2+1 \tag{18}$$

where C_{zl} is the diameter of the fairing envelope with the unit of m.

2.3.4. Satisfaction index of separation accuracy. The attitude accuracy index of the carrier rocket is given as:

$$C_{fl}=\frac{\frac{|v_{fl}|}{0.5}+\frac{2}{p_{zt}}+\frac{1}{|\omega_\theta|}+\frac{1}{|\omega_\psi|}+\frac{0.5}{|\omega_\varphi|}}{5} \tag{19}$$

where $\mathbf{v_{fl}}$ is the relative separation velocity in the unit of $m\cdot s^{-1}$, p_{zt} is position accuracy in the unit of °, $\boldsymbol{\omega_\theta}$,$\boldsymbol{\omega_\psi}$, and $\boldsymbol{\omega_\varphi}$ are the pitch, yaw, and roll direction per unit time under the deviation of attitude angular velocity, respectively, with the unit of $°\cdot s^{-1}$.

2.3.5. Satisfaction index of the satellite-rocket interface. The satisfaction index of the satellite-rocket interface is given as:

$$C_{xj}=\frac{\frac{6.1}{g_k}+\frac{4000}{k_d}+\frac{k_j}{2}}{3} \tag{20}$$

where g_k is the maximum overload factor, k_d is the separation shock response with the unit of g, k_j refers to the type of mechanical interface, including standard interface and special-shaped interface.

2.4. Cost efficiency evaluation

Although the benefit estimation of carrier rocket transportation is comprehensive, multiple, and difficult to quantify, cost efficiency is an important indicator to characterize the performance of carrier rockets. In this paper, the cost efficiency R is achieved based on the following equation:

$$R=\frac{E}{P} \tag{21}$$

where E is the effectiveness of the carrier rocket system. In addition, P is the cost with the unit of 100 million CNY, which is obtained from the following relationship [14]:

$$P=P_{rocket}+P_{launch}+P_{others} \tag{22}$$

where P_{rocket} is the cost of the rocket. The purchase price of a single rocket is jointly determined by the purchaser and the manufacturer. P_{launch} is the investment for organizing one launch, including the payments for the launch site, the propellant, the gas consumption, the manpower, the measurement and control, the measurement ship, the landing area evacuation, etc. In addition, P_{others} is other costs covering the landing safety control, the evacuation compensation, the launch insurance, etc.

3. Case analysis

According to the demand analysis of the launch mission, typical mission scenarios are designed based on different target orbits. This paper mainly analyzes three scenarios, including the LEO, GTO, and SSO. Based on those three orbits, an evaluation of the system contribution rates of three typical rockets in service is conducted in this work[15]. [1]

3.1. Availability evaluation

Data of the MTBF and the MTTR of measurement and control system and data of the reliability of the load system and the rocket system in the carrier rocket system are summarized in Table 2 and Table 3, respectively.

Table 2. Data of the MTBF and the MTTR of the measurement and control system.

Rockets	MTBF (h)	MTTR (h)
Model I	21000	221.6
Model II	25000	220.5
Model III	22000	194

[1] Note: Specific information related to the rockets is obtained from online open-accessed sources.

Table 3. Data on the reliability of subsystems in different types of rockets.

Rockets	Reliability of load system	Reliability of rocket power system
Model I	0.980	0.990
Model II	0.980	0.995
Model III	0.980	0.995

Availability matrixes of different load rockets are given as follows.

1) Model I: **A**=[0.969, 0.0102, 0.0102, 0, 0.0102, 0, 0, 0]

2) Model II: **A**=[0.974, 0.0086, 0.0086, 0, 0.0086, 0, 0, 0]

3) Model III: **A**=[0.975, 0.0082, 0.0082, 0, 0.0082, 0, 0, 0]

3.2. Dependability evaluation

According to Equations (5-9), the dependability matrix **D** of different types of carrier rockets during the task execution is achieved as follows:

$$1)\ \text{Model I}: \mathbf{D} = \begin{bmatrix} 0.94 & 0.018 & 0.04 & 0 & 0.002 & 0 & 0 & 0 \\ 0 & 0.959 & 0 & 0.04 & 0 & 0.001 & 0 & 0 \\ 0 & 0 & 0.984 & 0.014 & 0 & 0 & 0.002 & 0 \\ 0 & 0 & 0 & 0.906 & 0 & 0 & 0 & 0.094 \\ 0 & 0 & 0 & 0 & 0.98 & 0.017 & 0.003 & 0 \\ 0 & 0 & 0 & 0 & 0 & 0.9622 & 0 & 0.0378 \\ 0 & 0 & 0 & 0 & 0 & 0 & 0.981 & 0.019 \\ 0 & 0 & 0 & 0 & 0 & 0 & 0 & 1 \end{bmatrix}$$

$$2)\ \text{Model II}: \mathbf{D} = \begin{bmatrix} 0.9315 & 0.0125 & 0.0526 & 0.0004 & 0.003 & 0 & 0 & 0 \\ 0 & 0.943 & 0 & 0.053 & 0 & 0.004 & 0 & 0 \\ 0 & 0 & 0.984 & 0.013 & 0 & 0 & 0.003 & 0 \\ 0 & 0 & 0 & 0.996 & 0 & 0 & 0 & 0.004 \\ 0 & 0 & 0 & 0 & 0.955 & 0.0213 & 0.0237 & 0 \\ 0 & 0 & 0 & 0 & 0 & 0.9469 & 0 & 0.0531 \\ 0 & 0 & 0 & 0 & 0 & 0 & 0.987 & 0.013 \\ 0 & 0 & 0 & 0 & 0 & 0 & 0 & 1 \end{bmatrix}$$

$$3)\ \text{Model III}: \mathbf{D} = \begin{bmatrix} 0.9821 & 0.005 & 0.0129 & 0 & 0 & 0 & 0 & 0 \\ 0 & 0.987 & 0 & 0.011 & 0 & 0.002 & 0 & 0 \\ 0 & 0 & 0.995 & 0.004 & 0 & 0 & 0.001 & 0 \\ 0 & 0 & 0 & 0.908 & 0 & 0 & 0 & 0.092 \\ 0 & 0 & 0 & 0 & 0.989 & 0.002 & 0.009 & 0 \\ 0 & 0 & 0 & 0 & 0 & 0.987 & 0 & 0.013 \\ 0 & 0 & 0 & 0 & 0 & 0 & 0.996 & 0.004 \\ 0 & 0 & 0 & 0 & 0 & 0 & 0 & 1 \end{bmatrix}$$

3.3. Capability evaluation

The statistic of the current medium-sized carrier rocket system is shown in Table 4.

Table 4. Statistic of the current medium-sized carrier rocket system.

Rockets	LEO carrying capacity(ton)	GTO carrying capacity(ton)	SSO carrying capacity(ton)
Model I	3.8	5.5	3.0
Model II	8.2	14.0	9.0
Model III	5.0	7.0	4.5
$\overline{C}$	10	20	10

The orbit entry accuracies of the three rockets are shown in Table 5.

Table 5. Statistic of entry accuracy of different orbits.

Rockets	LEO			GTO			SSO		
	Semi-major axis deviation (km)	Perigee altitude deviation (km)	Orbit inclination deviation (°)	Semi-major axis deviation (km)	Perigee altitude deviation (km)	Orbit inclination deviation (°)	Semi-major axis deviation (km)	Eccentricity	Orbit inclination deviation (°)
Model I	53	35	0.17	75	30	0.21	7	0.003	0.08
Model II	53	35	0.17	75	30	0.21	5	0.003	0.08
Model III	53	35	0.17	75	30	0.21	5	0.003	0.08

The statistic of the envelope diameters of fairings is shown in Table 6.

Table 6. Statistic of the envelope diameters of fairings of carrier rockets.

Rockets	Maximum fairing diameter (m)
Model I	4.2
Model II	5.2
Model III	4.2

The separation attitude accuracies of rockets in different orbit missions are shown in Tables 7-9.

Table 7. Statistics of separation accuracy in LEO.

Rockets	Relative separation speed (m·s^{-1})	Attitude pointing accuracy (°)	Angular velocity deviation (°·s^{-1})		
			Pitch	Yaw	Roll
Model I	0.6	2	1.0	1.0	0.6
Model II	0.8	2	1.0	1.0	0.6
Model III	1.0	2	1.0	1.0	0.6

Table 8. Statistics of separation accuracy in GTO.

Rockets	Relative separation speed (m·s^{-1})	Attitude pointing accuracy (°)	Angular velocity deviation (°·s^{-1})		
			Pitch	Yaw	Roll
Model I	0.5	2	0.5	0.5	0.5
Model II	0.6	2	0.5	0.5	0.5
Model III	0.5	2	1	1	0.5

Table 9. Statistics of separation attitude accuracy in SSO.

Rockets	Relative separation speed (m·s^{-1})	Attitude pointing accuracy (°)	Angular velocity deviation(°·s^{-1})		
			Pitch	Yaw	Roll
Model I	0.5	2	0.5	0.5	0.5
Model II	0.5	2	0.5	0.5	0.5
Model III	0.5	2	1	1	0.5

Table 10 shows the statistic of the satellite-rocket interface of rockets.

Table 10. Statistic of the satellite-rocket interface.

Rockets	Satellite-rocket interface		
	Maximum overload factor	Separation shock response (g)	Number of interface types (types)
Model I	5.3	4000	2
Model II	6.5	6000	2
Model III	6.1	4000	2

1) Capability matrix **C** of different types of carrier rockets executing LEO missions are shown as:
Model I: $\mathbf{C}^T$=[0.3693, 0, 0, 0, 0, 0, 0, 0]
Model II: $\mathbf{C}^T$=[0.7897, 0, 0, 0, 0, 0, 0, 0]
Model III: $\mathbf{C}^T$=[0.519, 0, 0, 0, 0, 0, 0, 0]
2) Capability matrix **C** of different types of launch vehicles executing GTO missions are shown as:
Model I: $\mathbf{C}^T$=[0.38176, 0, 0, 0, 0, 0, 0, 0]
Model II: $\mathbf{C}^T$=[0.963, 0, 0, 0, 0, 0, 0, 0]
Model III: $\mathbf{C}^T$=[0.512, 0, 0, 0, 0, 0, 0, 0]
3) Capability matrix **C** of different types of launch vehicles executing SSO missions are shown as:
Model I: $\mathbf{C}^T$=[0.28468, 0, 0, 0, 0, 0, 0, 0]
Model II: $\mathbf{C}^T$=[0.8462, 0, 0, 0, 0, 0, 0, 0]
Model III: $\mathbf{C}^T$=[0.46, 0, 0, 0, 0, 0, 0, 0]

3.4. Cost efficiency evaluation

The statistic of the cost of rockets performing conventional launch missions is shown in Table 11.

Table 11. Statistics of cost of rockets performing conventional launch missions.

Rockets	**Cost**		
	Manufacturing cost (10,000 CNY)	**Launch cost (10,000 CNY)**	**Other costs (10,000 CNY)**
Model I	28,000	10,000	2,000
Model II	67,000	26,000	4,900
Model III	39,500	10,000	2,600

4. Analysis of system contribution rate

According to the launch demand of three selected medium-sized carrier rockets, this paper evaluates their effectiveness and calculates the corresponding system contribution rates.

4.1. As for LEO Mission

When the carrier rockets execute the LEO mission, the evaluation results of all types of rockets are shown in Figure 2.

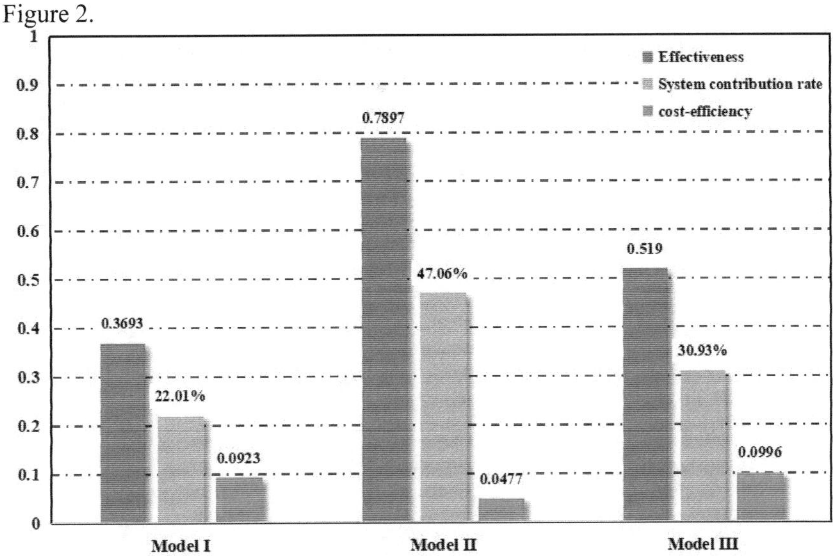

Figure 2. Evaluation results of various rockets executing LEO mission.

As for LEO Mission, it can be seen that (1) the effectiveness of Model III is smaller than that of Model II but higher than that of Model I; (2) the cost efficiency of Model II is the lowest among all the three models. It proves that although Model II is more suitable for the LEO mission, its cost performance needs to be improved.

4.2. As for GTO Mission

When the carrier rockets execute the GTO mission, the evaluation results of all types of rockets are summarized in Figure 3.

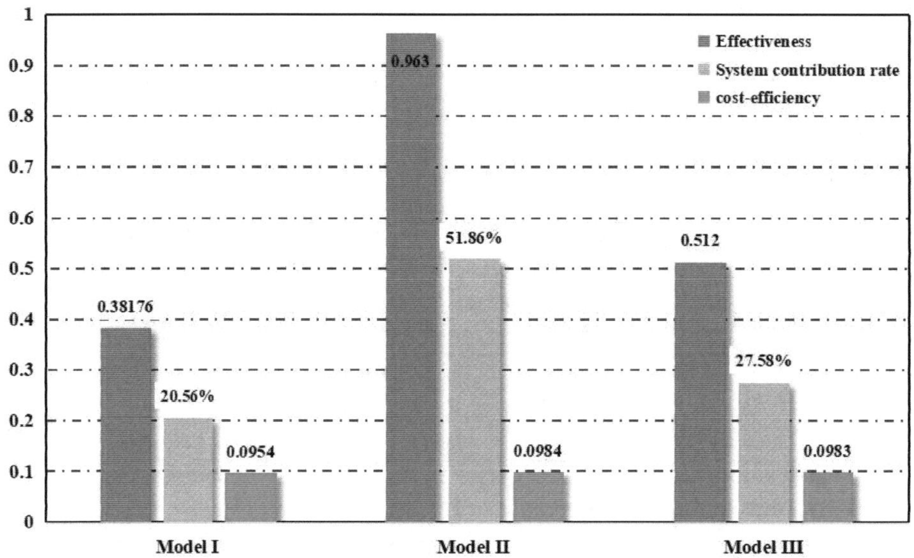

Figure 3. Evaluation results of various rockets executing GTO mission.

As for the GTO mission, it can be seen that (1) the effectiveness of Model I is only 0.38176, with the lowest capability. Although Model III is still not high, it already meets the launch requirements. The effectiveness of Model II is as high as 0.963, which is the highest. Therefore, Model II has a very good foundation to complete the mission; (2) the cost efficiencies of Model II and Model III are almost the same, which are slightly higher than that of Model I; (3) the contribution rate of Model II is the highest, reaching up to 51.86%, which reflects its strongest delivery capacity.

4.3. As for SSO Mission

When the carrier rockets execute the SSO mission, the evaluation results of all types of rockets are drawn in Figure 4:

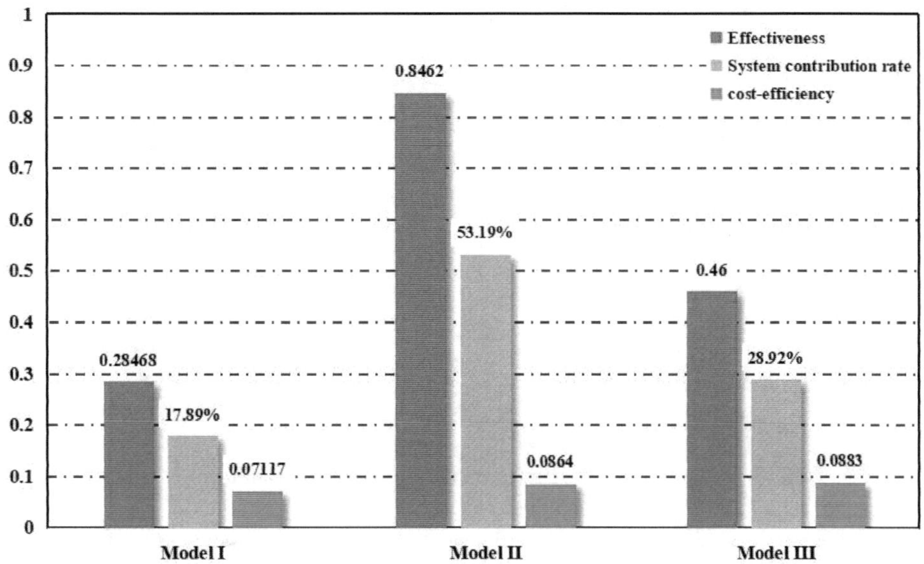

Figure 4. Evaluation results of various rockets executing SSO missions.

As for the SSO mission, it can be seen that (1) the capability of Model I is the poorest with an effectiveness of only 0.28468. The capability of Model III with an effectiveness of 0.46 also lacking in the SSO mission. In comparison, Model II has the highest effectiveness of 0.8462; (2) the system contribution rate of Model III is 28.92%, which is 11.03% higher than that of Model I. In addition, Model III has the highest cost efficiency, indicating that Model III has strong adaptability to orbital launch missions and can complete different launch missions better and faster.

5. Conclusion

Based on the evaluation requirements of the scientific algorithm and the prospect of application scenarios, the ADC method is employed in this work to evaluate the whole launch process of medium-sized carrier rockets. The availabilities, dependabilities, and capabilities of the measurement and control systems, the load systems, and the rocket systems of three types of rockets during executing launch missions in LEO, GTO, and SSO missions are calculated to evaluate the effectiveness, system contribution rates and cost efficiencies of those three rockets. The thorough process of evaluation and analysis provides a full view of the performance indexes of the new generation rockets and quantifies the mission satisfaction of various types of equipment in the medium-sized rocket system. In the future, we should build a more detailed and comprehensive evaluation index system, and adopt different methods to examine rocket capabilities from different aspects.

References

[1] Chen S M 1996 Evaluating weapon systems using fuzzy arithmetic operations *J. Fuzzy Sets and Systems* **77** 265-76

[2] Jiang T, Wang S and Wei R 2007 Study on tradeoffs between weapon system cost and performance based on Support Vector Machine *International Conference on Wireless Communications* **15** 5232–35

[3] İrfan E and Nilsen K 2009 Performance evaluation of Turkish cement firms with fuzzy analytic hierarchy process and TOPSIS methods *Expert Systems with Applications* **36(1)** 702-715

[4] Subrata C 2022 TOPSIS and Modified TOPSIS: A comparative analysis *Decision Analytics Journal* **2** 100-21

[5] Ertugrul E K and Ethem T 2001 Fuzzy multi-criteria decision-making procedure for evaluating advanced manufacturing system investments *International Journal of Production Economics* **69(1)** 49-64

[6] Shyur H J 2006 COTS evaluation using modified TOPSIS and ANP *Applied Mathematics and Computation* **177(1)** 251-59

[7] He Y J, Zhou S P, Liu Y 2017 Efficiency evaluation model of power communication network based on ADC method *Journal of North China Electric Power University (Natural Science Edition)* **44(01)** 76-81

[8] Yue J and Lu D W 2022 Effectiveness evaluation of security control system of the anti-ship missile in high-altitude cruise phase based on ADC method *Fire and Command and Control* **01** 156-61

[9] Zhu C W, Shi W H and Fu Z Q 2022 Firing effectiveness evaluation of ship-to-air missile weapon system with improved ADC method *Naval Electronics Engineering* **05** 104-08

[10] Zhang T Q, Lv W M and Li G 2021 Analysis of air defense system strike effectiveness evaluation based on improved ADC method *Fire and Command and Control* **10** 107-13

[11] Yu X H, Cai Y W, Yang R B 2005 Effectiveness evaluation method of launch vehicle *Missile and Space Launch Technology* **03** 19-22

[12] Yang R B, Cai Y W 2002 Research on the effectiveness evaluation system of launch vehicle *Journal of Equipment Command Technology College* **01** 50-2

[13] Zhong W A, Zhang J X, Li Z B, Zhu L P and Yan Z 2021 Strategy for optimizing the test and launch process of a large launch vehicle *Missile and Space Launch Technology* **05** 85-8

[14] Si Q Y, Ren J T, Zhang L and Fan S M 2022 Exploration on cost management of new-generation launch vehicles *Missile and Space Launch Technology* **02** 144-47

[15] Lu Y 2017 Space Launch Vehicle's Development in China *Aerospace General Technology* **03** 1-8

Fatigue Life Analysis of Bearing in Oscillatory Applications based on Ncode DesignLife

Hailun Ren [1,2], Aiguo Zhou [1], *Jianjun Jia [2];Zhaohui Li [2]

[1] School of Mechanical Engineering, Tongji University, Shanghai 200092, China;

[2]Shanghai Institute of Technical Physics Chinese Academy of Sciences, Shanghai 200083, China

[1]m18435135167@163.com;[2]zhouaiguo@tongji.edu.cn; [3]jjjun10@mail.sitp.ac.cn;[4] lizhaohui@mail.sitp.ac.cn

Abstract: Aiming at the fatigue life of bearing in oscillatory applications, through the analysis of the equivalent load of each part of bearing in oscillatory applications, the equivalent load of each part is calculated. Taking the 7005 angular contact ball bearing as the research object, the stress changes of each component are obtained through the transient dynamic analysis of the model, the fatigue of the bearing is analyzed based on the stress damage theory, and the damage and fatigue life of the bearing is obtained. The results show that the rolling element of the 7005 bearings is the weak part of failure under the given working condition, and it is more reasonable to design the service life of the shafting structure with an equivalent load of the rolling element, which provides reference measures for improving the service life of the oscillated bearing.

1. Introduction

The bearing motion of aerospace optical load is mostly in oscillatory applications, such as in the field of precision measuring instruments, two-dimensional tracking turntable of acquisition, tracking and pointing system (ATP system), laser galvanometer optical scanning system, and optical focusing mechanism. To prolong the service life of space mechanical devices and improve the reliability of mechanisms, it is necessary to study the fatigue reliability and life prediction of bearings in oscillatory conditions.

L. Houpert analyzed the inner and outer rings of the oscillated bearing and calculated the life of the oscillated bearing with the equivalent load of the rotating ring and the life correction coefficient table of the oscillated bearing [1]. At present, some scholars have established an adhesive wear model based on Archard theory and frictional work density for life analysis of solid lubricated bearings under vacuum swing conditions [2][3]. For bearings with good lubrication and sealing under vacuum conditions, the bearing life should still be calculated based on the fatigue failure model. In [4], the Ncode is used to simulate the damage evolution of the thrust bearing raceway. The results were very consistent with the laboratory tests under the same conditions. Still, at the same time, there were a large number of literature on the integral bearing tests, which showed that there were rolling element failures in multiple groups of samples [5][6].

A large number of domestic works of literature on oscillated bearings mainly study the life of solid lubricated bearings and spherical plain bearings, and there is no systematic analysis method for fatigue life prediction of rolling bearings due to fatigue failure. In this paper, through the analysis of the equivalent load of each part of the bearing in oscillatory applications, considering the failure of the rolling element, the calculation formula of the equivalent load of each part is given. Based on the fatigue analysis software Ncode DesignLife, the fatigue life of the bearing is predicted, and the failure fatigue life of the rolling element of the oscillated bearing is verified, which provides a method for the prediction of the fatigue life of the bearing.

2. Bearing life theory and equivalent load analysis

The discussion on the fatigue life model of oscillated bearing in this paper is based on the following assumptions:

1) The failure form of each bearing unit is rolling fatigue failure, without considering the wear and friction corrosion;

2) The motion condition of the bearing is low-speed oscillating, and the rolling body motion follows the inner ring control theory;

2.1. Fatigue failure life theory of bearing

The failure of traditional steel bearings is caused by the maximum shear stress under the surface depth. The crack extends to the surface. The L-P theoretical life calculation formula is:

$$ln\frac{1}{S} = A\frac{{\tau_0}^c N^e V}{{z_0}^h}$$

Where S is the reliability of N million stress cycles of the material; Is the maximum orthogonal shear stress under the material surface; Is the generated depth; V is the stress-bearing volume, e is the Weibull distribution coefficient, representing the discrete degree of the life of a batch of bearings. For steel rolling elements bearings, the value is generally 10/9; c. H is the coefficient related to material and shape

Under the given service probability (such as S=0.9), the bearing life is recorded as L10. After the bearing structural parameters are determined, the life is mainly related to the load and the number of stress cycles. For bearings of the same model, the relationship between their service life and rolling element load [5] is:

$$L \propto (\frac{1}{Q})^{(c-h+1)/2e}$$

After the bearing bears the axial load and the radial load, each rolling element is loaded differently. When the bearing is running, the outer ring is usually fixed and the inner ring rotates with the shaft. Therefore, when calculating the bearing life, it is necessary to use an equivalent load to simplify the load on the inner ring under different working conditions.

According to the classical L-P theory of bearing life, when the rolling element contacts with the inner and outer rings, the ratio of the rated dynamic load of the rolling element to its average load is equivalent to the ratio of the rated dynamic load of the inner and outer rings to its equivalent dynamic load. This approximately assumes that the rated dynamic load and equivalent dynamic load of the rolling element are the same as those of the inner and outer rings and that the rolling element is only used as an intermediate medium to calculate the life of the inner and outer rings when calculating the whole bearing life, and the life of the rolling element is not considered separately.

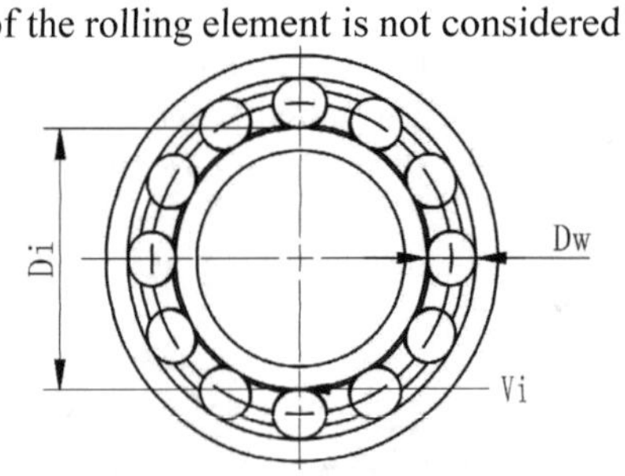

Figure 1. Schematic Diagram of Bearing

According to the analysis of the speed of each part in the bearing schematic diagram in Figure 1, the revolution speed of the bearing rolling element, that is, the speed of the rolling element center, is equal to the average speed of the two contact points of the inner and outer rings,

$$v_{pw} = \frac{v_e + v_i}{2}$$

When the outer ring of the bearing is fixed and the inner ring rotates completely, $v_e = 0$, the revolution speed of the rolling element is only half of the inner circle speed.

When the included angle between rolling elements is γ, the relationship between rolling element rotation amplitude $2\theta_w$ and inner ring oscillatory amplitude 2θ is:

$$\theta_w = \frac{D_i}{D_w}\theta \tag{1}$$

As shown in Figure 1, D_i is the diameter at the contact point of the inner ring; D_w is the diameter of the rolling element;

When the full-bearing ring is loaded, the number of stress cycles of inner and outer rings and rolling elements is as follows:

- When $2(n-1)\gamma \le \theta < 2n\gamma$, the number of stress cycles applied to the contact area of the inner ring in a motion cycle is 2n, where, $n \ge 1$ and n is an integer;
- When $n\pi \le \theta_w < (n+1)\pi$, the number of stress cycles applied to the contact area of the outer ring in a motion cycle is n+1, where $n \ge 0$ and n is an integer;
- The number of stress cycles in the rolling element contact area in a motion cycle is $\frac{2\theta_w}{\pi}$.

When the number of stress cycles on the inner ring and each rolling element in a cycle is the same, the revolution amplitude of the rolling element is smaller than the inner ring, the equivalent load of the rolling element bearing a larger load is higher than the inner ring, and the rolling element has a higher failure probability. In specific cases, the number of cycles on the corresponding unit can be obtained through bearing parameters and load distribution under working conditions for analysis.

2.2. *Equivalent load analysis*

When the reliability of each point is the same, the calculation of equivalent load is mainly based on Miner's linear cumulative damage criterion; When the reliability of each point is different, it is usually necessary to calculate the product or logarithmic sum of the reliability of each point. When the bearing rotates and oscillates, the calculation method of the equivalent load of the rolling element and outer ring is the same, but the working condition parameters differ. Therefore, this section analyses the equivalent load of the inner ring and outer ring of the bearing with continuous rotation and the inner ring and rolling element of the oscillated bearing, respectively.

2.2.1. *Equivalent load of continuously rotating bearing inner ring*

The inner ring of the bearing with continuous rotation uniformly bears the action of cyclic variable load, so any point on the inner ring has the same reliability during its service life. According to Miner's linear cumulative damage criterion, the total number of stress cycles of the inner ring under different loads is equal to the number of stress cycles of an equivalent load under the same service life:

$$Q_i Z L = \sum Q^s(\varphi) L \tag{2}$$

Where Q_i is the equivalent load acting on the bearing integral inner ring and rolling element; The **s** is the exponential coefficient related to material fatigue.

$$Q_i = [\frac{\sum Q^s(\varphi)}{Z}]^{\frac{1}{s}} \tag{3}$$

$$Q_i = Q_{max} J_1 \tag{4}$$

$$J_1 = \left\{\frac{1}{2\pi}\int_{-\varphi_0}^{\varphi_0} [1 - \frac{1-cos\varphi}{2T}]^{ts} d\varphi\right\}^{\frac{1}{s}} \tag{5}$$

Where J_1 is the integral of the equivalent load acting on the inner race and rolling element of the bearing with continuous rotation

2.2.2. Equivalent load of continuously rotating bearing outer ring

The bearing outer ring is usually fixed in the bearing seat and does not rotate with the shaft. Its load distribution is certain. The reliability of each point of the inner and outer rings is different during the bearing life. When calculating the equivalent load acting on the continuously rotating bearing outer ring and the rolling element, the sum of the reliability logarithms of each point of the outer ring is required.

$$\int_{-\varphi_0}^{\varphi_0} Q(\varphi)^{\omega} L^{e} \frac{D_e}{2} d\varphi = \pi Q_e{}^{\omega} L^{e} D_e \tag{6}$$

$$Q_e = [\frac{1}{2\pi}\int_{-\varphi_0}^{\varphi_0} Q(\varphi)^{\omega}\, d\varphi]^{\frac{1}{\omega}} \tag{7}$$

$$Q_e = Q_{max} J_2 \tag{8}$$

$$J_2 = \left\{\frac{1}{2\pi}\int_{-\varphi_0}^{\varphi_0} [1 - \frac{1-cos\varphi}{2T}]^{t\omega} d\varphi\right\}^{\frac{1}{\omega}} \tag{9}$$

Q_e is the equivalent load acting on the bearing integral inner ring and rolling element; and for point contact, $\omega = \frac{c-h+2}{3}$; J_2 is the integral of the equivalent load acting on the outer ring and rolling element of the bearing with continuous rotation.

2.2.3. Equivalent load of oscillated bearing inner race

When the outer ring is fixed, and the inner ring oscillates, the load and stress cycles on each part of the cycle are shown in Table 1:

Table 1. Load and Stress Cycle of Oscillated bearing

	Inner ring		Outer ring		Rolling elements	
	Load	Number of stress cycles at each point	Load	Number of stress cycles at each point	Load	Number of stress cycles at each point
Full revolution	Variable load and all points are the same	Same	Constant load	Determined by load distribution	Variable load and all points are the same	Same
Swing	Variable load and all points are different	Determined by load distribution	Constant load	Determined by load distribution	Variable load and all points are different	Determined by load distribution

Therefore, the equivalent load and life of the oscillated bearing cannot be calculated in the same way as that of the bearing with continuous rotation. Taking the time when the swing speed is maximum as the initial time, 2θ is the total oscillatory angle of the inner race. First, according to Miner's fatigue damage accumulation hypothesis, the variable load of each point on the inner ring of the oscillatory bearing in each cycle is:

$$Q_{si} = [\frac{\sum Q^s(\varphi)}{\frac{\theta}{\pi}Z}]^{\frac{1}{s}} \tag{10}$$

Any under fixed coordinates, the equivalent load on the inner ring at φ is:

$$Q_{si}(\varphi) = Q_{max}\left\{\frac{1}{2\theta}\int_{\varphi-\theta}^{\varphi+\theta} [1 - \frac{1-cos\varphi}{2T}]^{ts} d\varphi\right\}^{\frac{1}{s}} \tag{11}$$

The inner ring keeps swinging and the equivalent load on the inner ring $Q_i(\varphi)$ at φ can be regarded as a fixed value, so the reliability of each point is different, and the reliability of the inner ring is equal to the logarithm sum of the reliability of each point:

$$\int_{-\varphi_0}^{\varphi_0} Q(\varphi)^{\omega} L^{e} \frac{D_e}{2} d\varphi = \pi Q_e{}^{\omega} L^{e} D_e \tag{12}$$

$$Q_{si}{}^{\omega}L^{e}D_{i}\pi = L^{e}\frac{D_{e}}{2}\int_{-\varphi_0}^{\varphi_0} Q(\varphi)^{\omega}d\varphi \tag{13}$$

Q_{si} is the equivalent load acting on the oscillatory inner ring of the bearing and the rolling element;

$$Q_{si} = Q_{max}J_{s1} \tag{14}$$

$$J_{s1} = \left\{\frac{1}{2\pi}\int_{-\varphi_0}^{\varphi_0}\left[\frac{1}{2\theta}\int_{\varphi-\theta}^{\varphi+\theta}[1-\frac{1-cos\varphi}{2T}]^{ts}d\varphi\right]^{\frac{\omega}{s}}d\varphi\right\}^{\frac{1}{\omega}} \tag{15}$$

J_{s1} is the integral of the equivalent load acting on the inner ring and rolling element of the oscillatory bearing

2.2.4. Equivalent load of the rolling element of oscillatory bearing

Taking the rolling element bearing the maximum radial load as an example, the theoretical life of the rolling element is the minimum. When calculating the L_{10} life of all rolling elements, the overall equivalent stress can be calculated by the sum of the reliability logarithms of each rolling element. The rolling body is simultaneously subjected to the load of the inner and outer race raceways, and the number of stress cycles on the contact area in a motion cycle is $\frac{2\theta_w}{\pi}$. According to Miner's rule:

$$Q_{sw} = \left[\frac{2\theta_w}{\pi}\sum Q_w{}^{s}(\varphi)\right]^{\frac{1}{s}} \tag{16}$$

Each cycle is regarded as an equivalent load Q_{sw} cycle once, $Q_w(\varphi)$ cycle 2n+2 times.

Q_{sw} is the equivalent load of the rolling element when the bearing oscillates

$$Q_{sw}(\varphi) = Q_{max}J_{w} \tag{17}$$

$$J_{w} = \left\{\frac{2\theta_w}{\pi}\int_{\varphi-\frac{\theta}{2}}^{\varphi+\frac{\theta}{2}}[1-\frac{1-cos\varphi}{2T}]^{ts}d\varphi\right\}^{\frac{1}{s}} \tag{18}$$

J_{w} is the integral of an equivalent load of a rolling element of oscillatory bearing

3. Mechanical Model and Analysis

Transient dynamic analysis of 7005 angular contact ball bearing is carried out by using the Transient Structural module in the workbench. Its geometric parameters are shown in Table 2.

Table 2. Model Parameters

Inner diameter（mm）	25
Outer diameter（mm）	47
Rolling elements diameter（mm）	6.325
Thickness（mm）	12
Contact angle（°）	25
Number of rolling elements	12

Bearing assembly usually adopts the assembly method of fixing the outer ring and rotating the inner ring with the shaft. The inner ring is the main part of bearing failure. The inner ring grid unit size is set to be 1.5mm, and the rolling body is divided into grids according to the surface, with a size of 1mm.

During the transient dynamic analysis of the bearing, the friction contact between each rolling element and the inner and outer race raceways is set, and the coefficient is 0.5; The friction contact coefficient with the cage is 0.1; The cage and inner ring distribution, and outer ring are equipped with a rotary motion pair. The radial force is usually used as the accelerated stress in the bearing life acceleration test, so a radial force of 500N is applied to the inner ring of the bearing, and the oscillatory displacement of the inner ring is set as a time function in rad, that is, the oscillatory period of the bearing is 1s, and the oscillatory amplitude is 1rad. According to the analysis of the number of stress cycles in 2.1, under the set working conditions, the inner ring is subjected to 2 stress cycles, and the rolling elements are subjected to 10/π stress cycles in a motion cycle. The stress cloud diagram of the bearing is shown in Figure 2.

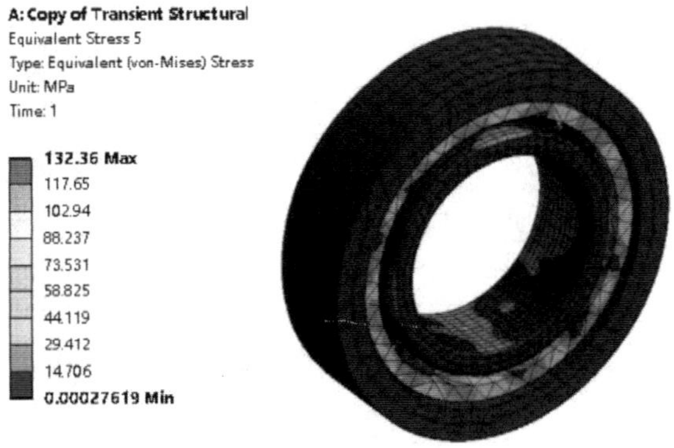

Figure 2. Overall Stress Diagram of Bearing

4. Fatigue Analysis Results

NCode DesignLife is a module integrated into the Workbench software [7], which can directly read the stress changes of each unit in a cycle obtained from the transient dynamic analysis. This paper uses Workbench 2021, and the project data sharing created is shown in Figure 3. This version combines the time step load analysis module with the time series load analysis module, and you need to modify the loading method in the load settings.

The time step load is the cyclic loading of the input one-cycle stress change, and the stress fatigue life of each unit is obtained according to the material S-N curve [8].

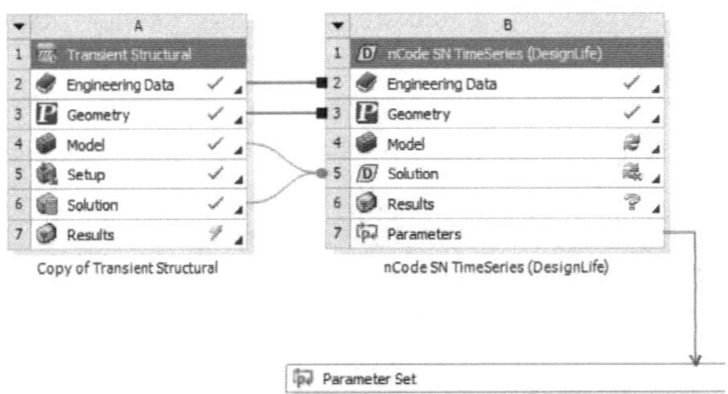

Figure 3. Creating Project Data Sharing

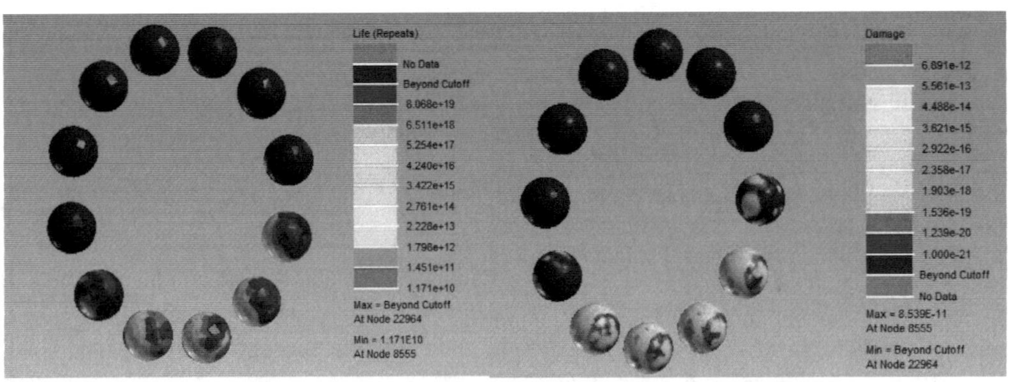

Figure 4. Cloud Picture of Life and Damage of Rollers

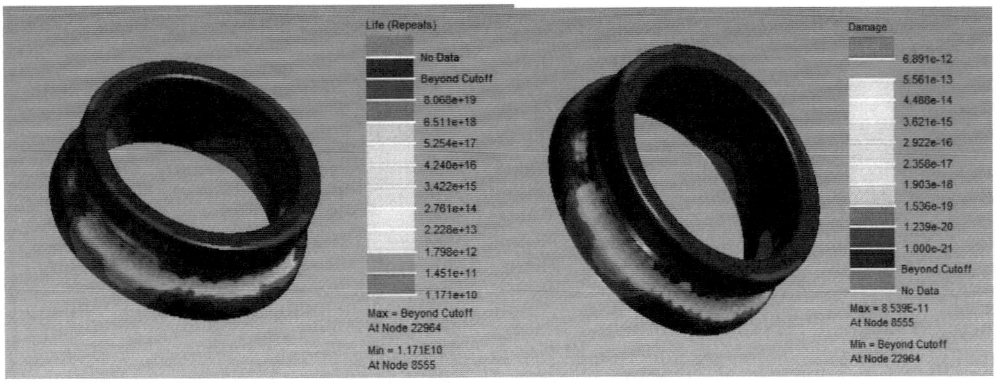

Figure 5. Cloud Picture of Life and Damage of Inner Ring

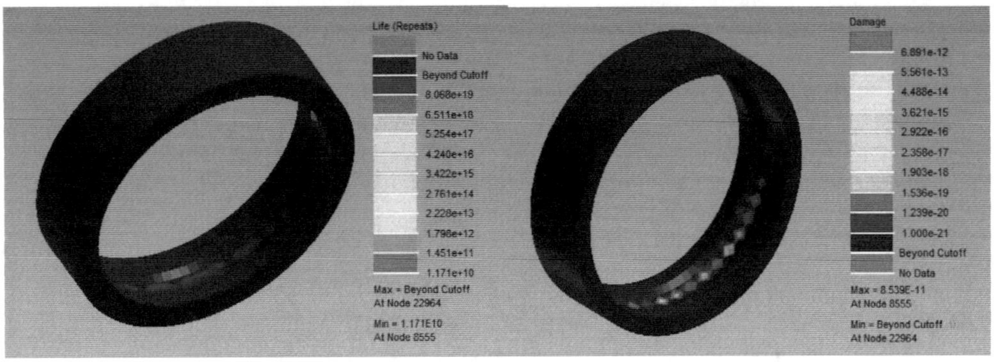

Figure 6. Cloud Picture of Life and Damage of Outer Ring

Figures 4, 5 and 6 show the life and damage distribution of bearing rolling element, inner ring and outer ring respectively, which directly shows the life distribution and damage distribution of each node of the part. The greater the load on each point, the smaller the life. If you want to obtain the overall reliability of a single part, you need the reliable product of each point.

5. Conclusion

In this paper, through the analysis of the equivalent load of each part of the bearing in oscillatory applications, it is concluded that the weak part of the fatigue failure of the bearing in oscillatory applications is related to the structural parameters of the bearing (pitch diameter of the contact point of the inner and outer rings, diameter of the rolling element) and oscillatory amplitude, and the calculation formula of the equivalent load of each part is given;

For 7005 bearing, a three-dimensional bearing model was established by UG. Combined with Workbench and nCode DesignLife, the fatigue simulation analysis of the bearing was carried out, and the bearing's stress distribution, damage, and life distribution were obtained. The results show that the weak part of the fatigue damage failure of the bearing is the rolling element in the load distribution area. The service life of the outer ring of the bearing is higher than that of the inner ring in oscillatory applications, and the service life of the inner ring is higher than that of some rolling elements. The analysis of the structure shows that this is mainly because the equivalent load of the rolling element in the load distribution area is greater than that of the inner ring in a cycle, and it bears more stress cycles than that of the outer ring.

To improve the fatigue strength of shafts and bearings, structural optimization can be carried out to improve the service life of bearings. For example, large angle rotation can be properly added in the oscillation to rotate the rolling elements in the load distribution area; Materials with better SN curves are selected, appropriate surface heat treatment measures are adopted to improve surface hardness, and lubrication of bearing parts is maintained.

The equivalent load calculation and fatigue simulation analysis in this paper can provide a theoretical basis for the fatigue design of the oscillated bearing and provide reference measures for the improvement of the durability, which has practical engineering significance.

Acknowledgments
This work was financially supported by the Quantum information technology integration test platform of the Shanghai Municipal Science and Technology Major Project (2019SHZDZX01-ZX08).

References

[1] L. Houpert. (1999). Bearing life calculation in oscillatory applications. Tribology Transactions, 42(1), 136-143.
[2] Jiang Ziqing, Jia Jianjun. Life analysis of space solid lubricated ball bearings under low-speed and small-angel-swing[J]. Science Technology and Engineering, 2018,18(31):17-22.
[3] Fabian Schwack, Volker Schneider, Sebastian Wandel, Román José de la Presilla, Gerhard Poll, Sergei Glavatskih. On the critical amplitude in oscillating rolling element bearings[J]. Tribology International, 2021.107154
[4] M.Y. Toumi, S. Murer, F. Bogard, F. Bolaers. Numerical simulation and experimental comparison of flaw evolution on a bearing raceway: Case of thrust ball bearing[J]. Journal of Computational Design and Engineering, 2018,5(4): 427-434.
[5] Zhao Qing. Fatigue life test and failure behavior analysis of bearing in space motor [D]. Harbin Institute of Technology.2010.
[6] Wang Chunliang. Remaining Life Prediction and Experimental Research of Metro Traction Motor Bearing[D]. Henan University of Science and Technology.2020.
[7] Liu Wenwei, Cheng Keqiang, WangYuanhang, etc. Research on the Durability Life Simulation of the motor based on nCode DesignLife[J]. Electronic Product Reliability and Environmental Testing, 2021,37(1).
[8] Lin Shaojiang, Li Shun, Dai Jiangxiong, etc. Fatigue Life Analysis of Compressor Connecting Rod Based on nCode DesignLife[J]. Journal of Hubei Polytechnic University, 2019,37(3):3-4.

 doi:10.1088/1742-6596/2489/1/012028

Experimental study on fuselage fuel hose guillotine test of civil aircraft

Haibin XIAO[a*], Zhibin WU[b] and Degang LIU[c]

Shanghai Aircraft Design and Research Institute, Shanghai, China

[a*]xiaohaibin@comac.cc, [b]wuzhibin@comac.cc, [c]liudegang@comac.cc

Abstract—The crash-worthiness design of civil transport aircraft fuel systems is significance to improve the safety of the aircraft in the case of a crash. CCAR25.993 (f) requires that "fuel feed line in the fuselage shall be assembled to allow reasonable deformation and tension without fuel leakage". Fuel system fuselage fuel supply line with the appropriate hose configuration can take the advantage of the ductility of the material, and the guillotine test is accepted by worldwide examiners to validate the deformation and tensile capacity of the fuel hose to keep it from leaking fuel.

1. Introduction

CCAR25.993 (f) requires that "fuel feed line in the fuselage shall be assembled to allow reasonable deformation and tension without fuel leakage". The aircraft body fuel feed line consists metal pipe outside and hose material inside . For the metal tube, material parameters is clear, and the simulation method can be used to show compliance. The hose material parameter is unknown, guillotine test is one of method accepted in the industry. For the test, the choice decision of test configuration should take the actual diagram of the aircraft into consideration [1].

To ensure the fuselage fuel feed line followed CCAR25.993 (f) and to avoid the fire caused by the leakage of the fuel conduit after aircraft crashing, fuel hose guillotine test has been studied by many aircraft manufacturers worldwide. Normally the aircraft structures and fuel hose materials are not the same, while the fuel requirements of hose deformation is similar. Figure 1 shows the fuel hose distribution of one aircraft.

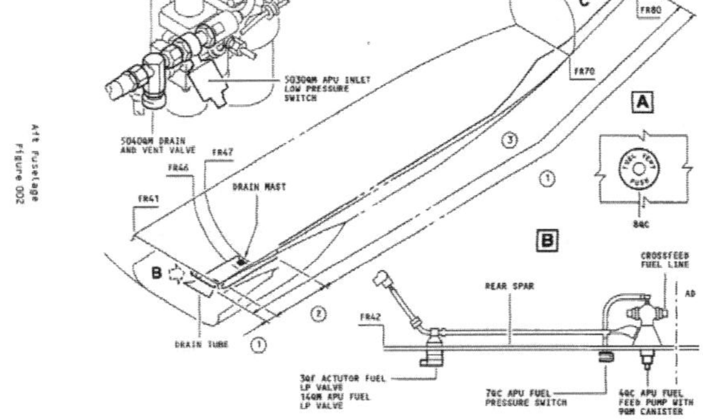

Fig. 1 Distribution of fuel pipe of aircraft industry of overseas.

Accordingly, to harmonize the test requirements and after extensive research, the FAA released a draft Issue Paper IP-23 in 2011, recommending that Section 25.993 (f) be verified by the guillotine test. The test keeps the consistency of the fuel pipe and the mounting style, and the test piece configuration to be the same as the mounting style on the aircraft.

The fuel line should be full of liquid to the operating pressure. The weight of the drop was 34.5 kg of a 3.18 mm thick edge. The drop-weight free-fall from a distance of 11.43 m and struck the weakest point of the test piece (according to the analysis). There was at least 0.457 m of deformation along the vertical direction of the test tube after impacted, as shown in Figure 2. without liquid leakage.

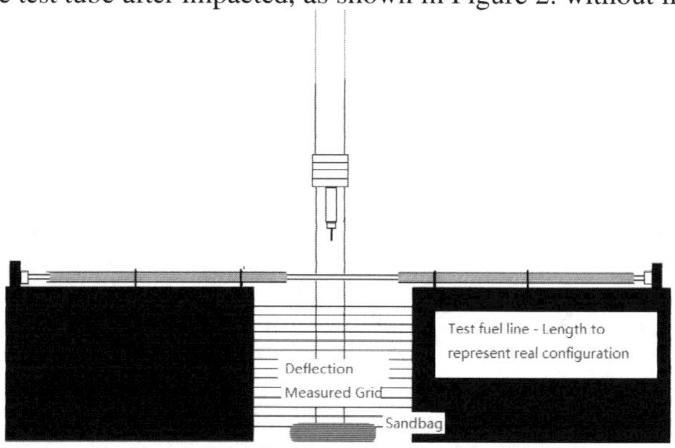

Fig. 2 Distribution of the guillotine test

2. Introduction for the test configuration of aircraft fuel hose guillotine

The test key points including the hammer-weight, the falling distance, the thickness of tool edge, the test piece, and the distance between the test piece and the buffering material. This is in line with the "Fuselage fuel line design and installation in the event of a crash, by the maximum drop weight of the aircraft deformation without fuel leakage" requirements.

2.1 Selection of Test Schematic

The fuselage system arrangement of a civil aircraft is shown in Figure 3, the fuselage fuel hose arrangement is shown in Figure 4, the fuselage fuel supply pipeline is the double-layer casing, and the outer tube is aluminum alloy material, an inner tube used to protect fuel. The inner tube is a kind of hose material whose ductility reaches 20%. The house is divided into three sections, which are of the same material, and the lengths are 4.93 m, 7.52 m, and 6.4 m respectively. The 3-segment inner tube is connected in a non-flared way. Under the protection of the outer metal tube, the 3-segment hose effectively utilizes the ductility of the hose material to ensure that there is no fuel leakage under the condition of large tensile deformation.

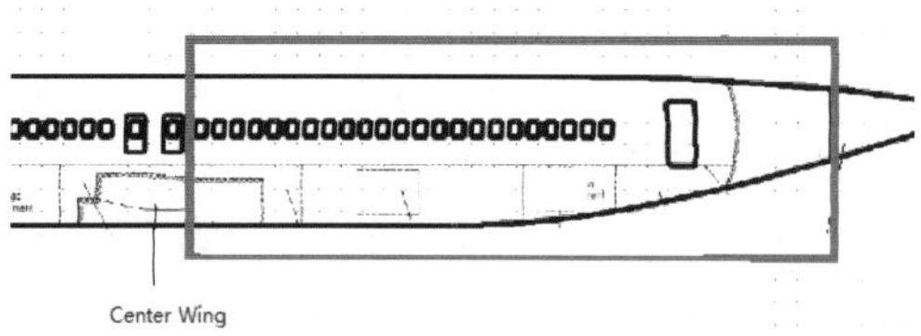

Fig. 3 Distribution of the fuel system of aircraft

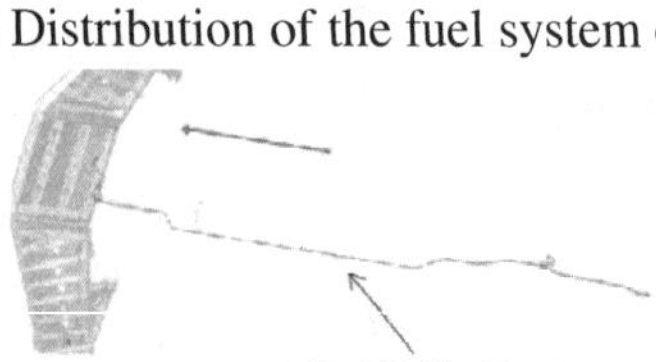

Fig. 4 Fuselage Feed Line Layout

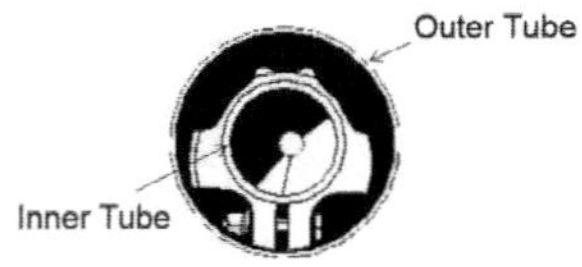

Fig. 5 Section of Fuselage Feed Line

2.2 Selection of Load Condition

According to the ARAC-TACDWG-FAA-Report-Final Transport Aircraft Crash-worthiness and Ditching Working Group Report to FAA [2], the vertical impact velocity-mass curve corresponds to a sinking velocity of 22 ft/s, i.e., for a large airliner, the maximum vertical impact velocity is 22 ft/s. From this, we obtain the simulation analysis condition as shown in Table 1.

Table 1 Simulation analysis of working conditions

Load case	Analysis Model
Full-22ft/s-00degree	Velocity = 22 ft/s (6.71 m/s), pitch 0 degree

2.3 Deformation analysis of fuselage hose

The structure of the aircraft is shown in Figure 6, with fuel lines and supports arranged below the cabin floor. In this paper, the deformation of the fuel pipeline under impact is calculated by the finite element method. During calculation, the fuel pipeline is only modeled on the outer metal pipeline, and the maximum deformation of the fuel pipeline is obtained by finite element analysis.

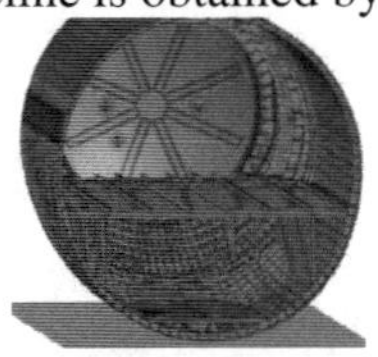

Fig. 6 Structure of Fuselage

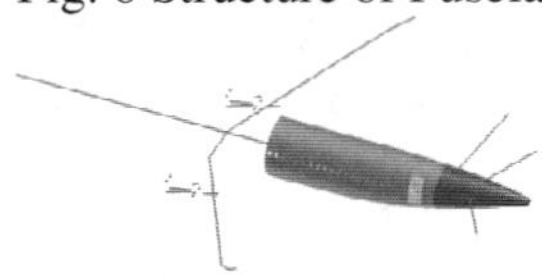

Fig. 7 Finite Element Model of Fuselage

The simulation results show that the fuel conduit is protected by main aircraft body, and the structure does not collide with the fuel pipeline. The displacement information of the fuel pipeline bracket and the joint is shown in Figure 8. The maximum deformation of the pipeline is 30.48 mm (see Figure 9 for details), and the total deformation of the hose is less than 762mm (the hose length is 18.85 m), which is the deformation of the hose in the circumstances of the aircraft crash speed at 22ft/s.

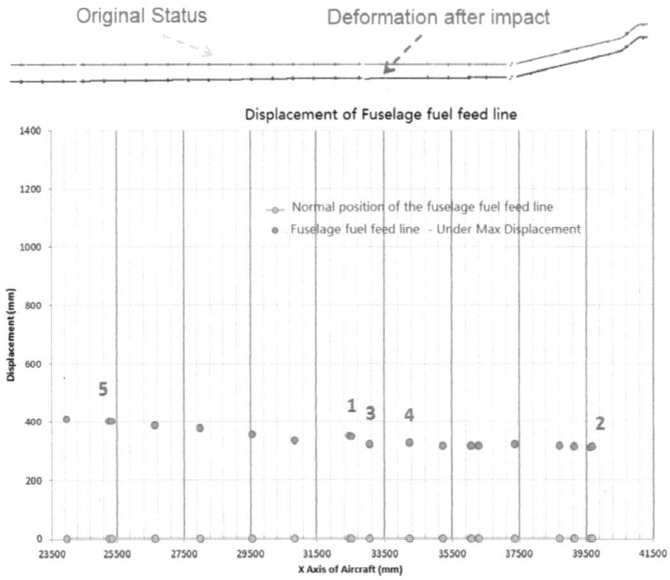

Fig. 8 Feed Line Deformation Under Crash Condition

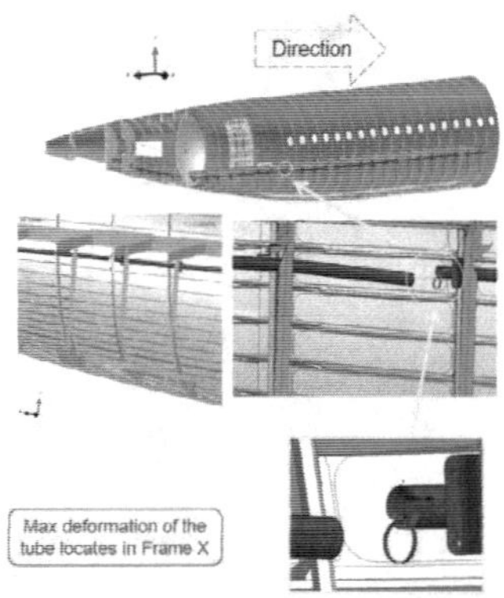

Fig. 9 Maximum Feed Line Deformation Under Crash Condition

The results show that the maximum tensile capacity of the hose is 762mm, and the tensile ratio is about 5% according to the principal full length of the test. This tensile ratio is 20% lower than the tensile capacity of the hose. However, considering the material parameters of the aluminum alloy pipeline used in the simulation, the deformation ability is less than that of the hose, so the simulation results only show that the minimum tensile capacity of the fuel hose (inner tube) is 762mm.

2.4 Impact Energy Screening

The simulation results show that the leakage cannot occur under the condition of a 22ft/s vertical crash.

For the test, the impact energy was chosen considering the industry-recognized IP-23, which was issued by the FAA specifically for this type of aircraft. The document states that if aircraft uses fuel feed hose, the hose will be impacted by the survivable crashing [3]. This article is raised from the accident B727 aircraft crashing (shown in Figure 10 of the crash scene, the requirement is the aircraft feed line after crashing will not leak to cause a fire). The fuel lines should be flexible enough to reduce the risk of fuel leakage during a severe crash. Crash scenarios include the landing gear striking on the tube, the tensile action of the structural deformation influenced on the tube, and the shear force from the structure and/or separation on the tube. The guillotine test is accepted to be one of the conformance method to demonstrate that the fuel conduit is strong and tensile enough to avoid fuel leakage [4].

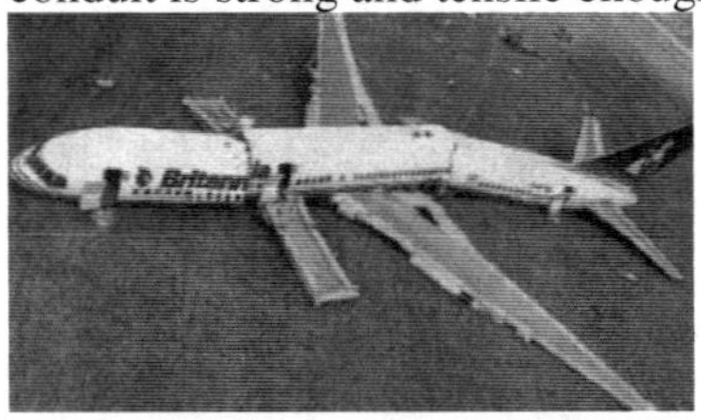

Fig. 10 Scene after crash

To validate of the fuel line design for this problem, guillotine test is accepted by authorized consideration. The drop weight should be 34.47 kg, including a steel plate 1/8 inch thick, which fell from a height of 11.28 m and strike the weakest point of the pipe test piece. The fuel line should be installed with sufficient tensile strength. Fig. 11 is the test equipment principle of the test.

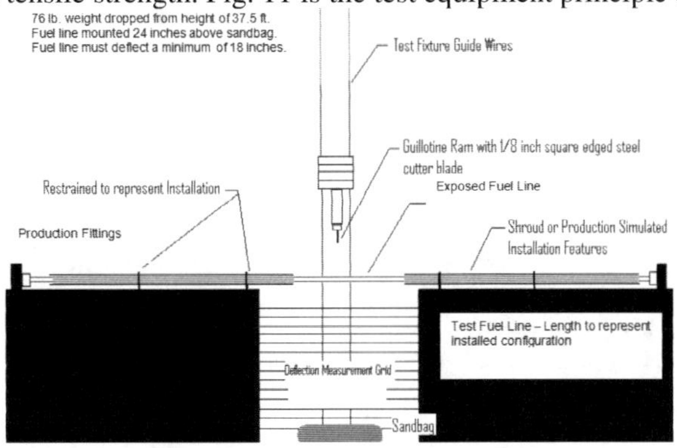

Fig. 11 Guillotine test installation

3. Configuration of test fixture

Take IP-23 as the requirement, guillotine test should be carried under the most critical case, the weight, height, and the transfer hose of 3.7 m in length was used in the test. The assembling of the test piece should be same as the real scenario, the impacting location is at the midpoint of the edges of the outer hard tube and the outer hard tube, and it impacts the inner transfer hose in this position, simulating the most critical case of aircraft situation.

In addition, to ensure that the test is conservative enough, the shortest 4.93 m hose from the three inner tubes was selected as the test piece.

4. Test Validation

The test is shown clear in Fig. 12. The 34.5 kg drop-weight of the hammer used to impact a 4.93 m specimen from 11.43 m height distance. Test piece is installed according to the real condition, and several clamps are fixed on the fixture. The outer metal pipe and inside hose of the test fixture are fixed together, the other end of the test fixture fixed to the flange by stiffing connection. The test bench and test piece installation dimension control requirements shown in Figure 12.

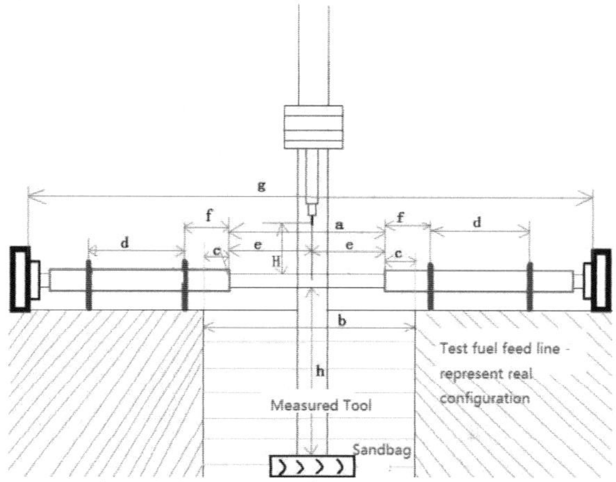

Fig. 12 Controlled dimension of test article

After the test fixture finishes, the test piece pipeline is filled with pressured liquid, and the pressure should be kept for 5 minutes to ensure no leakage. When the weight of the falling hammer is 34.7 kg and the thickness of the tool head is 3.18 mm, the free fall is raised by 11.43 m, the falling hammer hitting the hose is recorded with high-speed camera. Within 15 minutes after the impact is completed, the leakage of the pipe is observed, and the liquid pressure in the pipe is recorded.

After 15 minutes from impacting specimen deflection is 614.5 mm without fuel leakage occurs during the impact process, meeting criteria. The max. deflection in the test is shown in Fig. 13, and the appearance of the impact position of the pipeline.

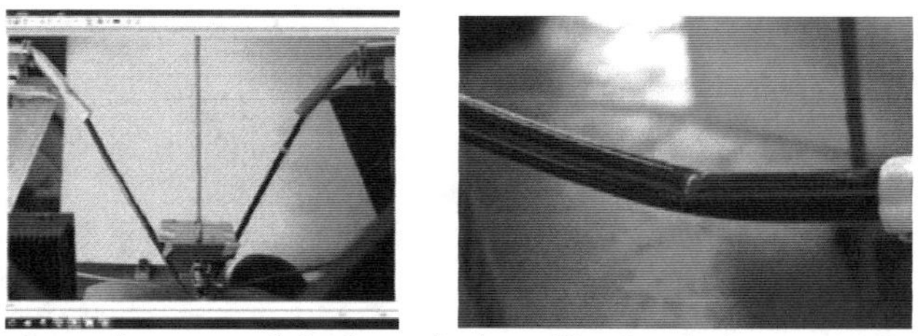
Fig. 13 Max deformation state of the hose during affecting & Appearance of impact position

5. Simulation versus Test

The results of the simulation and test is shown in Table 2.

Table 2 Simulation results and test results

Verification Method	Load Case	Length of Test Article (mm)	Max Hose deflection (mm)	Leakage Condition
Simulation	34.5	18850	762	Metal tube break
Test	34.7	4930	614.5	N

6. Conclusion

Based on the survivable crash-worthiness condition of ARAC-TACDWG-FAA-Report- -Final, the minimum tensile capacity of the fuselage fuel hose under this condition is analyzed by using the finite element method, while the experimental result shows that the deformation of the hose is far away from what obtained by the method of simulation analysis. To ensure a conservative design for the crashworthiness of civil aircraft fuselage fuel hoses, the use of a guillotine test should be preferred for review. So, before the simulation method is improved to get the same results as the test, the simulation can be used to judge the deformation trend and the position of maximum deformation only from an economic perspective.

Reference

[1] H B XIAO, Yang WANG, X Y LIAO, N A LI, D G LIU, T L GUAN. Research on guillotine test of a fuselage fuel hose of civil aircraft auxiliary fuel system, Journal of Physics: Conference Series Volume 2030.

[2] ARAC-TACDWG-FAA-Report-Final. Transport Aircraft Crash-worthiness and Ditching Working Group Report to FAA.

[3] Zhang Xiaomin, Feng Zhenyu, Xie Jiang Data Analysis and Research on Civil Aircraft Crash Accidents, Science and Technology Frontier, 2013.

[4] FAA Technical Center, Structural Response of Transport Airplanes in Crash Situations[R], DOT/FAA/CT-83/42 1983.

Journal of Physics: Conference Series 2489 (2023) 012029 doi:10.1088/1742-6596/2489/1/012029

Research on flight path planning method based on genetic ant colony algorithm and Maklink diagram

Xijian Bian[1a] and Kezhi Zhang[1*]

[1]COMAC Shanghai Aircraft Design and Research Institute, Shanghai 201200, China

[a]xijianbian@163.com, [b]zhangkezhi@comac.cc

Abstract. To reduce the cost of a flight, a calculation method is proposed in this paper, which takes the shortest distance as the goal and avoids the obstacle zone, and controls the turning angle as a constraint. Firstly, the Maklink diagram is used to build the obstacle zone model and establish the relationship between the obstacle zones. Then Dijkstra algorithm is used to construct the initial flight path to form a suboptimal path. Then the genetic ant colony algorithm is used to optimize the obtained suboptimal path. When there is a conflict risk between the two planned paths, the flight altitude of the aircraft on one path is adjusted, and the constraints are checked to form the optimal path. The simulation results show that the method can realize flight path optimization and verify the feasibility of the method.

1. Introduction

Fuel costs have become an increasingly important component of airline operating costs in recent years. There are a variety of techniques that can be used to control fuel consumption, among which flight trajectory optimization has the advantages of short time and low cost [1]. The objective function of the flight trajectory problem is complex and involves the processing of a large number of different information. It is generally divided into two levels: one is the overall reference trajectory planning, and the other is the local trajectory dynamic planning. The overall reference trajectory planning is performed on the ground before flight. The advantages and disadvantages of the reference trajectory are determined according to the predetermined performance index, which is generally determined according to the combination of mission requirements, safety requirements, flight time, and other factors of the aircraft. Based on the optimal performance index, an optimal reference trajectory is generated according to a certain algorithm.

This paper takes the route planning of two aircraft in a static obstacle zone environment as the research object. Firstly, the mathematical model of route planning is established, the genetic ant colony algorithm is used to optimize the route, the collision possibility analysis is carried out on the obtained route, and finally, the optimized path is obtained.

2. Flight path modeling

2.1 Establishment of obstacle zone

Maklink lines are used to establish the connection between obstacle zone in this paper [2]. In 1991, Habib M.K. constructed a vertex diagram method called Maklink for generating collision-free paths. This method can make connections between obstacle zone and provide a basis for establishing collision-free paths.

2.2 Constructing the initial solution of the flight path

The Dijkstra algorithm is used to construct the initial solution of the flight path in this paper. The Dijkstra method was proposed by Dijkstra in 1959, and its basic idea is to find the shortest path between the starting point and the ending point [3]. The main feature of Dijkstra's algorithm is that it starts from the starting point and adopts the strategy of a greedy algorithm.

2.3 Flight path constraints

The flight path is the "highway" connecting the departure airport and the destination airport. Good flight path planning can minimize fuel consumption and improve the economy of operation. Therefore, the shortest total length of the flight path is set as the objective function, so as to achieve the best economy.

The purpose of flight path planning is to avoid restricted zone, dangerous zone, restricted zone, etc., in addition to taking into account the impact of flight costs, flight procedures, aircraft performance, and workload. For safety and comfort considerations, the flight segment turning angle limit that needs to be considered in planning a path [4]. That is, limited by the aircraft's own turning performance and considering the pilot's maneuverability, the deflection angle of the aircraft's nose should not exceed 90 °.

2.4 Flight path optimization algorithm

In this paper, the genetic ant colony algorithm is used to optimize the initial path. The genetic algorithm (GA) is derived from the random search technique of natural selection and the genetic evolution mechanism in biological evolution, which has a good global search property. Ant colony algorithm (ACO) is a heuristic optimization algorithm based on population. It can find the optimal solution by selecting the shortest path in the process of simulating ant foraging. The ant colony algorithm uses pheromones to control the actions of ants.

However, both algorithms have their own shortcomings. The ant colony algorithm has the disadvantages of a large amount of computation, a slow search process, and easy stagnation [5]. The improvement points of the genetic algorithm are mainly that it relies too much on the initial population, the population diversity is difficult to maintain, and the convergence speed is difficult to control [6]. The genetic ant colony algorithm can improve the solving efficiency and convergence efficiency by combining the two algorithms [7].

3. Algorithm design

According to the model established above, the solution process is divided into three stages: obstacle zone establishment, initial path planning, and path optimization. The overall flow diagram of the algorithm is shown in Figure 1.

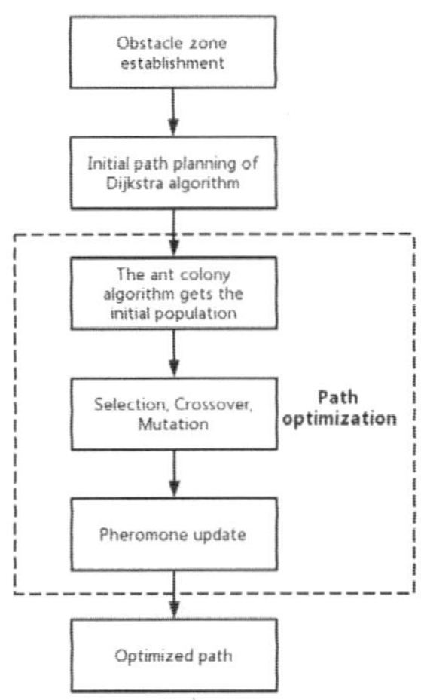

Figure 1. Overall flow diagram of flight path planning method

3.1 Obstacle zone establishment

Firstly, a convex polygon model is established for the restricted zone, danger zone, restricted zone, and other obstacles in the selected airspace. Graham algorithm is applied to build a convex polygon model, and then the boundary of the convex polygon is extended outward by *h*, as shown in Figure 2. The expanded portion is called the protected zone so that the minimum threshold for safe passage can be met during the path planning process.

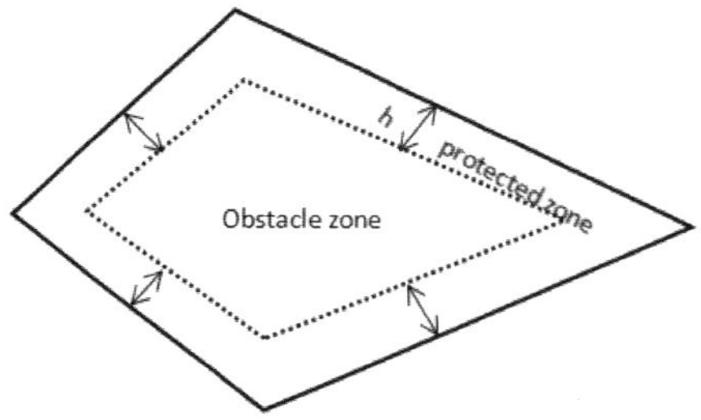

Figure 2.Schematic of the extended obstacle zone

On the basis of convex polygons, Maklink lines are generated to build a Maklink diagram. Its construction steps are as follows:

Step 1 Number the restricted zone in the airspace as R_1, R_2, R_3... R_i, R_n.

Step 2 Select restricted zone R_i and number its vertices D_1, D_2,..., D_j, D_n.

Step 3 Draw vertical lines from vertex D_j of the convex polygon to the boundary, connect all vertices in pairs, and number all lines as L_1, L_2, L_3... L_m, L_n. Store it in set *L*.

Step 4 Select a line segment in set *L* and check whether line segment *L*i intersects the limit zone. If yes, delete the line segment and repeat Step 4.

Step 5 Check whether the outer Angle between line segment L_i and the vertex is less than 180 °. If not, go to Step 6. If not, delete the other lines of this vertex, keep this line, and go to Step 2.

Step 6 Set L_i as a candidate line segment.

Step 7 Check whether the Angle between the cable segments is less than 180 °. If not, select the next line segment of *L* and go to Step 3; if not, go to Step 8.

Step 8 Reserve candidate line segments.

Repeat steps 1-8 until all vertices are complete.

After building the Maklink diagram, see Figure 3. Select midpoints on all Maklink lines and connect adjacent midpoints, starting points, and ending points to form an undirected network diagram. As shown in Figure 4.

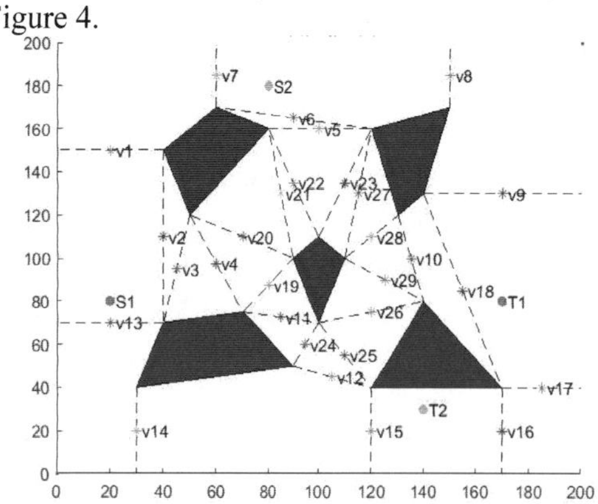

Figure 3. Maklink diagram and Matlab model

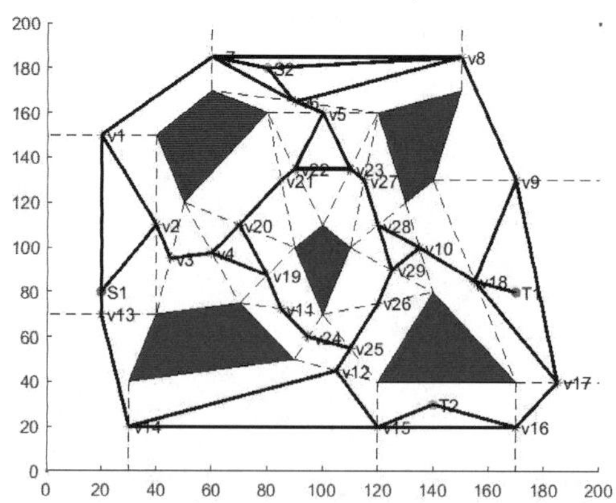

Figure 4. Maklink undirected network diagram

3.2 Initial Path Planning

Dijkstra algorithm is used to construct the initial solution of the path. The basic idea of the Dijkstra method is to find the shortest path between the starting point and the ending point. The passable path of each point is shown in the figure, and the solid black line in the figure is the passable path. In the figure, point S is taken as the starting point, point T as the ending point, and point vi as the midpoint of the Maklink line. Dijkstra algorithm starts from the starting point, and each time, it finds the node with the shortest distance from the starting point among the unlabeled nodes, marks it, and collects it into the set of optimal paths. Calculate the distance between the newly added node *A* and its neighboring node *B*. If it is less than the distance of node *B*, update the distance between node *B* and its previous point until the endpoint. So, we have the shortest path from S to T through point *vi*.

3.3 Optimization of genetic ant colony algorithm

The genetic ant colony algorithm is used to optimize the initial path of Dijkstra. The idea of the genetic ant colony algorithm is to make use of the pheromone of the ant colony algorithm to influence the gene recombination and mutation of the genetic algorithm. The steps are as follows:

Step 1 On the basis of Dijkstra's initial solution, the Maklink line *Li* that the path passes through is obtained, and the two endpoints P_{i1} and P_{i2} are obtained. Then other points on *Li* can be represented by Equation (1) below,

$$P_i(h_i) = P_{i1} + h_i(P_{i2} - P_{i1}), \quad h_i \in [0,1], \quad i = 1,2,..., d \tag{1}$$

In Equation (1), h_i is the proportion parameter; d is the number of nodes on the path.

Step 2 Define the number of ants m, pheromone selection threshold θ, pheromone importance factor α, node elicitation value β, and pheromone volatilization factor ρ. The initial pheromone τ and the initial elicitation value η, and the iteration number *NC*.

Step 3 The k ant selects the node. Generate a random number q, if $q <= \theta$, select the node corresponding to the largest pheromone; if $q > \theta$, adopt roulette $p_{hoz} = \frac{\tau_i^{\alpha}\eta_i^{\beta}}{\sum_{z \in Li} \tau_l^{\alpha}\eta_l^{\beta}}$ to obtain the next node.

Step 4 Update the local pheromone, $\tau_i = (1-\alpha)\times\tau_i + \alpha\times\rho$.

Step 5 Repeat Step 3 until k is equal to the population m.

Step 6 Calculate the total length of each ant's path, select the four ants with a short distance, take the nodes they pass as chromosomes, and take the distance as the fitness value to select, cross and mutate them.

Step 7 Merge the obtained new population with the original population, calculate the fitness value, and select M populations with good fitness as the new population.

Step 8 Update the local pheromone, $\tau_i = (1-\alpha)\times\tau_i + \alpha\times\rho$.

Step 9 Calculate the fitness of the new population, and select the track with the best fitness as minL.

Step 10 Update the pheromone corresponding to the shortest distance : $\tau_i = (1-\alpha)\times\tau_i + \alpha\times(\frac{1}{\min L})$

Step 11 Return to Step 3 until the number of iterations reaches *NC* to obtain the optimal trajectory.

4. Algorithm simulation

MATLAB software was used to generate convex polygonal obstacles, and the Maklink diagram was completed to construct a wireless network diagram. Dijkstra algorithm was used to generate a suboptimal path, and then the optimized path and turning point were obtained by using the algorithm. As shown in Figure 5, S1, T1, S2 and T2 are the starting point and ending point of the two lines respectively, the shaded part is the obstacle zone, the dotted line is the Maklink line, and the solid line is the passable line. Two suboptimal routes S1→v2→v3→v4→v20→v20→v22→v23→v27→v28→v10→T1 、

S2→v6→v5→v23→v27→v28→v29→v26→v25→v12→v15→T2 are generated by Dijkstra algorithm.

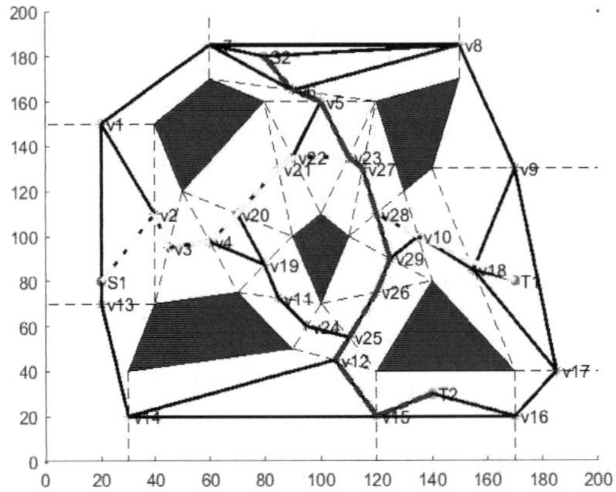

Figure 5.Dijkstra algorithm suboptimal path

On the basis of the obtained suboptimal path, the genetic ant colony algorithm is used to optimize, and the optimal path after the shortest path is obtained as shown in Figure 6. The length of the two paths calculated by the Dijkstra algorithm is 213.535 for S1→T1 and 198.01 for S2→T2. The optimized path lengths are 168.093 and 166.003. The optimized path in the vertical direction is shown in Figure 7. According to MATLAB software, the segment turning Angle of the optimal trajectory can meet the segment turning Angle constraint requirements and meet the segment lateral safe interval constraint requirements.

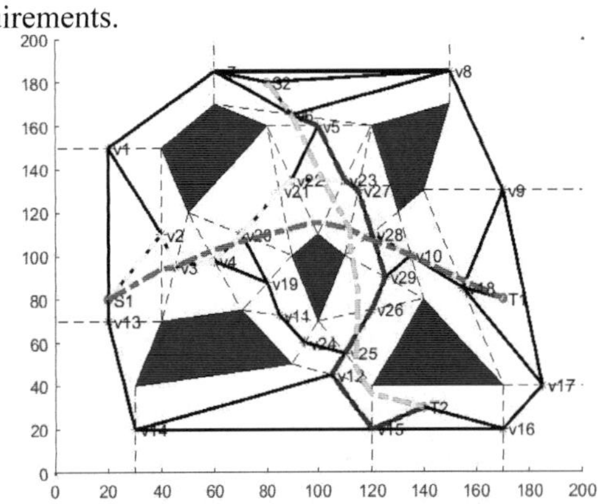

Figure 6. Final optimization path

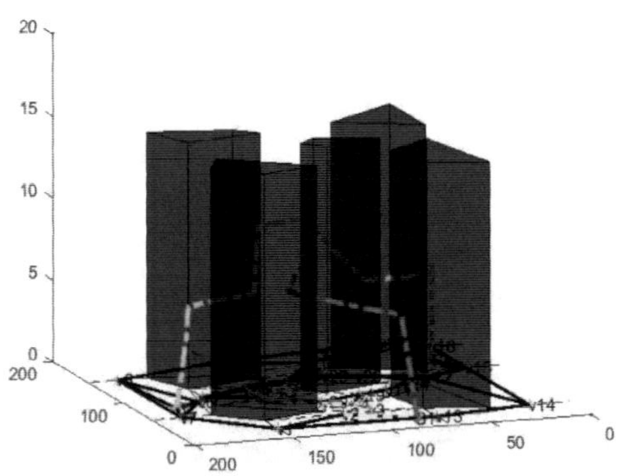

Figure 7. Optimize the path in the vertical direction

5. Conclusion

In this paper, flight trajectory optimization is realized from three aspects: obstacle zone establishment, initial path construction, and path optimization. Firstly, the Maklink diagram and Dijkstra algorithm are used to plan the feasible path, and then the genetic ant colony algorithm is used to optimize the path. Finally, we can verify whether the optimized path meets the restriction conditions.

The simulation results show that the proposed method can safely avoid the flight obstacle zone, meet the constraints of safety interval and turning angle, and has certain reference significance for flight path planning. In this paper, the genetic ant colony algorithm is chosen for the design of the scheme, but because of the continuous emergence of new algorithms, it is believed that there will be a better optimization algorithm. In terms of experimental design, few impact factors are considered, which are still different from the actual flight situation, so a lot of experiments and analysis are needed.

Reference

[1] Murrieta-Mendoza A, Botez R M. Aircraft Vertical Route Optimization Deterministic Algorithm for a Flight Management System[C]// SAE 2015 AeroTech Congress & Exhibition. 2015.
[2] Cheng Chunhua, Wu Jinhua, Zhou Dawang, Qi Yuhua. Route Planning Based on Link diagram Method and Genetic Algorithm [J]. Ship Electronic Engineering, 2010(1):85-87
[3] Chuan-xiang REN, Xin-gang HAO, Ying-rui WANG, Guang-hui PAN (College of Information and Electrical Engineering, Shandong University of Science and Technology, Qingdao 266510, China). Research on the Optimization and Simulation of the Shortest Path Based on Algorithm of Dijkstra[J].2010,0(S1):199-20137
[4] LI Shaohua. 3D Flight Path Planning for Airline Operation [D]. Civil Aviation University of China.
[5] WANG L J. Fusion of ant Colony Algorithm and Genetic Algorithm and its application Research [D]. Nanjing University of Posts and Telecommunications,2005.
[6] Wang Fei, Wang Hongyong. Research on Route Planning Method Based on Maklink diagram and Genetic Algorithm [J]. Transportation systems engineering and information,2014,14(5):154-160
[7] LIU X L. Research on Traveling Salesman Problem based on Genetic Ant Colony Algorithm [D]. Shandong: Shandong University of Science and Technology,2017.

Pseudo-Theory-Based Optimization Design Method for Lunar Gravity-Assisted Human Earth-Mars Transfer Trajectory

Yiting Yuan, Zichen Zhao, Yue Dong

Beijing Institute of Technology, Beijing, 100081, China

yuanyiting0206@163.com

Abstract. The lunar gravity assist (LGA) is recognized as one of the potential ways to reduce the energy demand of human Mars exploration. Nevertheless, it indeed increases the difficulty of designing the exploration trajectory due to its inevitable considerations of a more complex gravitational environment and additional flying stages. Focusing on this issue, an efficient pre-design method is proposed on the basis of the pseudo-state theory. By embedding the pseudo-sweep back step into the dynamical propagation, the accuracy of two-body trajectories can be recovered to a high level, which brings significant convenience of initially designing high-precision trajectory, rapidly searching the departure window, and fast evaluating the LGA effect. A large amount of simulation results show that the application of the pseudo theory could enhance at least 88% trajectory accuracy, increase the 60% success rate, and decrease 5% the necessary cost when optimizing the Earth-Mars transfer trajectory, with comparisons to the two-body trajectory.

1. Introduction

Human Mars exploration is of great value for exploring extra-terrestrial life, interstellar migration, and promoting science and technology development and human social progress. At the same time, the Earth-Mars transfer trajectory design is essential in realizing manned Mars exploration. Due to the significant velocity increment required for human Mars exploration trajectory transfer, the mass of payload to be carried by human Mars exploration is higher than unmanned exploration, so the mass scale of the human Mars exploration mission is enormous, and the mission has higher requirements for the rocket's carrying capacity. Lunar gravity assist can increase the escape energy of the spacecraft and is a potential strategy to reduce the incremental Earth-Mars transfer velocity and decrease the departure mass scale [1]. However, the LGA has exacerbated the difficulty of optimizing the Earth-Mars transfer trajectory. On the one hand, the simplified LGA model cannot simulate the real dynamics environment well, and the trajectory accuracy of the initial design is poor. On the other hand, the high-precision LGA model is nonlinear, and the trajectory optimization problem is highly dependent on reasonable initial guesses and low computational efficiency. Both fast and high-precision design has been a hot issue for research.

Early studies usually used the multiple patched conics (MPC) method to initialize the LGA trajectory. For example, R.S. Wilson assumed that the spacecraft dynamics outside the lunar influence sphere were a spacecraft-Earth two-body model, and the dynamics inside the lunar influence sphere were a spacecraft-moon two-body model [2]. He eliminates the velocity mismatch at the connection point of the segmented design trajectory outside the lunar influence sphere by numerical differential correction method. Based on this, Lorenzo Casalino [3] used an indirect optimization method to extend the MPC method to interplanetary trajectory design, and a design and optimization procedure for LGA Earth escape maneuvers for interplanetary missions was proposed. This process first optimizes the LGA trajectory to determine the escape hyperbolic velocity and subsequently optimizes the interplanetary trajectory to match the hyperbolic hypervelocity. The method optimizes the interplanetary and Earth escape trajectory separately, without global optimization of the entire transfer trajectory. The MPC method considers the presence of multiple gravity fields and enables a preliminary design of the LGA trajectory but ignores the gravitational influence of the spacecraft flyby near the Moon, which differs significantly from the actual model.

Scholars have been working on the third-body gravitational influence in the trajectory optimization model to optimize the LGA trajectory with high accuracy. Csaaell has studied the optimization problem of the double LGA Earth escape trajectory under the model considering the solar gravitational influence [4,5]. By introducing the principal vector, the optimization problem of the double LGA is converted into an optimal control problem and is performed by a nonlinear programming solution. The method is fast and straightforward to calculate. Still, the solved LGA trajectory to reduce the velocity increment is not ideal, and the solved trajectory, as the initial guess, cannot predict the optimal solution well. R. V. Ramanan and V. Adimurthy [6] proposed a numerical search technique for the high-precision LGA trajectory using an improved genetic algorithm, but the search time cost is high.

Many scholars have shown that the sensitivity of the design parameters for optimizing the LGA high-precision trajectory is greatly enhanced due to the introduction of the third-body gravitational uptake. Luo Zongfu studied the existence mechanics mechanism and modeling design method of the binary lunar side steering trajectory and clarified the parameter sensitivity mechanics of the LGA from a particular perspective [7]. Peng Qiju analyzed the effect of the Moon on the first-order partial derivative matrix of the Earth-Moon transfer trajectory through the rapid nonlinear amplification of the orbital parameters which is analyzed by the first-order partial derivative matrix of the Earth-Moon transfer trajectory [8]. The greatly enhanced sensitivity of the design parameters leads to the optimization of high-precision LGA Earth-Mars transfer trajectories highly dependent on accurate initial guesses. So, the initial design of the trajectories should be performed to provide reasonable reference trajectories before the numerical optimization of high-precision trajectories. Considering the influence of third-body gravitational uptake in the initial trajectory design can provide more accurate initial guesses for high-precision LGA Earth-Mars transfer trajectories.

To address the above problems, this paper applies the pseudo theory to the LGA human Earth-Mars trajectory optimization design problem. Besides, it proposes a fast search strategy for the LGA departure window. Through comparative simulations, it is found that the pseudo trajectory design method can provide better initial guesses for high-precision trajectory design than the MPC design method, which is conducive to the convergence of high-precision models and better performance indexes. The window search strategy proposed in this paper can realize the fast evaluation of the LGA effect and launch window length in any year, and the search results of the LGA launch window from 2031 to 2037 are given. The structure of this paper is as follows: Chapter 2 presents the optimization model of the LGA human Mars exploration transfer trajectory based on the pseudo method, Chapter 3 gives the departure window search strategy based on the pseudo method, Chapter 4 shows the simulation numerical results and analysis, and Chapter 5 concludes the paper.

2. Trajectory optimization method using pseudo theory

2.1. pseudo theory

Wilson proposed an approximate method for solving the restricted three-body problem [9,11], which he called the "pseudo theory " [9]. This method can reduce the error of the two-body trajectory by 90% without increasing the complexity and cost of the operation [10]. This section briefly introduces the pseudo theory, which gives its concepts and basic formulas without providing a detailed description of its derivation process, which can be found in the reference [9].

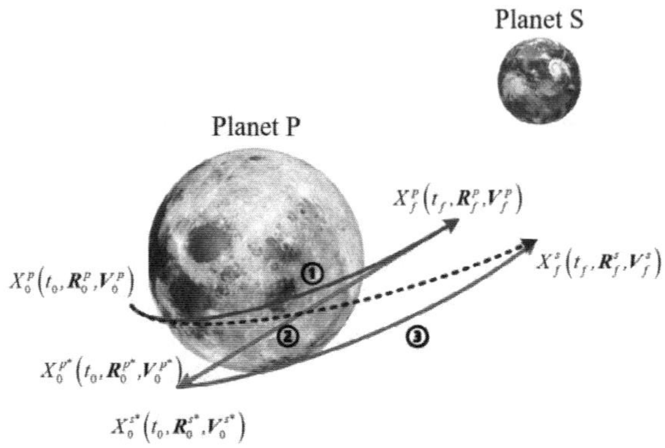

Figure 1. Steps for approximating the restricted three-body problem by the pseudo theory.

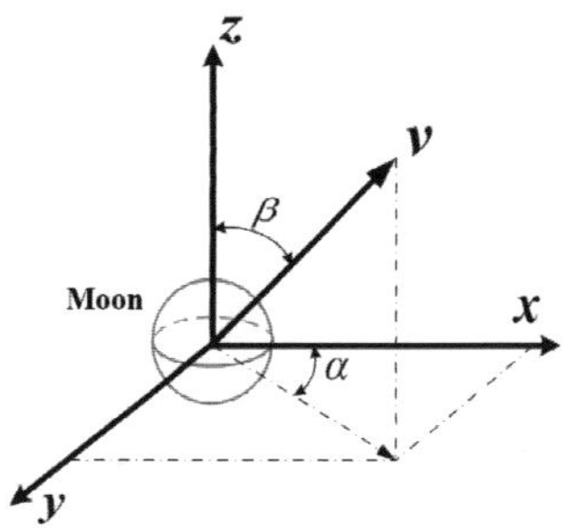

Figure 2. The directional angle of the velocity increment at the Lunar perigee in the equatorial coordinate system of the lunar.

Assuming that the primary gravitational object of the spacecraft is P-body and the secondary gravitational object is S-body during the flight, the steps of the pseudo theory method to approximate the restricted three-body problem are as follows:

1) Propagate a P-centred conic forward from the position and velocity state $X_0^p(t_0, \boldsymbol{R}_0^p, \boldsymbol{V}_0^p)$ at time t_0 to the $X_f^p(t_f, \boldsymbol{R}_f^p, \boldsymbol{V}_f^p)$ at the time t_f.
2) Propagate a field-free segment backward from $X_f^p(t_f, \boldsymbol{R}_f^p, \boldsymbol{V}_f^p)$ to $X_0^{p^*}(t_0, \boldsymbol{R}_0^{p^*}, \boldsymbol{V}_0^{p^*})$, and the state $X_0^{s^*}(t_0, \boldsymbol{R}_0^{s^*}, \boldsymbol{V}_0^{s^*})$ is obtained after coordinate transformation to the S celestial body.

$$\boldsymbol{R}_0^{p^*} = \boldsymbol{R}_f^p - (t_f - t_0)\boldsymbol{V}_f^p \tag{1}$$

$$\boldsymbol{V}_0^{p^*} = \boldsymbol{V}_f^p \tag{2}$$

3) Propagate an S-centred conic from the position and velocity state $X_0^{s^*}(t_0, \boldsymbol{R}_0^{s^*}, \boldsymbol{V}_0^{s^*})$ forward to obtain the approximation to the final S-body relative state $X_f^s(t_f, \boldsymbol{R}_f^s, \boldsymbol{V}_f^s)$.

Knowing the state of the spacecraft $X_0^p(t_0, \boldsymbol{R}_0^p, \boldsymbol{V}_0^p)$ at t_0, we can quickly find the state $X_f^s(t_f, \boldsymbol{R}_f^s, \boldsymbol{V}_f^s)$ at t_f when it is subjected to both primary and secondary gravity by the above steps shown in **Figure 1** [10]. When the spacecraft performs LGA, the primary gravitational object P is the Moon, and the secondary gravitational object S is the Earth. When the spacecraft flies in the Earth-influenced sphere, the primary gravitational object P is the Earth, and the secondary gravitational object S is the Sun. When the spacecraft flies in the Mars-influenced sphere, the primary gravitational object P is Mars, and the secondary gravitational object S is the Sun.

To verify the accuracy, the trajectories were propagated from the Lunar perigee for two-body propagation, pseudo theory method propagation, and high-precision propagation of the same flight time to reach the vicinity of the lunar impact sphere, respectively. Taking the high-precision trajectory as the reference trajectory, the terminal errors of the two-body trajectory and the pseudo trajectory are shown in **Table 1**. According to the simulation results, the pseudo-theory method can correct about 90% of the trajectory errors compared to the two-body propagation.

Table 1. The terminal errors of the two-body trajectory and the pseudo trajectory.

propagation error	2031year	2033year	2035year	2037year
Two-body trajectory	198.39 km	103.19 km	201.37 km	231.63 km
Pseudo trajectory	19.28 km	11.38%	17.15 km	14.54 km
Error correction amount	90.28%	88.97%	91.48%	93.72%

2.2. Trajectory optimization model based on the pseudo theory

For a human Mars exploration mission, the spacecraft should not cross the Van Allen belts multiple times to consider the health of the astronauts, so this paper designs the number of lunar gravitational assists as a single.

The specific flight process is as follows: the spacecraft applies an impulse maneuver in Earth trajectory, meets the Moon, applies an impulse maneuver at the Lunar perigee (usually less than 1 km/s), then escapes the Earth for interplanetary flight, applies a deep-space maneuver (DSM) midway, enters the Mars impact sphere, applies an impulse maneuver at the perigee of the target trajectory, and finally enters the target Mars trajectory.

The moments, position vectors, and velocity vectors of the spacecraft at the Earth's perigee, the Moon's perigee, and the Mars' perigee are the key design parameters that determine the entire Earth-Mars transfer trajectory, and the height of the lunar leap is the key parameter that determines the effect of the LGA. For the position and velocity vectors, using slow variables like orbital elements to express them instead can reduce the nonlinearity of the model. Combined with the above analysis, the optimization variables are

$$\boldsymbol{X} = \left[e_d, i_d, \Omega_d, \omega_d, t_d, t_{dl}, r_p, e_m, i_m, \Omega_m, \omega_m, T_{EM}, e_M, \Omega_M, \omega_M, i_M, \Delta v_p, \beta, \alpha\right]^T \tag{3}$$

where e_d, i_d, Ω_d, and ω_d are the eccentricity, inclination, ascending nodal equinox, and perigee amplitude of the Earth-Moon transfer hyperbolic trajectory, respectively. t_d is the moment of departure from NEO, and t_{dl} is the transfer time from the NEO to the lunar pseudosphere. r_p is the height of the lunar leap. e_m, i_m, Ω_m, and ω_m are the orbital eccentricity, inclination, ascending nodal equatorial longitude, and perigee amplitude of the spacecraft in the lunar equatorial system before the maneuver, respectively. T_{EM} is the flight time from the Earth's pseudosphere to the Mars pseudosphere. Δv_p is the magnitude of the velocity increment at the Lunar perigee. The α and β are the directional angle of the velocity increment applied by the spacecraft at the Lunar perigee, and the specific relationship is shown in **Figure 2**. e_M, i_M, Ω_M, and ω_M are the orbital eccentricity, orbital inclination, ascending node declination, and perihelion amplitude of the spacecraft before the maneuver is applied at the Mars perigee.

This paper considers a multi-segment splicing strategy to reduce the difficulty of trajectory optimization design, as shown in **Figure 3**. The trajectory splice point is designed in two places, the first is at the spacecraft's entry to the lunar pseudosphere, and the second is at the Mars pseudosphere.

1) The trajectory propagates positively from the perigee to the lunar pseudosphere, and the position $\boldsymbol{r}^e_{pse\_f}$ and velocity vectors $\boldsymbol{v}^e_{pse\_f}$ of the spacecraft at the lunar pseudosphere are obtained.
2) The trajectory propagates backward from the lunar perigee to the lunar pseudosphere, and the position $\boldsymbol{r}^e_{pse\_b}$ and velocity vectors $\boldsymbol{v}^e_{pse\_b}$ are obtained. Spliced position and velocity with the trajectory from step 1) at the entry point to the lunar pseudosphere.

3) An impulse maneuver is applied at the lunar perigee, and the trajectory propagates positively from the lunar perigee to the position of the deep space maneuver to obtain the position vector $\boldsymbol{r}^{s}_{dsmbefor}$ and velocity vector $\boldsymbol{v}^{s}_{dsmbefor}$.
4) The trajectory propagates backward from the Mars perigee to enter the Mars pseudosphere to obtain the position vector $\boldsymbol{r}^{s\_RA\_M}_{pse}$ and velocity vector $\boldsymbol{v}^{s\_RA\_M}_{pse}$ at the Mars pseudosphere.
5) Solve the Lambert trajectory using $\boldsymbol{r}^{s}_{dsmbefor}$ and $\boldsymbol{r}^{s\_RA\_M}_{pse}$ to obtain the velocity vector after the deep space maneuver $\boldsymbol{v}^{s}_{dsmafter}$ and the velocity vector at the pseudosphere of Mars $\boldsymbol{v}^{s\_RA\_M}_{lamb}$.
6) Splice the velocity vector $\boldsymbol{v}^{s\_RA\_M}_{pse}$ with the velocity vector $\boldsymbol{v}^{s\_RA\_M}_{lamb}$ at the Mars pseudosphere.

The splicing constraint at the entry to the lunar pseudosphere is

$$\boldsymbol{f} = \begin{bmatrix} \boldsymbol{r}^{e}_{pse\_f} - \boldsymbol{r}^{e}_{pse\_b} \\ \boldsymbol{v}^{e}_{pse\_f} - \boldsymbol{v}^{e}_{pse\_b} \end{bmatrix} \tag{4}$$

The splicing constraint at the Mars pseudosphere is

$$\boldsymbol{f} = \boldsymbol{v}^{s\_RA\_M}_{pse} - \boldsymbol{v}^{s\_RA\_M}_{lamb} \tag{5}$$

The magnitude of the incremental maneuver velocity at the LEO perigee, lunar perigee, DSM point, and Mars perigee is calculated by

$$\begin{cases} \Delta v_1 = \left\| \boldsymbol{v}^{e}_{dl} \right\| - \sqrt{\dfrac{\mu_e}{R_E + r_d}} \\ \Delta v_2 = \Delta v_p \\ \Delta v_3 = \left\| \boldsymbol{v}^{s}_{dsmbefor} - \boldsymbol{v}^{s}_{dsmafter} \right\| \\ \Delta v_4 = \left\| \boldsymbol{v}^{m}_{p} - \sqrt{\dfrac{\mu_m \left(1 + e^{target}_m\right)}{a^{target}_m \left(1 - e^{target}_m\right)}} \right\| \end{cases} \tag{6}$$

Where $\boldsymbol{v}^{e}_{dl}$ is the magnitude of the velocity after maneuvering at LEO perigee and $\boldsymbol{v}^{m}_{p}$ is the magnitude of the velocity before maneuvering at Mars perigee, calculated from the root of the hyperbolic trajectory at perigee, respectively. μ_e and μ_m are the gravitational constants of Earth and Mars, respectively. R_E is the radius of the Earth. r_d is the altitude of the Earth's departure circle trajectory. a^{target}_m and e^{target}_m are the semi-long axis and eccentricity of the Mars target trajectory, respectively.

The performance metric for the entire optimization problem is the minimum total speed increment:

$$J = \Delta v_1 + \Delta v_2 + \Delta v_3 + \Delta v_4 \tag{7}$$

3. Departure Window Search Strategy

The start of the departure window search relies on the MPC trajectory to provide a reference departure moment t_0 and the trajectory parameters x_0. The reference MPC trajectory is used as the initial guess for the pseudo-theory model to obtain the t_0 momentary pseudo-trajectory optimization results X_{t_0}. The optimization result X_{t_0} is used as the initial guess of the trajectory at the time of $t_0 + \Delta t$ and $t_0 - \Delta t$ to optimize the trajectory $X_{t_0+\Delta t}$ and $X_{t_0-\Delta t}$, and then the optimization result is continued as the initial guess at the next time. When all the optimal tracks in the time range T have been searched, the starting time with the best performance target t' is found and used as the reference time t_0 for the next search round, as shown in **Figure 4**. After repeated iterative searches, the departure window with the lowest speed increment is finally determined, and the specific search process is shown in **Table 2**.

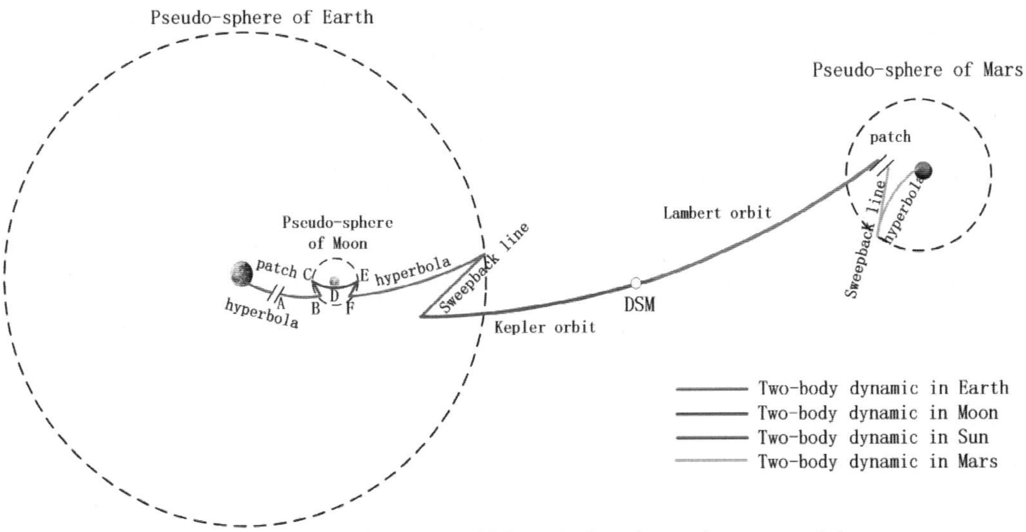

Figure 3. The LGA manned Mars exploration trajectory model.

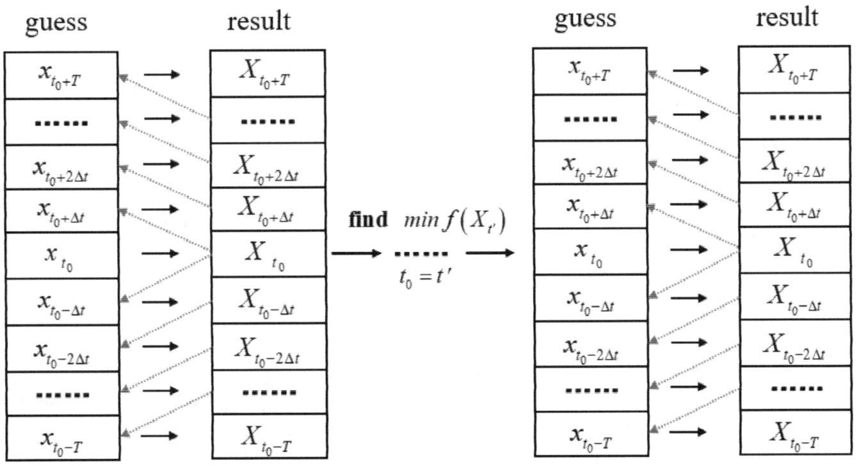

Figure 4. Launch window search strategy.

Table 2. Departure window search steps.

Step1	Set the number of cycles for the entire process, the time range T of the window search, and the time interval Δt of the search. The departure moment of the MPC trajectory is used as the window search time initial value t_0, and its trajectory data is used as the initial value guess $x(t_0)$, let $t = t_0$.
Step 2	Optimize the pseudo trajectory to obtain trajectory optimization result $X(t)$.
Step 3	Determine whether the window search time range is exceeded at the moment t. If yes, go to step 4. If no, the trajectory optimization result at the moment t becomes the initial guess at the next moment, i.e. $x(t+\Delta t) = X(t)$ ($x(t-\Delta t) = X(t)$), let $t = t + \Delta t$ ($t = t - \Delta t$), and return to step 2.

Step 4	Determine whether the process is cycled to the specified number of times. If yes, output the window search results of the last cycle. If no, find the departure moment t' with the lowest performance index in the whole search time range, update the initial value of the departure time, let $t_0 = t'$ and $t = t_0$, and return to step 2.

4. Numerical results and analysis

4.1. Comparison of convergence results of initial values for high-precision models

Set the LEO to a 500 km altitude circular trajectory, the Mars target trajectory to 250×33813 km elliptical trajectory, and the LGA altitude to 40 km.

Table 3. Convergence results of high precision models.

Initial Guesses	t_d y/m/d	t_{dl} d	T_{EM} d	Δv_1 km/s	Δv_2 km/s	Δv_3 km/s	Δv_4 km/s	Total ΔV km/s
Pseudo	2028/12/05	1.6008	314.47	3.2540	0.3539	7.61e-05	1.1121	4.7201
MPC	2028/12/06	1.3015	306.53	3.3959	0.7931	0.0036	1.1700	5.3817
Pseudo	2031/01/13	1.5986	342.75	3.2833	0.9998	0.2298	0.8310	5.3439
MPC	2031/01/13	1.6490	302.65	3.2523	0.7881	0.3366	1.5038	5.8865
Pseudo	2033/04/18	1.2474	182.20	3.4992	0	0	1.3105	4.8097
MPC	2033/04/17	1.5497	215.80	3.3106	0	0.5834	1.2646	5.1586
Pseudo	2035/06/25	1.3118	199.98	3.4562	5.25e-05	1.45e-04	0.8764	4.3326
MPC	2035/06/25	1.2803	190.38	3.4795	0.9992	4.53e-05	0.9126	5.3913
Pseudo	2037/08/05	0.9219	355.78	3.9361	7.78e-15	4.59e-04	0.9357	4.8718
MPC	2037/08/04	0.9799	353.09	3.8370	6.52e-6	0.2852	0.8760	4.9982
Pseudo	2039/09/09	1.1260	340.99	3.6756	0	0.0022	0.7807	4.4585
MPC	2039/09/10	1.0452	345.85	3.7857	0.0150	1.22e-06	0.8837	4.6844
Pseudo	2041/10/13	1.3183	341.10	3.4386	0	2.99e-04	0.8581	4.2967
MPC	2041/09/10	1.4802	314.72	3.4442	0.0141	0.0021	0.8174	4.2778
Pseudo	2043/11/20	1.2519	314.22	3.4427	4.70e-04	2.71e-06	1.0079	4.4506
MPC	2043/11/20	1.2815	295.37	3.4142	0.1895	1.2711	0.9645	5.8393

Table 3 shows the convergence results of the high-precision ingestion model with MPC trajectories and pseudo trajectories as initial values, respectively. The simulation results show that the optimization results of using MPC trajectories as initial guesses are not satisfactory, which usually show larger velocity increments at the Lunar perigee and the DSM stitching point. Compared with the simulation results of MPC trajectories as the initial value, the optimization performance targets of high-precision trajectories are better when using pseudo trajectories as the initial value.

4.2. Results and analysis of the LGA window for human Mars exploration

Under the MATLAB environment with a computer CPU of 2.90GHz, the optimization time of a single trajectory does not exceed 1.5 s by using the pseudo theory model for the optimization calculation of the human Mars transfer trajectory. In this paper, the window search interval is set to 0.05 days, the search time range is 7 days, and the number of external cycles is 10 times. As shown in **Figure 5**, the window search results from 2031 to 2037 are used as an example to show the results of the window search method in this paper.

As shown in **Figure 5**, there is a longer window of 7 days in 2031, and the LGA effect is better; using the LGA strategy saves about 400 m/s of velocity increment than the direct transfer of the Earth-Mars transfer trajectory. The LGA window of 2033 and 2035 is typically about three days, and using the LGA strategy can reduce the velocity increment by about 200 m/s. The 2037 LGA window is more sensitive to departure time, and the optimal departure window can save 400 m/s of velocity increment over a direct transfer.

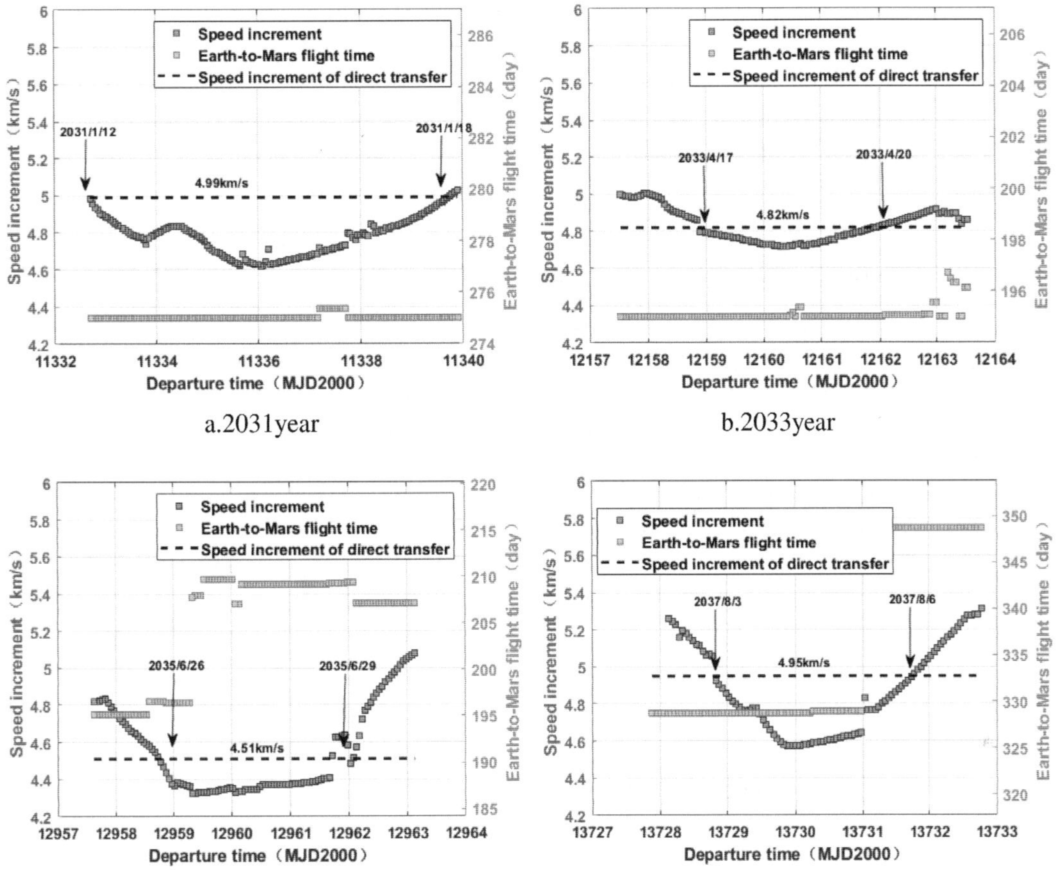

a.2031year

b.2033year

c.2031year

d.2033year

Figure 5. 2031 to 2037 LGA Earth-Mars transfer departure window search.

5. Conclusion

In this paper, a pseudo-theory-based method is developed for the initial design of the LGA human Earth-Mars transfer trajectory. The method is fast and easy to calculate, and the optimized trajectory of the method is beneficial to the convergence of the high-precision model as the initial guesses. The window search method designed in this paper can quickly evaluate the departure time, window length, and performance metrics of the human LGA Earth-Mars transfer trajectory.

References

[1] A. P. Paul, F. B. David, R. C. Christopher. Multiple Lunar Swingbys for Small Body and Planetary Mission. Proceedings of Spaceflight Mechanics Meeting, Albuquerque, NM, Feb, 1995. AAS 95-147;

[2] R. S. Wilson, " A Design Tool for Constructing Multiple Lunar Swingby Trajectories," M.S. Thesis, Purdue University, December 1993;

[3] Lorenzo Casalino. Design Of Lunar-Gravity-Assisted Escape Trajectories. The Journal of the Astronautical Sciences (2020) 67:1374-1390;

[4] C. R. Cassell, P. A. Penzo. LAMDA: a Software Tool for Lunar-Assist Escape Missions. Proceedings of AAS/AIAA Astrodynamics Specialists Conference, Halifax, Nova Scotia, Canada, Aug, 1995. AAS 95-405;

[5] C. R. Cassell. Optimization of Solar-perturbed Double Lunar Swingby Escape Trajectories. Proceedings of the AAS/AIAA Spaceflight Mechanics Meeting, Monterey, CA, Feb, 1998. AAS 98-165;
[6] R. V. Ramanan∗ and V. Adimurthy. Precise Lunar Gravity Assist Transfers to Geostationary Trajectories. Journal Of Guidance, Control, And Dynamic Vol. 29, No. 2, March-April 2006;
[7] LUO Z F, MENG Y H, TANG G J, et al. Dynamics and modeling of double lunar swing-by trajectories[J]. Journal of Astronautics, 2012, 33(10): 1361-1369(in Chinese);
[8] PENG Q B, HE B Y, ZHANG H L. Analysis of deviation propagation for trans-lunar free return trajectory[J]. Journal of Deep Space Exploration, 2016, 3(1): 56-60(in Chinese);
[9] S. W. Wilson, "A Pseudo state Theory for the Approximation for Three-Body Trajectories," AIAA Paper 70-1061, AAS/AIAA Astrodynamics Conference, Santa Barbara, California, August 1970;
[10] Yang B., Yang H. & Li S., "Pseudo state theory based iterative preliminary design method for powered gravity-assist interplanetary trajectories," Acta Astronautica, Vol.165(2019), pp.139-149;
[11] V. Ramanan R. & Adimurthy V., "Nonimpact Lunar Transfer Trajectories Using the Pseudo state Technique," *Journal of guidance, control, and dynamics*, Vol.28, No.2(2005), pp.217-225;

Nozzle Thrust Eccentricity Calculation Based on FLUENT Software

Yongzheng Hu [1*], Jun Chen [2], Xiaodong Shen [3] and Jiantao Dang [4]

[1]School of Mechanical Engineering, Nanjing University of Science and Technology, 210094, Nanjing, People's Republic of China.

[2]School of Mechanical Engineering, Nanjing University of Science and Technology, 210094, Nanjing, People's Republic of China.

[3]School of Mechanical Engineering, Nanjing University of Science and Technology, 210094, Nanjing, People's Republic of China.

[4]Sichuan Institute of Aerospace Systems Engineering, 610100, Chengdu, People's Republic of China.

e-mail: 1934413133@qq.com

Abstract. Thrust eccentricity refers to the phenomenon that the thrust vector deviates from the axis of symmetry. This phenomenon will produce lateral force and torque around the aircraft, which affects the normal flight of the aircraft, increases the required control torque, and reduces the density of uncontrolled rockets. There are three sources of thrust eccentricity: mass eccentricity, geometric eccentricity, and aerodynamic eccentricity. The aerodynamic eccentricity is caused by the asymmetrical flow of combustion gas in the nozzle. This paper uses FLUENT software to carry out a numerical simulation of a cone-shaped nozzle. By changing different factors such as the structure of the inner surface of the nozzle and the convergence radius, the influence law of the inner surface structure of different nozzles on the lateral force of the pneumatic eccentricity of the nozzle is explored. The following conclusions were drawn: With the expansion of the expansion half-angle, the second zero point moves toward the nozzle outlet. The relative radius of curvature has little effect on the main thrust of the nozzle but has a certain influence on the eccentricity of its thrust, and as the relative radius of curvature increases, the first and second zeros move toward the throat.

1. Introduction

The Thrust eccentricity is the phenomenon that the thrust vector deviates from the symmetry axis. It produces a lateral division and torque around the vehicle's centroid, causing greater dispersion and reduced density of field rockets. Thrust eccentricity is divided into three categories: mass eccentricity, geometric eccentricity, and pneumatic eccentricity. Thrust eccentricity caused by mass eccentricity and geometric eccentricity can be reduced by improving the design and manufacturing accuracy of the nozzle. Pneumatic eccentricity refers to the phenomenon of lateral force caused by the incident of gas and the coordinate axis at a certain angle [1-3]. Through theoretical analysis and experimental research, the technology designed to make the theoretical value of the thrust eccentricity of the Laval nozzle tend to zero is called micro-push bias nozzle technology, and the nozzle designed with this technology is called micro-push bias nozzle [4].

In this paper, fluent uses the computational fluid dynamics software to simulate the asymmetrical inflow conditions of the tapered nozzle, and by changing the geometric parameters of the nozzle (expansion half-angle, laryngeal radius, etc.), the influence of these nozzle geometric parameters on the lateral force of the nozzle and its Zero-Point position is explored by comparing the cloud map, contour plot, and lateral force value under different factors, and the relevant laws are obtained.

2. Numerical calculation method

2.1. Geometry

Here is a primitive tapered nozzle, and subsequent explorations are made on this nozzle to change factors such as the expansion half-angle, the length of the expansion section, the laryngeal radius, and the relative radius of curvature. The parameters are shown in Table 1, and the internal profile and mechanical relationship of the nozzle are shown in Figure 1.

Table 1. Nozzle geometry parameters

Parameter	Value
Laryngeal radius rt	7.0 mm
Laryngeal curvature radius Rt	3.5 mm
Convergence half-angle β	30 °
Expand by half an angle α	9.5 °
Expansion segment length X	70 mm
The convergence radius	18 mm

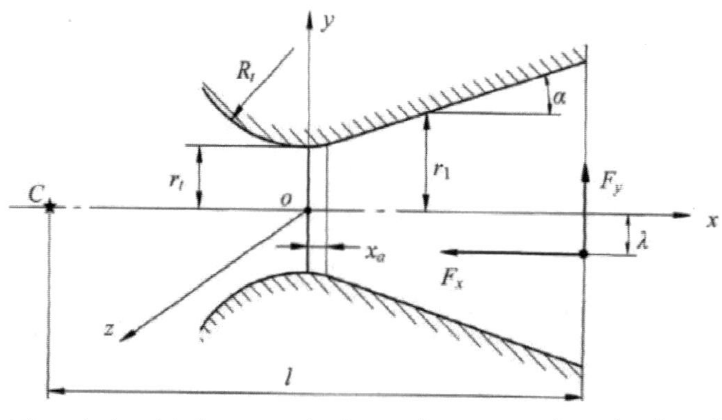

Figure 1. The relationship between the internal pattern and mechanics of the nozzle

2.2. Mesh model

The mesh adopts O-type slicing, grid encryption is carried out near the throat and boundary layer, the grid near the nozzle outlet is the sparsest, the number of meshes is 131774, and the mesh quality is as low as 0.764, which is of good quality and meets the calculation requirements. The grid diagram is shown in Figure 2.

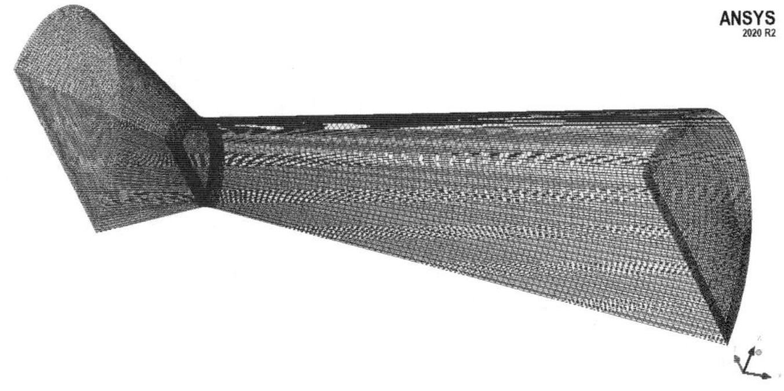

Figure 2. Tapered nozzle mesh

2.3. Calculation conditions
The gas conditions are as follows: $Rg = 320J/(kg \cdot K)$, $\gamma = 1.2216$. The gas is the ideal gas. Boundary conditions: pressure inlet (the pressure is 7 MPa, the total temperature is 3000 K), pressure outlet (the pressure is atmospheric pressure 101325 Pa, the total temperature is 300 K), the velocity direction perpendicular to the y-axis and z-axis angle is 89 ° and x-axis angle is 1 °. The wall surface adopts the slip boundary condition, and the symmetrical surface adopts the symmetrical boundary condition.

3. Calculation results and analysis

3.1. Nozzle thrust and lateral force calculation
Cloud and contour plots in Fluent can be directly processed by using the post-processing functions that come with Fluent, but they cannot directly give the thrust and lateral forces on different sections [5-6]. The important parameters such as pressure, temperature, velocity, density, and other important parameters on any cross-section can be known, and the calculation formula of thrust and two lateral forces (1, 2, 3) is given in [2], and from the formula, it can be seen that the speed, density, and pressure can be calculated on the cross-section to indirectly derive the main thrust and lateral force on each cross-section.

$$F_x(x) = \int_A (p + \rho u^2) dA \quad (1)$$

$$F_y(x) = \int_A \rho uv dA \quad (2)$$

$$F_y(x) = \int_A \rho uv dA \quad (3)$$

3.2. Effect of the expansion half-angle on thrust eccentricity
To explore the influence of different expansion half-angles, two conical nozzles with expansion half-angles of 10° and 10.25° are designed, and the remaining conditions are the same as the 9.5 ° expansion half-angle nozzle, and the other conditions applied are the same as above. To know exactly the specific relative thrust and lateral forces of these two nozzles need to be calculated by integrating Equations (1, 2, 3). The main thrust and lateral force curves are shown in Figures 3-4.

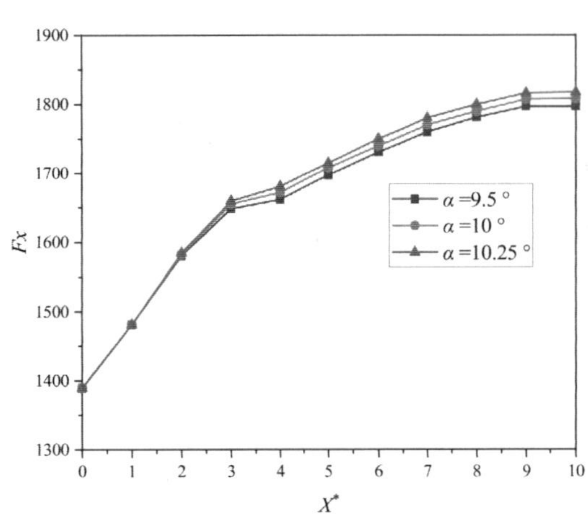

Figure 3. The relative main thrust of different expansion half angles

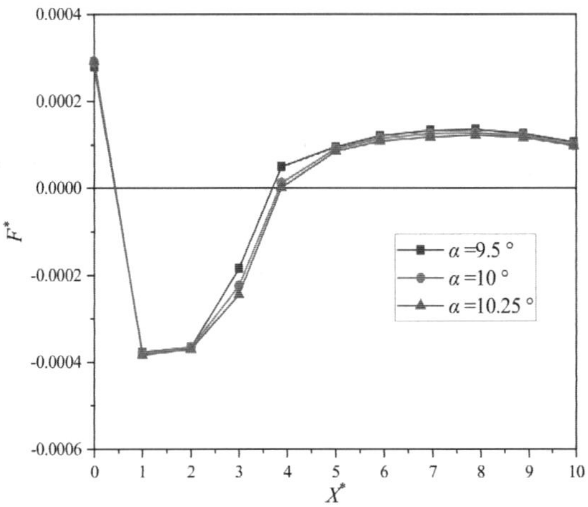

Figure 4. The relative lateral forces of different expansion half angles

From Fig 3, the main thrust of the nozzle increases slightly with the increase of the expansion half-angle. The first zero of the three curves in Fig 4 almost overlaps and the second zero point is different, with the second zero point at 10.25 ° being the closest to the larynx and the second zero at 9.5 ° being closest to the larynx. This indicates that the second zero point moves toward the nozzle outlet as the expansion half angle is expanded. As the expansion angle increases, the degree of expansion of the airflow at the same axial position increases, that is, the Mach number increases, resulting in a smaller Mach cone angle, which slows down the propagation of pressure waves downstream and laterally [7-8].

3.3. Effect of different radii of curvature on nozzle thrust eccentricity

The relative radius of curvature is the radius of curvature divided by the radius of the larynx, which has little effect on the geometry of the nozzle. When other factors such as the expansion of the half-angle and gas parameters are limited, the relative lateral force can be adjusted by adjusting the relative radius of curvature. To eliminate the influence of the laryngeal radius, the relative radius of curvature is controlled by controlling the radius of curvature unchanged by 7 mm. Two new nozzles are redesigned with a radius of curvature of 7 mm and 10.5 mm, according to the definition of their relative curvature radii of 1 and 1.5. They are put together with the nozzle in the previous 2.1 (relative curvature radius of 0.5), still, the same cross-section is taken as above, the relative thrust is calculated and compared, and the curve chart is as follows.

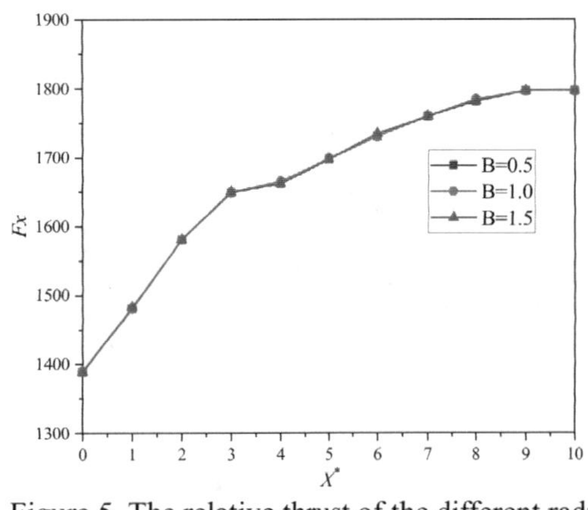

Figure 5. The relative thrust of the different radii of relative curvature

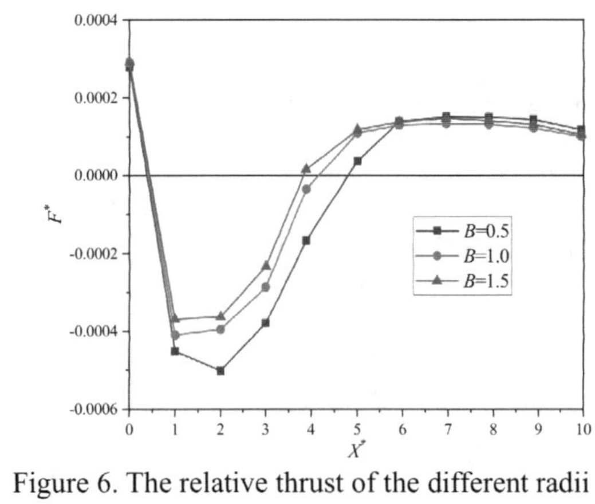

Figure 6. The relative thrust of the different radii of relative curvature

As can be seen from Figure 5, the relative principal thrust is almost unchanged with the increase of the relative radius of curvature, and it can be assumed that the relative thrust does not change with the change of the relative radius of curvature. In Figure 6, the relative radius of curvature has a significant effect on the lateral force, and as the relative radius of curvature increases, the first and second zero points both move toward the larynx. The movement of the first zero point is small, while the movement of the second zero point is obvious. Since the relative radius of curvature has little effect on the overall size of the nozzle, its influence on the size of the nozzle is more concentrated in the larynx, so the relative radius of curvature is an effective way to change the position of the lateral force zero point without making large changes to the nozzle, and changing the relative radius of curvature does not affect the relative thrust. The amplitude of the three curves does not follow a very obvious law, and the oscillation decreases with the expansion segment. In terms of the absolute value of the lateral force, the relative lateral force curve with a relative radius of curvature of 0.5 is higher than the other two curves [9-11].

4. Conclusions

In this paper, the numerical simulation calculation of the tapered nozzle is carried out by Fluent software, the calculation results of the thrust and lateral force of the nozzle are obtained, and the relative thrust and relative lateral force extension x-axis change curve of the nozzle are depicted, and the change law of relative thrust and relative lateral force is observed. Then, by changing the different geometric parameters of the tapered nozzle, the influence of these geometric factors on the ejection characteristics is explored, and the following conclusions are obtained:

- The half-width of the nozzle expansion has a certain influence on its thrust eccentricity characteristics, and the second zero point moves toward the nozzle outlet as the expansion of the expansion half-angle is expanded.
- The relative radius of curvature has little effect on the main thrust of the nozzle, and there is a certain influence on the eccentricity of its thrust, and as the relative radius of curvature increases, the first and second zero points move to the larynx.

After this study, it is found that the geometric parameters of the nozzle have a certain influence on its thrust eccentricity characteristics, which plays a decisive role in the design of the micro-thrust eccentric nozzle. For future researchers, the thrust eccentricity characteristics of gas-solid two-phase flow nozzles can be investigated by considering nozzle flow including particle phases.

References

[1] Chen, J., Wang, D., Feng, F. (2013) The principle and progress of modern aircraft propulsion. Tsinghua University Press.

[2] Wu, X.S., Chen, J., Wang, D. (2006) Numerical Simulation of Solid Rocket Motor Working Process. Higher Education Press.

[3] Chen, J., Wang, D., Feng, F. (2013) The principle and progress of modern aircraft propulsion[M]. Tsinghua University Press.

[4] Wang, D. (2012) Theoretical and experimental study of the deflection characteristics of Laval nozzles. The Nanjing University of Technology.

[5] Wu, X.S., Wang, Z.S., J, Z.D. (1998) Thrust Eccentricity characteristics of double-arc supersonic nozzles. Military Engineering Journal, 19 (2):126-129.

[6] Qiu, G.Z., Wu, X.S. (1999) New Methods for Microdepresential Nozzle Design [J]. Journal of Military Engineering, 20 (1):24-27.

[7] Wu, X.S., Wang, Z.S., J, Z.D. (1998) Thrust Eccentricity characteristics of double-arc supersonic nozzles Military Engineering Journal, 19 (2): 126-129.

[8] Zang, Q.M., Wu, X.S., Wang, D. (2005) Study on the eccentricity of rocket thrust. Journal of the Nanjing University of Technology: Natural Science Edition, (03): 257-259.

[9] Zhou, C.S., Xu, B.Q. (1998) Study on Error Sources Affecting Thrust Eccentricity Characteristics of Solid Rocket Motor[J]. Journal of Nanjing University of Science and Technology, (004):293-296.

[10] Erdem, E., Kontis, K., Albayrak, K., (2008) Numerical Investigation of Flowfield Inside a Conical Rocket Nozzle in the Presence of Transverse Injection. AIAA - 359.
[11] Knauber, R.N. (2015) Thrust misalignments of fixed-nozzle solid rocket motors[J]. Journal of Spacecraft & Rockets, 33(6):794-799.

Simulink-Based Simulation of Gas Pilot Servo Actuator for Aviation Ejection Seat

Congcong Zhou[1*], Jun Chen[1], Jianing Duan[1], Ruiqiang Wu[2]

[1]School of Mechanical Engineering, Nanjing University of Science and Technology, 210094, Nanjing, People's Republic of China.

[2]Shanxi Beifang Xingan Chemical Industry CO., LTD., 030000, Taiyuan, People's Republic of China.

E-mail: zhoucongcong_c@163.com

Abstract. The ejection seat needs the ability of rapid attitude adjustment to save lives, and the gas servo steering engine is a feasible scheme for correcting the seat position. The three closed-loop control system of the permanent magnet synchronous motor is built to realize the rotor position loop control of the motor; the working principle of the gas steering gear is analyzed, and the mathematical model of the gas servo system is constructed; Simulink is used to simulate the position of the motor and gas valve; the simulation results show that the steering gear can respond quickly to the control command, and the torque to the rudder surface generated by gas is large enough to meet the needs of the rudder blade to overcome the torque generated by the erosion of the rudder surface and the timeliness of the ejection seat.

1. Introduction

The ejection seat is the main carrier and tool for military aircraft pilots. To escape from the aircraft under low altitude adverse attitude and high-speed flight conditions, the ejection seat requires rapid attitude adjustment and life-saving capability [1]. As a position (angle) servo controller, the steering gear is suitable for those control systems that require constant angle changes and can be maintained [2]. According to different power sources driving rudder blade deflection, it can be divided into pneumatic steering gear, hydraulic steering gear, and electric steering gear [3]. Among them, the gas steering gear is characterized by a relatively small mass, compact structure, short working time, and the need for a filter device to obtain clean gas [4]. The working time required for attitude correction of the ejection seat is short, so it is a feasible scheme to use a gas servo steering gear as an ejection seat to correct the position of the ejection seat during adverse attitude ejection. Therefore, the research on the whole dynamic process of the gas steering engine, combined with the specific structure, has a certain practical significance and value for the development of the ejection seat vector control system.

In this paper, the permanent magnet synchronous motor is combined with the gas steering gear to control the motor rotor in the gas source part, distribute high-pressure gas to the pneumatic actuator at a certain working frequency, and the gas drives the rudder blade to deflect. The mathematical models of the three-phase permanent magnet synchronous motor and gas servo valve are constructed respectively, and the dynamic response curve of the rudder deflection angle is obtained through simulation using Simulink, and the curve is analyzed to provide some reference for the research of ejection seat attitude control system.

2. Modeling and Simulation of Electric Servo Valve

A permanent magnet synchronous motor has the advantages of high efficiency, high precision, a small moment of inertia, etc., and is widely used in steering gear control [5]. The permanent magnet synchronous motor is used to distribute high-temperature gas to the pneumatic actuator so that the working pressure of the cavities at both ends of the piston is dynamically balanced. The piston drives the mechanical connecting rod to drive the rudder blade to deflect, so as to realize the control of the gas steering engine. Permanent magnet synchronous motor adopts PID three closed-loop control, namely current loop, speed loop, and position loop three closed-loop control system. The rotor is controlled to deflect at the specified angle by giving the positioning ring position signal. The rotor shaft is nested with a swing block structure, which deflects within ± 8 ° to control the distribution of gas in the actuator cavity.

2.1. Establishment of the mathematical model of permanent magnet synchronous motor

To simplify the mathematical model of a permanent magnet synchronous motor, the following assumptions are made:

- Ignore motor core saturation.
- Neglect eddy current and hysteresis loss in the motor.
- Neglect the space harmonic effect in the magnetic field.
- Each phase of winding is symmetrical [6].

The mathematical model in the q and d-axis coordinate system is as follows:

$$\frac{d}{dt}\mathrm{i}_d = \frac{1}{L_d}v_d - \frac{R}{L_d}i_d + \frac{L_q}{L_d}p\omega_m i_q \tag{1}$$

$$\frac{d}{dt}\mathrm{i}_q = \frac{1}{L_q}v_q - \frac{R}{L_q}i_q + \frac{L_d}{L_q}p\omega_m i_d - \frac{\lambda p\omega_m}{L_q} \tag{2}$$

$$T_e = 1.5p\left[\lambda i_q + \left(L_d - L_q\right)i_d i_q\right] \tag{3}$$

Where, $\mathrm{L_d}$ and L_q are d-axis and q-axis inductances; R is the resistance of stator winding; i_d and i_q is d-axis and q-axis current; v_d and v_q is the voltage of axis d and axis q; ω_m is the angular velocity of the rotor; λ is the magnetic flux amplitude induced by the rotor permanent magnet in the stator phase; P is the polar logarithm; T_e is electromagnetic torque.

2.2. Simulation of the electric servo valve

SVPWM control is mainly composed of a sector judgment module, space voltage vector working time module, switching time module, and PWM generation module. Its simulation flow chart is shown in Figure 1.

The permanent magnet synchronous motor servo system based on $i_d = 0$ MATLAB/Simulink is built. The current i_d and i_q output by the current controller and the control voltage signal of the drive circuit are obtained through inverse Park transformation. The three-phase inverter control signal is obtained through seven-segment SVPWM algorithm modulation, and the three-phase electric motor is output to control the operation of the motor. The three-phase current obtained from the motor sensor is fed into the feedback current i_d and i_q PI controller through Clarke transformation and Park transformation to achieve closed-loop control. The selected motor pole pair is 4, and the sensor output is a mechanical angle. Therefore, in the current feedback, the angle input when the three-phase current i_a, i_b and i_c are converted into i_d and i_q, for which coordinates must be multiplied by the motor pole pair number 4 to turn into an electrical angle. The simulation diagram of zero d-axis current control is shown in Figure 2.

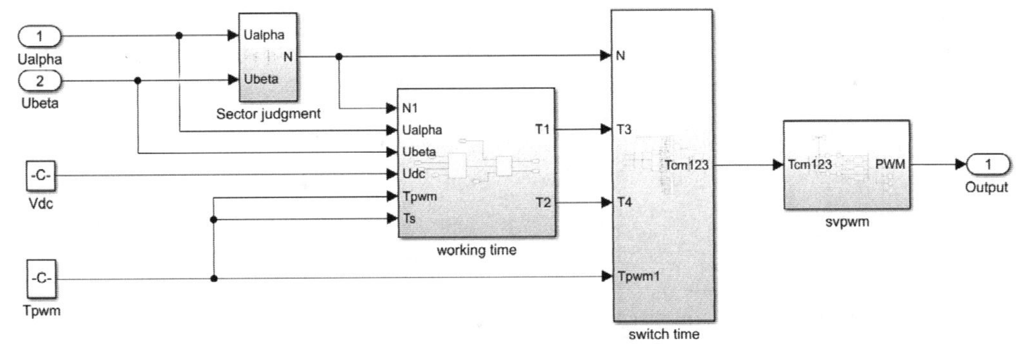

Figure 1 SVPWM control simulation

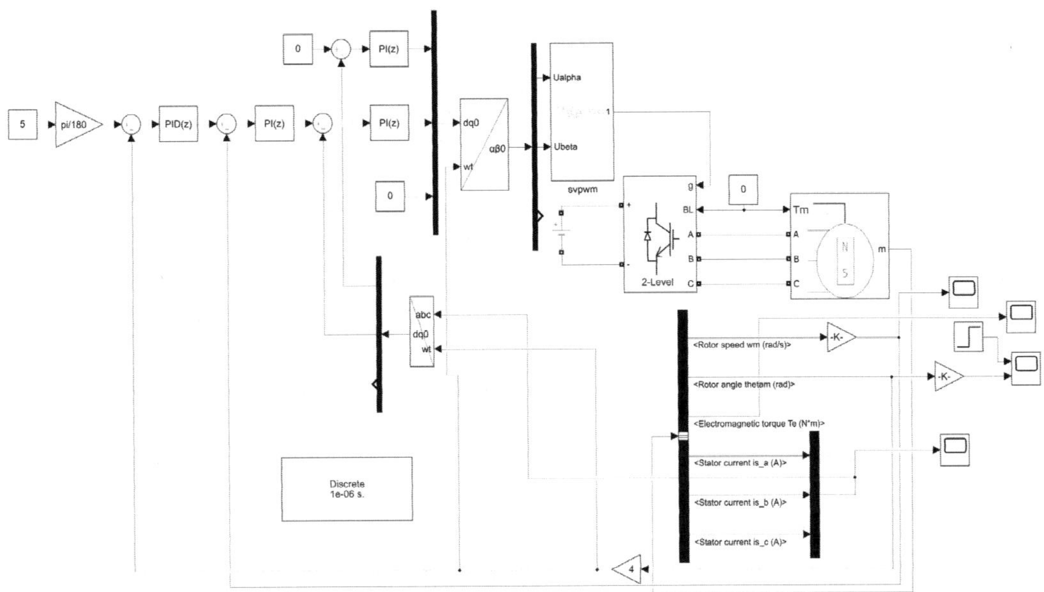

Figure 2 Zero d-axis current position servo simulation

3. Modeling and Simulation of the gas servo valve

The fuel gas generated by the rocket engine passes through the electric servo valve to control the airflow into the two ends of the actuator cylinder and acts on the piston. The piston is driven to produce displacement by generating pressure difference at both ends of the piston. The piston is connected with the rudder blade support through the connecting rod mechanism, and the linear displacement of the piston is converted into the angular displacement of the rudder blade. When the swing block of the electric servo valve deflects to the right, the gas inlet of the gas splitter seat is connected with the right channel of the gas outlet, and the gas cleaned by the filter screen is introduced into the right chamber of the actuator. The left chamber of the actuator is combined with the atmosphere, generating a pressure difference to drive the piston to move to the left, driving the spoiler to produce a deflection angle. The potentiometer on the spoiler seat feeds back the deflection signal, and the controller compares the control signal with the feedback signal of the potentiometer. When the spoiler deflects to the specified angle, the swing block of the electric valve will swing to zero, which will make the pressure at both ends of the actuator produce a new dynamic balance, force the piston to stop moving, and realize the spoiler angle control.

3.1. Establishment of the mathematical model of the gas valve

To establish the mathematical model of the gas servo valve, the following assumptions are made:

- Gas is an ideal gas with a viscosity-free, pure gas phase and constant specific heat capacity.
- The flow process of gas in the gas chamber is quasi-static, and the state parameters of the working medium at each point in the control chamber are the same at any instant.
- Ignore the influence of gravity and other penetrating forces on the flow [7].

The whole working process of a gas servo valve can be represented by a set of thermodynamic differential equations, including a gas state equation, mass conservation equation, and flow equation [8-9].

According to the equation of the state of the ideal gas,

$$\mathrm{pV} = mRT \tag{4}$$

By time differentiation of Equation (4), we can get

$$\frac{d}{dt}\left(\frac{pV}{T}\right) = R\frac{dm}{dt} = \left(q_i - q_e\right)R \tag{5}$$

Where q_i and q_e are the flow rates at the inlet and outlet of the actuator respectively.

If the cross-sectional area of the piston is S, the displacement of the piston is x, the volume of one side air chamber is V, and the pressure in the chamber is p, the above formula can be changed to

$$\dot{p} = \frac{k}{V}\left(q_i RT - q_e RT_e - p\dot{x}S\right) \tag{6}$$

$$\dot{T} = \frac{T\dot{x}S}{V} + \frac{T\dot{p}}{p} - \frac{RT^2\left(q_i - q_e\right)}{pV} \tag{7}$$

The mass flow entering the actuator can be expressed as

$$Q_m(\mu, f, p_0, p) = \begin{cases} \dfrac{\mu A p_0 C}{\sqrt{T_0}} & \dfrac{p}{p_0} \le C_{t1} \\ \dfrac{\mu A p_0 B}{\sqrt{T_0}} \Phi\left(\dfrac{p}{p_0}\right) & \dfrac{p}{p_0} \ge C_{t1} \end{cases} \tag{8}$$

The mass flow rate leaking from the piston is

$$Q_n = \frac{\pi \mathrm{D}' \Delta_D^3 \left(P_1 - P_a\right)}{12 vl} \tag{9}$$

The outlet flow is laminar flow, where n is taken as 2 and 4, D' is the outer diameter of the clearance, Δ_d is the radial clearance of the piston, P_1 is the pressure in the cavity, P_a is the atmospheric pressure, v is the kinematic viscosity coefficient, $0.15\times10^{-4} m^2/s$ and l are the piston clearance length, taken as 4 to 12 mm.

The motion equation of the rudder blade is

$$\begin{cases} F_d L = I\dfrac{d^2\theta}{dt^2} - k_\theta\left(\theta_0 - \theta\right) \\ y = L\theta \end{cases} \tag{10}$$

Where $F_d L$ is the moment of the piston connecting the swivel arm, I am the moment of inertia of the rudder blade, k_θ is the load force proportional coefficient, $\theta_0 = 15\times\dfrac{\pi}{180}$.

3.2. The Simulation of the gas servo valve

The pressure simulation is shown in Figure 3. The piston is initially calculated in the intermediate state, and the initial temperature T_s Takes 1000 k, and the gas constant is 287. The cross-sectional area of the piston is $50.2mm^2$, and the initial volume of the left and right chambers is $201.1mm^3$.

Sector selection module calculation

$$N = A + 2B \tag{11}$$

Where A and B are

$$A = \begin{cases} 1 & x \geq angle \\ 0 & x < angle \end{cases} \quad B = \begin{cases} 1 & x \leq angle \\ 0 & x > angle \end{cases} \tag{12}$$

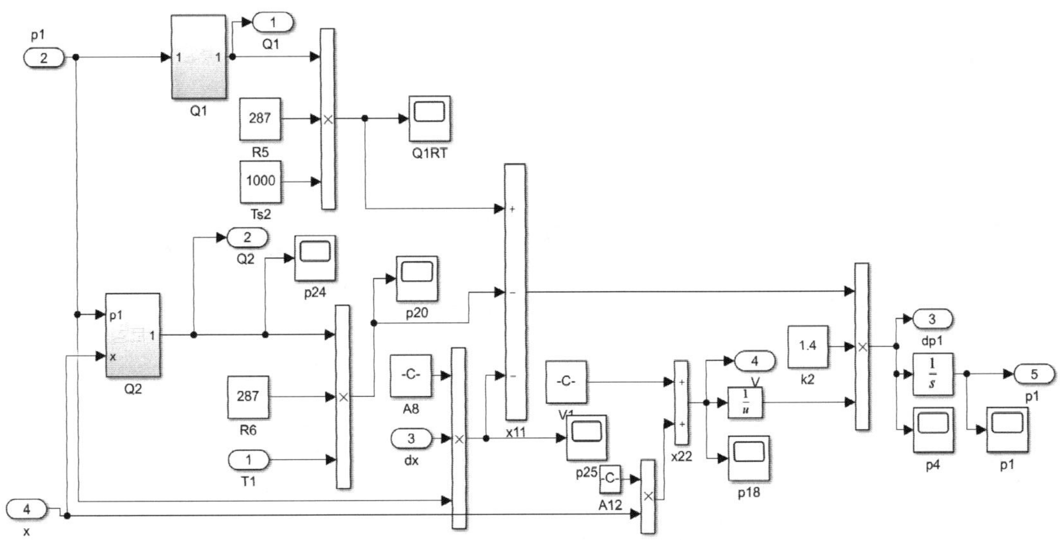

Figure 3 Pressure simulation module

Figure 4 shows the gas servo simulation model, including the pressure simulation module, temperature simulation module and displacement output module at both ends of the piston.

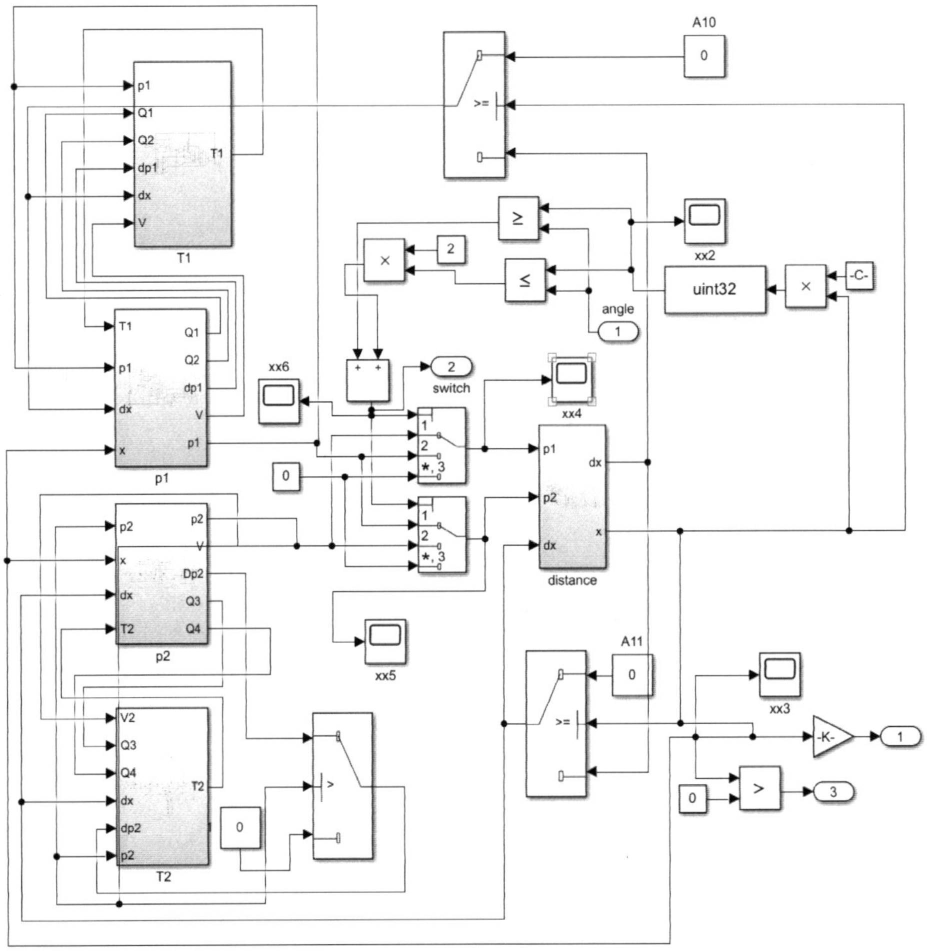

Figure 4 Gas servo simulation model

4. Analysis of simulation results

A step signal with a step time of 0 s and a final value of 5 is given as the motor position signal, and the simulation time is set as 0.1 s to simulate the position servo of the motor with a deflection angle of 15 degrees. As can be seen from Figure 5, the motor has a fast response speed, short adjustment time, and no overshoot.

Given the rudder blade angle of 16 degrees, the gas servo valve is simulated, and the simulation curve is shown in Figure 6. After the motor deflects, the gas begins to enter the actuator under the action of the splitter plate, which drives the piston to move. It can be seen that there is a certain vibration in the curve, which is caused by only three modes of gas distribution at both ends of the piston. However, the response of gas action is fast, the frequency response is high, and the torque generated is large, which can enable the rudder blade to overcome the torque generated by 2.7 gas flushing the rudder surface.

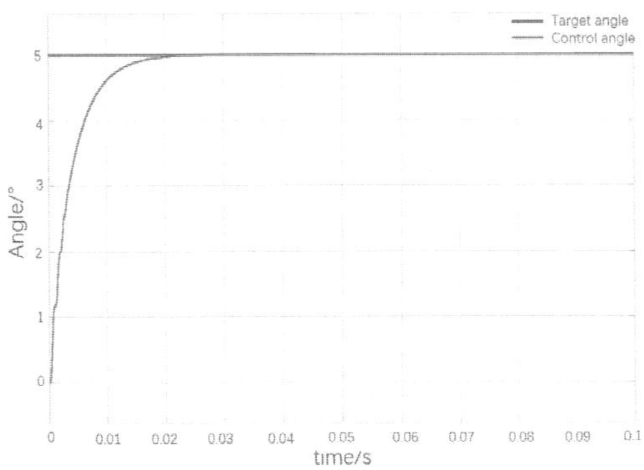

Figure 5 Simulation of motor position servo angle control

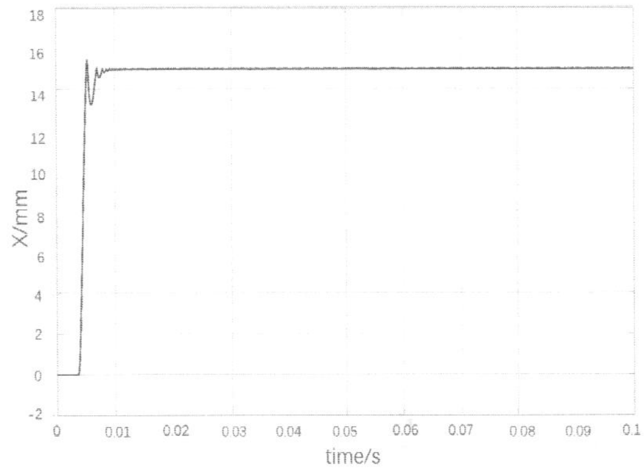

Figure 6 Simulation of rudder blade

5. Conclusion

In this paper, the basic mathematical model of the permanent magnet synchronous motor is expanded by combining the permanent magnet synchronous motor with the gas valve. The angle deflection of the motor is controlled by the three closed-loop nested structures and the SVPWM control algorithm. After that, the mathematical model of the gas steering engine is derived, and the gas servo system is built to realize the deflection of the spoiler angle. Starting from the attitude adjustment system of the rocket engine, this paper studies the working process of the gas valve obtains the working curve of the gas valve from power on to the control surface response and obtains the dynamic response parameters of the gas valve, which provides a certain reference and practical value for the specific structural design and development of the vector control steering gear of the subsequent aviation ejection seat. The influence of gas pressure, power supply voltage, and other layers on the motion response of the rudder blade can be further studied later.

References

[1] Fan, H., Lian, B., Wang Z.T. (2012) Analysis of the Development on the Rocket Motor for Ejection-seat, J. Advances in Aeronautical Science and Engineering, 03:279-283.

[2] Cai, R.Y. (2012) Design of Servo Control System Based on Arduino. J. Computer Knowledge and Technology, 15:3719-3721.

[3] Su, X., Gui, C.B., Zhou, Y.J. (2008) Research Status and Key Technology Analysis of Steering Gear of Guided Ammunition. J. Aerodynamic Missile Journal, 11:48-50.

[4] Ran, H.K. (2012) Improved design and research of a pneumatic servo. D. Nanjing University of Science and Technology.

[5] Wang, X., Ma, R.Q., Ji, P.P. (2010) Simulation of PMSM position servo system based on fuzzy PI control. J. Micromotors, 03:52-55.

[6] Zhao, S.B., Bi, Y.L., Zhang, Y. (2020) Research on Speed Control System of Permanent Magnet Synchronous Motor Based on SVPWM. J. Journal of Heilongjiang Hydraulic Engineering College, 01:63-68.

[7] Wu, X.S., Chen, J., Wang D., (2015) Principles of Solid Rocket Motor. M. Ordnance Industry Press.

[8] Wu, X.S., Chen, J., Wang D. (2005) Gas dynamics of solid rocket engines. M. National Defense Industry Press.

[9] Wu, Q. (1995) Pneumatic Engineering Manual. M. Beijing National Defense Industry Press.

A Space-ground Joint Target Detection System Based on Artificial Intelligence

Jinghua Wang[1], Junqing Shi[1,*], Yong Lei[1], Xinyu Yao[1], Chen Chen[1], Xiyan Dong[2]

[1]Institute of Remote Sensing Satellite, Beijing, China

[2]Beijing Institute of Technology, Beijing, China

[*]shijunqing1996@gmail.com

Abstract: Due to the low efficiency and poor adaptability of template matching in target detection, this paper represents a space-ground joint detection system based on artificial intelligence, which takes advantage of the high computation power of the ground system to train neural networks and utilize the measurement and control channel to update and synchronize the on-orbit system. Based on YOLOv5, an oriented bounding box is used for labeling and training, the angle is considered as a class and the loss for the angle is added to the code. After training the network with the ground system, the weight will be sent to the satellite. This process would repeat once there are new training data that worse the update. Based on the new routine for target detection, the accuracy and robustness are significantly higher than before.

1. Introduction

Target detection by remote sensing satellites usually refers to the use of different image process methods to acquire and label regions of interest from images, including roads, aircraft, ships, and vehicles. Target detection is an important part of remote sensing analysis and application, and an important basis for monitoring areas such as harbors and airports.

In recent years, with the launch of modern satellites and the rapid development of remote sensing technology, getting high spatial resolution images has become much easier. These images contain rich texture information of the target, which provides the possibility of locating targets in a large scene. Due to the increase in background complexity and target quantity, traditional template-matching methods are becoming increasingly inefficient to extract targets from the large scene [1, 2].

Due to the support of big data, artificial intelligence has become the future of target detection. Currently, target detection methods fall into two main categories: two-stage detection represented by R-CNN, and one-stage detection represented by YOLO and SSD. In the long run, as single-stage detection greatly simplifies the framework of two-stage detection, it not only speeds up the inference process but also maintains relatively high precision, making it more suitable to use in systems that require a high degree of effectiveness [3].

This paper builds a joint space-ground target detection system with a rotated bounding box that can predict target orientation based on YOLOv5. The system relies on the ground server to repeatedly train the constantly updated training data, the latest weight is then uploaded to the satellite to improve the preciseness of the on-orbit target detection system.

This system can not only significantly speed up the detection process and improve the detection rate, but also have the capability of increasing the recognition rate of any region of interest by simply updating the weight information from the ground server. Compared to the traditional image processing template matching method [4], this system greatly improves the efficiency of the target detection process. The major contributions of this paper are listed below:

- Trian model based on YOLOv5 with rotated bounding box with remote sensing images data from NWPU;
- Add a new function to the ACUITY toolkit to make it possible to quantify different layers into different precision while exporting the ONNX framework to the OpenVX framework;
- Rewrite the NMS in C++ that support rotated IoU;
- Compare the execution time and detection rate of the Yulong 810A AI chip running the quantified model in the OpenVX framework with a ground server with Tesla P100 running the original Pytorch model.

2. System platform framework

The hardware setup of the space-ground joint system mainly consists of three parts: ground GPU server, satellite on-orbit high computing power platform, and calibration personnel. As shown in Figure 1, the ground GPU server acts as the center of the whole system, and its main function is to receive new data sets from the manual calibration of the target of interest, train them and finally update the optimal weights to the satellite in orbit through the measurement and control channel.

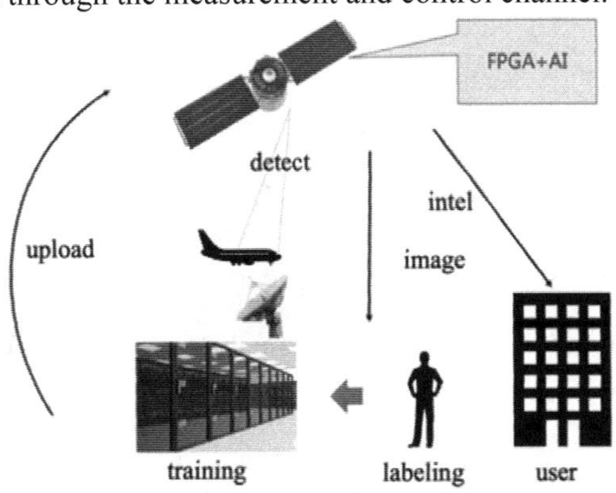

Figure 1. Space-ground joint system

The high-performance processing unit on the satellite is the Yulong 810A from Orbit. The chip has a heterogeneous architecture. The processor (CPU) is equipped with a four-core ARM A9 and four-core SPARC V8, an eight-core graphics processing unit (GPU), and an eight-core neural network acceleration unit (NNA), and the peak processing capacity of INT8 can reach 12TOPS. The chip has astronaut-level anti-radiation performance and has been used by satellites in orbit.

The open-source tool "labeling" is used to label the objects in the newly acquired remote sensing images with a rotated bounding box (OBB) [5]. The generated DOTA dataset was converted into YOLO dataset by a format conversion tool in MATLAB and then imported into the server together with the image for training.

The ground GPU servers are based on the NVIDIA DGX™ series, which features four different levels of solutions to meet different needs, including the NVIDIA DGX™ A100, which is compact and easy to deploy and contains eight NVIDIA A100 GPUs with a total of 640GB of video memory, enough performance for average-sized intensive computing tasks and the world's most secure AI system solution in the world. Its peak INT8 performance is up to 10petaOPS, providing up to 83 times throughput compared to CPU alone. The high-performance ground server significantly reduces the convergence time of the network, thus improving the cycle time of on-orbit updates and always keeping the on-orbit detection system at its peak performance

3. Artificial intelligence-based target detection and recognition

3.1. YOLOv5 detection model

YOLOv5 was released in June 2020, at just 27MB in size (compared to 244MB for darknet YOLOv4), it is a very lightweight model while still matching the YOLOv4 benchmark in terms of accuracy. Like other YOLO detection models, YOLOv5 can be divided into four main parts: input, backbone, neck, and prediction. Figure 2 illustrates the main structure of YOLOv5:

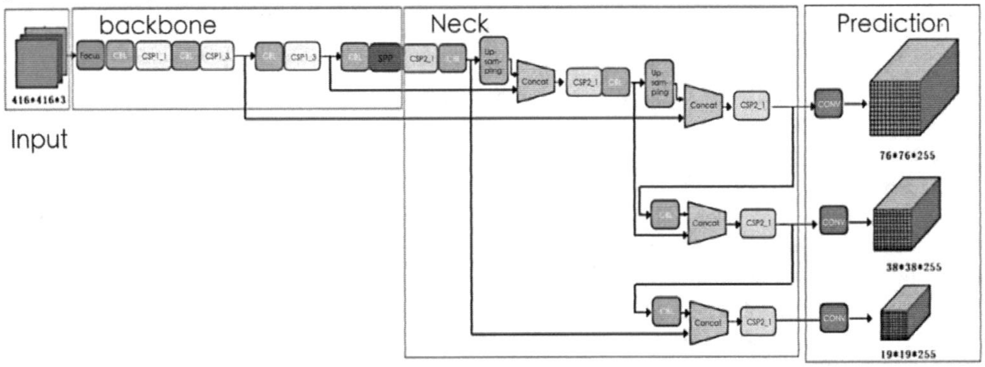

Figure 2. Main structure of YOLOv5

The input part can realize data enhancement, adaptive anchor calculation, and adaptive image scaling. The feature extraction part mainly uses the focus structure, which can complete the slicing and winding machine, and the CSP structure to enhance the learning ability of the network. Because the Focus and CBL of different networks have different convolution kernels and the number of residual modules, the performance of the model can be improved by changing the width and depth of the network. The neck uses FPN and PAN structures to enhance the fusion ability of the network using information extracted from the backbone. The output layer is divided into three volume base channels, and the final prediction result containing target information in the output image is suppressed by calculating the loss function and non-maximum value [6].

3.2. Arbitrary rotation of the bounding box

The application of the original YOLOv5 model is aimed at the target under the natural field of view and uses a horizontal bounding box (HBB). When the viewing angle changes, the feature of the object will change. To better match the image features, the labeling method needs to be changed. From a remote sensing perspective, orientation features are more obvious than in the natural field of view, so the rotated bounding box can be used to calibrate the region of interest more accurately. In this paper, the definition method of arbitrary rotation box is point-based ($[x_1y_1, x_2y_2, x_3y_3, x_4y_4]$), and the leftmost points are arranged counterclockwise as the starting point, as shown in Figure 3.

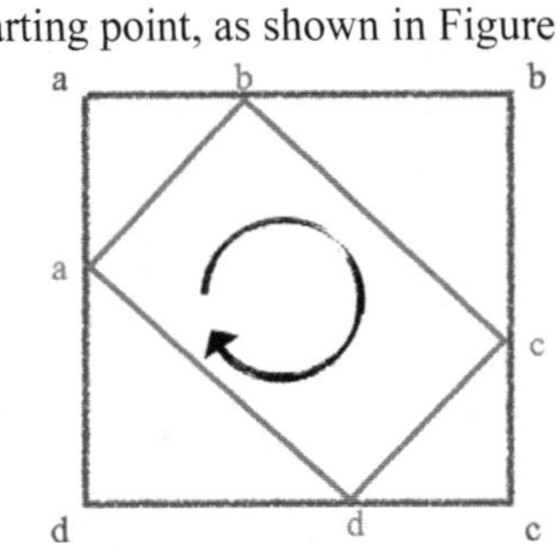

Figure 3. Oriented bounding box

The problem with using a rotated bounding box is that the angle is continuous while the classification result is limited to a certain number. One solution is to set every degree into different classes using the one-hot label as shown in Fig 4 below:

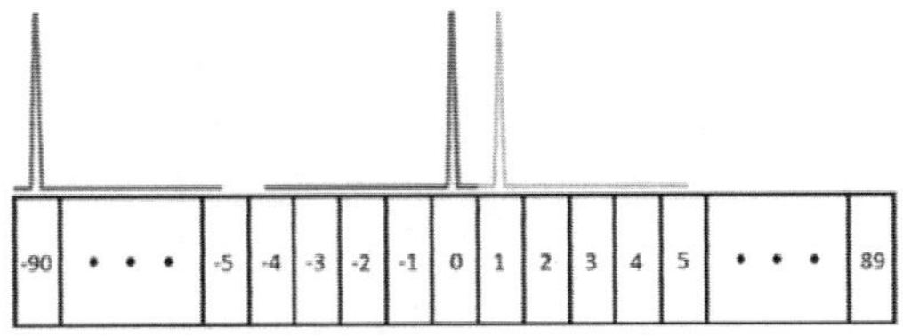

Figure 4. One-hot label

Expect accuracy loss ($E(loss)$) and max accuracy loss ($Max(loss)$) using this method can be calculated with Equations (1) and (2):

$$Max(loss) = \frac{\omega}{2} \tag{1}$$

$$E(loss) = \int_a^b x \times \frac{1}{b-a} = \int_0^{\frac{\omega}{2}} x \times \frac{1}{\frac{\omega}{2}-0} = \frac{\omega}{4} \tag{2}$$

It is feasible to convert angle prediction into a classification problem since the accuracy loss is acceptable. However, if the ground truth is 0°, the accuracy loss for predicting it as 1° or -90° is the same, which is problematic. To solve this problem, Circular Smooth Label (CSL) is applied (Fig.5):

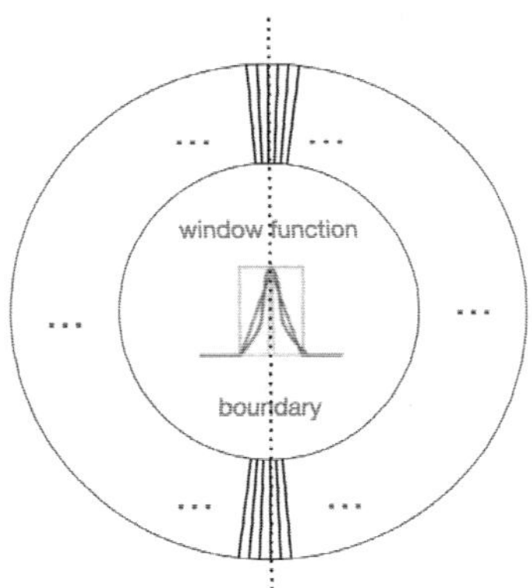

Figure 5. Circular Smooth Label

The mathematical expression of CSL is as below:

$$CSL(x) = \begin{cases} g(x), \theta - r < x < \theta + r \\ 0, otherwise \end{cases} \tag{3}$$

where $g(x)$ is window function, and r is window size.

Due to the use of the window function, to some degree, accuracy loss can gradually become smaller when the prediction gets closer to the ground truth. Notice is the window size is too small, Circular Smooth Label is equivalent to a one-hot label.

3.3. Data set preparation and preprocessing

In this paper, the remote sensing image scene NWPU-RESISC45 created by Northwestern Polytechnical University is used as training data [7], which contains a total of 315,000 images with a size of 256*256, covering 45 scenes.

Using the aircraft category as the experimental object, the images are divided into a training set and a test set according to the ratio of 7:3. The aircraft target was manually calibrated using the rotation box in the "rolLabelImg" tool. Compared with the traditional horizontal frame, it can better distinguish the target from the background, and provide less redundant information for the network, which is helpful to constrain the training director of the network and reduce the convergence time of the network.

Before training, images in the training set are randomly selected to be flipped left and right, then flipped up and down. Mosaic and aliasing image enhancement methods are also done to increase the richness of samples and to improve the robustness of the detection network. The pre-process result is shown in Figure 6.

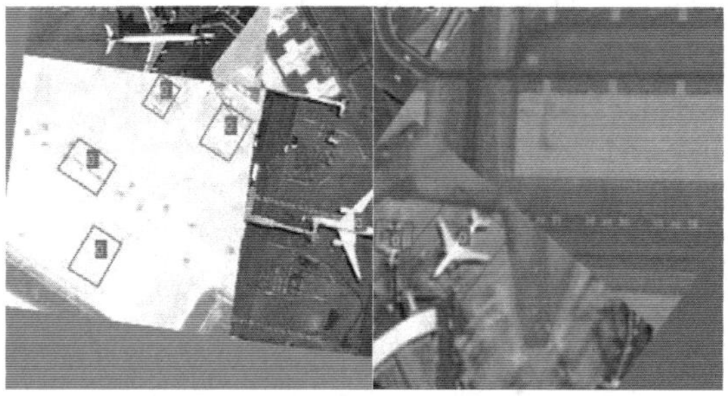

Figure 6. Mosaic and Mixed-up comparison

4. Experiment and analysis

The YOLOv5 source code containing the detection of rotating anchor frame adopts the YOLOV5_OBB project of Kaixuan Hu on GitHub, which uses the GPL-3.0 protocol and can be used and modified as an open-source project for free [8].

A laptop with a six-core, 12-thread I7-9750H CPU and a Tesla P100 GPU running Linux 18.04 was selected as the test device.

Experiment 1 used only 110 images as training data and paused the training after 187 epochs to run detection on test data to simulate the on-orbit system performance that has not been updated by the ground server. As the training data only contains one category of aircraft and YOLOv5 does not work well for a single classification. However, this problem can be solved by setting the category to "nc=2" and giving the dummy category any name during parameter setting. Because the training data was small and the network did not converge well, the target recognition rate was 91.2% and the accuracy of the rotating frame angle was poor (Figure 8(a)). In addition to this, it was found that in scenes with small and dense targets, the recognition rate dropped particularly significantly.

Experiment 2 retrained the model with more images. The training data was expanded to 980 images with a total of nearly 3000 targets covering aircraft and ships, and a dense scene of small targets was added as training samples to simulate the on-orbit update of the satellite by the ground server. Because the learning rate of this model is adjusted during the training process through a method called Stochastic Gradient Descent with Warm Restarts (SGDR) [9], a preset learning rate is adopted after certain epochs of Warm-up (4).

$$\eta_t = \eta_{min}^i + \frac{1}{2}\left(\eta_{max}^i - \eta_{min}^i\right)\left(1 + \cos\left(\frac{T_{cur}}{T_i}\pi\right)\right) \quad (4)$$

where η_{min}^i and η_{max}^i are ranges for learning rate, T_{cur} is how many epochs have been performed since last restart, T_i is total epochs. T_{cur} is updated at each iteration t, thus $\eta_t = \eta_{max}^i$ when $t = 0$ and $T_{cur} = 0$.

SGDR could prevent problems that may occur during training such as a sudden increase in error, and accelerate the convergence of the network. In the case of using the default learning rate, the network appeared to be misfitting, so the learning rate was adjusted to "lr0=0.0015, lrf=0.08". After 300 epochs, the training was paused to manually reduce the learning rate hyperparameters to "lr0 =0.001, lrf =0.01". Then restart the training process and run 50 epochs based on the last training to fine-tune the model. The first 300 training sessions are shown in Figure 7.

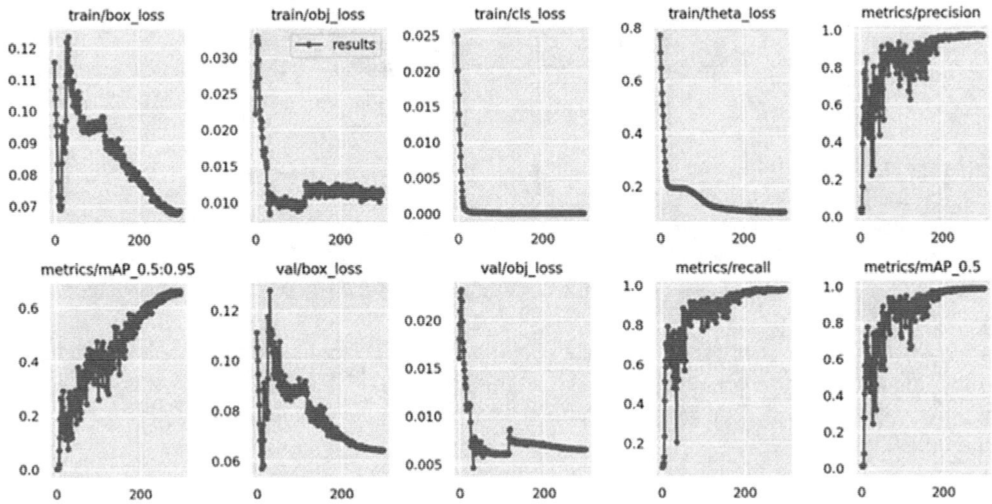

Figure 7. lr0=0.0015, lrf=0.08, epochs=300

After the training was completed, the new weights file was used to run detection in the test data. Since Experiment 2 used more images and applied a more suitable learning rate during the training process, the detection rate increased from 91.2% to 99.04% (Table 1). Moreover, the detection rate of the small target was significantly improved with new types of training data being fed to the model, and the frame angle could now be properly represented. (Fig. 8(b)). This experiment shows that it is feasible to increase the on-orbit detection rate of certain targets by training more related data with the ground server.

TABLE I. UPDATED AIRCRAFT TARGET RECOGNITION RATE

Category	*P*	*R*	*Tp*	*Fp*	*Fn*
Aircraft	99.04%	98.27%	625	6	11
Ship	97.65%	96.73%	415	14	10

Precision ***P*** and recall rate ***R*** can be calculated with Equations (5) and (6) :

$$R = \frac{T_p}{T_p + F_n} \quad (5)$$

$$P = \frac{T_p}{T_p + F_p} \quad (6)$$

where T_p, F_n and F_p represent true positive, false negative and false positive, respectively.

Figure 8. Before update and after the updated comparison

Experiment 3 performs model lightweight. The layer of the model was reduced from 290(50.2GFLOPS) to 217(17.3GFLOPS) with 65% fewer parameters to create a model that could speed up the inference process. The training was restarted to reduce the volume of weight files. As result, the average inference time of a 256*256 image using a lightweight model is shortened from 10.7ms to 8.5ms, and the size of the weight file is also reduced from 43.5MB to 15.2MB. With training epochs set to 150, the detection rate of aircraft is 89.06%.

In Experiment 4, since the lightweight version with reduced layers has huge impacts on the detection rate, the original model was converted into ONNX framework in PyTorch, then converted into OpenVX framework with INT8+FP32 mixed quantization method through a customized ACUITY-toolkit to run in Yulong 810A AI chip (Fig 9).

Figure 9. Yulong 810A development board

NMS in C++ with rotated IoU (7) was added to the project to collect sensor data from the model and output the target information.

$$IoU = \frac{Intersection\ area}{Whole\ area} \quad (7)$$

GPU and NNA were used to accelerate the inference process. Run the graph ten times and get an average execution time of 12.10ms (Fig.10), which costs about the same processing time as Tesla P100 running the original model at FP32. A detailed comparison between the two systems and running on a CPU is listed in Table 2.

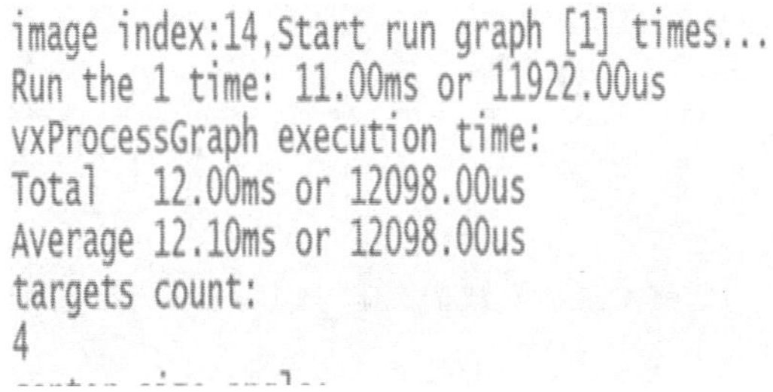

Figure 10. On-orbit system performance

TABLE II. COMPARISON BETWEEN THE TWO SYSTEM

Hardware	*Quantization*	*Framework*	*Inference time*	*P*
Tesla P100	FP32	PyTorch	10.7ms	99.04%
Intel i7-9750H	FP32	PyTorch	64ms	97.36%
Yulong 810A	INT8+FP32	OpenVX	12.1ms	93.99%

5. Conclusions

The space-ground joint detection system based on artificial intelligence can not only greatly improve the detection rate and effectiveness, but also can be rapidly updated by the ground server to improve the detection rate for a particular scenario, solving problems that traditional template matching methods meet that constantly affected by many uncertain factors which lead to low detection rate; hence the robustness of the system is enhanced. This paper provides a method based on YOLOv5, which uses rotated bounding boxes to identify aircraft and ship targets, compares the use of different degrees of training samples for detection to simulate on-orbit update, and runs the quantified lightweight version model in Yulong 810A to simulate the actual satellite hardware setup. The experimental results show that the method has good practical application prospects.

References

[1] X. Zhang, B. Jiang, Y. Yang, Y. Fu, and W. Yue, "A Survey on Object Detection Technology in Optical Remote Sensing Images", Spacecraft Recovery & Remote Sensing, Vol.40, no.4, pp. 95-104, August 2019
[2] H. Wang, "A research of Real-time Object Detection and Tracking Based on Shape-based Template Matching", Hangzhou Normal University, 2019
[3] L. Ma, X. Liu, X. Zhang, Y. Ye, G. Yin, and B. Johnson, "Deep learning in remote sensing applications: A meta-analysis and review", Journal of Photogrammetry and Remote Sensing, vol.152, pp. 166-177, 2019
[4] F. Ren, "Research on real-time target detection and tracking algorithm based on shape template matching", Huazhong University of Science and Technology, 2017
[5] relabeling [Online]. Available: https://github.com/cgvict/roLabelImg#readme
[6] J. Zhou, P. Jiang, A. Zou, X. Chen, and W. Hu, "Ship Target Detection Algorithm Based on Improved YOLOv5", Journal of Marine Science and Engineering, vol. 9, no. 8, pp. 908, 2021
[7] G. Cheng, J. Han, and X. Lu, "Remote Sensing Image Scene Classification: Benchmark and State of Art", Proceeding of the IEEE, vol. 105, no. 10, pp. 1865-1883, 2017
[8] YOLOv5 for Oriented Object Detection [Online]. Available: https://github.com/hukaixuan19970627/yolov5_obb
[9] L. Ilya and H. Frank, "SGDR: Stochastic Gradient Descent with Warm Restarts", arXiv, 2016

Application and research of linear ADRC in Buck converter

Wang Shen[1*], Xieyan Zhang[1], Yixue Xie[1]

[1] College of Information Science and Engineering, Hunan Normal University, Changsha 410081, China

[*]xiaosheng@hunnu.edu.cn

Abstract: This paper takes PWM modulated DC-DC Buck converter as the object to improve the traditional LADRC method. Firstly, the LESO module directly feeds back an error differential signal, which can improve its disturbance observation performance. Secondly, the disturbance differential term is imported to the LESO module to enhance the observation speed of disturbance. Finally, a first-order inertial link is joined to the disturbance estimation channel to suppress the high-frequency noise disturbance caused by the increase in observation bandwidth. By analyzing the frequency response characteristic curves of the improved LADRC method and the conventional LADRC method, it can found that improved LADRC method has preferable disturbance resistance performance. Using Simulink model, experimental consequence confirm the availability and correctness of the optimization method.

1. Introduction

The Buck converter is an indispensable part of the power conversion of power electronics, which is responsible for voltage conversion. However, in order to achieve accurate control of the output voltage of the Buck converter, the main problems are: voltage regulation can be affected by external disturbances (changes in load resistance, input voltage jumps, etc.) as well as internal disturbances (changes in system parameters, etc.) [1]-[2]. The quality and efficiency of voltage conversion can be greatly reduced by various uncertain factors, which poses great challenges to the high-quality voltage regulation for Buck converters [3]-[5].

To acquire a fine control result and simplify controller design, linear active disturbance rejection control (LADRC) is came up with in reference [6]. The LESO module is the kernel of influencing LADRC control performance and is also the current research hotspot [7-8]. In reference [9], the dynamic capability of T-LESO (traditional LESO) from the perspective of a frequency domain was analyzed. In reference [10], the method of switching between linear and nonlinear ESO was proposed, which integrated the advantages of the two ESO methods but also increased the complexity of the control's method in response.

In this paper, the traditional LESO was optimized. To improve the observation bandwidth and accuracy, a state variable was added to LESO to estimate the perturbation differential signal and the observation error differential signal was used as feedback. To restrain the amplification of high-frequency noise, adding a first-order inertial link to the perturbation estimation channel can effectively restrain the high-frequency noise. The experimental consequence confirm the availability of the I-LADRC (improved LADRC) method.

2. Model of Buck converter

The topology of the Buck converter based on PWM modulation is shown in Figure 1, including input voltage V_{in}, PWM gate-driven switch VT, diode VD, inductance L, capacitance C, and load resistance R.

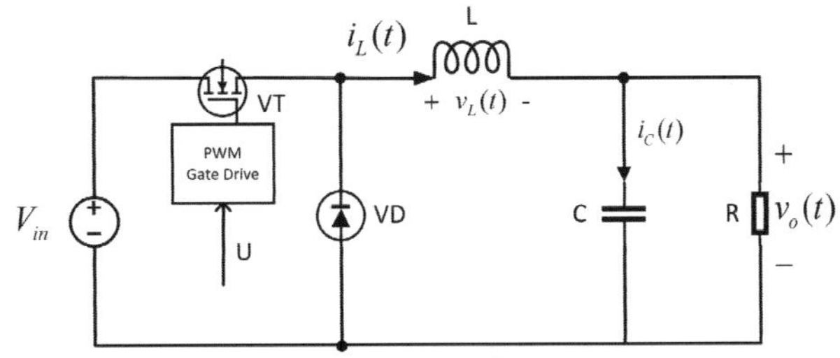

Figure.1 Buck converter topology.

According to the figure above, the dynamic model of the Buck converter is shown below:

$$\begin{cases} \dfrac{dv_o(t)}{dt} = \dfrac{1}{C}\left(i_L(t) - \dfrac{v_o(t)}{R} \right) \\ \dfrac{di_L(t)}{dt} = \dfrac{1}{L}\left(V_{in}\mu(t) - v_o(t) \right) \end{cases} \tag{1}$$

v_o is average output voltage, i_L is average input current, and the duty ratio is $\mu(t) \in (0,1)$ of the said control signal.

3. Improved LADRC for Buck converter

3.1 LESO improvement and disturbance rejection performance analysis

The traditional LESO module controls the state variable through the feedback of the output error, but for the Buck circuit, the output voltage can be directly measured, and the output voltage differential signal can be indirectly obtained from the capacitance current value. Therefore, the feedback of the differential of the output error can be directly introduced to obtain the state variable, and the estimation of the differential of the disturbance term can be increased. The new LESO structure is obtained as follows:

$$\begin{cases} \dot{a}_1 = l_1(\dot{y} - a_1) + a_2 + b_0 u \\ \dot{a}_2 = l_2(\dot{y} - a_1) + a_3 \\ \dot{a}_3 = l_3(\dot{y} - a_1) \\ \hat{\dot{v}}_o = a_1, \hat{f} = a_2, \hat{\dot{f}} = a_3 \end{cases} \tag{2}$$

Accordingly, the improved LSEF design is:

$$u = \frac{\mathrm{k}_p(r - y) - k_d a_1 - a_2}{b_0} \tag{3}$$

a_1, a_2, a_3 respectively represents the estimated value of $\hat{\dot{v}}_o, \hat{f}, \hat{\dot{f}}$. Parameters l_1, l_2, l_3, k_p, k_d in Equations (2) and (3) are configured by bandwidth method as $3w_o, 3w_o^2, w_o^3, w_c^2, 2w_c$.

According to equation $f = \ddot{y} - b_0 u$ and the Laplace transform of Equation (2), the disturbance observation transfer function in improved LESO is:

$$G_R(s) = \frac{a_2}{f} = \frac{3w_o{}^2 s + w_o{}^3}{(s + w_o)^3} \tag{4}$$

Compared with the perturbation observation loop of T-LESO [5], Equation (3) is equivalent to adding one zero point. From the analysis in the frequency domain, adding a leading network to the loop alleviates the problem of LESO observation disturbance phase lag, improves the observer bandwidth, and obtains better disturbance observation performance. However, the above-improved method can increase the control bandwidth but reduce the suppression effect of high-frequency signals. Based on the idea of lead-lag, a first-order inertial link is added in the perturbation observation loop to mine the system's own phase Angle margin, which can suppress the high-frequency noise disturbance:

$$a_4 = \frac{(\beta Ts+1)}{(Ts+1)} a_2, \beta \in (0,1) \tag{5}$$

T is the correction time constant and β is the correction link coefficient. The frequency domain analysis of the I-LESO module was conducted, the comparison with the previous two methods was made, and the comparison analysis diagram of disturbance rejection in Figure 2 was obtained.

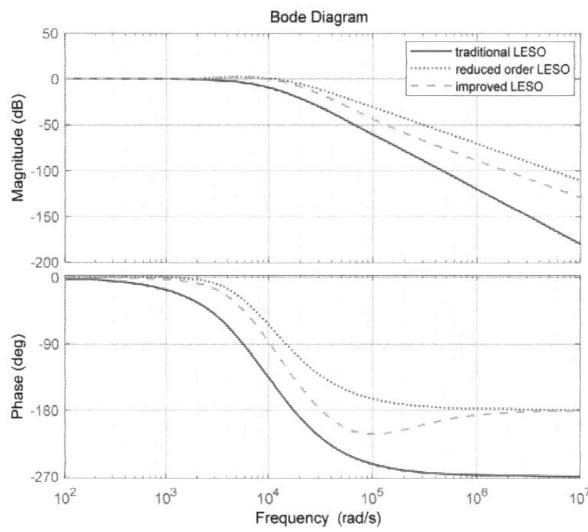

Figure.2 Three LESO disturbance Observations diagram.

According to the analysis in Figure 2, compared with the improved LESO, the improved LESO method with the addition of the first-order inertia link retains the feature of the low and middle-frequency band, which increases bandwidth of the improved LESO. Meanwhile, due to the influence of the first-order inertia link, the suppression ability of high-frequency signals has increased. The final improved LADRC control block diagram is shown in Figure 3.

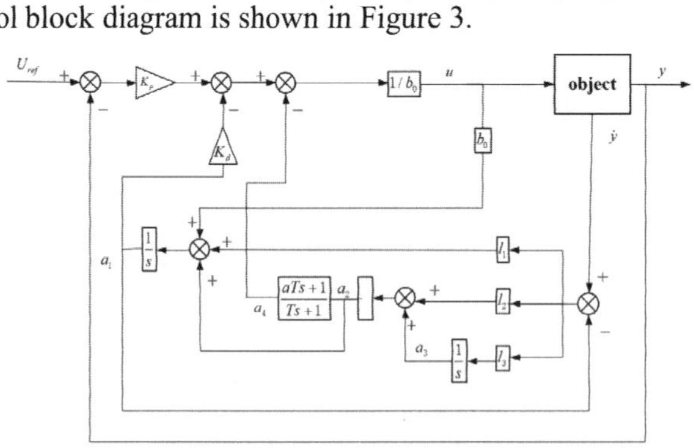

Figure.3 Improved LADRC block diagram.

3.2 Improved LADRC disturbance resistance performance analysis

Through the analysis of the control block diagram, a_4 is obtained from a_3 through the filter module, which acts on the LSEF module as the estimation of the overall perturbation of the system as follows:

$$u = \frac{k_p(r-y) - k_d a_1 - a_4}{b_0} \tag{6}$$

By connecting Lines (2), (5), and (6), we can obtain:

$$u(s) = \frac{k_p\left(s+w_o\right)^3}{b_0\left[\left(s+w_o\right)^3 + k_d s^2 - \frac{(aT_e s+1)}{(T_e s+1)}(3w_o^2 s + w_o^3)\right]} * U_{ref}$$

$$- \frac{k_p\left(s+w_o\right)^3 + k_d\left(3w_o s^3 + 3w_o^2 s^2 + w_o^3 s\right) + (3w_o^2 s^3 + w_o^3 s^2)\frac{(aT_e s+1)}{(T_e s+1)}}{b_0\left[\left(s+w_o\right)^3 + k_d s^2 - \frac{(aT_e s+1)}{(T_e s+1)}(3w_o^2 s + w_o^3)\right]} * y(s)$$

$$= \frac{B(s)}{b_0 A(s)} * U_{ref} - \frac{C(s)}{b_0 A(s)} * y(s) \tag{7}$$

According to Equation (6), the system control block diagram can be simplified from Figure 3 to Figure 4:

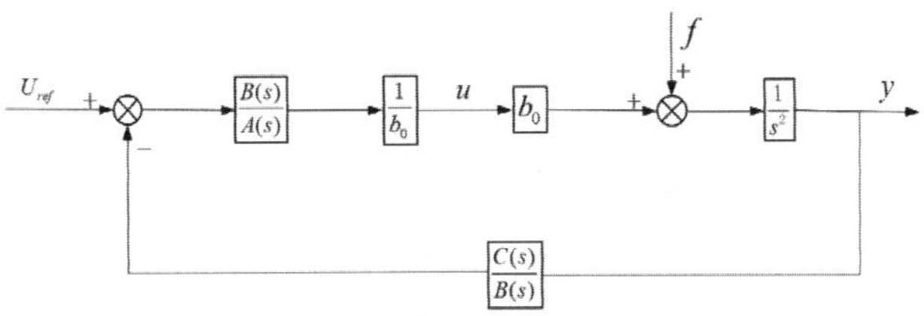

Figure.4 LADRC simplified control block diagram

From this, we can write the closed-loop transfer function of the system can be listed:

$$y = \frac{k_p}{s^2 + k_d s + k_p} U_{ref} + \frac{A(s)}{A(s)s^2 + C(s)} f \tag{8}$$

According to Equation (8), the system output will be affected by tracking terms and disturbance terms, and by comparing the closed-loop transfer function of T-LADRC, it can be found that improved suppression of LADRC disturbance signals affects the control precision and accuracy of LADRC. Therefore, the disturbance transfer function is analyzed and in comparison with the T-LADRC, and the corresponding curve in the frequency domain is shown in Figure 5.

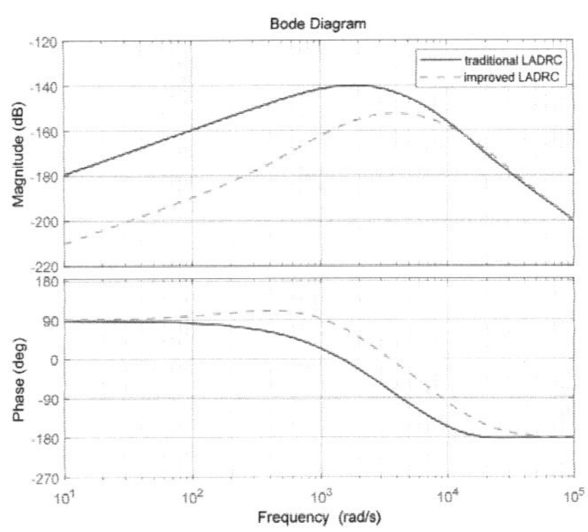

Figure.5 Disturbance rejection analysis diagram of two LADRC methods

By comparison, it can be seen that the I-LADRC method has a preferable suppression effect than the T-LADRC method in the low and middle-frequency bands. In the high-frequency band, due to the limitation of observer bandwidth, the curves of the two methods are close and the disturbance rejection performance is almost the same.

4. Experimental analysis and performance verification

To further prove the availability and feasibility of the I-LADRC method, a DC-DC converter model of 24V to 12V is established. The specific system parameter settings are shown in the following Table 1:

Table 1. Parameters Setting.

DC-DC converter parameters	Valve	Controller parameters	Value
Reference Input voltage	$24V$	**Controller gain** b_0	1.2×10^8
Reference Output Voltage	$12V$	**Observer bandwidth** w_o	1×10^4
Inductance	$2mH$	**Controller bandwidth** w_c	2000
Capacitance	$10\mu F$	**Correction coefficient** a	0.5
Nominal Resistance	12Ω	**Correction time constant** T_e	1×10^{-4}

The robustness of the improved LADRC method for DC-DC Buck converters under load resistance and input voltage transformation is tested.

(a) Robustness analysis of load resistance change. The load specific Settings are as follows:

$$R=\begin{cases}12\Omega(=R_0) & ,t\in(0,0.4)\\ 6\Omega(=0.5R_0) & ,t\in[0.4,0.8)\\ 24\Omega(=2R_0) & ,t\in[0.8,1)\end{cases}$$

The output voltage curves of the two-control method under load disturbance are shown in Figure 6.

(b) Robustness against input voltage change. The input voltage transformation is as follows:

$$V_{in}=\begin{cases}24V(=V_{in0}) & ,t\in(0,0.4)\\ 18V(=0.75V_{in0}) & ,t\in[0.4,0.8)\\ 36V(=1.5V_{in0}) & ,t\in[0.8,1)\end{cases}$$

The output voltage response curve of the two-control method under input voltage disturbance is shown in Figure 7.

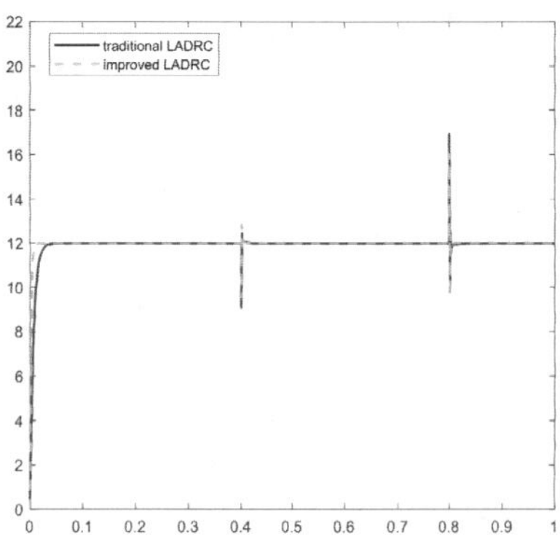

Figure 6. Output voltage curve under load resistance transformation

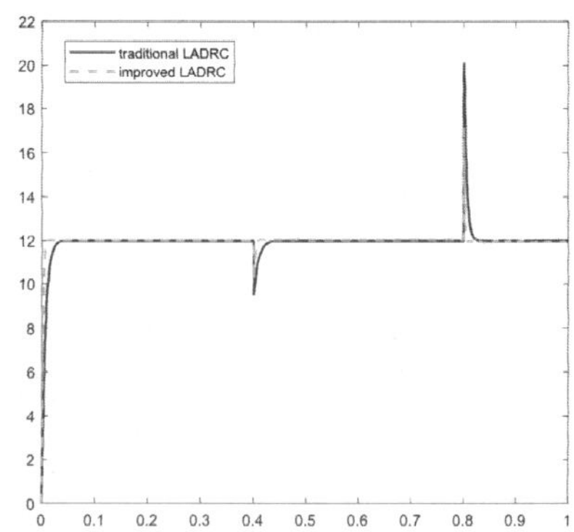

Figure 7. Output voltage curve under input voltage transformation

As shown in Figs. 6, and Figs. 7, the improved LADRC(I-LADRC) method proposed in this paper further improves the static performance in the front of various perturbation and uncertainties. Similarly, the dynamic performance of the I-LADRC method is better than that of the traditional LADRC(T-LADRC) method after load mutation and input voltage change. For completeness, the performance indicators (the maximum output voltage raise/drop (MOVR/MOVD) and the maximum recovery time (MRT)) of the two control methods are compared as shown in Table 2.

Table 2. Experimental performance index under two LADRC methods.

Test Type	Performance	Controllers	
		T-LADRC	I-LADRC
Case a	MOVR (V)	4.32	3.36
	MOVD (V)	2.92	2.46
	MRT (sec)	0.026	0.009
Case b	MOVR (V)	8.02	5.03
	MOVD (V)	3.02	1.83
	MRT (sec)	0.031	0.009

5. Conclusion

Taking Buck converter as the research object, an optimized LESO method is put forward in this article. First of all, the traditional LESO idea of controlling the state error according to the feedback of the output voltage error is no longer adopted in the design. Instead, the feedback is carried out through the differential of the output error, so as to increase the bandwidth of the observer, and improve the observation accuracy. Secondly, according to the thought of adjustment and amendment of lead-lag, a first-order inertia link is joined to the perturbation term estimation channel to suppress the interference of high-frequency noise. In the end, the estimation of disturbance differential signal is added to LESO to promote the observation speed of LESO. The simulation results show that compared with the T-LADRC method, the I-LADRC method has faster speed and a smaller adjustment range in the output voltage conversion, both load resistance disturbance and input voltage disturbance, and improves the dynamic performance and stability of the system.

References

[1] Wang J, Rong J and Li Y. (2019) Design and implementation of reduced-order extended state observer and sliding mode control for DC-DC buck converter. Control Theory & Applications, 36(09):1486-1492

[2] Samir A, Ali D, Azzeddine B and Maamar B. (2020) Design and experimental implementation of a new robust observer-based nonlinear controller for DC-DC buck converters. Energy,213(15): 118816.1-118816.11

[3] Gao Z. (2003) Scaling and bandwidth-parameterization based controller tuning. Proceedings of the 2003 American Control Conference. Denver, USA. 4989-4996.

[4] Gao Z. (2013) Research on the idea of active disturbance rejection control. Control Theory and Application, 30 (12): 1498-1510.

[5] Chen W, Yang J, Li G and Li S. (2016) Disturbance-observer-based control and related methods - an overview. IEEE Transactions on Industrial Electronics, 63(2): 1083–1095.

[6] Han J. (2007) Auto Disturbance Rejection Control Technology. Frontier Science,1(1):24-31.

[7] Yang L, Zeng J, Ma W, etc. (2019) Voltage control of microgrid inverter based on improved second-order linear active disturbance rejection control. Automation of Electric Power Systems, 43(4):146-153.

[8] Wang X, Wu F, Pan Z, et al. (2018) Application of linear active disturbance rejection control in full-bridge DC-DC converters. Control Theory & Applications,35(11):1610-1617.

[9] Yuan D, Ma X, Zeng Q and Qiu X. (2013) Research on frequency-band characteristics and

parameters configuration of linear active disturbance rejection control for second-order systems. Control Theory & Applications,30(12): 1630-1640.

[10] Li J, Qi X, Xia Y and Gao Z. (2016) On Linear/Nonlinear Active Disturbance Rejection Switching Control. Acta Automatica Sinica, 42(2): 202-212.

Journal of Physics: Conference Series **2489** (2023) 012035 doi:10.1088/1742-6596/2489/1/012035

Dual-antenna GPS-based UAV Attitude Determination Method and Accuracy Analysis

Yilun Wu[1, 2a*], Chunsheng Sun[1b]

[1] Department of Weapon Engineering, Naval University of Engineering, Wuhan, China

[2] No. 92038 Troops of PLA, Qingdao, China

[a*] e-mail: 1131444744@qq.com; [b] e-mail: scs96581@163.com

Abstract. Accurate attitude information of small-sized unmanned aerial vehicles (UAVs) is required to obtain absolute azimuth information of a target via an airborne electro-optical imaging system. The direct attitude solution method of dual-antenna Global Position System (GPS) is investigated in this paper to address this need. We analyzed the attitude measurement principle of the direct method, gave the error estimation model, and estimated the inner and outer precision through simulation. The simulation results demonstrated that to achieve the requirement of UAV yaw angle error within 1°, the baseline length should be no less than 1 m. The applicability of the model and algorithm was confirmed.

1. Introduction

Airborne electro-optical imaging tracking and measurement systems can accurately measure the relative azimuth and pitch angles between ground or space targets and carriers [1]. To receive the viewing indicator axis' absolute orientation angle with high accuracy and, consequently, the target's absolute orientation information, the carrier must have high-accuracy attitude information. Traditional attitude information acquisition methods include the high-precision inertial navigation system or the combined INS/GPS navigation system, magnetometer, etc. The high-precision INS equipment is high-cost, high-load, accumulates attitude error over time, and has significant usage limitations [2]; magnetometers are susceptible to interference from the surrounding hard and soft magnetic environment and are complicated to install and calibrate [3]. None of the traditional methods apply to small rotorcraft UAVs in complex electromagnetic environments.

GPS attitude measurement technique, as a significant branch of Global Navigation Satellite System in the application research, has the strengths of being inexpensive, tiny in size, great accuracy, convenient installation, no need for initial calibration, and no time accumulation error [4-6]. It is set up on the technique of carrier phase difference. The short baseline is measured using multiple GPS antennas attached to the carrier, public error interference is removed, basis vectors are resolved, and carrier attitude information is acquired.

Over the years, various experts have studied the multi-antenna attitude solution methods, the most common ones being the direct method as well as the least squares method. [7-9]. Moreover, the quantity and configuration of antennas and the longness of baselines can impact the attitude solution accuracy. However, the antenna configuration is often limited by the carrier to meet the ideal conditions. Therefore, an appropriate attitude solution method must be sought depending on the applied environment. The UAV carrier applied in this paper has a limited load, and small installation space and does not require roll angle information, furthermore, the yaw angle error is required to be within 1 °. Under the short baseline condition, both attitude-solving methods can obtain high-accuracy attitude information. Therefore, the baseline solution basics, the direct method attitude solution model, and the accuracy estimation model are all examined in this work. By simulating the accuracy study, the viability of using the dual-antenna GPS direct attitude solution method is confirmed.

2. The rationale of the dual-antenna attitude solution

2.1. Short baseline carrier phase difference principle

The short baseline carrier phase differential principle is shown in Figure 1. A GPS antenna is placed at each end of the baseline, a set of the same satellites are measured at the same time, and the baseline vector is determined after differential operation.

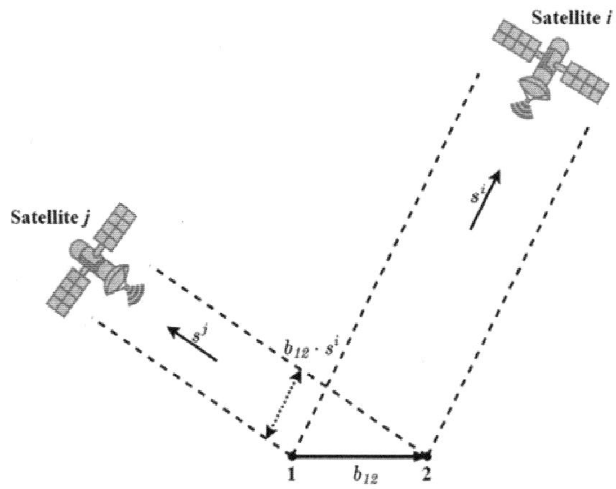

Figure 1. Carrier phase differential schematic

The satellite signal's carrier phase measurement equation is

$$\varphi_1^i + N_1^i = \left(\rho_1^i + c\delta t_1 - c\delta t^i - \left(\delta\rho^i\right)_{ion} + \left(\delta\rho^i\right)_{\text{trop}}\right)/\lambda + \varepsilon_1^i, \tag{1}$$

In Equation (1), φ_1^j represents the carrier phase observation value between antenna $1(A_1)$ and satellite $i(S_i)$, N_1^i represents integer ambiguity, ρ_1^i represents the geometrical distance between A_1 and S_i, δt_1 represents the A_1 clock-bias, δt^i represents the S_i clock-bias, $\left(\delta\rho^i\right)_{ion}$ represents the ionosphere delay, $\left(\delta\rho^i\right)_{\text{trop}}$ represents the troposphere delay, and ε_1^i represents the rest of the error, which is generally the measurement noise.

When A_1 and A_2 observe S_i simultaneously, the satellite signal carrier phase observation equations for the two antennas are

$$\varphi_1^i = \lambda^{-1}(\rho_1^i - \delta\rho^i{}_{ion} + \delta\rho^i{}_{\text{trop}}) + f(\delta t_1 - \delta t^i) - N_1^i + \varepsilon_1^i, \tag{2}$$

$$\varphi_2^i = \lambda^{-1}(\rho_2^i - \delta\rho^i{}_{ion} + \delta\rho^i{}_{\text{trop}}) + f(\delta t_2 - \delta t^i) - N_2^i + \varepsilon_2^i, \tag{3}$$

Since the distance from the carrier to the satellite is much greater than the baseline length, the two antenna positions are the same concerning the satellite, The single difference operation from Equations (2) and (3) to eliminate the ionospheric delay, tropospheric delay, satellite orbit error and satellite clock difference, the carrier phase single difference equation between A_1 and A_2 is obtained as

$$\Delta\varphi_{12}^i = \lambda^{-1}\Delta\rho_{12}^i + f\Delta\delta t_{12} - \Delta N_{12}^i + \Delta\varepsilon_{12}^i, \tag{4}$$

In Equation (4), Δ represents the single difference operator.

Similarly, at the same moment when the two antennas observe S_j, the carrier phase single difference equation is as follows:

$$\Delta\varphi_{12}^j = \lambda^{-1}\Delta\rho_{12}^j + f\Delta\delta t_{12} - \Delta N_{12}^j + \Delta\varepsilon_{12}^j, \tag{5}$$

Performing a single difference operation on Equation (4) and Equation (5), the carrier phase double difference equation is obtained as follows:

$$\nabla\Delta\varphi_{12}^{ij} = \lambda^{-1}\nabla\Delta\rho_{12}^{ij} - \nabla\Delta N_{12}^{ij} + \nabla\Delta\varepsilon_{12}^{ij}, \tag{6}$$

In Equation (6), $\nabla\Delta$ represents the double difference operator,$\nabla\Delta\varepsilon_{12}^{ij}$ represents the error, which is negligible. From Equation (6), it can be seen that the receiver clock bias and other receiver-related errors are eliminated after the double difference operation.

As the baseline length is much smaller than the distance between the antenna and the satellite, the carrier signal from the satellite can be considered a plane wave when it arrives at the two antennas and can be obtained as

$$\Delta\rho_{12}^{i} = |b_{12}| \cdot \cos\alpha = \boldsymbol{b}_{12} \cdot \boldsymbol{s}^{i}, \tag{7}$$

$$\Delta\rho_{12}^{j} = |b_{12}| \cdot \cos\beta = \boldsymbol{b}_{12} \cdot \boldsymbol{s}^{j}, \tag{8}$$

$\mathbf{b}_{12}$: the baseline vector; $\mathbf{s}^{i}$: the antenna-to-satellite unit direction vector.

Taking Equations (7) and (8) into Equation (6) yields

$$\nabla\Delta\varphi_{12}^{ij} = \lambda^{-1}(\boldsymbol{s}^{i} - \boldsymbol{s}^{j}) \cdot \boldsymbol{b}_{12} - \nabla\Delta N_{12}^{ij} + \nabla\Delta\varepsilon_{12}^{ij}, \tag{9}$$

From Equation (9), the baseline vector in the navigation coordinate system can be achieved when the integer ambiguity is determined.

2.2. Coordinate system and coordinate transformation

The rotation angle of the carrier coordinate system relative to the navigation coordinate system represents the attitude angle of the carrier. The dual-antenna attitude solution is a coordinate system conversion using the azimuthal connection of two coordinate systems to obtain the attitude angle of the carrier.

Navigation coordinate system (n-system): it is generally a local horizontal geographic coordinate system with the origin at A_1, the X-axis pointing east horizontally, the Y-axis pointing north horizontally, the Z-axis pointing straight up, together with the X and Y axes, and the Z-axis creates a right-handed coordinate system, i.e., East-North-Up (ENU) coordinate system.

Carrier coordinate system (b-system): the origin is located at the carrier center, the Y-axis is parallel to the longitudinal axis of the carrier and points to the forward direction of the carrier, the X-axis is parallel to the transverse axis of the carrier and perpendicular to the Y-axis, the Z-axis creates a right-handed coordinate system with the X and Y axes and is located in the longitudinal plane of the carrier, i.e., Right-Front-Up (RFU) coordinate system.

When A_1 and A_2 are symmetrically installed in the carrier center along the Y-axis of the carrier coordinate system, the baseline formed by them is in the same direction as the Y-axis of the carrier coordinate system, and A_1 is located at $\mathrm{b}_{12}/2$ on the Y-axis of the carrier coordinate system, as shown in Figure 2.

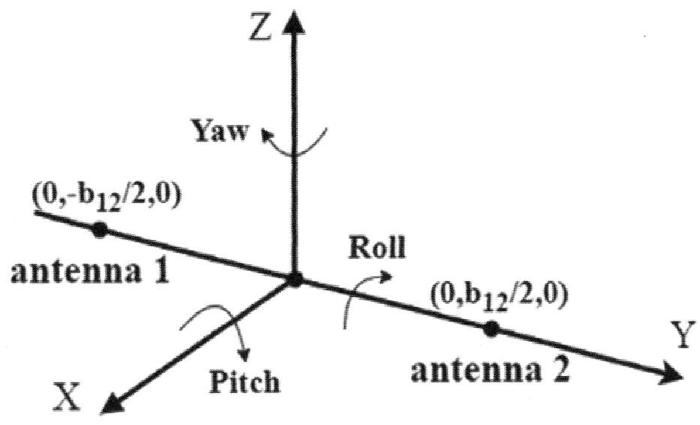

Figure 2. Carrier coordinate system and antenna schematic

Since the n-system is ENU and the b-system is RFU, after first translating $-\mathrm{b}_{12}/2$ along the Y-axis and then rotating in the order of Z-axis, X-axis, and Y-axis, n-system can be transformed into a b-system.

The direction of the pitch angle (θ) and roll angle (γ) is the same as the corresponding axial rotation direction, and the yaw angle (ψ) is opposite to the corresponding axial rotation direction. So, the attitude matrix is

$$C_n^b = C(\gamma)C(\theta)C(-\psi) = \begin{bmatrix} \cos\gamma & 0 & -\sin\gamma \\ 0 & 1 & 0 \\ \sin\gamma & 0 & \cos\gamma \end{bmatrix} \begin{bmatrix} 1 & 0 & 0 \\ 0 & \cos\theta & \sin\theta \\ 0 & -\sin\theta & \cos\theta \end{bmatrix} \begin{bmatrix} \cos\psi & -\sin\psi & 0 \\ \sin\psi & \cos\psi & 0 \\ 0 & 0 & 1 \end{bmatrix}$$
$$= \begin{bmatrix} \cos\gamma\cos\psi + \sin\gamma\sin\theta\sin\psi & -\cos\gamma\sin\psi + \sin\gamma\sin\theta\cos\psi & -\sin\gamma\cos \\ \sin\psi\cos\theta & \cos\psi\cos\theta & \sin\theta \\ \sin\gamma\cos\psi - \cos\gamma\sin\theta\sin\psi & \cos\psi\cos\theta - \sin\gamma\sin\psi - \cos\gamma\sin\theta\cos\psi & \cos\gamma\cos \end{bmatrix} \quad (10)$$

3. Direct attitude solution method and error analysis

3.1. The direct attitude solution method

As shown in Figure 2, assuming the baseline length is d_{12}, the coordinates of main antenna *1* and slave antenna *2* are $(0, -d_{12}/2, 0)$, $(0, d_{12}/2, 0)$, so the coordinates of the baseline vector $\boldsymbol{b}_{12}$ in the b-system can be noted as

$$b_{12}^b = (0 \quad d_{12} \quad 0)^T, \quad (11)$$

The coordinates under the n-system obtained by solving the GPS observation data are

$$b_{12}^n = (x_{12}^n \quad y_{12}^n \quad z_{12}^n)^T, \quad (12)$$

The baseline vector's transformation related to the two reference frames is

$$\boldsymbol{b}_{12}^b = (C_n^b)\boldsymbol{b}_{12}^n, \quad (13)$$

Taking Equations (10), (11), and (12) into Equation (13), the yaw angle (ψ) and pitch angle (θ) are derived as

$$\psi = arctan\left(\frac{x_{12}^n}{y_{12}^n}\right), \quad (14)$$

$$\theta = arctan\left(\frac{z_{12}^n}{\sqrt{x_{12}^n + y_{12}^n}}\right), \quad (15)$$

3.2. The error estimation model of the direct method

Differentiating Equation (14) yields

$$\delta\psi = \frac{y_{12}^n \delta x_{12}^n - x_{12}^n \delta y_{12}^n}{{x_{12}^n}^2 + {y_{12}^n}^2}, \quad (16)$$

Assuming that the coordinate components are uncorrelated, it is obtained from the error transfer formula that the standard deviation of yaw angle error is

$$\sigma_\psi = \sqrt{{y_{12}^n}^2 \sigma_{x_{12}}^2 + {x_{12}^n}^2 \sigma_{y_{12}}^2 / {x_{12}^n}^2 + {y_{12}^n}^2}$$
$$= \sqrt{cos^2(\psi)\sigma_{x_{12}}^2 + sin^2(\psi)\sigma_{y_{12}}^2 / d_{12}cos\theta}, \quad (17)$$

Similarly, the error of the pitch angle is

$$\sigma_\theta = \sqrt{sin^2(\theta)cos^2(\psi)\sigma_{x_{12}}^2 + sin^2(\theta)sin^2(\psi)\sigma_{y_{12}}^2 + cos^2(\theta)\sigma_{z_{12}}^2 / d_{12}}, \quad (18)$$

For the convenience of analysis, it is assumed that the errors of each coordinate component for each antenna under the navigation coordinate system are independent and identically distributed, so it can be considered that

$$\sigma_{x_{12}} = \sigma_{y_{12}} = \sigma_{z_{12}} = \sigma, \quad (19)$$

Taking Equation (19) into Equation (17) and Equation (18) respectively, we derive

$$\sigma_\psi = \sigma / d_{12} \cos\theta, \quad (20)$$
$$\sigma_\theta = \sigma / d_{12}, \quad (21)$$

Analysing Equation (20) and Equation (21), we can see that the baseline length has a significant impact on the precision of the yaw and pitch angle measurements.

4. Simulation results and analysis

4.1. External conformity accuracy analysis

The current GPS dual-frequency combined short baseline relative positioning accuracy reaches 0.47 cm in the E-direction, 0.48 cm in the N-direction, and 1.17 cm in the U-direction [10]. According to the distribution law of positioning errors, we took the baseline lengths of 0.8 m, 1 m, and 1.2 m to generate 500 sets of simulated positioning data, calculated the yaw and pitch attitude information by the direct method, and obtained the attitude error results in Figure 3; the statistics of attitude errors are shown in Table 1.

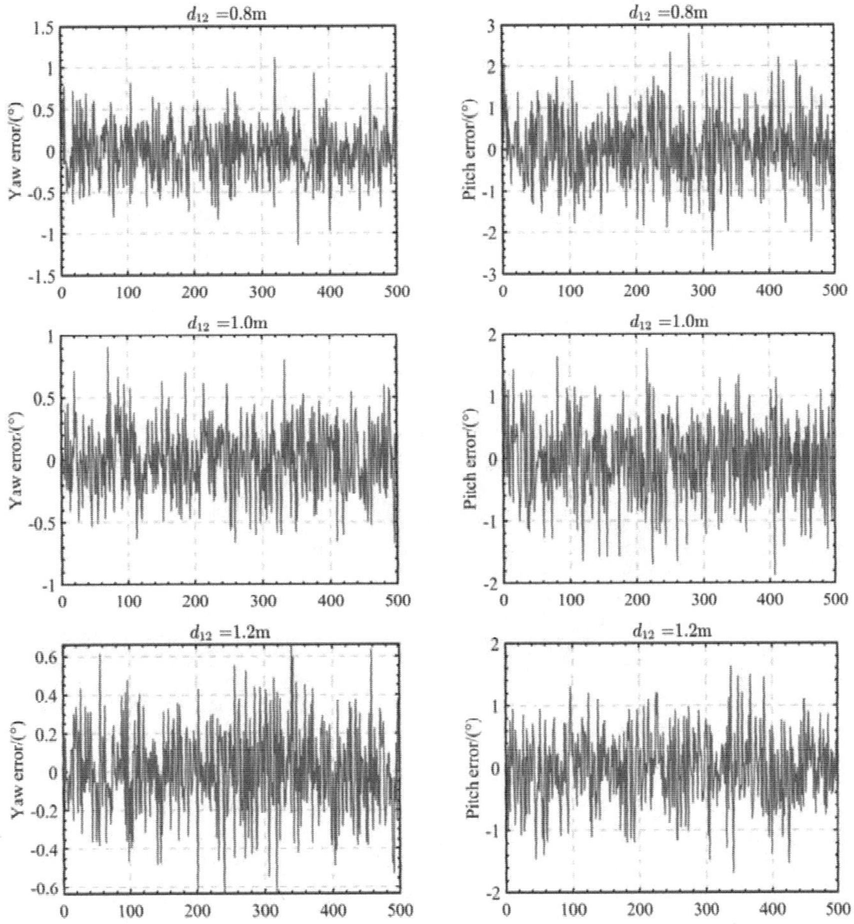

Figure 3. Carrier coordinate system and antenna schematic

Figure 3 and Table 1 demonstrate that the simulation is performed for different baseline lengths and attitude angles, and the yaw angle has greater precision than the pitch angle under the same conditions. When the baseline length is 0.8 m, the error range of the yaw angle is between ±1.2 ° and the pitch angle between ±2.8 °; when the baseline length is 1m, the error range of the yaw angle is between ±1.0 ° and pitch angle is between ±1.9 °; when the baseline length is 1.2 m, the error range of yaw angle is between ±0.7 ° and pitch angle is between ±1.7 °. The yaw angle and pitch angle accuracy showed a negative correlation with baseline length, and the accuracy of the attitude angle increased with baseline length.

Table 1. Consequences of attitude errors in statistics.

Baseline Length (m)	Attitude angle	Maximum (°)	Minimum (°)	Average (°)	Root-mean-square error (°)	Standard Deviation (°)
0.8	Yaw	1.1257	-1.1289	-0.0163	0.3274	0.3273
	Pitch	2.7857	-2.4329	0.0300	0.8108	0.8110
1.0	Yaw	0.9089	-0.6679	0.0018	0.2731	0.2734
	Pitch	1.7626	-1.8655	-0.0258	0.6224	0.6225
1.2	Yaw	0.6603	-0.6344	0.0032	0.2203	0.2205
	Pitch	1.6255	-1.6793	-0.0078	0.5591	0.5596

4.2. Internal conformity accuracy analysis

Using Equation (20) to calculate the yaw angle error, σtook the GPS dual-frequency combination of short baseline relative positioning maximum error value, i.e., the positioning accuracy of 1.17 cm in the U-direction; taking into account factors such as the axial length of the UAV platform, the influence of wing rotation on the placement of the GPS antenna, and the flight speed of the UAV in the application environment, the range of the baseline length d_{12} was taken to be 0.8 ~ 1.2 m, and the pitch angle of the UAV was taken in a range of 0 ~ 30 °, with yaw angle error as shown in Figure 4.

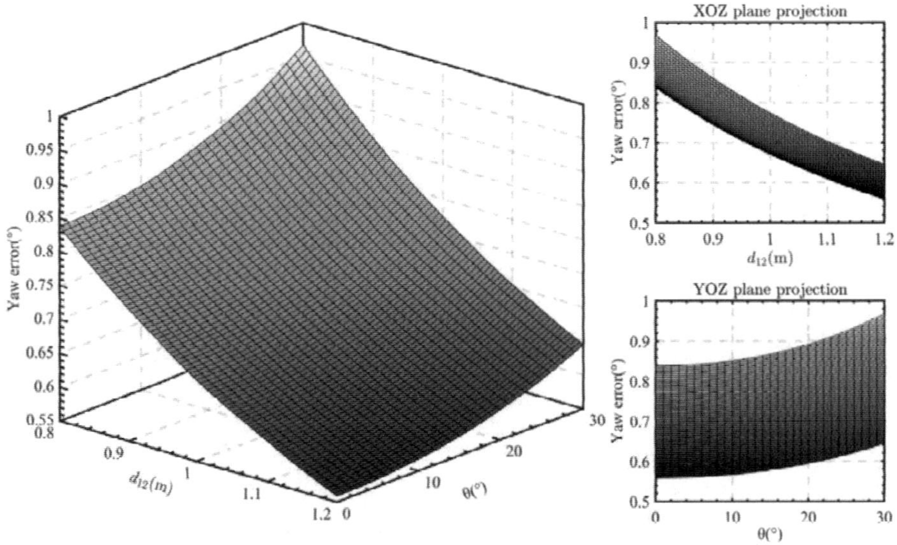

Figure 4. Standard deviations of attitudes from the direct method

From Figure 4, under a certain location error of the GPS carrier phase, yaw angle accuracy is negatively correlated with the baseline length and positively correlated with the absolute value of pitch angle. As the length of the baseline is longer, the pitch angle is closer to 0, and the accuracy of the yaw angle is higher; Pitch angle accuracy is negatively correlated with baseline length, and the longer the baseline length is, the higher the precision of pitch angle.

From the above analysis, it is obvious that the inner precision of the yaw angle is better than the outer precision, which may be caused by the deviation between the simulation data and the actual data. Comprehensive inner and outer precision, to meet the requirements of the yaw angle error of less than 1 °, the baseline length should not be less than 1 m, within the scope of the conditions allowed, increasing the baseline length can improve the attitude accuracy.

5. Conclusions

The dual-antenna GPS direct attitude determination method overcomes the shortcomings of INS and magnetometer, such as large load, difficult installation, and error accumulation. It is essential for attitude measurement of small UAVs. The target's absolute orientation can be established utilizing the UAV attitude information in conjunction with the relative orientation provided by the airborne optoelectronic pods. This makes it possible to do on-site performance inspections of specific equipment, which has important practical ramifications.

In this paper, we studied the mathematical model and accuracy estimation model of the dual-antenna GPS direct attitude determination method and analyzed the internal and external conformity accuracy of the direct method through simulation. Through simulation results, the precision of dual-antenna GPS attitude measurement mainly depends on GPS positioning accuracy and baseline length. In addition, when the baseline length is not less than 1 m, the direct method attitude determination can satisfy the index requirement of yaw angle error within 1 °, which verifies the feasibility of the application of the dual-antenna GPS direct attitude solution method. Future research can compare the three-antenna attitude solution approach and the Least Squares method for various application scenarios and accuracy demands.

References

[1] Jin G 2018 Analysis of Error and Image Motion in Airborne Electro-optical Imaging and Tracking Measurement System (Beijing: National Defense Industry Press) p 2
[2] Liu S, Sun F, Zhang L, Li W and Zhu X 2016 Tight integration of ambiguity-fixed PPP and INS: model description and initial results *Gps Solutions* **20** 39-49
[3] Caruso M J 2000 Applications of magnetic sensors for low-cost compass systems *IEEE 2000 Position Locat. Navig. Symp.* 177-184
[4] Li Y, Zhang K, Roberts C and Murata M 2004 On-the-fly GPS-based attitude determination using single- and double-differenced carrier phase measurements *Gps Solutions* **8** 93-102
[5] Raskaliyev A, Pate S H, Sobh T M and Ibrayev A GNSS-Based Attitude Determination Techniques-A Comprehensive Literature Survey *Ieee Access* **8** 24873-24886
[6] Zhang P, Zhao Y, Lin H, Zou J, Wang X and Yang F 2020 A Novel GNSS Attitude Determination Method Based on Primary Baseline Switching for A Multi-Antenna Platform *Remote Sens.* **12**
[7] Chen D and Lachapelle G 1995 Comparison of the FASF and least-squares search algorithms for on-the-fly ambiguity resolution *Navig. J. Inst. Navig.* **42** 371-390
[8] Zhao S, Huang C, Qi X and Lu M 2015 Application of BeiDou navigation satellite system on attitude determination for Chinese space station *China Satellite Navigation Conference* 13-25
[9] Liu X, Ballal T, Chen H and Al-Naffouri T Y 2022 Constrained Wrapped Least Squares: A Tool for High-Accuracy GNSS Attitude Determination *IEEE Trans. Instrum. Meas.* **71**
[10] Yang C 2022 Analysis of GPS Positioning Accuracy in Different Modes *Geomatics & Spatial Information Technology* **45** 139-142

Anomaly detection of satellite telemetry data based on extended dominant sets clustering

Xin Jin, Hui Quan Wang* , Zhong He Jin

Micro-satellite Research Centre, Zhejiang University, Hangzhou 310000, China

*Corresponding author's e-mail: hqwang@zju.edu.cn

Abstract: To mine out anomalies in satellite telemetry data under unsupervised conditions, a cluster-based method is proposed in this paper. Firstly, an extended dominant sets clustering algorithm is proposed to cluster the telemetry data with arbitrary shapes. Secondly, objects that do not belong to any cluster or belong to small clusters are traditionally identified as anomalies. Thirdly, the anomalies in large clusters are detected according to the relative similarity. Finally, the information on anomaly windows in the telemetry sequence is obtained according to the local anomaly rate, which provides more characteristics of the anomalies. Experimental results show that: 1) The proposed extended dominant sets clustering algorithm can deal with the dataset containing multiple and arbitrarily shaped clusters; 2) The introduction of relative similarity increases the AUC values of anomaly detection by 3%~10%; 3) The proposed anomaly detection method can effectively detect the anomalies in magnetometer telemetry data of Tianping-2B satellite. Therefore, the proposed method can achieve anomaly detection of satellite telemetry data under unsupervised conditions, and provide support for improving satellite safety and reliability.

1. Introduction

Once a spacecraft like a satellite is launched into orbit, the telemetry data becomes the only basis for the ground to judge its status. Therefore, the anomaly detection of telemetry data becomes an important way to enhance the reliability and safety of spacecraft in-orbit [1]. Since a spacecraft system is very complex and its telemetry data is highly relevant, professional, multi-modal, and unpredictable, it is extremely difficult and costly to generate an accurate and comprehensive labelled training set for anomaly detection. Therefore, unsupervised anomaly detection algorithms are gradually becoming mainstream in the field [2][3].

Among the many unsupervised anomaly detection algorithms, cluster-based methods are often used for spacecraft telemetry data. For example, NASA has developed a spacecraft system health monitoring software — IMS (Inductive Monitoring System) [4][5] based on cluster-based anomaly detection technology. IMS performs clustering processing on the achieved system data and characterizes nominal interactions between selected parameters [6]. When the system behaviour deviates significantly from the nominal IMS models, an alarm is provided to remind people of potential system failures or precursors of major failures. Inspired by IMS, Gao [7] proposed a normal behaviour clustering-based anomaly detection method and used this method to detect the fault of the satellite power system. Yairi [8] proposed a satellite health monitoring method based on probabilistic clustering and dimensionality reduction and successfully applied it to the small demonstration satellite 4(SDS-4) of the Japan Aerospace Exploration Agency (JAXA). Wang [9] proposed a density peak clustering-based two-stage anomaly detection method and mined out the anomaly of the analogue solar sensor on the ZDPS-1A satellite.

The above cluster-based anomaly detection methods consider objects belonging to large clusters to be normal objects. However, since the goal of clustering is to mine clusters in the dataset, there is no guarantee that all anomalies are excluded from large clusters. In addition, the output of the above method isolated anomalies without further mining the association between them. For these phenomena, we propose a novel cluster-based anomaly detection method, which will be described in detail in Section 2. We will discuss the experimental results and conclude the paper in Sections 3 and 4, respectively.

2. Description of the proposed method

The telemetry data anomaly detection algorithm proposed in this paper mainly includes four parts: sequence segmentation, clustering, anomaly detection, and anomaly window searching:

(1) **Sequence segmentation**: Split telemetry sequence into multiple sub-sequences. For a telemetry sequence with length L, a non-overlapping sliding window with size N is used for segmentation to form $\lceil L/N \rceil$ sub-sequences. For the last sliding window, we fill the first $\lceil L/N \rceil \cdot N - L$ part of it with the last $\lceil L/N \rceil \cdot N - L$ samples of the penultimate subsequence.

(2) **Clustering:** Clustering the objects in each subsequence.

(3) **Anomaly detection**: Find out all anomalies in each subsequence.

(4) **Anomaly window searching**: Search the starting positions and lengths of all anomaly windows in the entire telemetry sequence.

The content of clustering, anomaly detection, and anomaly window searching will be described in detail below.

2.1. Extended dominant sets clustering

According to the principle of dominant sets clustering [10], for a dataset $\boldsymbol{D}$ of N size, the similarity matrix on $\boldsymbol{D}$ is defined as $\boldsymbol{A}=(a_{pq})$, $p,q \in \boldsymbol{D}$, where a_{pq} represents the weighted similarity between object p and q, and a_{pq} is given by:

$$a_{pq} = \begin{cases} e^{-\frac{d(p,q)}{\sigma}} & p \neq q \\ 0 & p = q \end{cases} \tag{1}$$

In (1), $d(p,q)$ means the Euclidean distance of object p and q; σ is the attenuation factor, which is taken as σ=0.5median(d). The dominant set of $\boldsymbol{D}$ is represented by $\boldsymbol{S}$, where $\boldsymbol{S}$ is a non-empty data set and $\boldsymbol{S} \subseteq \boldsymbol{D}$. To save space, the solution process of $\boldsymbol{S}$ is not described here, and readers can be referred to in [10]. In this paper, we take $\boldsymbol{S}$ as the cluster centre set, and name p as the centre member if $p \in \boldsymbol{S}$. To complete a cluster, we define the ε-similarity neighbourhood of p as:

$$N_\varepsilon(p) = \{q \mid q \in \boldsymbol{D}, a_{pq} \geq \varepsilon\} \tag{2}$$

Here ε is given by:

$$\varepsilon = \max\big(\text{median}(\boldsymbol{A}(p,\boldsymbol{S})), p \in \boldsymbol{S}\big) \tag{3}$$

If q is an ε-similarity neighbour of p, then p is an ε-similarity neighbour of q. Furthermore, p and q are considered to have the ε-similarity relationship. Based on the dominant set $\boldsymbol{S}$ and the ε-similarity relationship, a cluster is defined as the largest object set derived from the centre set $\boldsymbol{S}$ through the ε-similarity relationship. Formula (8) gives the recursive definition of the cluster:

$$\boldsymbol{C}^{[i]} = \begin{cases} \boldsymbol{S} & i = 0 \\ \boldsymbol{C}^{[i-1]} \cup N_\varepsilon\left(p \in \boldsymbol{C}^{[i-1]}\right) & otherwise \end{cases} \tag{4}$$

Here i is the growing step of cluster $\boldsymbol{C}$, and $\boldsymbol{C}$ will stop growing once $\boldsymbol{C}^{[i]}=\boldsymbol{C}^{[i-1]}$ occurs.

The pseudocode in Table 1 describes the process of extended dominant set clustering.

Table 1. Extended dominant set clustering

Input	***D***	/* data set*/
Output	***ClusteId***	/* cluster id of each object in ***D*** */
	$\boldsymbol{S}_{size}$	/*size of each dominant set */
	k=1; N=|***D***|; ***Cluster Id*** = -1*ones(N,1)	
	Calculate ***A*** according to formula (1);	
	While |***D***|>0.0001*N **do**	
	Get the current dominant set: ***S***;	
	Record dominant set size: $\boldsymbol{S}_{size}(k)$=|***S***|;	
	Get k^{th} cluster according to formula (4): ***C***;	
	Record the cluster id of each cluster member: ***ClusterId***(***C***)=k;	
	k++;	
	end While	

2.2. Anomaly detection

The cluster-based anomaly detection algorithm considers objects that do not belong to any cluster or belong to a small cluster as anomaly objects [11][12]. According to Table 1, an object with a cluster id of -1 is an object that does not belong to any cluster, and thus it can be treated as an anomaly.

Suppose $\boldsymbol{C} = \{C_1, C_2, ..., C_k\}$ is the set of clusters in the sequence that $|C_1| \geq |C_2| \geq |C_k| > 1$, and k is the number of clusters in ***D***. To identify small clusters, we define b as the boundary of a large and small cluster if one of following formulas holds:

$$\begin{cases} |C_{b-1}|/|C_b| \geq 5 \\ |C_b| \leq \boldsymbol{S}_{size}(b-1) \end{cases} \tag{5}$$

Here $\boldsymbol{S}_{size}$ records the size of the dominant set of each cluster. Then the set of a large cluster is defined as: $\boldsymbol{LC} = \{C_i, 1 \leq i < b\}$; and the set of the small cluster is defined as: $\boldsymbol{SC} = \{C_i, b \leq i \leq k\}$. Objects belonging to ***SC*** are treated as anomalies.

To detect the anomalies in ***LC***, we calculate the similarity of p about its m-distance neighbourhood:

$$sim(p) = \frac{1}{m} \sum_{q \in \boldsymbol{M}_m(p)} a_{pq} \tag{6}$$

Here $\boldsymbol{M}_m(p)$ represents the set of m objects with the highest similarity to p in ***LC***, and m is calculated by Formula (7):

$$m = \max\left(\left\lceil \frac{|C_i|}{\boldsymbol{S}_{size}(i)} \right\rceil\right),\ 1 \leq i < b \tag{7}$$

Based on (6), we define the relative similarity of p about its m-distance neighbourhood as:

$$rsim(p) = \frac{\frac{1}{m} \sum_{q \in \boldsymbol{M}_{m(p)}} sim(q)}{sim(p)} \tag{8}$$

The $rsim$ value reflects the local anomaly degree of large cluster members: the more p deviates from its neighbours, the larger the $rsim(p)$ value is.

According to the above anomaly detection process, the anomaly label of object p is given by:

$$AnomalyLabel(p) = \begin{cases} 0 & rsim(p) < thresh; p \in \boldsymbol{LC} \\ 1 & otherwise \end{cases} \tag{9}$$

Here $thresh$ is the anomaly filtering threshold, and is calculated from the mean and standard deviation of $rsim$ values:

$$thresh = rsim_{mean} + 5 \cdot rsim_{std} \tag{10}$$

2.3. Anomaly window searching

Suppose the set of anomalies in the telemetry sequence is formed as $\boldsymbol{O} = \{o_1, o_2, \ldots, o_n\}$, where o_i represents the serial number of the ith anomaly in the telemetry sequence, $1 \leq i \leq n$ and n are the total numbers of anomalies. If i, j holds Formula (11), then $[o_i, o_j]$ is treated as an anomaly window.

$$\begin{cases} \dfrac{j-i+1}{o_j - o_i + 1} \geq Rate_{th} \\ \dfrac{j+1-i+1}{o_{j+1} - o_i + 1} < Rate_{th} \end{cases} \tag{11}$$

Here $Rate_{th}$ represents the threshold value of the local anomaly rate, with the range of (0, 1]. And the larger $Rate_{th}$ is, the narrower the anomaly window generated; especially when the $Rate_{th}$ is set to 1, the objects in the anomaly window generated will be all anomalies.

3. Experimental results and analysis

3.1. Synthetic data sets

To intuitively display the clustering and anomaly detection process, and evaluate the impact of the introduction of relative similarity on the detection results, this experiment uses three synthetic data sets (see Figure 1 and Table 2 for details). And the AUC (Area Under Curve) value under the ROC (Receiver Operating Characteristic) curve is used as the evaluation index.

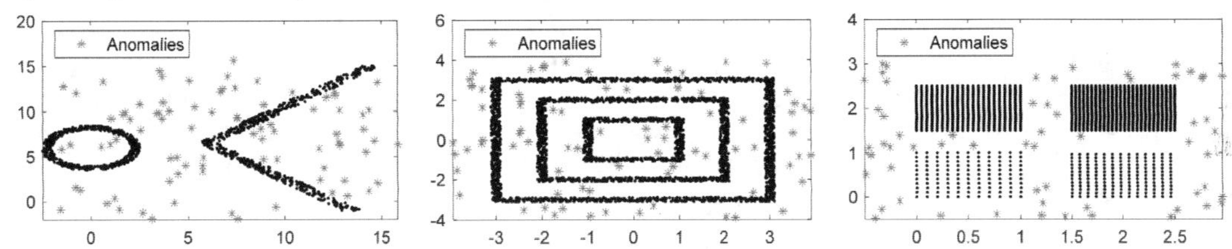

(a) D1 (b) D2 (c) D3

Figure 1. Synthetic data sets

Table 2. Description of synthetic data sets

Name	Size	Number of large clusters	Number of anomalies	Anomaly rate
D1	1137	2	85	7.48%
D2	3031	3	79	2.61%
D3	1467	4	60	4.09%

Figure 2 shows the clustering, *rsim* calculation, and anomaly detection results of the proposed method on synthetic data sets. In Figure 2, 'Anomalies 1' indicates the anomalies detected traditionally, which are objects that do not belong to any cluster or belong to small clusters, while 'Anomalies 2' indicates the anomalies detected according to the relative similarity index.

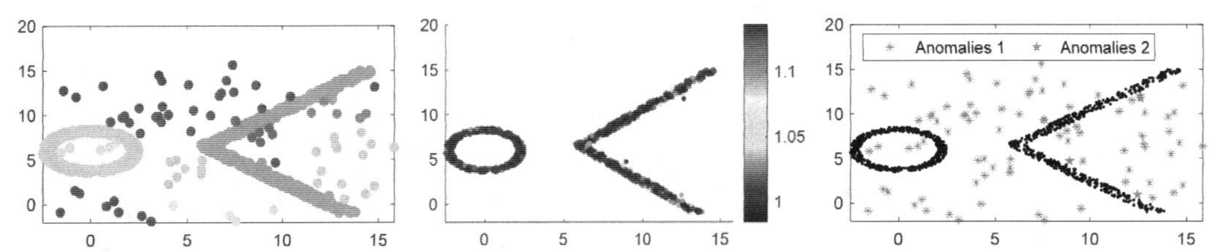

(a) D1: Clusters (b) D1: *rsim* values (c) D1: Anomalies

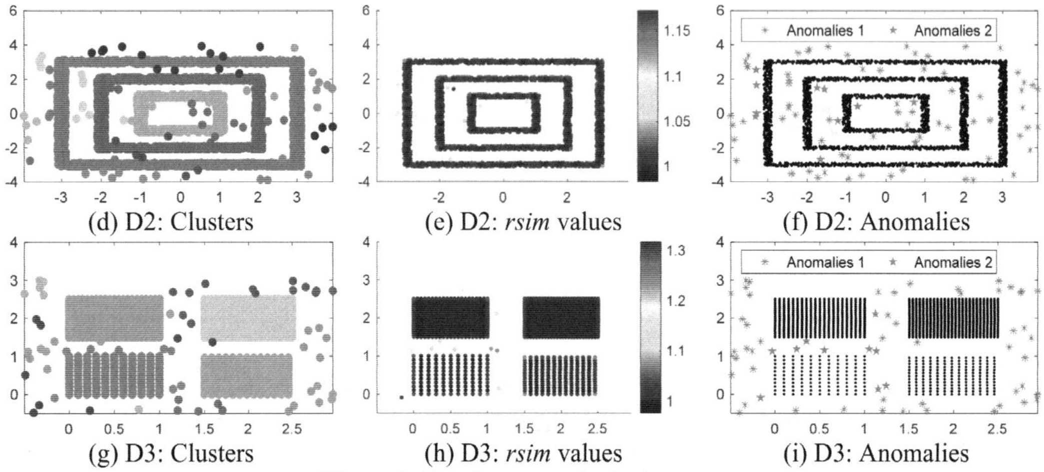

(d) D2: Clusters (e) D2: *rsim* values (f) D2: Anomalies

(g) D3: Clusters (h) D3: *rsim* values (i) D3: Anomalies

Figure 2. Results on synthetic data sets

Experimental results show that: 1) The extended dominant sets clustering method proposed in this paper can handle data sets with different cluster numbers and cluster shapes; 2) The AUC values obtained by traditional means are 0.9706, 0.9241 and 0.9083; 3) The introduction of relative similarity increase AUC values by 3.03%~10.09%.

3.2. Magnetometer telemetry data

This experiment uses the magnetometer telemetry data of the real in-orbit satellite named Tianping-2B, as shown in Figure 3. The data is generated from 16:09:00~22:00:00 on July 5, 2022, with a total of 21,010 samples. Tianping-2B satellite has a 600km sun-synchronous orbit, and the orbital period is about 96 minutes. Besides, the magnetometer telemetry data has 3 dimensions, namely X-axis, Y-axis and Z-axis; and the update frequency is 1Hz. According to the above physical characteristics, the sliding window size is set as N=5760, that is, the divided subsequence contains the data amount of one orbital period.

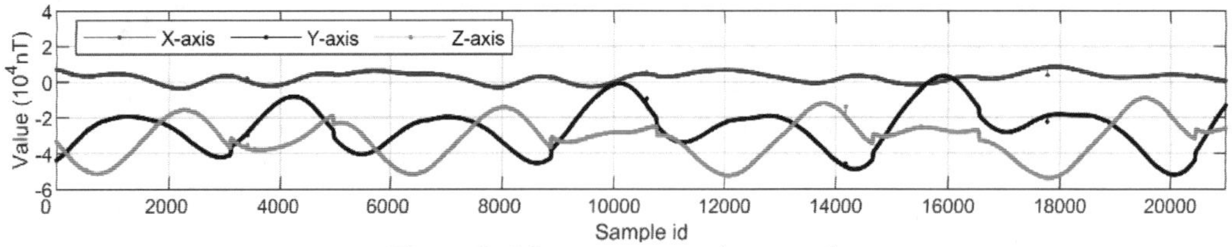

Figure 3. Magnetometer telemetry data

Figure 4 shows the result of the anomaly detection method proposed in this paper on magnetometer telemetry data. According to the result, this magnetometer data contains 1.07% of anomalies, which form 19 anomaly windows. Information on these anomaly windows is shown in Table 3.

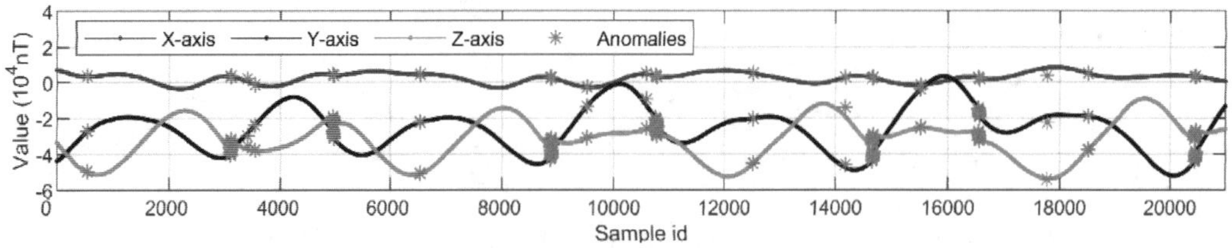

Figure 4. Anomaly detection result on Magnetometer telemetry data

Table 3. Information on each anomaly window

Id	Window type I		Window type II		Window type III	
	Start position	Length	Start position	Length	Start position	Length
1	3398	1	536	7	3089	18
2	8861	1	3531	6	4969	27
3	10587	1	6530	9	8869	21
4	14182	1	9525	6	10749	32
5	17776	1	12524	5	14652	24
6	-	-	15514	8	16542	25
7	-	-	18508	10	20439	26

(1) 5 anomaly windows of Type I: This type of window has a length of 1, and anomalies appear randomly. These anomalies are considered to be wild values.

(2) 7 anomaly windows of Type II: The length of such windows is between 5 and 10, and the average difference between the starting positions of adjacent windows is about 2995 sample points. Figure 5(a) shows the Z-axis data of the magnetometer around these windows. It is speculated that the magnetometer will be subject to fixed interference every 49 minutes and 55 seconds on average. Furthermore, according to the expert knowledge of the system, it is determined that the interference comes from the reconfiguration operation of the TT&C transponder A. To resist the space single-particle effect, the TT&C transponder of the Tianping-2B satellite is designed to automatically reconfigure the main FPGA every 50 minutes. During the reconfiguration process, the transponder device has a small current, and the magnetic field generated by this current is weak, which leads to a decrease in the amplitude of magnetometer data.

(3) 7 anomaly windows of Type III: The length of such windows exceeds 15, and the starting positions of adjacent windows are separated by about 1883 and 3900 sample points, corresponding to the interval of 31 minutes 23 seconds and 65 minutes 0 seconds, respectively. Figure 5(b) shows the Z-axis data around these windows. Since the orbital period is about 94 minutes, it is speculated that this phenomenon may occur during the movement of the satellite in and out of shadow areas. According to the analysis of the satellite mechanical structure and the output of the solar sensors, it is determined that the interference comes from the input current of the expanded solar panel, which is manifested: 1) When the solar panel switches from the illuminated to the unilluminated state, the input current decreases rapidly, and the magnetic field strength generated by the current weakens rapidly, which will cause the amplitude of the X-axis output of the magnetometer to increase rapidly, while the amplitude of the Y-axis and Z-axis output decreases rapidly; 2) When the solar panel switches from the unilluminated to the illuminated state, the strength of magnetic field changes in the opposite way.

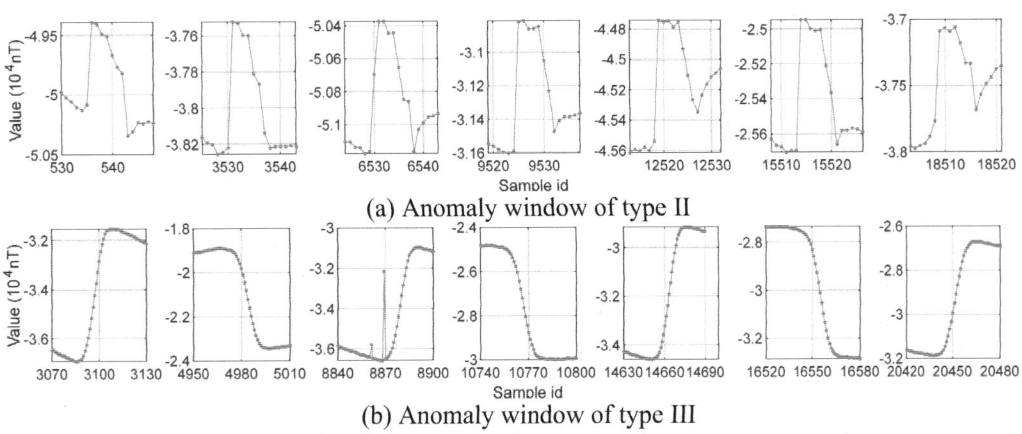

(a) Anomaly window of type II

(b) Anomaly window of type III

Figure 5. Details of Z-axis data of magnetometer in anomaly windows

The above results show that the anomaly detection method based on extended dominant sets clustering proposed in this paper can effectively mine the anomalies in the magnetometer telemetry data of Tianping-2B, and can provide data support for the ground to grasp the health status of the magnetometer device, and the changes of the magnetic field inside the satellite.

4. Conclusions

This paper proposes a clustering-based anomaly detection method for telemetry data, which mainly includes extended dominant sets clustering, anomaly detection and anomaly window searching. Experimental results show that this method can effectively mine anomalies in satellite telemetry data under unsupervised conditions, and provide data support for improving the safety and reliability of satellites in orbit.

Acknowledgments

This research was supported by the Fundamental Research Funds for the Central Universities (2021XZZX038).

References

[1] Peng X Y, Pang J Y, Peng Y and Liu D T. (2016) Review on anomaly detection of spacecraft telemetry data. *Chinese Journal of Scientific Instruments*, 9: p1929-1945.

[2] Hundman, K, Constantinou V, Laporte C, Colwell I and Soderstrom T. (2018) Detecting Spacecraft Anomalies Using LSTMs and Nonparametric Dynamic Thresholding. *In Proceedings of the 24th ACM SIGKDD international conference on knowledge discovery & data mining*, London, p387-395.

[3] Nalepa, J, Myller M, Andrzejewski J, Benecki P and Piechaczek S. (2022) Evaluating algorithms for anomaly detection in satellite telemetry data. *Acta Astronautical*, 2022, 198: p689-701.

[4] Iverson D L. (2004) Inductive system health monitoring. *International Conference on Artificial Intelligence*. p1-7.

[5] Iverson D L. (2008) System Health Monitoring for Space Mission Operations. *IEEE Aerospace Conference*, Montana, p1-8.

[6] Iverson D, Martin R, Schwabacher M, Spirkovska L and Taylor W. (2009) General Purpose Data-Driven System Monitoring for Space Operations. *AIAA Infotech @ aerospace conference*, Seattle, p1-11.

[7] Gao Y, Yang T S, Xu M Q, Xing N. (2012) An Unsupervised Anomaly Detection Approach for Spacecraft. *Fifth International Conference on Intelligent Computation Technology and Automation,* Zhangjiajie, p478-481.

[8] Yairi T, Takeishi N, Oda T, Nakajima Y, Nishimura N and Takata N. (2017) A Data-Driven Health Monitoring Method for Satellite Housekeeping Data Based on Probabilistic Clustering and Dimensionality Reduction. *IEEE Transactions on Aerospace & Electronic Systems*, 53: p1384-1401.

[9] Wang C, Wang H Q, Jin Z H. (2018) Pico-satellite telemetry anomaly detection through clustering. *Journal of Harbin Institute of Technology*, 50: p110-116.

[10] Pavan M, Pelillo M. (2007) Dominant Sets and Pairwise Clustering[J]. *IEEE Transactions on Pattern Analysis and Machine Intelligence*, 29: p167-172.

[11] He Z, Xu X, Deng S. (2003) Discovering cluster-based local outliers. *Pattern Recognition Letter*, 24: p1641-1650.

[12] Huang J L, Zhu Q S, Yang L J, Cheng D D, W Q W. (2017) A novel outlier cluster detection algorithm without top-n parameter. *Knowledge-Based Systems*, 121: p32-40.

Simulation and analysis of drag for inflatable aerodynamic decelerator

Chenghu Yun, Danghui Liu

Department of Aerospace Science and Technology, Space Engineering University, Beijing, 101416, China

Liudanghui@sohu.com

Abstract. Inflatable aerodynamic decelerator, as a novel deceleration and recovery technology, has received attention due to its lightweight and freedom from the launch vehicle fairing size constraints for the space mission. Aimed at the aerodynamic characteristics of the inflatable aerodynamic decelerator, this work studied the effect of the cone angle on its drag and discussed the drag varied with the angle of attack under the different cone angles. The simulation showed that the drag was affected by the cone angle and increased locally with the angle of attack. Furthermore, the larger cone angle was beneficial to the stability of the flow field on its leeward in the case of a high Mach number. The math model of drag with the Mach number and the angle of attack was obtained by fitting the data, which provided favorable support for the prospective discussion of the flight dynamics characteristics.

1. Introduction

With the vigorous development of the aerospace industry, the effective carrier capability of the launch vehicle is expected to enhance and decrease the launch cost [1]. In recent years, SpaceX has realized the recycling of the first stage through the retro-propulsion and attitude control technology of the launch vehicle, which effectively reduces the launch cost and the launch cycle and promotes the development of the reusable launch vehicle (RLV). But a certain amount of propellant is reserved for the reentry process, which results in the loss of carrier capability [2]. In this situation, the inflatable aerodynamic decelerator (IAD) is noticed gradually due to its small size, foldable, and lightweight [3]. The primary explorations have been carried out to apply the IAD in the Mars landing [4], the payload reentry [5], the launch vehicle recovery [6], and so on. However, some complex effects, such as aerodynamic loads, aerothermal, and aeroelasticity during the reentry process of the IAD caused by its flexible and inflatable, has to be studied deeply [7].

Researchers have made an in-depth study of computational fluid dynamics (CFD) [8], wind tunnels [9], and flight experiments [10] to solve the above problems. Restricted by the actual conditions, CFD is the most common method that lots of studies have been performed. For example, Wei [11] compared and analyzed the difference between the cone angle and the shape of the IAD surface (straight and concave surface) by simulating the aerodynamic characteristics of the flow field during the re-entry. The results showed that the drag of the IAD had a significant relationship with its cone angle, and the shape of the IAD would have an influence on the aerodynamic characteristics. Takahashi [12] studied the aerodynamic coefficients of the IAD by transonic wind tunnel experiment and CFD and obtained the mechanism of differences between the above two methods and flight experiments. The results showed that the leeward of the test model affected the drag at the transonic and supersonic regimes. Bhaskar [13] researched the influence of the cone angle, the length of the load on the leeward, and the position of the center of gravity for the aerodynamic coefficient of the IAD. The researcher obtained a statically stable configuration according to the results. However, the above research had a little discussion on the relationship between the cone angle and the aerodynamic characteristics of the IAD. The relationship between the cone angle and the drag needs to be studied further.

In this work, based on the IAD model adopted by IRVE (Inflatable Reentry Vehicle Experiment) [14], the effect of different cone angles (Unless otherwise specified, the cone angle below refers to the IAD.) on drag was studied in detail. Then, the tendency of the drag at different angles of attack (AoA) and cone angle, and the influence of cone angle on the stability of the leeward flow field at different Mach numbers (Ma) were studied. Finally, by fitting and analyzing the above simulation data, the math model between the drag and the Ma and AoA was obtained for the IAD.

2. Computational model

2.1. Math model

The CFD experiment was performed by Fluent, and the physical field of the computational domain was solved by the Reynolds-averaged N-S equation based on the finite volume method, and the SST model was selected as the turbulence model. The governing equations expressed by the Einstein summation symbol are listed as follows [15]:

$$\begin{aligned} &\frac{\partial \rho}{\partial t}+\frac{\partial}{\partial x_i}(\rho u_i)=0 \\ &\frac{\partial}{\partial t}(\rho u_i)+\frac{\partial}{\partial x_j}(\rho u_j u_i)=-\frac{\partial \mathrm{p}}{\partial x_i}+\frac{\partial \tau_{ij}}{\partial x_j} \\ &\frac{\partial}{\partial t}(\rho \mathrm{E})+\frac{\partial}{\partial x_j}(\rho u_j H)=\frac{\partial}{\partial x_j}(u_i\tau_{ij}-\dot{q}_j) \end{aligned} \quad (1)$$

where u_i represents the i-th velocity component ($i, j = 1,2$), p for the pressure of the flow field, E for the total energy of the flow field, H for the total enthalpy of the flow field, τ_{ij} for the fluid viscous stress tensor and is the sum of the laminar flow and turbulent flow components, and $\dot{q}$ for the fluid heat flux density.

2.2. Boundary conditions

Referring to the test data of IRVE [16], different altitudes were selected for simulation, and the detailed free stream parameters were shown in Table 1. The conditions were assumed as follows: the reentry altitude was less than 91 km, the free incoming flow was continuous flow, no chemical reaction occurred between the incoming components, the incoming gas was an ideal compressible gas, and the viscosity condition adopted Sutherland's law. For the incoming flow, Ma and static conditions were known for the flow field, the velocity inlet condition was set as the pressure-far-field boundary, and the IAD wall was adiabatic with non-slip.

2.3. Method validation

The model of the IAD used in this work was shown in Figure 1(a). The outer flow field mesh was generated by ICEMCFD (The Integrated Computer Engineering and Manufacturing code for Computational Fluid Dynamics). The wall mesh satisfies that the y^+ was less than 1, which met the calculation requirements, and the mesh was shown in Figure 1(b). The method of use in this work was compared with the published literature [11]. As shown in Figure 1(c), the maximum error was 6.1% and the minimum error was 1.2%. The main error was from the mesh generation and boundary conditions, and the error range was acceptable. The result proved that the simulation method used in this work was feasible.

Table 1. Free stream parameters.

altitude/km	Ma	pressure/pa	temperature/k
38	1	377	244
48	3	102	270
68	4.5	5	219

(a)

(b)

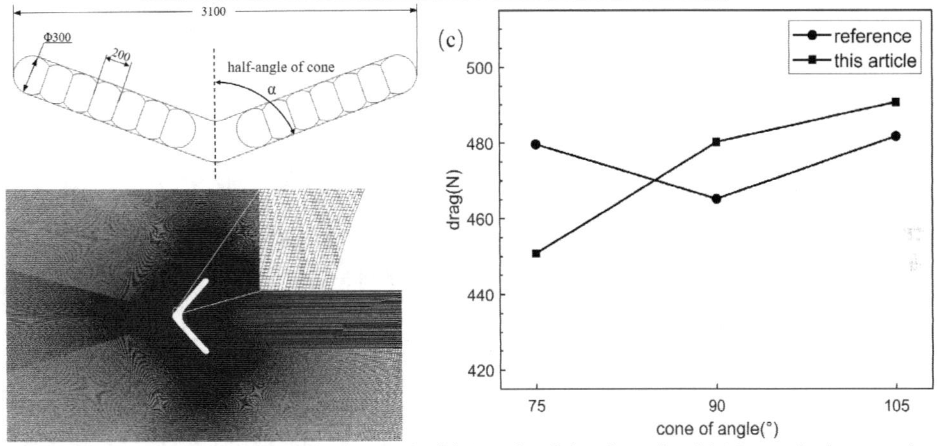

Figure 1. (a) The model of the IAD (unit: mm). (b) Mesh of the domain. (c) Numerical examples validation.

3. Results and analysis

3.1. Influence of the cone angle on drag

The cone angle ranged from 90 ° to 160 ° with every 10 ° as an interval, and other conditions remained unchanged. The drag of the IAD was simulated experimentally at Ma = 1 and AoA = 0. Figure 2(a) showed that the drag increased with the cone angle. The drag increased by 61.76 N when the cone angle varied from 90 ° to 120 °. However, the drag increased by only 35.25 N when the cone angle varied from 120 ° to 150 °, which was less than the additional drag when the cone angle varied from 90 ° to 120 °. The drag had an increasing trend with the increase of the cone angle, but the trend slowed down when the cone angle was greater than 140 °. This was not enough to determine the optimum cone angle, and further research was needed on the aerodynamic characteristics under different cone angles and AoA.

3.2. Compound effects of cone angle and AoA on drag

3.2.1. Ma=1

The drag of the IAD was positively related to the size of its cone angle at AoA = 0. But in the reentry process, there would be an AoA due to its reentry speed, aerodynamic characteristics, flight stability, and other requirements. Therefore, it was necessary to simulate the drag under the AOA. The cone angles were 100 °, 120 °, 140 °, and 160 ° respectively, and the AoA ranged from 0 to 50 °, with an interval of every 5 °.

It could be seen from Figure 2(b) that when the AOA was less than 20 °, the drag decreased with the growing AoA. The drag of the IAD changed nonlinearly when the AoA exceeded 20 °. There was little difference of about 1.1% in drag between the IAD with the cone angle of 160 ° and 140 ° at the same AoA. When the cone angle was 140 °, the drag decreased with the increase of the AoA, but when the AoA increased from 35 ° to 40 °, the drag increased slightly, and then decreased sharply. To explain the phenomenon, the distributions of flow properties were studied when the cone angle was 140 ° and AoA was 35 °, 40 °, and 45 °. As shown in Figure 3(a-c), with the increase of the AoA, the stagnation points gradually moved down, and the angle between the free stream and the surface of the IAD gradually decreased. The free stream was perpendicular to the surface at the critical AoA (30 ° - 40 °), which led to a decrease in the velocity and dynamic pressure, resulting in a decrease in drag.

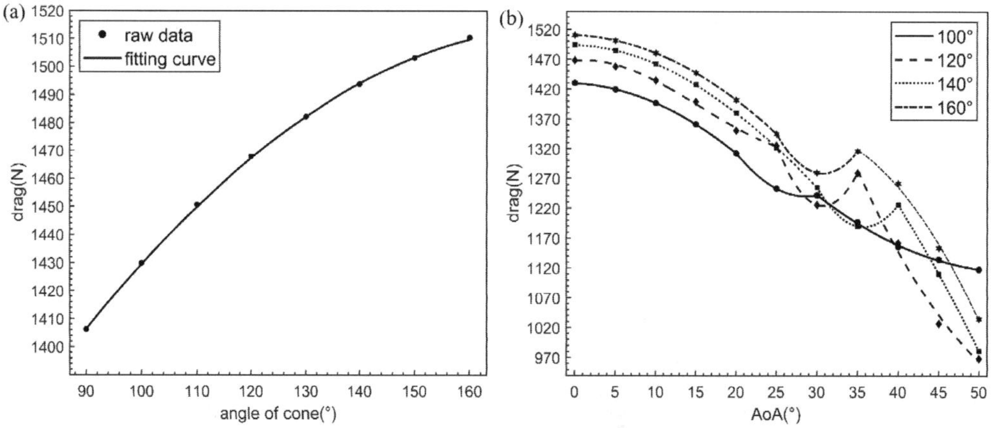

Figure 2. (a) Drag changing with the cone angle. (b) Drag changing with the AoA.

The alike phenomenon was observed with other cone angles. The corresponding range of critical AoA varied with the cone angle. For the cone angle of 100 ° (Figure 3(d-f)), the critical AoA range was reduced to 20 ° - 30 °. And for the cone angle of 120 ° (Figure 3(g-i)), the critical AoA ranged from 25 ° to 35 °. The range of critical AoA gradually increased with the cone angle. However, when the cone angle was 160 °, the critical AoA range was reduced to 25 ° - 40 °. It revealed that under the AoA, the maximum critical AoA range appeared when the cone angle was 140 ° that better than the larger cone angle.

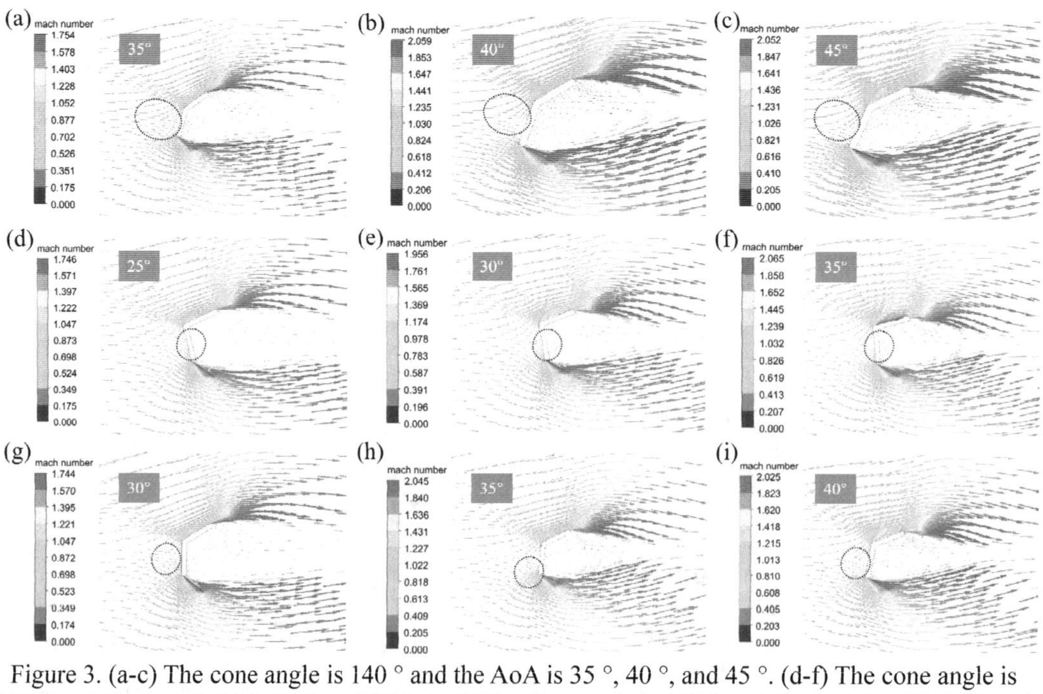

Figure 3. (a-c) The cone angle is 140 ° and the AoA is 35 °, 40 °, and 45 °. (d-f) The cone angle is 100 ° and the AoA is 25 °, 30 °, and 35 °. (g-i) The cone angle is 120 ° and the AoA is 30 °, 35 ° and 40 °.

3.2.2. Different Ma flight conditions (AoA = 0)

The above discussions were all at Ma = 1. The flow field of the IAD became complicated under the high Ma due to the combined effects of shock and boundary layers. To study the influence of cone angle for the IAD at high Ma, the flow field of the IAD was simulated when the cone angle was 100 °, 120 °, 140 °, and the Ma was 1, 3, 4.5 respectively.

There was an obvious expansion wave that appeared in front of the IAD (Figure 4(a)) at Ma = 1, while an obvious bow shock appeared in front of the IAD (Figure 4(b-c)) when the Ma was 3 and 4.5. These were consistent with the wind tunnel experimental results in the literature [12]. With the increase of the cone angle, the pressure of the external flow field increased slightly at Ma = 1 (Figures 4 (b)(e)(f)) and changed nonlinearly. However, with the cone angle increased, the radial area of the affected leeward became smaller. It revealed that the leeward of the IAD with a larger cone angle was less affected, and its leeward flow field was more stable. In practical application, the payload would be connected to the tail of the IAD and the large cone angle was beneficial to the stability of the IAD. Based on the above analysis results, when the cone angle was 140°, its general characteristics of aerodynamics were better than other cone angles.

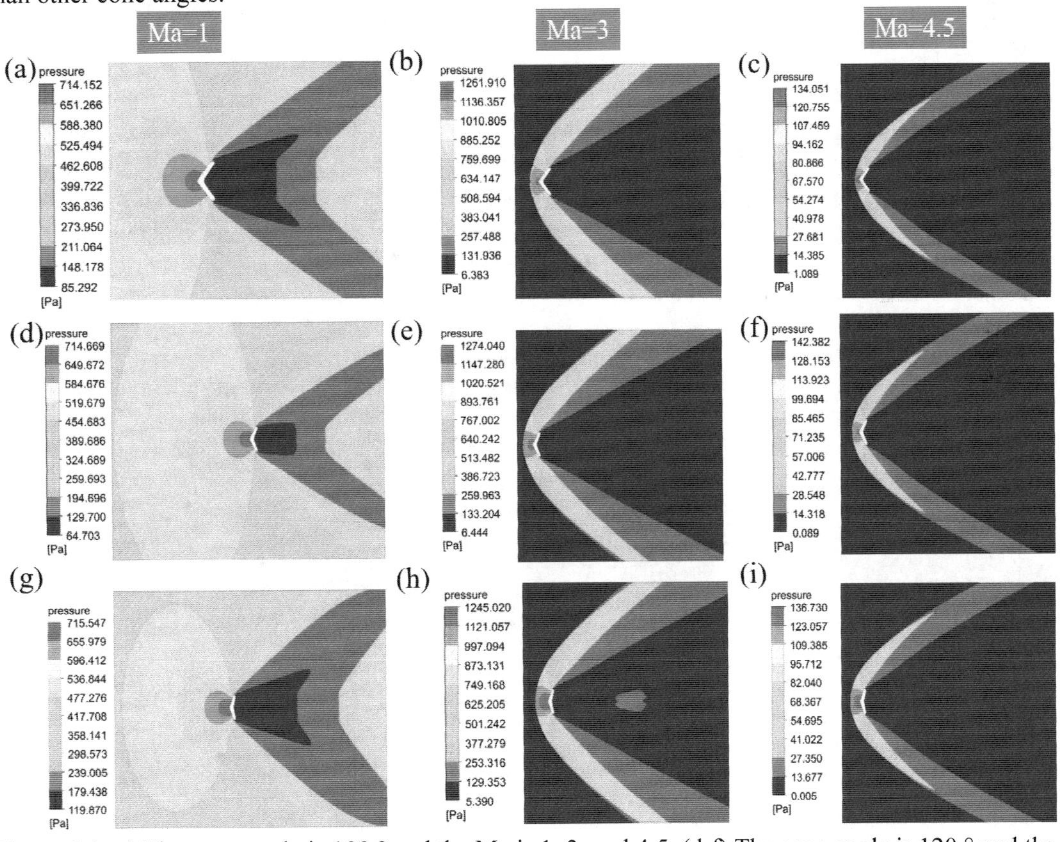

Figure.4 (a-c) The cone angle is 100 ° and the Ma is 1, 3, and 4.5. (d-f) The cone angle is 120 ° and the Ma is 1, 3, and 4.5. (g-i) The cone angle is 140° and the Ma is 1, 3, and 4.5.

4. Fitting analysis of drag with AoA and Ma

Based on section 3.2, the IAD with the cone angle of 140 ° was selected for the reentry process. In this section, the relationship among drag, AoA, and Ma was obtained through fitting analysis.

4.1. Fitting curve of drag and AoA (Ma=1)

Figure 5(a) showed the relationship between the drag and the AoA that was obtained by fitting the data.

Due to the change of the drag was obviously different from the data of other segments, the piecewise fitting was used when the AoA was 30 ° - 40 °, and the function of the drag with respect to the AoA (in radian, $rad = \pi \times AoA/180^{\circ}$) in Equation (2) was taken as $F_{AoA}(x)$. It could be seen in Figure 5(a) that in the range of small AoA (AoA less than 10 °), the drag changed little, and it gradually decreased as the AoA increased. When the AoA was 30 ° - 40 °, the drag started to decrease and then subsequently increase, and then decrease obviously when the AoA was greater than 40 °.

$$
\begin{aligned}
F_{AoA} &= -767.3013\times x^2 - 57.6314\times x + 1494.6(0 \le x < 0.5236)\\
F_{AoA} &= 6643.1\times x^2 - 8286.2\times x + 3772.9(0.5236 \le x \le 0.6981)\\
F_{AoA} &= -878.479\times x^2 - 20.9703\times x + 1668.5(0.6981 < x \le 0.8727)
\end{aligned} \tag{2}
$$

4.2. Fitting curve of drag and Ma (AoA=0)

The experimental data of the drag at Ma of 0.3 - 4.5 were fitted by the polynomial and neural network, and the fitting curves were shown in Figure 5(b). Using polynomials of order 6 to 9 to fit, the goodness of fit R^2 increased with the independent variables, but it was due to the false growth caused by the increase of independent variables, and the adjusted R^2 was less than 0.99, which was under-fitting and did not meet the fitting requirements. Then the neural network was used for fitting, and the default number of hidden layers was 1 and the number of neurons was 6, and the Bayesian Regularization algorithm was used for training. The R^2 obtained by training was greater than 0.99, which met the fitting requirements. The function of drag with respect to Ma was $F_{Ma}(v)$. It could be seen from Figure 5(b) that in the reentry process of the IAD, the drag showed the trend of increasing first and then decreasing as the Ma decreased, and there is a trough and peak at Ma about 0.5 and 2.5. Through the fitting equation obtained in this work, the drag of the IAD at the Ma range could be easily calculated.

4.3. Models of drag on Ma and AoA

According to the formulas obtained in Sections 4.1 and 4.2, Equation (3) about the relationship between the drag and the Ma and the AoA was derived. Figure 5(c) showed the three-dimensional surface graph of the function.

$$
F_d = \frac{F_{AoA}(x)}{F_{AoA}(0)} \times F_{Ma}(v)(0 \le x \le 0.8723, 0.3 \le v \le 4.5) \tag{3}
$$

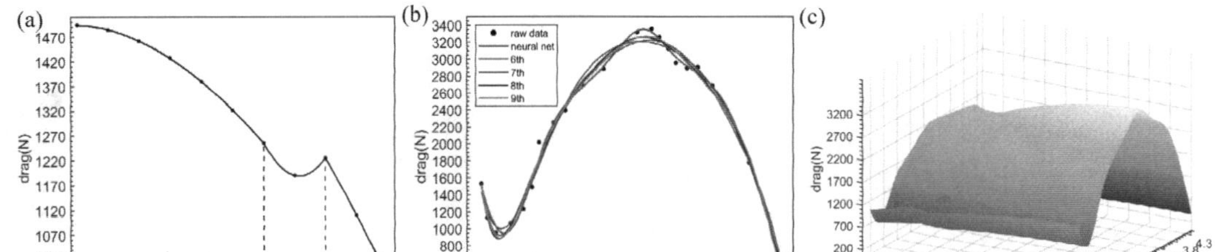

Figure 5. (a) The fitting curve of drag and AoA. (b) The fitting curve of drag and Ma. (c) 3-D surface plot of drag with respect to Ma and AoA.

5. Conclusion

This work discussed the influence of the cone angle on the drag of the IAD, including the change of the drag with the cone angle, the effect of the AoA on the IAD with different cone angles, and the influence of the cone angle on the IAD at different Ma. The results showed that the drag increased with the cone angle, however, the increased value of drag gradually decreased. With the increase of AoA, the drag decreased. When the AoA reached the critical value, the free stream was perpendicular to the surface of the IAD, and the drag changed abruptly it increased first and then decreased greatly. At high Ma, the cone angle would affect the radial area of its leeward side. Under the large cone angle, the leeward side of the IAD was smaller and the flow field was more stable. Overall, the results showed that the performance of the IAD was excellent when the cone angle was 140 °, and the fitting equations of drag with respect to Ma and AoA were obtained through the simulation and data fitting, which laid a foundation for further practical application of the IAD.

References

[1] Sivan K and Pandian S 2018 An overview of Reusable Launch Vehicle Technology Demonstrator *CURR. SCI. INDIA* **114** 38-47
[2] Mandalia A B and Braun R D 2015 Supersonic Retropropulsion Thrust Vectoring for Mars Precision Landing *J. Spacecraft Rockets* **52** 827-35
[3] Takahashi Y, Koike T, Oshima N and Yamada K 2019 Aerothermodynamic analysis for the deformed membrane of inflatable aeroshell in orbital reentry mission *Aerosp. Sci. Technol.* **92** 858-68
[4] Friz P D, Samareh J A and Hosder S 2019 New Method for Systems and Cost Analysis of Human Mars Entry Vehicles *J. Spacecraft Rockets* **56** 1742-56
[5] Finchenko V S, Ivankov A A, Golomazov M M and Shmatov S I 2020 On the Use of Inflatable Decelerators in the Design of Spacecraft Intended for the Study of Venus *Solar Sys. Res.* **54** 595-602
[6] Dek C, Overkamp J L, Toeter A, Hoppenbrouwer T, Slimmens J, Zijl J V, Ross I A, Machado R, Hereijgers S, Kilic V and Naeije M 2020 A recovery system for the key components of the first stage of a heavy launch vehicle *Aerosp. Sci. Technol.***100** 105778-86
[7] Wu J, Zhang Z, Hou A, Wang L W, Wang J and Cao Xu 2020 Dynamic Aerodynamic Load and Structural Characteristics of Inflatable Reentry Reducer *Journal of Astronautics* **41** 287-97
[8] Brune A J, West T K, Hosder S and Edquist K T 2015 Uncertainty Analysis of Mars Entry Flows over a Hypersonic Inflatable Aerodynamic Decelerator *J. Spacecraft Rockets* **52** 776-88
[9] Hollis B R 2018 Surface Heating and Boundary-Layer Transition on a Hypersonic Inflatable Aerodynamic Decelerator *J. Spacecraft Rockets* **55** 856-76
[10] Lichodziejewski D, Tutt B and Jurewicz D 2013 Ground and Flight Testing of a Stacked Tori Hypersonic Inflatable Aerodynamic Decelerator Configuration *54th AIAA/ASME/ASCE/AHS/ASC Structures, Structural Dynamics, and Materials Conference* Boston 1864-73
[11] Wei J, Ma R, Tan H and Xie Z M 2013 Analysis of Aerodynamic Characteristics of Inflatable Membrane Aeroshell *Spacecraft Recovery & Remote Sensing* **34** 49-55
[12] Takahashi Y, Matsunaga M, Oshima N and Yamada K 2019 Drag Behavior of Inflatable Reentry Vehicle in Transonic Regime *J. Spacecraft Rockets* **56** 577-85
[13] Bhaskar N U, Deependran B and Ashok V 2020 Aerodynamic Configuration Analysis of a Typical Inflatable Aerodynamic Decelerator *7th Asian Joint Workshop on Thermophysics and Fluid Science* Trivandrum 585-600
[14] Lindell M C and Hughes S J 2006 Structural Analysis and Testing of the Inflatable Re-entry Vehicle Experiment (IRVE) *47th AIAA/ASME/ASCE/AHS/ASC Structures, Structural Dynamics, and Materials Confere* Newport 1699-1717
[15] Guo J, Lin G, Bu X, Fu S M and Chao Y M 2017 Effect of static shape deformation on aerodynamics and aerothermodynamics of hypersonic inflatable aerodynamic decelerator *Acta Astronaut.* **136** 421-33
[16] Olds A, Beck R, Bose D M, White J, Edquist K T, Hollis B R, Lindell M, Cheatwood F M, Gsell V and Bowden E L 2013 IRVE-3 Post-Flight Reconstruction *AIAA Aerodynamic Decelerator Systems (ADS) Conference* Florida 1390-1410

Distributed rotorcraft coupling modal parameters influence analysis

Guibao Zhang*[a],Yan Zhu[a],Siwen Wang[a],Fengnan Sun[a], Qiyou Cheng[a], Zhizhuang Feng[a]

Science and Technology on Rotorcraft Aeromechanics Laboratory, CHRDI

Jingdezhen, China

e-mail: zhanggb602@163.com

ABSTRACT: In order to explore the influence mechanism of different parameters on Multi-rotor aircraft dynamic characteristics,Hamilton principle and the theory of medium deformation beam were used to derive the rotors/nacelle/wing coupling structure dynamic equation,and a solution method suitable for the analysis of coupled modal was established.By using literature as an example calculate rotor and wing modal,the error is less than 5% compared with the experimental results in the literature,indicating that the coupling model in this paper is accurate and effective.On this basis,the influences of ,nacelle offset,rotor collective pitch,blade center of mass on the coupling modal are studied.

1. INTRODUCTION

Distributed rotorcraft is a new rotorcraft using a distributed electric drive concept, the lift and propeller units can be designed for hovering low speed and high speed forward flight respectively for aerodynamic optimization, avoiding the traditional tilt-rotor aircraft due to both hovering landing and forward flight performance caused by the speed can not be further enhanced problems.By distributing multiple propellers at the front of the wing, which can effectively reduce the vibration and noise level of the whole aircraft and enhance the comfort of the members and the mission load environment[1].For the distributed rotorcraft NASA conducted the number "X-57" program to verify the feasibility of the distributed electric propulsion system and to predict the problems that may be encountered in the future[2,3].DARPA launched the Vertical Takeoff and Landing Test Aircraft (VTOL X-Plane, VXP) [4] project to develop a more capable vertical takeoff and landing vehicle, and Boeing, Airbus, and other research institutions have also conducted research on this configuration [5,6].

Distributed rotorcraft configurations combine the advantages of several technologies, but they also pose difficult technical problems [7].Due to the distribution of multiple rotors mounted on the wing, equivalent to multiple external hangers, there are multiple points of concentrated mass and inertia attached to the wing,each additional outboard adds six intrinsic vibration patterns and these distributed rotors increase the chances of modal coupling with the wing.At the same time, the wing in the process of tilting;the flight speed;inertia center of gravity of the vehicle is constantly changing,these factors lead to the motion mode of the wing-distributed multi-rotor dynamics system will be very complex.Therefore, this paper establishes the dynamic equations and coupled modal solutions for the distributed rotorcraft configuration to study the effects of nacelle offset, rotor collective pitch, and blade mass center position parameters on the coupled modalities of the distributed rotorcraft.

2. SYSTEM MODELING AND SOLUTION METHODS

In this paper we derive the dynamical equations of the distributed rotor/nacelle/wing coupled system based on the principle of Hamilton with the following expressions:

$$\delta\Pi = \int_{t_1}^{t_2} (\delta U - \delta T - \delta W)dt = 0$$

$$\delta U = \delta U_w + \sum_{i=1}^{n}\sum_{j=1}^{Nb} \delta U_b \qquad (1)$$

$$\delta T = \delta T_w + \sum_{i=1}^{n} \delta T_p + \sum_{i=1}^{n}\sum_{j=1}^{Nb} \delta T_b$$

For the modal parameter impact analysis the external virtual work δW is no longer considered and the nacelle is assumed to be a rigid structure.

Where the subscript ***R*** refer to the rotor, ***W*** to the wing, ***p*** to the nacelle, N_b to the number of blades and ***n*** to the number of rotors,The followings are the derivations of the energy terms of wing and rotor.

2.1 Coordinate system definition

To describe the spatial position of the wing, nacelle and propeller conveniently, the coordinate system and the transformation relationship are defined as follows:

As in Figure 1, There is an angle ***app*** between $X_IY_IZ_I$ *and* $X_pY_pZ_p$ coordinate system,the deformed coordinate system($X_WY_WZ_W$) of the wing is converted to the undeformed coordinate system with a difference of three rotation angles.Rotor coordinate system parametric helicopter[8],and the transformation matrix between each coordinate system expressed as :T_{PI}、T_{WP}、T_{HW}、T_{RH}、T_{UR}、T_{DU}.

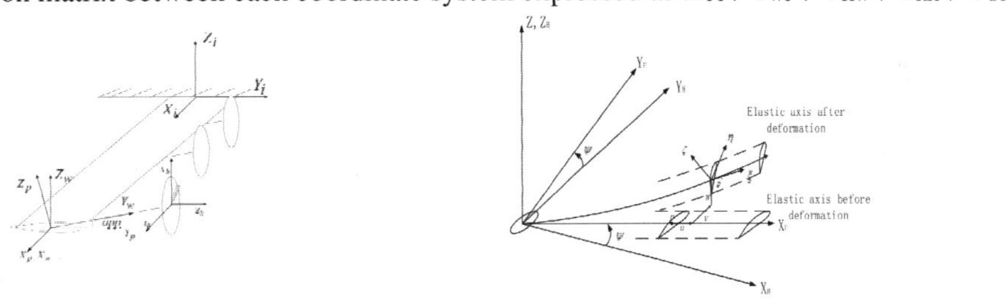

a) Sketch map of wing coordinate system b) Sketch map of rotor coordinate system

Fig.1 Sketch map of Multi-rotor aircraft coordinate

2.2 System Modeling

Taking the blade as an example, the derivation process is as follows:

According to the positional relationships of each component shown in Fig.1, the position vector ***R*** in inertial co-ordinate system of an arbitrary point on the cross-section at radius ***r*** of the ith blade can be expressed as:

$$R = \{yw_i + ut_i \quad vt_i \quad wt_i\}.[T_{pI}] + \{xh_i \quad yh_i \quad zh_i\}.[T_{wI}] + \{x+u_i \quad v_i \quad w_i\}.[T_{hI}] + \{0 \quad \eta_i \quad \zeta_i\}.[T_{DI}] \qquad (2)$$

Where yw_i is the installation position for distributed rotor,u_i, v_i, w_i is the displacement of the elastic axis of the wing at the installation,xhi, yhi, zhi is the rotor hub center position,$(x+u, v, w)$ is the coordinate of any point on the elastic axis of the deformed blade.

The velocity of the point is determined by taking the time derivative of the position vector (2)and is given as:

$$\vec{V} = R / dt \qquad (3)$$

The virtual kinetic energy variation of the ith blade is given by

$$T_b = \frac{1}{2}\int_0^L \left(\iint_A \rho\vec{V}\cdot\delta\vec{V}d\eta d\varsigma\right)dx \qquad (4)$$

where ***p*** is the blade density , ***L*** is the blade radius.

According to the description of kinematics, the virtual strain energy of the rotor is given by

$$U_b = \frac{1}{2}\int_0^L \iint_A (\sigma_{xx}\varepsilon_{xx} + \sigma_{x\eta}\varepsilon_{x\eta} + \sigma_{x\varsigma}\varepsilon_{x\varsigma}) d\eta d\varsigma dx \quad (5)$$

Where ε_{xx} and σ_{xx} is Strain and Stress of blade,the nacelles and wings can be derived by referring to the blades.

2.3 Solving method

Substituting the above derived virtual kinetic, strain energy expressions into Equation (1) and combining the terms with identical variables, the following expression in matrix form can be obtained

$$M\ddot{q} + C\dot{q} + Kq = F \quad (6)$$

where ***M***,***C*** and ***K*** are the mass, damping and stiffness matrices of the system,The right side ***F*** of Equation (5) includes the non-linear terms and constant terms independent of variables.

Define the vector: $Y = \{y \quad \dot{y}\}^T$,The system equation can be written as:

$$AY = BY$$
$$A = \begin{bmatrix} I & 0 \\ 0 & [M] \end{bmatrix} \quad B = \begin{bmatrix} 0 & I \\ -[K] & -[C] \end{bmatrix} \quad (7)$$

Where ***I*** is identity matrix.The imaginary part of the eigenvalue of the equation (7) represents the modal frequency of the system.

3 MODEL VALIDATION

Using the high-aspect-ratio-Tang[9]wing and SA349[10] rotor to verify the validity of this paper.Tables 2 and 3 show the comparison of the calculations in this paper with the literature,the maximum error is less than 5%, which proves that the model of this paper is effective.

Table.2 Tang Wing natural frequency

mode[Hz]	Ref.	paper	Ref. (With nacelle)	paper (With nacelle)
1flapwise	3.675	3.677	2.625	2.277
2flapwise	23.03	23.03	17.88	17.89
3flapwise	64.50	64.49	—	53.80
1lagwise	24.39	24.39	14.13	15.00
1twist	119.5	120.2	22.88	23.07

Table.3 SA349 natural frequency

Mode[1/rev]	Ref.	paper
1lagwise	0.5850	0.5851
1flapwise	1.0178	1.0181
1flapwise	3.7916	3.7923
1flapwise	4.8971	4.9971
1torsion	3.9313	3.8313

4 DISTRIBUTED ROTORCRAFT PARAMETER IMPACT ANALYSIS

The configuration of the model in this paper is: 3 rotors + 4m wing, each rotors consists of 4 blades and all kinds of parameters are dimensionless parameters.

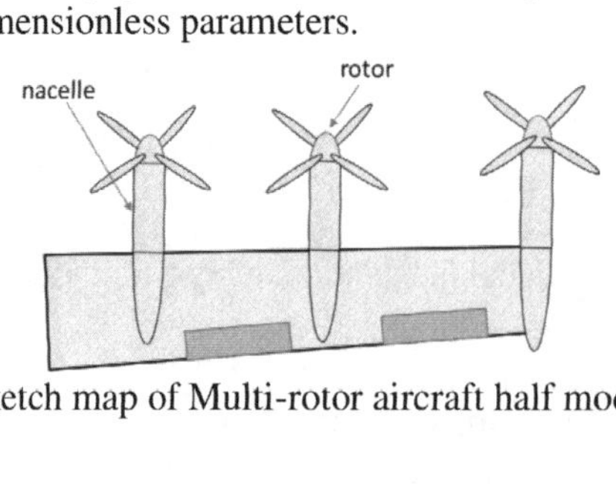

Fig.2 Sketch map of Multi-rotor aircraft half model

4.1Analysis Strategy

Multiple distributed rotors will make the rotor/wing modal coupling increase, resulting in a complex and frequency-dense system.In order to efficiently and accurately identify the complex modal frequencies, the coupled modal analysis strategy in this paper is to first plot the first 15 orders of the coupled modal of the system without identification, which includes the first (3*4) orders of the rotor blades + the first 3 orders of the wing frequencies. On this basis, further plot the frequency curves before and after the coupling of the rotor and wing respectively, and use the dashed line and the solid line (***Baseline***) to represent the coupled frequency and the uncoupled original frequency respectively, and use the uncoupled original frequency curve as the reference to compare and analyze the influence of the rotor/wing coupling on the rotor and wing modalities, In this way, the efficiency of the analysis can be balanced with the accuracy of the identification.

4.2 nacelle *Z*-axis offset effect analysis

Taking the intersection of the wing chord and the wing ***Z***-axis coordinate system as the base point (the position of the center of mass coincides with this point), the nacelle is moved along the ***Z***-axis and the frequency change curve is plotted in Figure 3.

Figure 3.a) plots the first 15th order frequency of the coupled system subject to the nacelle bias, It can be seen that most of the frequencies tend to decrease after the nacelle deviates from the base point, and the frequency curve is symmetrically distributed about the base point.The reason is that the cross-sectional moment of inertia will increase after the nacelle deviates from the base point, and thus the overall frequency tends to decrease,at the same time, the effect of gravity is not considered in this paper, so the frequency is symmetric about the base point.

Figure 3.b) shows the wing modal ratio diagram.It can be seen that the nacelle offset has a greater effect on the ***1twist*** and 2***Flap*** of the wing, and the ***1Flap*** has a relatively small effect; the wing frequency after coupling decreases in comparison with ***Baseline***, and the three frequencies are decreasing with the nacelle offset from the base point.The rotor modal variation is smaller compared to the effect on the wing.

In Figure 3.c) it can be seen that the rotor modal curve after coupling is the same as the ***Baseline*** curve.

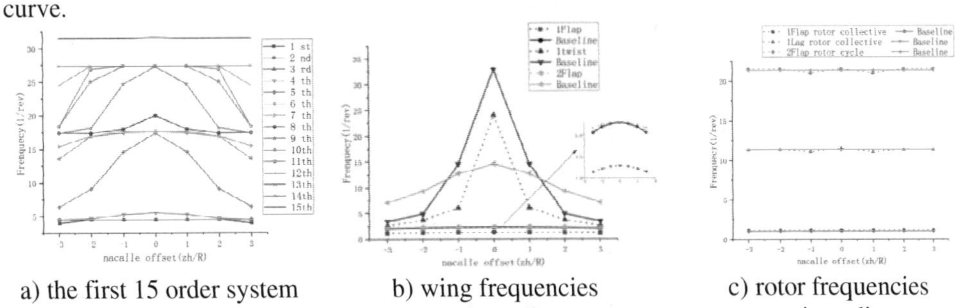

a) the first 15 order system frequency influence diagram b) wing frequencies comparison diagram c) rotor frequencies comparison diagram

Fig.3 influence of nacelle offset on coupling characteristics

4.3 Rotor collective pitch effect analysis

With reference to the tilt-rotor structure and maneuvering [11], this paper chooses to analyze the effect of the rotor collective pitch on the system in the range of variation from 0 to 30°.

Figure 4.a) plots the first 15 orders of frequency variation of the system, and it can be seen from it that the system shows 4 types of variation curves as the collective pitch increases:Decrease;increase; no change and decrease followed by increase.

In Figure 4.b), plot the wing frequency comparison curves. As the collective pitch increases, the ***1Lag*** is decreasing and the ***1Flap*** is increasing; comparing with the ***Baseline*** curve, the ***1Lag*** and ***2Flap*** are decreasing overall after coupling.Comparing with the ***Baseline***, the overall decrease of the ***1Lag*** and ***2Flap*** after coupling is due to the extra rotor mass on the wing after coupling, so the overall decrease of the wing frequency.Comparing the Baseline curves, ***1Lag*** and ***2Flap*** decrease after coupling because there is an additional rotor mass on the wing after coupling, so the frequency decreases.

Figure 4.c) shows the rotor frequency comparison. With the increase of collective pitch, the changes of ***Baseline*** and coupling curve basically keep the same, the ***2Flap rotor cycle*** frequency is basically unchanged, the ***1Flap rotor collective*** frequency increases, and the ***1Lag rotor collective*** frequency decreases with the increase of collective pitch.

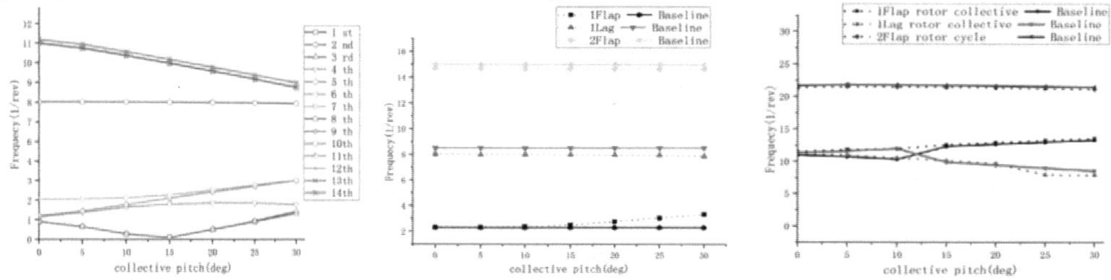

a) the first 15 order system frequency influence diagram
b) wing frequencies comparison diagram
c) rotor frequencies comparison diagram

Fig.4 influence of collective pitch on coupling characteristics

4.4 blade mass center effect analysis

Using the intersection of the blade elastic axis and the blade ***Y***-axis coordinate system as the origin, design the blade mass center to move along the ***Y***-axis, and plot the frequency variation curve in Figure 5.

Figure 5.a) plot the first 15 orders of frequency variation of the system, It can be seen that the frequency curve is symmetric about the origin and the curve changes in a complex way as the center of blade mass moves.

In Figure 5.b) plot the wing frequency comparison curves. With the offset of the blade center of mass position, the change of the wing frequency before and after coupling is small; comparing with the ***baseline*** curve, the wing frequency generally decreases after coupling.

Figure 5.c) shows the rotor frequency comparison. As the center of mass moves away from the origin, the ***1Flap*** rotor cycle and ***1Lag*** rotor collective frequencies are less affected, and the ***1torsion*** rotor cycle frequencies are decreases significantly .Zooming in on the ***1Flap*** rotor cycle and ***1Lag*** rotor collective curves, it can be seen that all three frequencies decrease as the center of mass moves away from the origin position, the ***1Flap*** rotor cycle and ***1Lag*** rotor collective curve basically overlap with the Baseline curve.

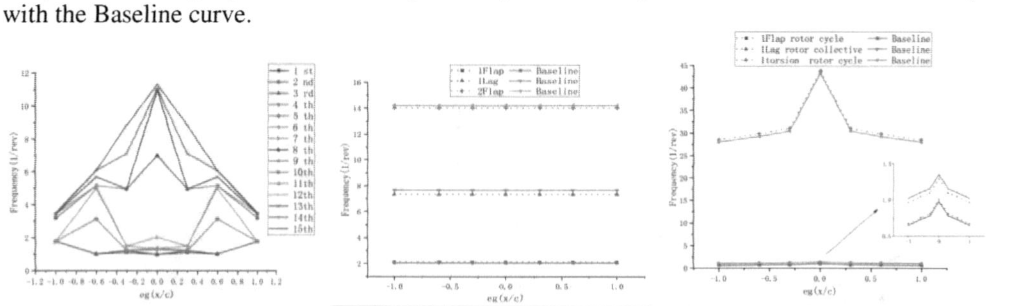

a) the first 15 order system frequency influence diagram
b) wing frequencies comparison diagram
c) rotor frequencies comparison diagram

Fig.5 influence of the center of blade mass offset on coupling characteristics

5. CONCLUSION

In this paper the structural dynamics model of the distributed rotorcraft is established based on the Hamilton principle and using the medium deformation beam theory.Analytical calculations of the system led to the following conclusions:

(1) For the established coupled structural model of distributed rotorcraft, verification of isolated rotor modules and high-aspect-ratio-wing is carried out to ensure the validity of the structural model;

(2) The nacelle **Z**-axis bias mainly affects the wing frequency of the coupled system, where the wing frequency decreases as the nacelle vertical offset increases;When increasing the rotor collective pitch , the ***1Flap*** rotor collective frequency frequency will increase and the ***1Lag*** rotor collective frequency will decrease;moving the position of the blade mass, the ***1torsion*** rotor cycle frequency decreases and the wing frequency remains basically unchanged;

(3) The distributed rotor/nacelle/wing coupling system presents the characteristics of dense structural vibration mode frequency and complex mode vibration pattern, and the frequency identification can be assisted by the strategy of this paper to identify the vibration pattern and improve the identification efficiency.

REFERENCES

[1] Steven C. Geuther, David D. North. Investigation of a Tandem Tilt-wing VTOL Aircraft in the NASA Langley 12-Foot Low-Speed Tunnel[C]. NASA/TM-2020-5003178, June, 2020 .

[2] Papachristos C , Alexis K , Tzes A. Hybrid model predictive flight mode conversion control of unmanned Quad-TiltRotors[C]. Control Conference. IEEE, 2013.

[3] Serhat Dikyar. Nonlinear Control of a Quad Tilt-Wing UAV[D], Sabancı University,2011.

[4] HARRINGTON J D, KAMLET M, KATHY B. NASA's X-57 electric research plane [R]. Washington, D.C.:NASA,2016.

[5] Paul M.Rothhaar, Patrick C.Murphy. NASA Langley Distributed Propulsion VTOL Tilt-Wing Aircraft Testing, modeling,Simulation,Control, and Flight Test Development. NASA Langley Research Center, Hampton, Virginia. https: //ntrs.nasa.gov/search.jsp?R=20140011413 2020-05-18T01:27:41+00:00Z.

[6] Serhat Dikyar. Nonlinear Control of a Quad Tilt-Wing UAV[D], Sabancı University,2011.

[7] Hironori Totoki, Yoshimasa Ochi and Masayuki Sato. Flight Testing of a Gain-Scheduled Stability and Control Augmentation System for a Quad-Tilt-Wing UAV[C]. AIAA SciTech 4-8 January 2016, San Diego, California, USA, AIAA Guidance, Navigation, and Control Conference.

[8] Sarker P, Chakravarty U K. On the dynamic response of a hingeless helicopter rotor blade[J]. Aerospace Science and Technology, 2021, 115: 106741.

[9] Deman Tang , Earl H. Dowell. Experimental and Theoretical Study of Gust Response for High-Aspect-Ratio Wing [J]. AIAA Journal, 2002, 40(3): 419—428.

[10] Gloria K, Yamauchi G K. Hub and Blade Structural Loads Measurements of an SA349/2Helicopter. NASA TM-101040, 1988.

[11] Wang X, Cai L. Mathematical modeling and control of a tilt-rotor aircraft[J]. Aerospace Science and Technology, 2015, 47: 473-492.

Engineering Application Method Of A New Satellite Simulation Test System

Tao LUO [1], Yuanmei JIA[1], Kai XUE[1], Le PAN[1] , Huamiao SUN[1], and Xianjie WANG[1]

[1.] Shenzhen of Aerospace Dongfanghong Satellite Ltd, Shenzhen, China, 51800

Corresponding author's e-mail:1858154 33@qq.com

Abstract. This paper studies a new simulation test system, which integrates the satellite management and attitude track control software test system, realizes the test configuration and equipment generalization, improves the system universality and automation capability, effectively saves human resources, and equipment costs, shortens the test cycle, avoids quality problems caused by human uncertainty, and improves the efficiency of software development on the satellite.

1. Introduction

In recent years, the microsatellites industry at home and abroad has entered a period of rapid development[1]. With the advantages of a high functional density ratio, short development cycle, low cost, and large batch size, small satellites have gradually played a huge role in the aerospace field of various countries. In the process of small satellite development, the satellite software determines whether the satellite is launched and operates properly in orbit, and the ground test of the satellite management software and attitude track control software is the key link in the development process[2]. As shown in Figure 1, the common test system is generally composed of a simulation control host computer and PXI real-time simulation computer[3], using their independent test systems. Most of them have problems, such as poor versatility, complex configuration process, insufficient data management, low production efficiency, low degree of automation, low utilization of equipment resources, etc. To meet the needs of batch automated testing of future satellite onboard software, based on the principle of the traditional simulation test system, this paper proposes a low-cost, high-performance, highly automated, scalable, and flexibly configured simulation test system design and implementation scheme to satisfy the satellite management software and attitude track control software test through the same set of equipment.

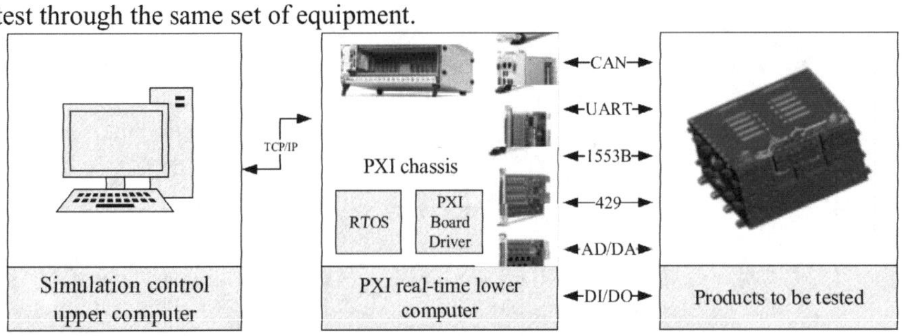

Figure 1. Traditional PXI simulation test system.

2. Overall system scheme

2.1 Overall system introduction

The simulation test system adopts the open-loop mode for the satellite management software test, and the attitude track control software test is based on the attitude and track closed-loop control mode, as shown in Figure 2. It can simulate the electrical and communication characteristics of the satellite components and realize the testing and verification of satellite software functions and performance. Users can configure management and system operation control through the host computer of the simulation test system and send the dynamic model executable file and operation configuration to the simulation test equipment and the dynamic computer, which realizes the interface interaction between satellite simulation components and satellite onboard computer. Real satellite components data can also be obtained through the interface adapter board. In addition, the simulation test system can interact with the integrated test system interface, which improves the ability of the system to automate testing and interpretation, data management, and data query and analysis. Its overall system block diagram is shown in Figure 3.

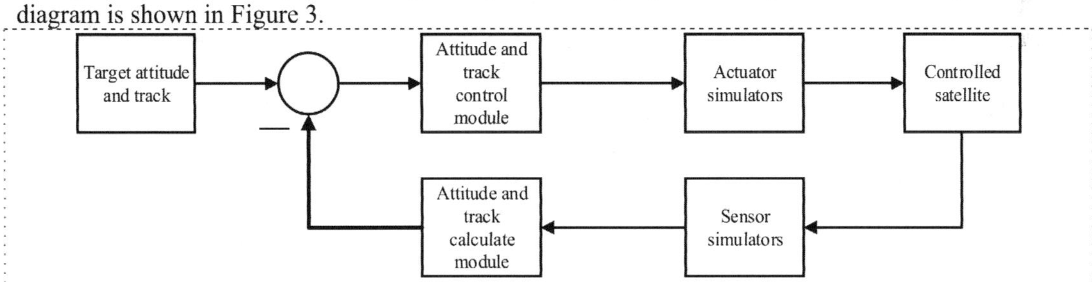

Figure 2. Attitude and track closed-loop control mode.

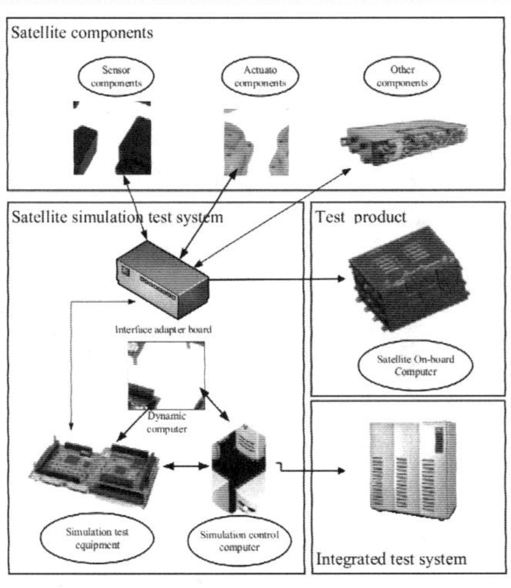

Figure 3. System block diagram.

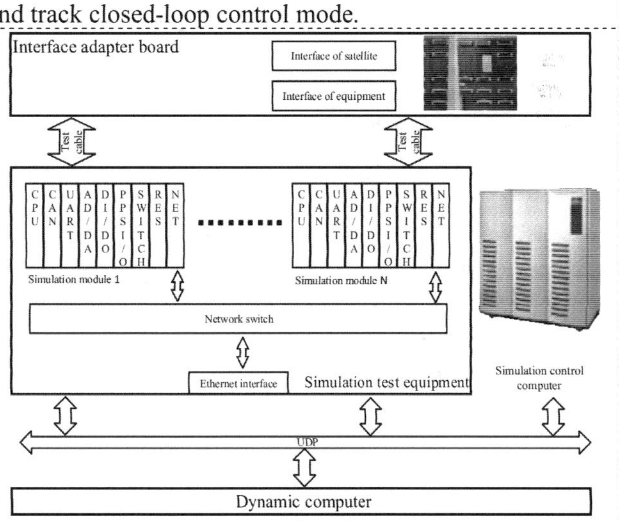

Figure 4. System hardware architecture.

2.2 System hardware

The requirements for hardware interface types are similar, but the number of requirements is different. To achieve full coverage of the test interface and independent control of hardware costs, a generic simulated test module is designed. The module provides different interface types and CPU processors to simulate the I/O processing of the peripheral interfaces of real satellite components. Through the integration of multiple modules and cable node transfer, the full coverage of the pilot test interface is realized. As shown in Figure 4, the system hardware includes simulated test equipment, an interface conversion board, a simulation control computer, and a dynamic computer.

2.3 System software

During the test, the simulation test system needs to respond to the user's requirements to modify the system configuration, real-time viewing of test data, log information, and other business functions. This function is designed and deployed on the non-real-time operating system for simulation control of the whole test process[4]. It also needs to meet the requirements of peripheral I/O and bus communication, acquisition and control, fast command request response, real-time data update of satellite simulation components, high-frequency solution of the dynamic model, etc[5]. This function is designed and deployed with high real-time embedded software. The system software consists of simulation control upper computer software, simulation test equipment lower computer software, and dynamics lower computer software[6]. The software architecture of the simulation test system is shown in Figure 5.

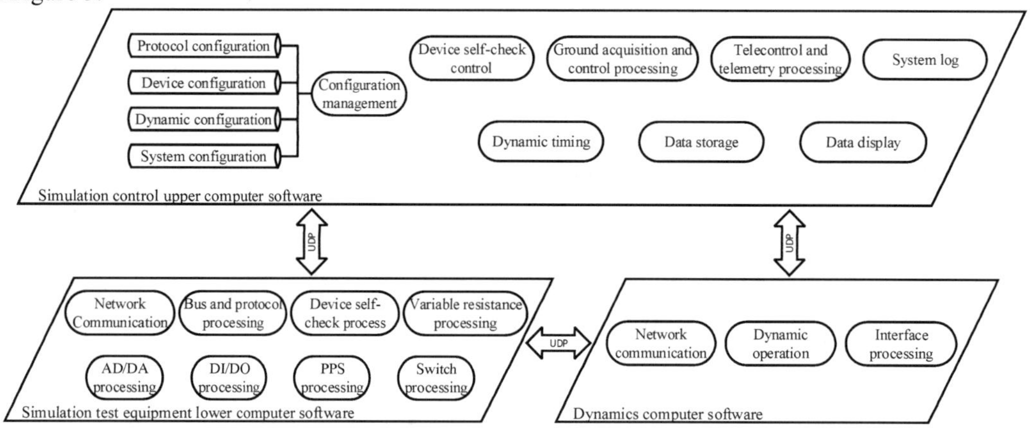

Figure 5. Software architecture.

3. Communication protocol design

As shown in the table 1, the simulation test system transport protocol consists of a protocol header and protocol body[7]. The protocol header includes protocol type, length, start, and terminal network information. The protocol body mainly transmits specific business processing content, including status settings (protocol configuration, model parameter configuration, digital/analog settings, etc.) and data broadcast (dynamic broadcast, protocol broadcast, digital/analog broadcast, etc.).

Table 1. Transport protocol contract.

Serial number	1	2	3	4	5	6
Field name	Sync words	Length	Data category	Network information	Reserve	Business information
Information composition	Protocol header					Protocol body

4. Database design

The simulation test system makes a general design for devices and protocols and realizes the binding between satellite components protocols, device I/O settings, simulation device modules, etc. The database storage information of the simulation test system includes satellite simulation components class, model, protocol, mapping relationships, channel configuration, etc.

Through the simulation device modules, device channel setting and satellite components communication protocol are related. The specific data relationship design is shown in Figure 6.

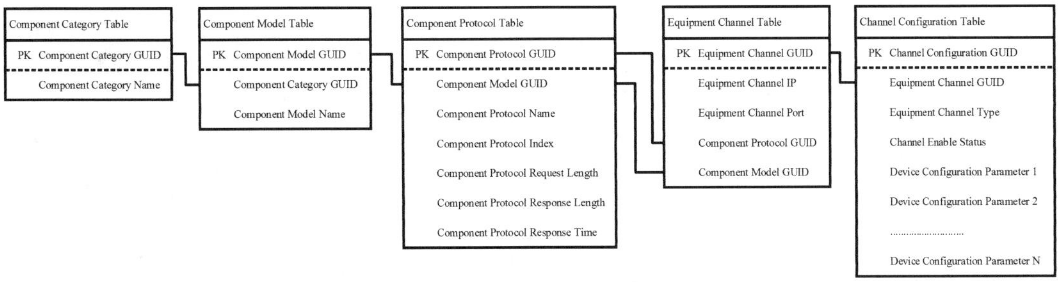

Figure 6. Database design.

5. System advantages

Compared with the traditional satellite management and attitude track control software test system, this system has the following advantages:

1) System application scenarios: This system can combine the protocol and interface commonality of the two application scenarios for testing, which can directly save 50% of the equipment cost.
2) Satellite components protocol simulation: By analyzing various types of satellite components protocol information, we design a universal protocol structure field, which can adapt to different satellite components protocol configuration requirements, establish a mapping between simulated satellite components protocols and devices, and conduct structural management through the database to ensure that the protocol configuration is standardized and reliable.
3) Test data management: By establishing interfaces with the integrated test system, it can transmit test data to the integrated test system to realize more functions, such as data storage, query, statistics, etc.
4) System time calibration method: This system acquires the dynamic model time, automatically generates and sends TC commands after the user sets the time calibration command, obtains satellite telemetry, and confirms the calibration results, realizing one-click fast time calibration of the satellite system and ground test system.
5) System processing performance: Compared with the traditional PXI test system, one device board simulates several satellite components, this system uses an independent simulation module to simulate one satellite component, which can process satellite protocol requests without being affected by other modules and has better data processing performance.

6. System integration and testing

The simulation test system uses ethernet to establish a LAN and designs a general communication protocol and a structured database to realize the information interaction of each module of the system. The workflow of the simulation test system is shown in Figure 7. After the user starts the software and hardware of the upper and lower computers of the system, the user performs the equipment self-test and tests the cable connection. The configuration is loaded, confirmed, and assigned according to the test task requirements.

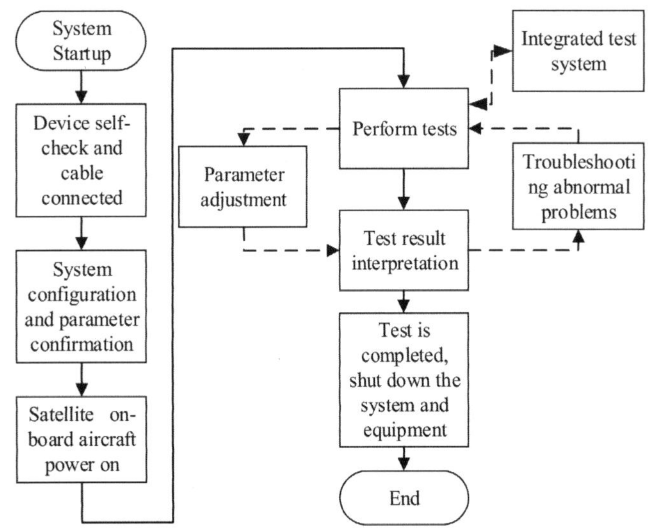

Figure 7. Test flow chart.

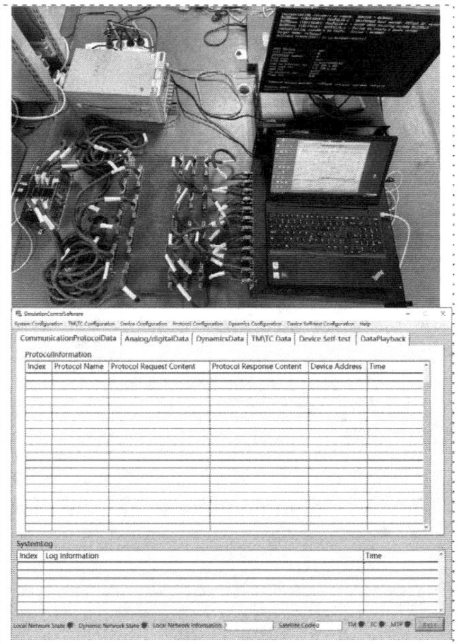

Figure 8. Test scene and software UI.

The satellite onboard computer is powered up to perform the test. During the test, as shown in Figure 8, the upper computer can display data (communication protocol, analog/digital data, dynamics data, remote control, telemetry data, etc.) to check and judge the test results or establish a communication connection with the integrated test system to achieve automatic test and interpretation. The configuration parameters can be dynamically adjusted during the test to view the test results. If an exception occurs, the exception needs to be investigated. After the test is finished, we close the test system and ground test devices.

7. Conclusion

Based on the development status of microsatellite testing technology, this paper proposes a new satellite simulation test system design and engineering implementation method, which supports function, performance, and interface testing of satellite onboard software. After testing and verification, when using this system for testing, the coverage of test items can reach 100%, and test performance is increased by about 30%, which has strong engineering application value.

Due to the wide variety of satellite component protocols, the optimal management methods of different satellite component protocols will be explored in the future. In addition, the serialization of test cases takes a long time to implement, and the improvement of system automation testing capabilities is also critical.

References

[1] Xiaojun WANG: The development and future of Chinese commercial aerospace. Missile and SpaceVehicles,1-6 (2020).

[2] N. M. H., Norsahperi, S., Ahmad, S. F., Toha, I. A., Mahmood. Analysis and practical validation on multi-linkage scissor platforms drive system for the satellite test facilities [J]. International Journal of Heavy Vehicle Systems, 2021, 28(1).

[3] Inigo Del Portillo, Bruce, G. Cameron, Edward, F., Crawley. A technical comparison of three low earth orbit satellite constellation systems to provide global broadband [J]. Acta Astronautica, 2019

[4] Electronics; School of Electrical Engineering and Automation Researchers Focus on Electronics (Onboard multi-lane parallel MLVDS bus design for in-orbit embedded satellite test device)[J]. Information Technology Newsweekly, 2020.
[5] Pan TIAN, Xingchen LI, Wenhao XUE, Zizheng BIAN, Tao HUANG, Yunjie LIU, A Docker-based LEO Satellite Network Testbed, Chinses Journal of Computers, 2022, 45(09).
[6] Xinwen GUO, Jiangguo PANG, Tiantian LI, Erfeng LI, Xiaorong JIN. Design and implementation of satellite hardware-in-the-loop simulation training equivalent, Computer Measurement & Control, 2020, 28(12).
[7] Satellite Today. Demystifying Software-Defined Networks for Satellite [J]. 2020.

Journal of Physics: Conference Series **2489** (2023) 012040 doi:10.1088/1742-6596/2489/1/012040

Research on Asynchronous Information Fusion Technology of the Distributed Active/Passive Imaging Detection System

Yifan Gu, Youwei Kang, Zeyuan Fang, Xin Liao, Shanjun Tang

Shanghai Electro-Mechanical Engineering Institute, Shanghai, 201109, China

loon_gyf@126.com

Abstract. In the process of target cooperative tracking of the distributed multi-imaging detection system, aiming at the problem of asynchronous measurement information from different detectors, a distributed Asynchronous and Heterogeneous Information Matrix Fusion algorithm based on the Cubature Kalman Filter (CKF-AHIMF) is proposed. According to the Local Detector Driven Communication (LDDC) method and CKF nonlinear mapping, IMF distributed fusion algorithm is generalized into nonlinear, asynchronous, and heterogeneous cases to integrate the information from a multi-imaging detection system and to obtain global high-precision target trajectory. Through the digital simulation, it is verified that the CKF-AHIMF algorithm under the LDDC method proposed in this paper is more effective and has higher tracking accuracy than other distributed fusion algorithms in dealing with the problem of target cooperative tracking under asynchronous information from local imaging detectors, which is also valid to implement in precision detection and guidance system.

1. Introduction

With the development of imaging detection technology, the requirements of missile-borne and airborne distributed platforms on the capability of precision guidance capability of imaging detection systems are gradually increasing [1]. Passive/active imaging detection technology, such as visible imaging technology, infrared imaging technology, and Synthetic Aperture Radar (SAR) imaging technology have gradually become important detection technologies of missile-borne and airborne detection systems [2]. However, imaging detection systems with distributed platforms and different imaging systems will face the problem of heterogeneous and asynchronous detection information, which requires the use of a high-precision information fusion algorithm to fuse the detection information from passive/active imaging systems to estimate more accurate target comprehensive information [3].

This paper first presents a distributed Information Matrix Fusion (IMF) algorithm, and according to the characteristics of the distributed active/passive imaging detection system, the IMF algorithm is extended to three cases: nonlinear, asynchronous, and heterogeneous. An asynchronous and Heterogeneous IMF algorithm based on Cubature Kalman Filter (CKF-AHIMF) is proposed and formed in this paper, and the problem of target state estimation and fusion under distributed asynchronous active/passive imaging detection system is solved. Finally, combined with the analysis of digital simulation results, the validity and robustness of the CKF-AHIMF algorithm proposed in this paper to estimate the target fusion state are verified, which provides a theoretical basis for achieving the precise guidance task of the distributed detection system.

2. Information Matrix Fusion (IMF) algorithm

The Information Matrix Fusion (IMF) algorithm is mainly used to solve the problem of linear distributed fusion under multiple detection systems with synchronous sampling rates. Since the IMF algorithm does not need to calculate the cross-covariance matrix between multiple detection systems, it is simple to be utilized in engineering applications [4].

The linear state and measurement equation of the target are as follows:

$$X_{k+1} = F_k X_k + w_k \tag{1}$$

$$z_{k+1} = H_k X_{k+1} + v_k \tag{2}$$

Wherein, F_k is the state transition matrix of the target, H_k is the measurement matrix, w_k is the process noise with the mean of zero and the variance of Q_k, and v_k is the measurement noise with the mean of zero and the variance of R_k.

The detection system uses a filtering algorithm to obtain the target estimation state $\hat{x}_k$ and its covariance matrix $\hat{P}_k$ at the local detector level. In the fusion center, the estimated state $\hat{x}_k$ and covariance matrix $\hat{P}_k$ of each detector is received and updated using the IMF algorithm. The calculation process is shown in Figure 1 below.

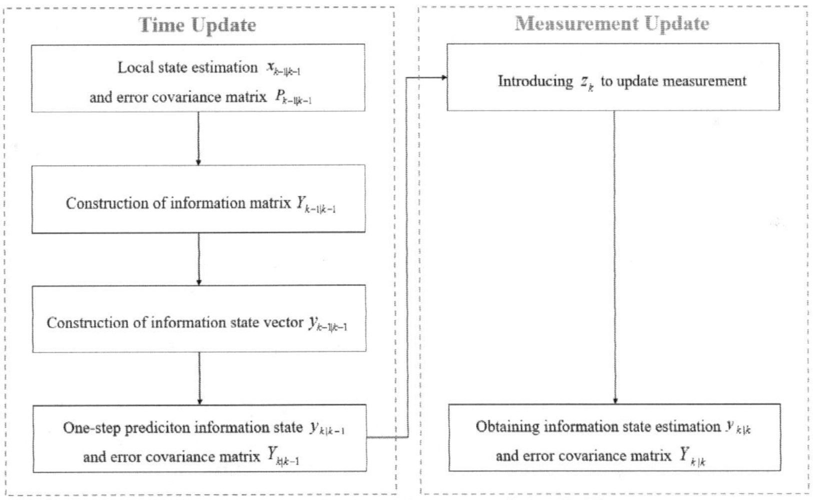

Figure 1. The architecture of the IMF

• ***Construction of information matrix***

The information matrix is the inverse matrix of the estimation error covariance matrix,

$$Y_{k-1|k-1} = P_{k-1|k-1}^{-1} \tag{3}$$

Wherein, $Y_{k-1|k-1}$ is the information matrix of the IMF algorithm, $P_{k-1|k-1}$ is the estimation error covariance matrix.

• ***Construction of information state vector***

The information state vector represents the information quantity of the target in the information space, which is defined as the product of the information matrix and the normal state space vector,

$$y_{k-1|k-1} = P_{k-1|k-1}^{-1} x_{k-1|k-1} = Y_{k-1|k-1} x_{k-1|k-1} \tag{4}$$

Wherein, $y_{k-1|k-1}$ is the information matrix, $x_{k-1|k-1}$ is the estimation error covariance matrix.

• ***Prediction equation***:

$$P_{k|k-1}^{-1} x_{k|k-1} = P_{k|k-1}^{-1} F_k P_{k-1|k-1} P_{k|k-1}^{-1} x_{k-1|k-1} \tag{5}$$

Wherein, F_k is the state transition matrix. Thus, the process of one-step prediction is:

$$y_{k|k-1} = Y_{k|k-1} F_k Y_{k-1|k-1}^{-1} y_{k-1|k-1} \tag{6}$$

Wherein, $Y_{k|k-1}$ can be derived by inversion:

$$Y_{k|k-1} = \left[F_k Y_{k-1|k-1}^{-1} F_k^T + Q_k \right]^{-1} \tag{7}$$

Wherein, Q_k is the process noise covariance matrix.

• ***Update equation***:

$$y_{k|k} = y_{k|k-1} + H_k^T R_k^{-1} z_k \tag{8}$$

$$Y_{k|k} = Y_{k|k-1} + H_k^T R_k^{-1} H_k \tag{9}$$

Wherein, z_k is the measurement of the detection system, R_k is the measurement noise covariance matrix, and H_k is the measurement transition matrix.

It can be seen from the above IMF update equation that the measurement z_k updates the information state vector $y_{k|k-1}$ in the form of $H_k^T R_k^{-1} z_k$. Therefore, $H_k^T R_k^{-1} z_k$ represents the information increment caused by the measurement of the detection system after sending it to the fusion center. Similarly, the uncertainty of measurement, that is, the measurement noise covariance matrix R_k, also updates the information matrix $Y_{k|k-1}$ in the form of $H_k^T R_k^{-1} H_k$.

In the IMF distributed fusion algorithm, the system state, and covariance for filtering iteration are the information matrix $Y_{k|k}$ and the information state $y_{k|k}$. No other inversion operations are required in the fusion center, which can reduce the computational load of the fusion center partly.

3. Asynchronous and Heterogeneous IMF algorithm based on Cubature Kalman Filter (CKF-AHIMF)

The IMF distributed fusion algorithm is a method under the condition that the distributed detection system is synchronous and linear. Since the passive imaging detection system can only estimate the nonlinear angle state of the target, while the active imaging detection system can estimate the three-dimensional position state of the target. The problem of asynchronous sampling rate between the passive and active imaging detection systems also needs to be solved, which requires the IMF algorithm to extend in the cases of nonlinear, asynchronous, and heterogeneous states [4].

3.1. Generalization of IMF algorithm in the nonlinear case

First, for the nonlinear measurement of the passive/active imaging detection system, the estimation state of the i-th local detector at the time t_k is:

$$\hat{x}_{k|k}^i = \hat{x}_{k|k-1}^i + P_{k|k}^i H^i(k, \hat{x}_{k|k-1}^i)'(R_k^i)^{-1} \left[z_k^i - h^i(k, \hat{x}_{k|k-1}^i) \right] \tag{10}$$

The measurement equation is:

$$z_k^i = h^i(k, x_k) + w_k^i \tag{11}$$

Wherein, w_k^i is the process noise of the i-th dete3ctor, which obeys the Gaussian distribution with the mean of zero and the covariance of R. $h^i(\square)$ is the nonlinear measurement function, which the Jacobian matrix is:

$$H'(k, \hat{x}_{k|k-1}^i) = \left[\nabla_{x_{k=}} h^i(k, x_k)' \right]' \Big|_{x_k = \hat{x}_{k|k-1}^i} \tag{12}$$

The update process of the covariance matrix using information matrix fusion is:

$$Y_{k|k}^i = (P_{k|k}^i)^{-1} = (P_{k|k-1}^i)^{-1} + H^i(k, \hat{x}_{k|k-1}^i)' \left(R_k^i\right)^{-1} H^i(k, \hat{x}_{k|k-1}^i) \tag{13}$$

The imaging detection system mentioned in this paper includes visible, infrared, SAR, etc. Different imaging detection systems have strong nonlinearity in measurement. Thus, in the traditional IMF algorithm, as shown in the above formula, it is necessary to calculate the Jacobian matrix of the measurement equation, and the calculation process will introduce linearization error. To avoid the linearization of the nonlinear measurement equation, this paper uses the following formula to calculate and simplify the computational complexity of the IMF algorithm under nonlinear local estimation, and meanwhile, the fusion estimation accuracy is also achieved.

Multiply the left and right formula of Equation (10) and the left and right formula of Equation (13) to obtain the i-th update equation of the information state $(P_{k|k}^i)^{-1}\hat{x}_{k|k}^i$ of the detection system:

$$\begin{aligned} y_{k|k}^i &= (P_{k|k}^i)^{-1}\hat{x}_{k|k}^i \\ &= \left[(P_{k|k-1}^i)^{-1} + H^i(k,\hat{x}_{k|k-1}^i)'(R_k^i)^{-1}H^i(k,\hat{x}_{k|k-1}^i)\right]\hat{x}_{k|k-1}^i \\ &\quad +(P_{k|k-1}^i)^{-1}(P_{k|k-1}^i)H^i(k,\hat{x}_{k|k-1}^i)'(R_k^i)^{-1}\left[z_k^i - h^i(k,\hat{x}_{k|k-1}^i)\right] \\ &= (P_{k|k-1}^i)^{-1}\hat{x}_{k|k-1}^i + H^i(k,\hat{x}_{k|k-1}^i)'(R_k^i)^{-1}\left[z_k^i - h^i(k,\hat{x}_{k|k-1}^i) + H^i(k,\hat{x}_{k|k-1}^i)\hat{x}_{k|k-1}^i\right] \end{aligned} \tag{14}$$

By sorting out the above formula, it can get:

$$(P_{k|k}^i)^{-1}x_{k|k}^i - (P_{k|k-1}^i)^{-1}x_{k|k-1}^i = H^i(k,\hat{x}_{k|k-1}^i)'(R_k^i)^{-1}[z_k^i - h^i(k,\hat{x}_{k|k-1}^i) + H^i(k,\hat{x}_{k|k-1}^i)\hat{x}_{k|k-1}^i] \tag{15}$$

In the above formula, the right formula is the innovation of the i-th detection system at the k-th time.

Therefore, for the distributed fusion center at synchronization time, its update equation under nonlinear measurement is:

$$\hat{x}_{k|k} = \hat{x}_{k|k-1} + P_{k|k}\sum_{i=1}^{N_s} H^i(k,\hat{x}_{k|k-1})'(R_k^i)^{-1}\left[z_k^i - h^i(k,\hat{x}_{k|k-1})\right] \tag{16}$$

$$P_{k|k}^{-1} = P_{k|k-1}^{-1} + P_{k|k}\sum_{i=1}^{N_s} H^i(k,\hat{x}_{k|k-1})'(R_k^i)^{-1}H^i(k,\hat{x}_{k|k-1}) \tag{17}$$

Wherein, $H^i(\Box)$ is the Jacobian matrix of the nonlinear measurement equation, and N_s is the number of detection systems. Based on the above, the left and right forms of the above two equations can be multiplied respectively to obtain:

$$P_{k|k}^{-1}\hat{x}_{k|k} = P_{k|k-1}^{-1}\hat{x}_{k|k-1} + \sum_{i=1}^{N_s} H^i(k,\hat{x}_{k|k-1})'(R_k^i)^{-1}\left[z_k^i - h^i(k,\hat{x}_{k|k-1}) + H^i(k,\hat{x}_{k|k-1})\hat{x}_{k|k-1}\right] \tag{18}$$

Through the calculation of the above formula, combining Formula (11), it can get two IMF distributed fusion methods under synchronous nonlinear measurement: the full Rate Communication (FRC) method and Reduced Rate Communication (RRC) method. Its definition and specific methods are as follows.[4]

3.1.1. Synchronous Nonlinear IMF Distributed Fusion Algorithm Based on FRC Method

Full Rate Communication (FRC) method [4] means that the Fusion Center (FC) does not have a fixed fusion rate, and the local estimated states of all Local Detectors (LD) need to be sent to FC sequentially for fusion. This method will not lose the local estimated state information of LD.

The information state update equation of the synchronous nonlinear IMF algorithm under the FRC method is:

$$P_{k|k}^{-1}\hat{x}_{k|k} = P_{k|k-1}^{-1}\hat{x}_{k|k-1} + \sum_{i=1}^{N_s}\left[(P_{k|k}^i)^{-1}\hat{x}_{k|k}^i - (P_{k|k-1}^i)^{-1}\hat{x}_{k|k-1}^i\right] \tag{19}$$

The information matrix update equation is:

$$P_{k|k}^{-1} = P_{k|k-1}^{-1} + \sum_{i=1}^{N_s}\left[(P_{k|k}^{i})^{-1} - (P_{k|k-1}^{i})^{-1}\right] \tag{20}$$

3.1.2. Synchronous Nonlinear IMF Distributed Fusion Algorithm Based on RRC Method

The rate Reduction Communication (RRC) method [4] means that FC has a fixed fusion rate, and LD only sends the local estimated state to FC for fusion which is the closest to the fusion time. This method will lose the local estimated state information of LD.

The information state and information matrix update equation of the synchronous nonlinear IMF algorithm under the RRC method are as follows:

$$P_{k|k}^{-1}\hat{x}_{k|k} = P_{k|k-m}^{-1}\hat{x}_{k|k-m} + \sum_{i=1}^{N_s}\left[(P_{k|k}^{i})^{-1}\hat{x}_{k|k}^{i} - (P_{k|k-m}^{i})^{-1}\hat{x}_{k|k-m}^{i}\right] \tag{21}$$

$$P_{k|k}^{-1} = P_{k|k-m}^{-1} + \sum_{i=1}^{N_s}\left[(P_{k|k}^{i})^{-1} - (P_{k|k-m}^{i})^{-1}\right] \tag{22}$$

Where, in the fusion period $[k-1,k]$, $k-m$ is the period of time when the local detector sent the information to FC which is closest to the fusion time k.

It can be seen from the above formula that the nonlinear synchronous IMF distributed fusion algorithm under FRC or RRC can avoid the calculation of the Jacobian matrix of the nonlinear measurement equation. It can reduce the calculation load at the fusion center and avoid the introduction of linearization errors.

3.2. Generalization of IMF algorithm in the asynchronous case

For the problem of asynchronous sampling rate, corresponding to the above FRC and RRC method, Fusion Center Driven Communication (FCDC) and Local Detector Driven Communication (LDDC) method are posed to solve the generation of IMF algorithm in the asynchronous estimation state of the detection system. The definition and specific method are as follows. [4]

3.2.1. Asynchronous Nonlinear IMF Distributed Fusion Algorithm Based on FCDC Method

The Fusion Center Driven Communication (FCDC) method means that the asynchronous distributed fusion is driven by FC, the target estimation state at the detector level (state $\hat{x}_{k|k}^{i}$, covariance $P_{k|k}^{i}$) of LD is obtained within the fixed fusion period of the FC, and the fusion is only performed at the specific fusion time, which will lose the local estimation state information of LD. From the perspective of FC, the FCDC method is an asynchronous RRC method.

The schematic diagram of the FCDC method is shown in Figure 2 below.

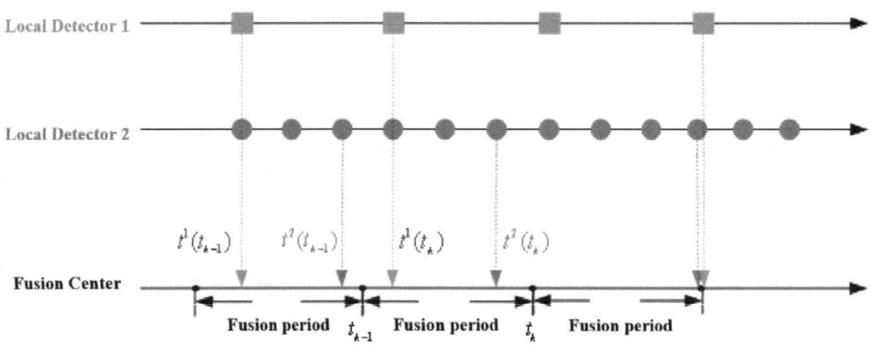

Figure 2. FCDC method dealing with asynchronous LD

FCDC method is driven by FC, that is, in the fusion period $[t_{k-1}, t_k]$ of FC, it needs to combine target estimation states at the detector level at the fusion time of $t^i(t_k)$ and $t^i(t_{k-1})$ to conduct IMF distributed fusion. Wherein, $t^i(t_k)$ is the local estimation time of the i-th LD which is closest to the fusion time t_k of FC, and $t^i(t_{k-1})$ is the local estimation time of the i-th LD which is closest to the fusion time t_{k-1} of FC.

The information state update equation of the nonlinear asynchronous IMF distributed fusion algorithm under the FCDC method is as follows:

$$\begin{aligned} P(t_k|t_k)^{-1}\hat{x}(t_k|t_k) &= P(t_k|t_{k-1})^{-1}\hat{x}(t_k|t_{k-1}) \\ &+\sum_{i=1}^{N_s}\left\{P^i\left[t_k|t^i(t_k)\right]^{-1}\hat{x}^i\left[t_k|t^i(t_k)\right]-P^i\left[t_k|t^i(t_{k-1})\right]^{-1}\hat{x}^i\left[t_k|t^i(t_{k-1})\right]\right\} \end{aligned} \tag{23}$$

Meanwhile, the updated equation of the information matrix is:

$$P(t_k|t_k)^{-1} = P(t_k|t_{k-1})^{-1} + \sum_{i=1}^{N_s}\left\{P^i\left[t_k|t^i(t_k)\right]^{-1} - P^i\left[t_k|t^i(t_{k-1})\right]^{-1}\right\} \tag{24}$$

3.2.2. Asynchronous Nonlinear IMF Distributed Fusion Algorithm Based on LDDC Method

The Local Detector Driven communication (LDDC) method means that asynchronous distributed fusion is driven by LD. Once LD obtains measurements, it conducts target local estimation (state $\hat{x}^i_{k|k}$, covariance $P^i_{k|k}$) at the detector level using filtering algorithms, and sequentially transmits the state estimation to FC for fusion without losing the information of the local estimation state from LD. From the perspective of FC, LDDC is an asynchronous FRC method.

The LDDC method is shown in Figure 3 below.

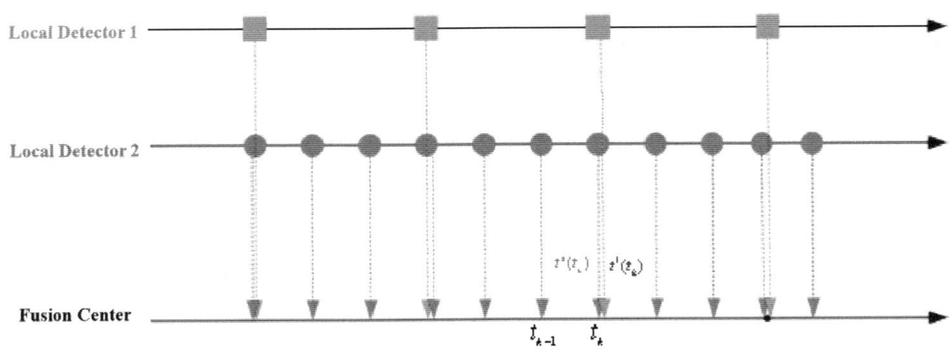

Figure 3. LDDC method dealing with asynchronous LD

The LDDC method is driven by LD, that is, it can be sent to FC for IMF distributed fusion, once the LD obtains the local estimation state of the target.

The information state update equation of the nonlinear asynchronous IMF distributed fusion algorithm under LDDC is as follows:

$$\begin{aligned} P(t_k|t_k)^{-1}\hat{x}(t_k|t_k) &= P(t_k|t_{k-1})^{-1}\hat{x}(t_k|t_{k-1}) \\ &+\sum_{i=1}^{N_s}\left\{P^i\left[t_k|t^i(t_k)\right]^{-1}\hat{x}^i\left[t_k|t^i(t_k)\right]-P^i\left[t_k|t^i(t_{k-1})\right]^{-1}\hat{x}^i\left[t_k|t^i(t_{k-1})\right]\right\}\chi^i(k) \end{aligned} \tag{25}$$

The information matrix update equation is:

$$P(t_k|t_k)^{-1} = P(t_k|t_{k-1})^{-1} + \sum_{i=1}^{N_s}\left\{P^i\left[t_k|t^i(t_k)\right]^{-1} - P^i\left[t_k|t^i(t_{k-1})\right]^{-1}\right\}\chi^i(k) \tag{26}$$

Wherein, t_k is the time when FC receives the state and covariance of LD, t_{k-1} is the last time when FC receives the state and covariance of LD, and $t^i(t_k)$ is the nearest local estimation time of the i-th LD. Wherein, $\chi^i(k)$ is defined as:

$$\chi^i(k) = \begin{cases} 1, \text{ LD sends local estimation to FC.} \\ 0, \text{ Otherwise.} \end{cases} \tag{27}$$

From the above asynchronous IMF algorithm, the nonlinear asynchronous IMF distributed fusion algorithm under the LDDC method can make full use of the local estimation state of the target at each measurement time from LD. The essence of the algorithm is to use the IMF algorithm at FC sequentially. Compared with the nonlinear asynchronous IMF distributed fusion algorithm under the FCDC method, it can make full use of the state information at the detector level, For the problem of target state estimation of distributed and asynchronous active/passive imaging detection system posed in this paper, it shows more application value.

3.3. Generalization of IMF algorithm in the heterogeneous case

The distributed fusion architecture proposed in this paper is that the passive/active imaging detection system forms the target state estimation at the local detector level based on the measurement information, that is, the passive imaging detection system will obtain the target angle state estimation, and the active imaging detection system will obtain the target three-dimensional position estimation. Each detection system sends the heterogeneous target estimation state to the Fusion Center (FC) for distributed fusion to form the target position state estimation.

Thus, the distributed Asynchronous Heterogeneous Information Matrix Fusion algorithm based on Cubature Kalman Filter [5] (CKF-AHIMF) is proposed in this paper. For the local target position estimation state of the active imaging detection system, IMF distributed fusion is performed at FC; for the local target angle estimation state of the passive imaging detection system, CKF is used to map the angle state estimation to obtain the three-dimensional position state estimation, and then IMF distributed fusion is performed. The Local Detector Driven Communication (LDDC) method is adopted to solve the asynchronous problem of state estimation at FC when conducting IMF distributed fusion. The core of the LDDC method is sequential processing. Therefore, the distributed fusion architecture of asynchronous and heterogeneous passive/active imaging detection systems in this section is shown in Figure 4 below.

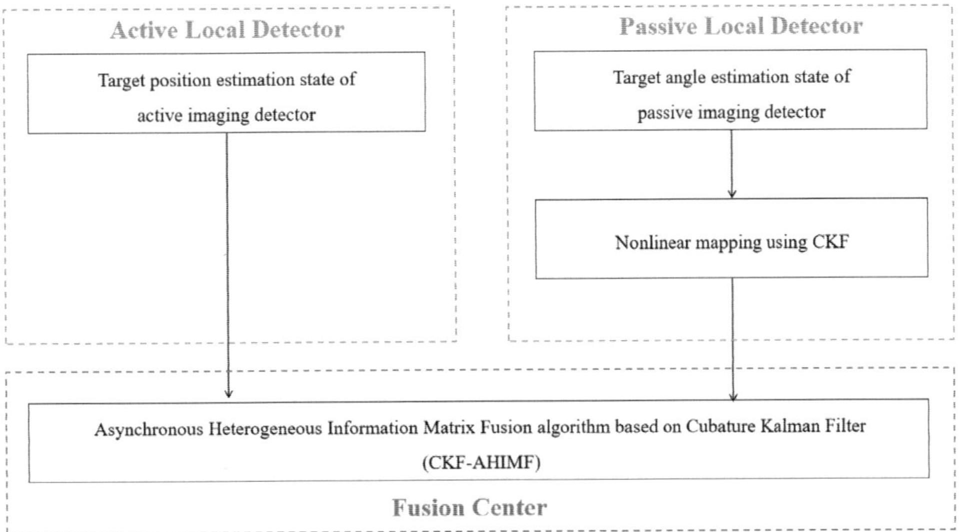

Figure 4. The distributed fusion architecture of asynchronous and heterogeneous passive/active imaging detection systems

Thus, it is necessary to unify heterogeneous states after the state estimation sending to FC. The relationship between low dimensional angle state estimation and high dimensional position state estimation can be expressed as [6]:

$$x^e = g\left(x^r, P^r, P^e\right) \tag{28}$$

Wherein, x^e is the angle state estimation, x^r is the three-dimensional position state estimation, P^r is the error estimation covariance matrix of the three-dimensional position state, P^e is the error estimation covariance matrix of angle state.

The mapping function $g(\cdot)$ is nonlinear mapping. It can map the high dimensional position state to the low dimensional angle state space, but $g(\cdot)$ is irreversible, that is, the state cannot be inversed from the function $g(\cdot)$ to map the low dimensional to the high dimensional.

Therefore, in this paper, angle state estimation is regarded as an "equivalent measurement", and position estimation state is regarded as a state variable, and its relationship can be described as:

$$z = g\left(x^r\right) + v \tag{29}$$

Wherein, v obeys the Gaussian distribution with the mean of zero and the variance of angle estimation error covariance matrix P^e.

In general, the estimation of angle state in three-dimensional position state space can be obtained by the Least Square Estimation $\hat{x}^E$, that is:

$$\hat{x}^E = (G'R^{-1}G)^{-1}G'R^{-1}\left(z - h\left(\hat{x}^r\right)\right) + \hat{x}^r \tag{30}$$

The error covariance matrix of the estimated state $\hat{x}^E$ is:

$$P^E = (G'R^{-1}G)^{-1} \tag{31}$$

Wherein, G is the Jacobian matrix of nonlinear transformation $g(x^r)$ at x^r.

$$G(x^r) = \left[\nabla_{x^r} g\left(x^r\right)' \right]' \Bigg|_{x^r = \hat{x}^r} \tag{32}$$

In the above formula, $G(x^r)$ is a $n^e \times n^r$ dimensional Jacobian matrix. Therefore, in Formula (31), $G'R^{-1}G$ is irreversible, that is, the covariance matrix P^E in the above formula does not exist. The above mapping from angle estimation to position estimation based on the Least Square Estimation method is not feasible. [7]

Thus, in literature [4], a mapping method based on Extended Kalman Filter (EKF) is proposed to map the nonlinear state to the linear state space by calculating the Jacobian matrix. However, this method will cause the problem of low fusion accuracy and even filtering divergence in the case of the distributed asynchronous fusion of passive/active imaging detection systems with strong nonlinearity in this paper.

Aiming at this problem, Cubature Kalman Filter (CKF) is used to solve the mapping problem from nonlinear state space to linear state space utilizing the distributed fusion algorithm of nonlinear asynchronous IMF under the LDDC method. The specific flow of the algorithm is shown in Figure 5 below.

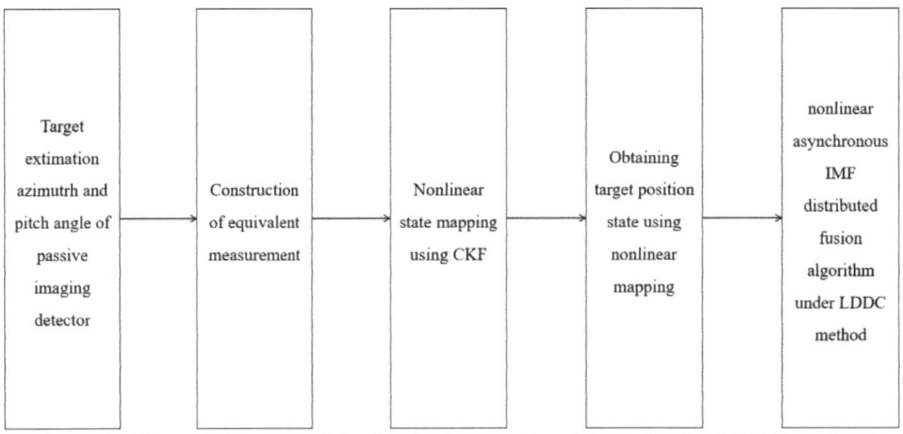

Figure 5. The architecture of the CKF-AHIMF algorithm with the LDDC method

- First, the angle state of the target (azimuth and pitch angle) estimated by CKF is used to construct the equivalent measurement and its measurement covariance matrix according to the parameter normalization method. The method is as follows:

$$\tilde{z}\left(t_k \middle| t_k, t_{k-1}\right) = \tilde{R}\left(t_k \middle| t_k, t_{k-1}\right) \cdot \left[\tilde{P}\left(t_k \middle| t_k\right)^{-1} \tilde{x}\left(t_k \middle| t_k\right) - \tilde{P}\left(t_k \middle| t_{k-1}\right)^{-1} \tilde{x}\left(t_k \middle| t_{k-1}\right) \right] \tag{33}$$

$$\tilde{R}\left(t_k \middle| t_k, t_{k-1}\right) = \left[\tilde{P}\left(t_k \middle| t_k\right)^{-1} - \tilde{P}\left(t_k \middle| t_{k-1}\right)^{-1} \right]^{-1} \tag{34}$$

Wherein, t_k is the time when the Local Detector (LD) obtains the local estimation, and t_{k-1} is the last estimation time of LD. According to this, angle equivalent measurement $\tilde{z}\left(t_k \middle| t_k, t_{k-1}\right)$ and measurement covariance $\tilde{R}\left(t_k \middle| t_k, t_{k-1}\right)$ can be obtained.

- Using the constructed equivalent measurement $\tilde{z}\left(t_k \middle| t_k, t_{k-1}\right)$ and measurement covariance $\tilde{R}\left(t_k \middle| t_k, t_{k-1}\right)$, based on the one-step prediction of the target spatial position state and error covariance matrix, with the measurement update part in CKF which includes the Cholesky decomposition of the one-step prediction error covariance matrix, cubature points propagation, measurement prediction, and filter gain calculation, the three-dimensional position estimation state of the target under angle equivalent measurement is finally obtained. The formulas involved are:

1) One-step prediction of global state and error covariance matrix:

$$\hat{X}\left(t_k \left| t_{k-1}\right.\right) = F\hat{X}\left(t_{k-1} \left| t_{k-1}\right.\right) \tag{35}$$

$$P\left(t_k \left| t_{k-1}\right.\right) = FP\left(t_{k-1} \left| t_{k-1}\right.\right)F^T + Q \tag{36}$$

Wherein, $\hat{X}\left(t_{k-1} \left| t_{k-1}\right.\right)$ and $P\left(t_{k-1} \left| t_{k-1}\right.\right)$ are the spatial position estimation state and covariance of LD respectively. F and Q are the state transition matrix and process noise matrix respectively.

2) Cholesky decomposition of one-step prediction error covariance matrix:

$$P\left(t_k \left| t_{k-1}\right.\right) = S\left(t_k \left| t_{k-1}\right.\right)S\left(t_k \left| t_{k-1}\right.\right)^T \tag{37}$$

3) Constructing cubature points:

$$x^i\left(t_k \left| t_{k-1}\right.\right) = S\left(t_k \left| t_{k-1}\right.\right)\xi_i + \hat{X}\left(t_k \left| t_{k-1}\right.\right) \tag{38}$$

4) Propagation of cubature points:

$$z^i\left(t_k\right) = h\left(x^i\left(t_k \left| t_{k-1}\right.\right), v\right) \tag{39}$$

5) Calculation of measurement prediction:

$$\hat{z}\left(t_k\right) = 1/\left(2n\right)\sum_{i=1}^{2n} z^i\left(t_k\right) \tag{40}$$

6) Calculation of measurement error covariance matrix and cross-covariance matrix:

$$P^z(t_k) = 1/(2n)\sum_{i=1}^{2n} z^i(t_k)\left[z^i(t_k)\right]^T - \hat{z}(t_k)\left[\hat{z}(t_k)\right]^T + R \tag{41}$$

$$P^{xz}(t_k) = 1/(2n)\sum_{i=1}^{2n} x^i(t_k \left| t_{k-1}\right.)\left[z^i(t_k)\right]^T - \hat{X}(t_k \left| t_{k-1}\right.)\left[\hat{z}(t_k)\right]^T \tag{42}$$

Wherein, R is the equivalent measurement error covariance matrix $\tilde{R}(t_k \left| t_k, t_{k-1}\right.)$.

7) Calculation of Kalman gain and updating state and error covariance matrix:

$$K(t_k) = P^{xz}(t_k)\left[P^z(t_k)\right]^{-1} \tag{43}$$

$$\hat{X}(t_k \left| t_k\right.) = \hat{X}(t_k \left| t_{k-1}\right.) + K(t_k)\left(z(t_k) - \hat{z}(t_k)\right) \tag{44}$$

$$P(t_k) = P(t_k \left| t_{k-1}\right.) - K(t_k)P^z(t_k)K(t_k)^T \tag{45}$$

Wherein, $z(t_k)$ is the angle equivalent measurement $\tilde{z}(t_k \left| t_k, t_{k-1}\right.)$.

• Use the updated state $\hat{X}(t_k \left| t_k\right.)$ and error covariance matrix $P(t_k)$ calculated by CKF as the local position estimation state of the nonlinear asynchronous IMF distributed fusion algorithm under the LDDC method and send them to FC for asynchronous IMF distributed fusion.

• If states from other LD need to be fused at this fusion time, repeat the first and second steps above for heterogeneous state estimation mapping from the local angle state and its error covariance matrix to the three-dimensional position state and its error covariance matrix.

Through the above steps from chapters 2.1 ~ 2.3 in this paper, the distributed Asynchronous Heterogeneous Information Matrix Fusion algorithm based on Cubature Kalman Filter (CKF-AHIMF) proposed in this paper can convert the nonlinear angle state of the passive imaging detection system to the position estimation state in the three-dimensional space without calculating the Jacobian matrix. On this basis, target state estimation and fusion of distributed multi-platform passive/active asynchronous and heterogeneous imaging detection systems can be finally completed, that is, the target estimation and fusion state of passive/active imaging detection systems in the condition of asynchronous sampling rate can be obtained.

4. Simulation results and analysis

The simulation in this section aims to verify the target state estimation accuracy of each distributed fusion algorithm[8]~[10] when the distributed passive/active imaging detection system obtains an asynchronous heterogeneous target estimation state, including distributed Asynchronous Heterogeneous Information Matrix Fusion algorithm based on EKF (EKF-AHIMF) under Local Detector Driven Communication (LDDC) method proposed in reference [4], the distributed Asynchronous Heterogeneous Information Matrix Fusion algorithm based on CKF (CKF-AHIMF) under Fusion Center Driven Communication (FCDC) method, and the distributed Heterogeneous Information Matrix Fusion algorithm based on CKF (CKF-AHIMF) under LDDC method proposed in this paper.

The simulation involves two imaging detection systems, including an active SAR imaging detection system and a visible/infrared passive imaging detection system. 50 times Monte Carlo simulations were conducted. The position estimation accuracy of each fusion algorithm in each coordinate axis is evaluated by using position estimation Root Mean Square Error (RMSE) of the X, Y, and Z axis and the overall position estimation RMSE.

The simulation scenario is as follows: an active SAR imaging detection system and a passive visible/infrared imaging detection system, which are respectively installed on two flight platforms. In the geodetic coordinate system, the longitudes, latitudes, and heights of the two flight platforms are $(31.0215°\,N,\ 112.1377°\,E,\ 5557.0\,m)$ and $(31.0216°\,N,\ 112.1168°\,E,\ 3554.8\,m)$respectively. In the earth's rectangular coordinate system, the positions of the two flight platforms are (8000 m, 5000 m, 5000 m), (6000 m, 5000 m, and 3000 m), and the speeds of the two flight platforms in each axis are (- 4 m/s, - 2 m/s, - 0.1 m/s), (- 2 m/s, - 1 m/s, - 0.1 m/s) respectively. The measurement noise of SAR imaging detection system 1 is $diag(25mrad^2, 25mrad^2, 10000m^2)$, and the sampling period is 5 s; the measurement noise of visible/infrared imaging detection system 2 is $diag(1mrad^2, 1mrad^2)$, and the sampling period is 2 s, that is, the two detectors are asynchronous and heterogeneous.

The true longitude, latitude, and height of the target is $(30.9765°\,N,\ 112.0540°\,E,\ 550.0\,m)$, and its position in the earth rectangular coordinate system is (0 m, 0 m, 0 m). The target moves in a Near Constant Velocity (NCV) in the positive direction of the X axis, and its speed is (1 m/s, 0 m/s, 0 m/s). The process noise covariance matrix Q of the target is:

$$Q = \begin{bmatrix} \Delta T^2/3 & \Delta T^2/3 \\ \Delta T^2/3 & \Delta T \end{bmatrix} \otimes diag(q_x, q_y, q_z) \tag{46}$$

Wherein, $\Delta T = 1s$ is the sampling period, (q_x, q_y, q_z) are the noise intensity of the X, Y, and Z axis respectively. In the simulation, $q_x = q_y = q_z = 0.001$.

The simulation scenario is shown in Figure 6 below.

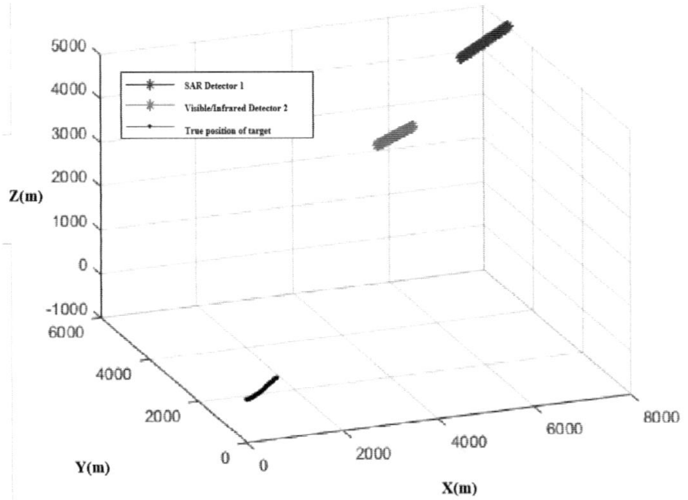

Figure 6. 3-D simulation scenario with the active and passive imaging detection system

In this simulation scenario above, the analysis results of the estimation state of the target from each distributed fusion algorithm are shown in Figures 7 ~ 9 below.

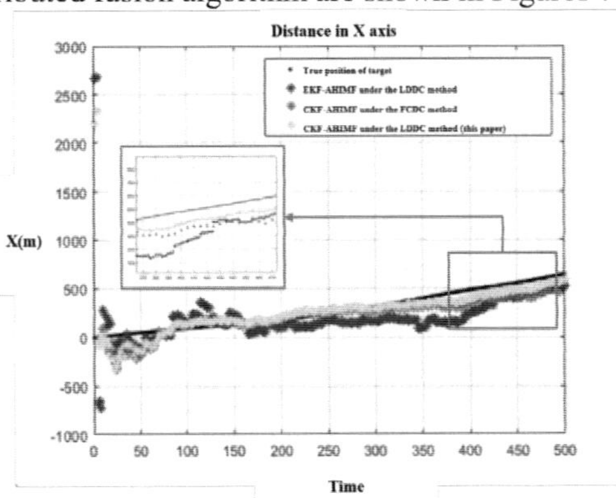

Figure 7. Position estimation in the X axis

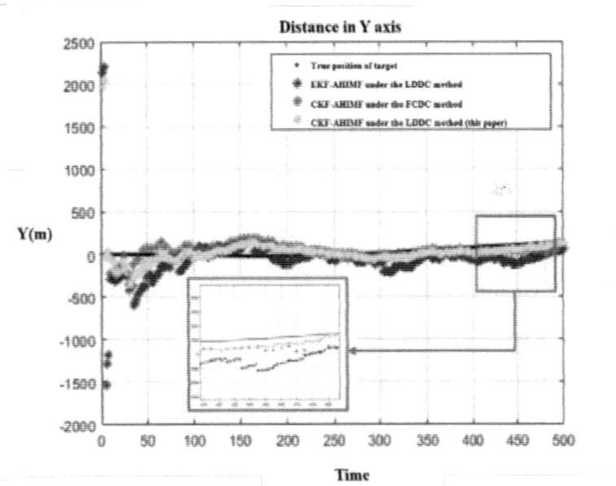

Figure 8. Position estimation in the Y axis

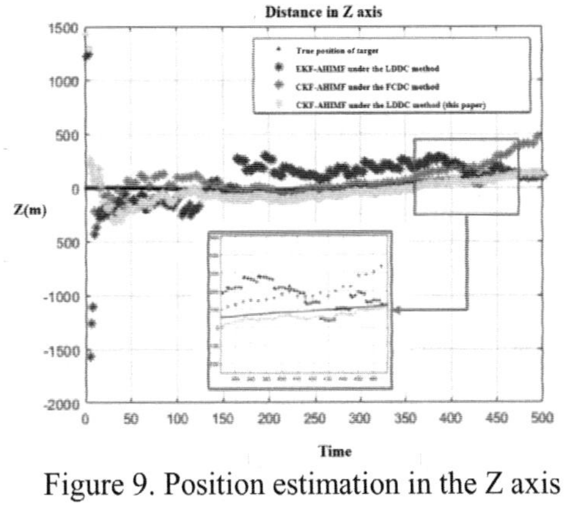

Figure 9. Position estimation in the Z axis

In the above figures, the blue dots represent the EKF-AHIMF algorithm under the LDDC method, the red dots represent the CKF-AHIMF algorithm under the FCDC method, and the green dots represent the CKF-AHIMF algorithm under the LDDC method proposed in this paper. The position estimation RMSE of each axis and the overall position estimation RMSE are calculated to evaluate the estimation accuracy of each algorithm. The error results are shown in Figures 10 ~ 13 below.

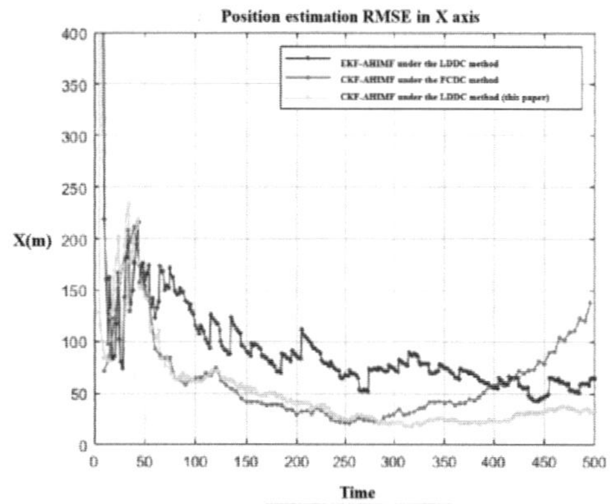

Figure 10. Position estimation RMSE in the X axis

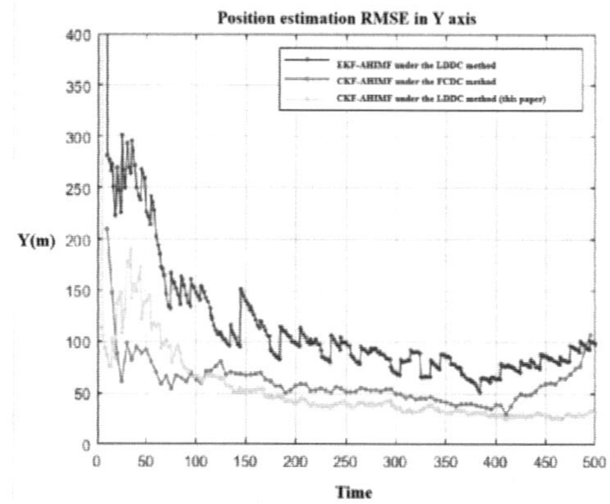

Figure 11. Position estimation RMSE in the Y axis

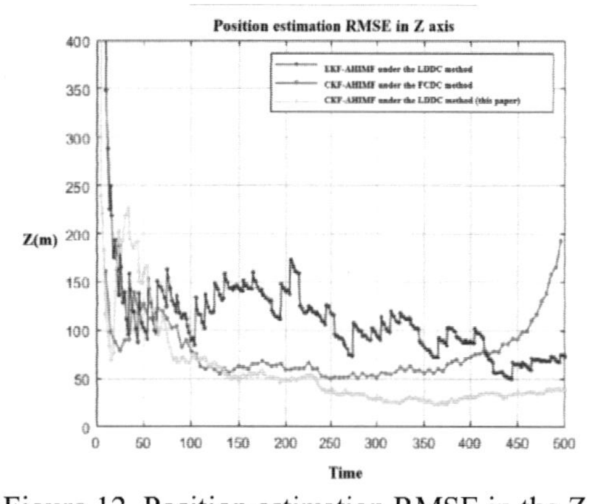

Figure 12. Position estimation RMSE in the Z axis

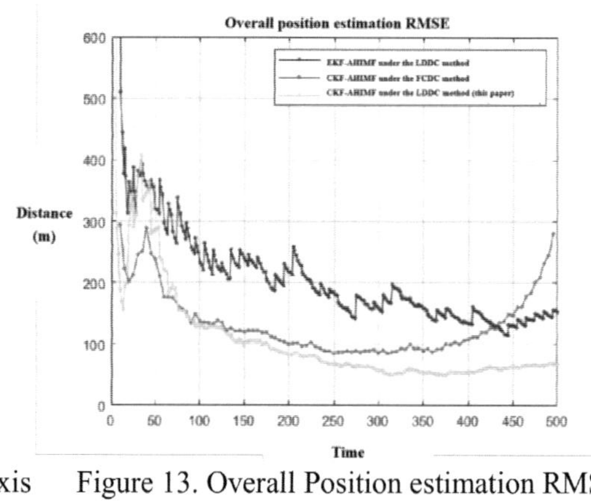

Figure 13. Overall Position estimation RMSE

In the above position state estimation RMSE results, the blue line represents the EKF-AHIMF algorithm under the LDDC method, the red line represents the CKF-AHIMF algorithm under the FCDC method, and the green line represents the CKF-AHIMF algorithm under the LDDC method proposed in this paper. The EKF-AHIMF algorithm under the LDDC method uses the EKF algorithm for nonlinear state mapping, and its filtering convergence effect is poor. Meanwhile, the CKF-AHIMF algorithm under the FCDC method will lose part of the local estimation state information of the detection system due to the fixed fusion period, leading to divergence in the later stage of filtering. The CKF-AHIMF algorithm under the LDDC method proposed in this paper has higher position state estimation accuracy than the EKF-AHIMF algorithm under LDDC; Compared with the CKF-AHIMF algorithm under FCDC, the algorithm has a better convergence effect, and the filtering divergence will not occur.

To further verify the effectiveness of the algorithm proposed in this paper, Table 1 shows the RMSE of the target position estimation of the above algorithms at the final filtering moment. It indicates that the CKF-AHIMF algorithm under the LDDC method proposed in this paper can achieve better precision of target state estimation in distributed fusion.

Table 1. Target position estimation RMSE under each distributed fusion algorithm

Distributed fusion algorithm	Position estimation error	
EKF-AHIMF algorithm under the LDDC method	RMSE_x (m)	64.2992
	RMSE_y (m)	97.9762
	RMSE_z (m)	73.1907
	RMSE_position (m)	152.9200
CKF-AHIMF algorithm under the FCDC method	RMSE_x (m)	137.9402
	RMSE_y (m)	106.6680
	RMSE_z (m)	192.9030
	RMSE_position (m)	280.0371
CKF-AHIMF algorithm under the LDDC method (proposed in this paper)	RMSE_x (m)	**32.9708**
	RMSE_y (m)	**31.4169**
	RMSE_z (m)	**38.0634**
	RMSE_position (m)	**66.4781**

The results in the above table also show that the CKF-AHIMF algorithm under the LDDC method in this paper has higher position state estimation accuracy than other distributed asynchronous and heterogeneous fusion algorithms.

5. Conclusion

This paper mainly aims at the problem of target state estimation and fusion when the distributed active/passive imaging detection system has an asynchronous sampling rate. The main method is to propose a distributed Asynchronous and Heterogeneous Information Matrix Fusion algorithm based on Cubature Kalman Filter (CKF-AHIMF). Combined with the generalization of the Information Matrix Fusion (IMF) algorithm in nonlinear, asynchronous, and heterogeneous conditions, the distributed fusion of asynchronous heterogeneous states in the fusion center is completed. The simulation verifies the efficiency of the CKF-AHIMF algorithm under the LDDC method proposed in this paper. compared with the EKF-AHIMF algorithm under the LDDC method and the CKF-AHIMF algorithm under the FCDC method, the algorithm proposed in this paper can obtain more accurate target position estimation which can help improve the accuracy of the precision-guided imaging system.

References

[1] Shuli Sun, Honglei Lin, Jing Ma, Xiuying Li. Multi-sensor distributed fusion estimation with applications in networked systems: a review paper [J]. Information Fusion. 2017.
[2] Yifan Gu, Wenlong Zhao, Shanjun Tang. Target cooperative positioning method of distributed active/passive imaging detection system[J]. Air & Space Defense. 2021.
[3] Tongde Li, Huaicheng Yan, Ge Zhou. Distributed fusion estimation of an asynchronous multi-rate multi-sensor network system in navigation guidance[J]. Air & Space Defense. 2021.
[4] Kaipei Yang, Yaakov Bar-Shalom. Information Matrix Fusion for Nonlinear, Asynchronous and Heterogeneous Systems[C]. 2019 International Conference on Information Fusion (FUSION). 2019.
[5] Arasaratnam I, Haykin S. Cubature Kalman Filters[J]. IEEE Trans on Automatic Control, 2009, 54(6): 1254-1269.
[6] Allig C, Wanielik G. Heterogeneous Track-to-Track Fusion Using Equivalent Measurement and Unscented Transform[C]. 2018 International Conference on Information Fusion (FUSION). 2018.
[7] Mallick M, Chang K C, Arulampalam S, et al. Heterogeneous Track-to-Track Fusion in 3D Using IRST Sensor and Air MTI Radar[J]. IEEE Transactions on Aerospace and Electronic Systems, 2019, 56(6):3062-3079.
[8] Zhu G M, Zhou F, Xie L, et al. Sequential Asynchronous Filters for Target Tracking in wireless sensor networks[J]. IEEE Sensors Journal, 2014, 14(9): 3174-3182.

[9] Guchong Li, Wei Yi, Suqi L. Asynchronous Multi-Rate Multi-Sensor Fusion Based on Random Finite Set[J]. Signal Processing, 2019, 160: 113-126.
[10] Alouani A T. Simplified Generalized Asynchronous Track Fusion Filter[C]. SPIE, 2006.

Journal of Physics: Conference Series **2489** (2023) 012041 doi:10.1088/1742-6596/2489/1/012041

Technical analysis on airworthiness compliance of lighting components for civil aircraft emergency evacuation slides

Hongyuan Song*, Feng Li and Xiaoliang Tang

COMAC Shanghai Aircraft Design and Research Institute(SADRI), 5188 Jinke Road, Pudong New Area, Shanghai, PRC

* Corresponding author: songhongyuan@comac.cc

ABSTRACT: The emergency landing safety of emergency evacuation slides can be improved greatly by using lighting components, especially in dark environments. To design a required lighting system, two problems need to be solved, which are also the focus of this paper. Firstly, what airworthiness compliance work should be planned and how to do it. Secondly, how to design to meet the airworthiness requirements should be considered. A complete and effective airworthiness compliance method was proposed by analyzing various performance characteristics such as self-opening function, lighting duration, illuminance, and installation environment. Airworthiness requirements are decomposed into design features, and the design requirements of lighting components are put forward. This study can provide verification methods and design ideas for the lighting components of civil aircraft emergency evacuation slides.

1. INTRODUCTION

The emergency evacuation slide is inflatable life-saving equipment for large and medium-sized civil aviation aircraft after an emergency land (water) emergency landing. After the emergency landing of the plane on land and water, the passengers on the plane can land quickly and safely or float on the water for rescue. According to airworthiness requirements, all cabin emergency exits with sill heights of more than 1.83 meters should be equipped with emergency evacuation slides.

The earliest slides were developed and manufactured by the American Air Cruisers Company. Based on becoming standard safety equipment on the flights of major airlines, the technology of emergency slides is also constantly updated. The design requirements for emergency evacuation slides mainly come from two parts, one is TSO (Technical Standard Order) requirements, such as TSO-C69c [1] and CTSO-C69c [2]. Another part is the requirements of installed airworthiness regulations, such as 14 CFR-25 [3] and CCAR-25-R4 [4]. The research on emergency evacuation and slides in the United States started earlier, and related research reports have been formed [5-6]. The research on slides in China started relatively late. Wang [7] discussed the general design process of civil aircraft emergency evacuation slides, including slide configuration, slide package installation, slide actuation form, and slide indication. Wang [8] and others used ADAMS software to study the slide release process and analyzed the design technology for the coordination of the slide pack and the movement of the cabin door. Li [9] and others used LS-DYNA software to simulate the evacuation of the slide, and compared the results of the test samples to analyze the influence of different internal working pressures on the evacuation rates. Zhu [10] and others used ANSYS WORKBENCH to analyze the influence of different design parameters on the ejection ratio of the high-pressure ejector in the slide, and the research results can be used to guide the improvement of the ejection ratio of the ejector.

Existing research mainly focuses on inflatable chambers, ejectors, motion simulation of slides, etc. In addition to these components, another important component of the slide is the lighting component, which is still lacking in research in this field. The lighting components are composed of a battery, slide lights, light harness, etc, to provide emergency lighting for emergency slides. The slide light is powered by a battery via a light harness, and the slide light version uses LED (Light Emitting Diode) for lighter weight and higher reliability. A schematic diagram of the arrangement of the lighting components in the slide is shown in Figure 1.

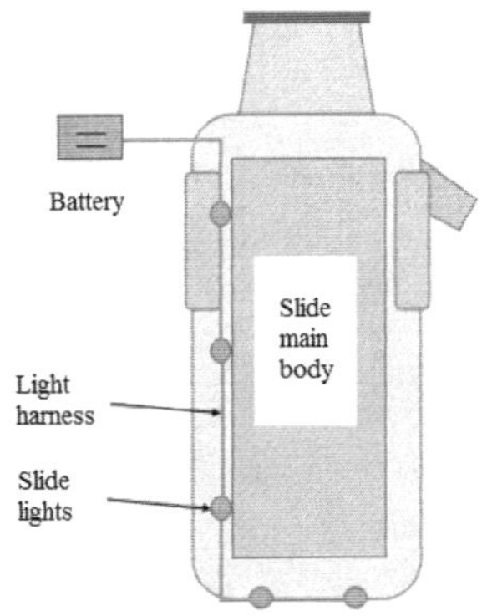

Figure 1. Composition of lighting components.

This paper takes the lighting components of emergency slides as the research object, and systematically studies the airworthiness compliance technology of lighting components. The airworthiness requirements related to lighting components are analyzed one by one, including self-opening function, lighting duration, illuminance, installation environment, etc.

2. AIRWORTHINESS REQUIREMENTS

Airworthiness requirement CCAR 25.812(h)(2)(i) puts forward airworthiness requirements for the self-opening function of the lighting system. If the emergency lighting subsystem of the assist means serves no other assist means, it is independent of the airplane's main emergency lighting system, and is automatically activated when the assist means is erected. The lighting provisions may not be adversely affected by stowage. Section 4.19 of CTSO-C69c, the integral device illumination must be designed so the illumination means is activated automatically during deployment or inflation and the level of illumination meets the appropriate requirements in CCAR 25.812 in effect at the time of application. The illumination means must not interfere with the safe evacuation of persons using the device in a normal manner.

Airworthiness requirement CCAR 25.812(i) specifies the requirements for the lighting duration of the slide lighting components. The energy supply to each emergency lighting unit must provide the required level of illumination for at least 10 minutes at critical ambient conditions after an emergency landing.

Airworthiness requirement CCAR 25.812(h)(2)(ii) put forward the requirements for illuminance. The lighting system must provide illumination of not less than 0.323 lx (0.03 foot-candle) (measured normal to the direction of incident light) at the ground end of the erected assist means where an evacuee would normally make first contact with the ground, with the airplane in each of the attitudes corresponding to the collapse of one or more legs of the landing gear.

Airworthiness requirements CCAR 25.1301 and 25.1309 are mainly about the requirements of the installation environment. It should be ensured that emergency evacuation slides function properly and are designed to perform their intended function under predictable operating conditions. The environment that can be expected here is generally screened and verified by the environment in DO160G.

3. VERIFICATION METHOD AND DESIGN

3.1 Self-opening function verification method and design

Combined with airworthiness requirements, a verification method can be analyzed and extracted. Through the consecutive deployment test of slides, the compliance of the terms is shown, and it is verified that the lighting components can be activated and lit at the same time during the deployment of the slide.

Verification conditions are as follows. In this test, the sill height of the test bench is set to the normal sill height and angle of the aircraft, and the slide test piece is installed on the test mockup for deployment to show compliance with the terms. The slide lighting assembly automatically activates when the slide is dropped.

Through the above analysis, the design idea of lighting components can be obtained to meet the airworthiness requirements. Slide lights and light wire harnesses should be securely mounted on the slide air column to maintain a stable connection during the slide deployment process. The battery maintains a reliable power supply during the slide release. The battery assembly has an emergency start function and cooperates with activating the lanyard of the slide as a switch to control the start of the battery. During deployment, when the battery is activated, the slide light will automatically illuminate. When the slide is not deployed, the battery should be kept off to save power and ensure battery life. A composition diagram of the battery components in the slide is shown in Figure 2.

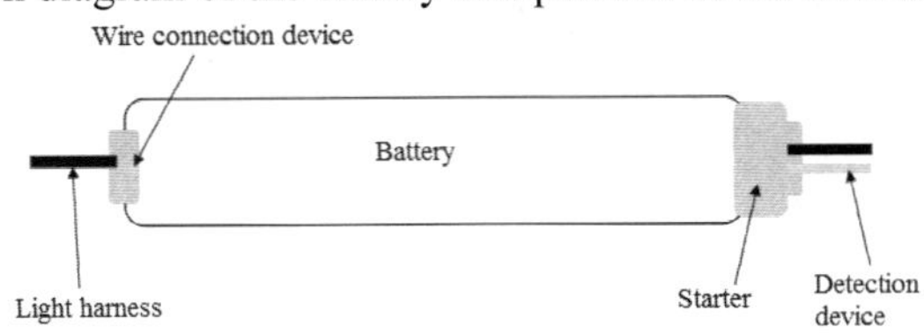

Figure 2. Composition diagram of battery components design.

3.2 Lighting duration verification method and design

A verification method is analyzed and summarized combined with airworthiness requirements. This clause shows compliance with the clause through the emergency lighting test in the slide TSO qualification test. Under critical ambient conditions, the operating time of the battery unit and the slide light assembly is detected.

Verification conditions are as follows. The capacity loss is determined by considering the aging of the battery, the service life of the battery is 6 years at least, and a certain safety factor is considered. By converting the loss into lighting time, the experiment will increase the lighting time. The lighting time is not less than 10min× (1+ battery annual attenuation ×life years × safety factor).

To assess critical ambient conditions, tests will be performed at normal, low, and high-temperature environments, operating the battery and slide light assembly for the required duration, and measuring voltage, current, and brightness every 30 seconds. According to the position of the emergency exit corresponding to the slide, it can be divided into the door slide and the off-wing slide. The door slides are generally located inside the cabin, and the off-wing slides are generally located outside the fuselage and in the belly fairing area at the root of the wing. The temperature requirements for the lighting components of the slides are shown in Table 1.

Table 1. Test temperature of slide lighting components.

Test Temperature	**Normal**	**Low**	**High**
Door slide	+21°C	-40°C	+71°C
Off-wing Slide	+21°C	-54°C	+71°C

Through the above analysis, the design idea of lighting components can be obtained to meet the airworthiness requirements. To ensure the lighting time of the slide lighting components, non-rechargeable batteries are generally used as part of the slide system. The battery assembly can maintain a stable connection with the light harness under the predetermined working environment to ensure normal operation. The battery pack has an external detection function, which can easily detect the remaining capacity of the battery.

3.3 Illuminance verification method and design

Combined with airworthiness requirements, an airworthiness verification method can be analyzed and extracted. The emergency lighting test in the TSO qualification test of the slide shows the compliance of the clauses, and the illumination of the typical measurement points in the landing area of the evacuee under different sill heights of the aircraft (simulating the collapse of one or more legs of the landing gear) is verified.

Verification conditions are clear and enforceable. By adjusting the position of the LED light on the slide body tube, the position of the LED light in different postures is simulated. It includes three working conditions, which simulate the minimum sill height, normal sill height, and maximum sill height respectively. The lighting positions under the three working conditions are shown in Figure 3. The test workshop is completely dark. At each sill height, the illuminance is measured at the typical measurement points in the landing area of the evacuees, to ensure that the illuminance of each point is not less than 0.323 lx. A schematic diagram of the measurement is shown in Figure 4.

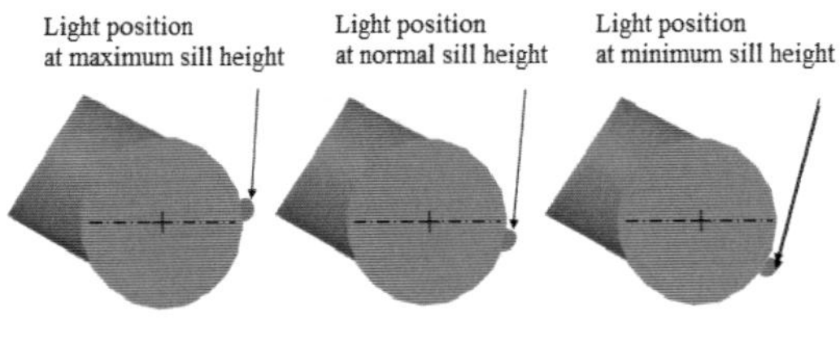

Figure 3. Lighting positions.

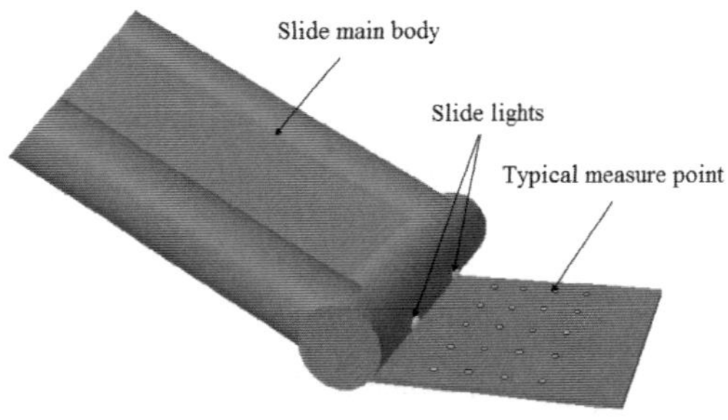

Figure 4. Schematic diagram of the measurement.

Through the above analysis, the design idea of lighting components can be obtained to meet the airworthiness requirements. The slide light has a lighting function and cooperates with the emergency lighting equipment on the aircraft to provide stable lighting for evacuees' evacuation. The slide light consists of a wick, a transparent shell, and a base, which are fixed on the slide body tube by bonding, as shown in Figure 5. Slide lights and light harnesses should be securely mounted on the slide body tube, and a stable connection should be maintained in the intended working environment (such as during slide dropping). Each module of the slide light should be connected in parallel to prevent the circuit failure of some modules from affecting other modules. The light harness adopts a ring-shaped positive wire and negative wire layout to ensure that the slide light can still light normally when a certain wire is disconnected. The light-emitting color of the lamp wick is white with low chromaticity and high penetration. The transparent shell and base are made of materials with better luminous flux, strength, and weather resistance.

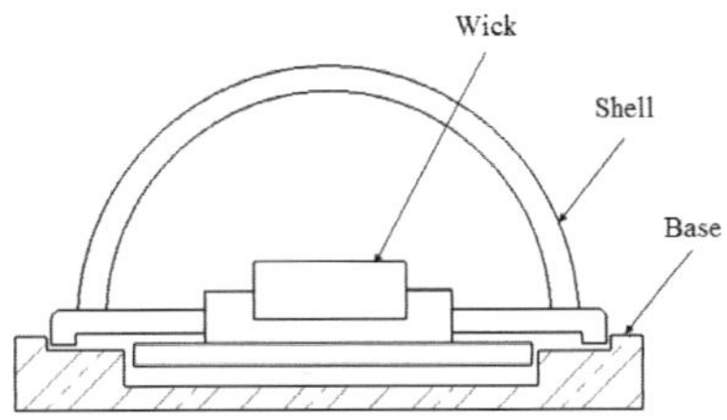

Figure 5. Composition diagram of slide light design.

3.4 Installation environment verification method and design

The temperature requirements are described in Table 1 and are verified along with the lighting duration. Among them, the verified temperature range should be extended according to the actual installed ambient temperature.

The altitude test is carried out according to DO-160G category A2. The lighting components of door slides should be able to work normally under air pressure conditions equivalent to 4600m (15,000 feet) above sea level and be able to work normally when exposed to a rapid decompression environment. By the provisions of category D2 in DO-160G, the lighting components of off-wing slides should be able to work normally under air pressure conditions equivalent to 15200m (50000 feet) above sea level.

The humidity test of the lighting components of door slides should meet Section 6 Category A (standard humidity environment) humidity requirements. The lighting components of off-wing slides should meet Section 6 Category C (external humidity environment) humidity requirements.

Electromagnetic compatibility requirements apply to electrical and electronic components in lighting assemblies. The slide lights should meet electrostatic discharge in DO-160G Section 25 Category A. Considering the power supply, operating environment, working preconditions, and equipment functions of the emergency slide, electromagnetic compatibility tests such as voltage spike, audio frequency conducted susceptibility - power inputs, induced signal susceptibility, radio frequency susceptibility (radiated and conducted), emission of radio frequency energy and lightning induced transient susceptibility should be added as appropriate.

The battery is generally verified according to the environmental requirements in the TSO, such as CTSO-C142a, and CTSO-C173a, or verified with the form of aircraft airworthiness certification.

Materials used to manufacture lighting assemblies must meet the requirements of CCAR 25.853(a), Appendix F, Section I(a)(1)(ii).

Combined with airworthiness requirements, complete and effective airworthiness verification ideals can be analyzed and extracted. The laboratory environmental tests and subsequent functional tests of lighting components or slides show the compliance of the clauses.

Verification conditions are as follows according to DO160G requirements to complete the laboratory test preparation. If it involves functional verification, the lighting components after the laboratory test shall be installed on the designated slide facilities to complete the functional verification.

Through the above analysis, the design index that meets the requirements of lighting components can be summarized. Lighting components in the design from material selection, and parts forming to assembly, should consider the adaptability to different environments, to ensure that the final product can meet the predictable operating environmental conditions.

4. GENERAL REQUIREMENTS AND DESIGN

The safety impact of the lighting components should meet the safety index requirements of the aircraft. The human factors that seriously affect safety must be reduced to a minimum during the design.

Lighting components should be designed to meet reliability requirements. Battery life and service intervals should not be less than 6 years. A functional inspection should be carried out every 6 years and any damaged slide lights should be replaced.

The lighting components design meets maintainability requirements. The design should ensure easy operation during installation and maintenance. The average repair or replacement time is no more than 0.2 h.

Components of the same part number shall be interchangeable in function and size. Attention shall be paid to the interchangeability details of parts during design.

The processing method used for the product should ensure that the appearance and quality of the product meet the specified requirements. There should be no burrs, cracks, bubbles, fine lines, pinholes, scratches, sharp corners, and other defects that affect usability or appearance. The outer packaging that protects the product should be to the customer's specifications or as defined by the manufacturer. Packaging shall identify the external markings of the product, including part number, warranty period requirements, and date of manufacture.

The identification requirements for lighting components are as follows: Manufacturer's identification and part number and/or customer's identification and part number should be marked on all components and replaceable elements as far as possible. Equipment labels should be visible. Serial numbers should be as permanent as possible marked on the product. The design of lighting components can only be marked by spraying according to the above requirements.

5. CONCLUSION

This paper studies the airworthiness compliance verification method for the emergency evacuation slide lighting assembly of civil aircraft. In this process, the airworthiness requirements that the design of slide lighting components should meet are studied, including self-opening, lighting duration, illuminance, and installation environment. General requirements and design are summarized. This paper systematically studies and analyzes the airworthiness compliance technology of slide lighting components, which has important guiding value for the design of civil aircraft slides. The research achievements in the design of lighting components are mainly reflected in the following aspects: the design composition and functions of the battery specially used for the slide, the lighting duration design and environmental resistance design requirements of the battery, and the design composition and installation requirements of the slide light, and the various adverse environmental factors and general design indicators that need to be considered for the lighting components.

REFERENCES

[1] U.S. Department of Transportation Federal Aviation Administration Aircraft Certification Service, "Emergency evacuation slides, ramps, ramp/slides, and slide/rafts," TSO-C69c, Washington, DC, (1999).

[2] Civil Aviation Administration of China, "Emergency evacuation slides, ramps, ramp/slides, and slide/rafts," CTSO-C69c, (2019).

[3] Federal Aviation Administration, "Airworthiness standards: transport category airplanes," 14 CFR-25, Washington, DC, (2011).

[4] Civil Aviation Administration of China, "Airworthiness standards: transport category airplanes," CCAR-25-R4, (2011).

[5] National Transportation Safety Board, "Emergency evacuation of commercial airplanes," Safety Study NTSB/SS-00/01, Washington, DC: NTSB, (2000).

[6] Es, G. V., "Analysis of evacuation slide problems," International Aircraft Cabin Safety Symposium, (2006).

[7] Wang, H., "Technical analysis of civil aircraft emergency evacuation slide system," Advances in Aeronautical Science and Engineering, 2(01):37-42, (2011).

[8] Wang, J., Feng, Y., Liu, J., et al, "Design technique of commercial aircraft emergency evacuation slide release," Mechanical Science and Technology for Aerospace Engineering, 33(08):1264-1267, (2014).
[9] Li, T., Wang, X., Huang, Y., "Numerical simulation and test verification of the evacuation rate of the emergency evacuation slide," Ship Electronic Engineering, 37(10):75-79, (2017).
[10] Zhu, S., Qiu, Y., Zhang, X., et al, "Performance Investigation of rapid ejecting inflation system," Aeronautical Science & Technology, 29(03):08-14, (2018).

An Improved Mickens Iterative Method for Nonlinear Aeroelastic analysis of Airfoils with an External Store

Feixin Chen[a, b], Jianxin Deng[b], Zhenghao Ding*[, c], Weihua Li[a] and Suhua Xiao[a]

[a.] Guangdong Polytechnic Normal University, 510665, Guangzhou, China;

[b.] Guangxi Key Lab of Manufacturing System and Advanced Manufacturing Technology, Guangxi University, 530004, Nanning, China.

[c.] Department of Civil and Environmental Engineering, The Hong Kong Polytechnic University, Kowloon, Hong Kong, China;

* Corresponding author: zhenghao.ding@polyu.edu.hk

Abstract: This paper aims to propose a modified Mickens iteration method for analyzing the flutter stability of the nonlinear airfoil with an external storage system. Unlike the regular Mickens approach, the modified form is established by substituting previous frequencies of the limit cycle oscillation (LCO) into unknown ones. Then the algebraic equations governing the LCO amplitude and frequency could keep linear, which might improve the calculation efficiency. The numerical example shows the LCOs obtained easily and accurately. Furthermore, the phase planes of the LCOs are investigated with the wind speed varying. The approach could be applicable to investigate mechanical systems with complicated nonlinearities.

1. INTRODUCTION

The airfoil with aerodynamic and structural nonlinearities has stimulated the curiosity of many researchers, and extensive literature has been published [1]. Due to the nonlinearities, the flutter vibration of an airfoil usually displays lots of nonlinear characteristics, such as the limit cycle oscillation (LCO) and bifurcation. Predicting the characteristics of the LCOs accurately and efficiently has always been an active research field in aeroelasticity [2]. The numerical methods, mainly including the Runge-Kutta integration and the finite difference scheme method, are often applied to study the stable solutions of the nonlinear system. To obtain LCOs more effectively, many analytical or semi-analytical techniques were used in nonlinear aeroelastic system, such as perturbation-incremental method, describing functions technique, harmonic balance method (HB), homotropy analysis method, and so on [3].

Among the abovementioned methods, the method of harmonic balancing is widely applied in determining analytical approximations [4]. However, because of complex nonlinear equations needing to be solved in the procedure, the method is difficult to extend into a higher-order nonlinear system. To address this issue, a type of iteration technique, initiated by Mickens, has been found valid in various nonlinear oscillators. Hu et al [5] obtained the periodic solutions by a modified iteration procedure. Another modified perturbation technique, that combined with Mickens iteration and variation iteration method (VIM), was proposed [6]. Mickens [7] also extended the iteration method to calculate a fractional-powers conservative oscillator. And, the iteration method has been applied to investigate the other strongly nonlinear system [8].

In this study, exemplified by the airfoil with an external store, a further modified Mickens iteration method will be proposed to solve nonlinear dynamical systems with multiple degrees of freedom (DOFs).

2. A 3-DOF AIRFOIL Motion

Figure1 is the sketch of an airfoil model. α denotes the pitch angle, which is positive when leading-edge rises. h denotes the plunge deflection, which is positive with the downward. At the aerodynamic center point ***F***, there is an external store, which has a central distance $\bar{L}b$ from the mid-chord. varying about ***F***, β is the pitch angle, which signifies the vibration of the external store. The length of total wing is $2b$. The elastic axis has a central distance ab from mid-chord. The distance from the mass center ***G*** to the elastic axis is $x_\alpha b$. $x_\beta b$ represents the distance from the mass center of the external store to ***F***.

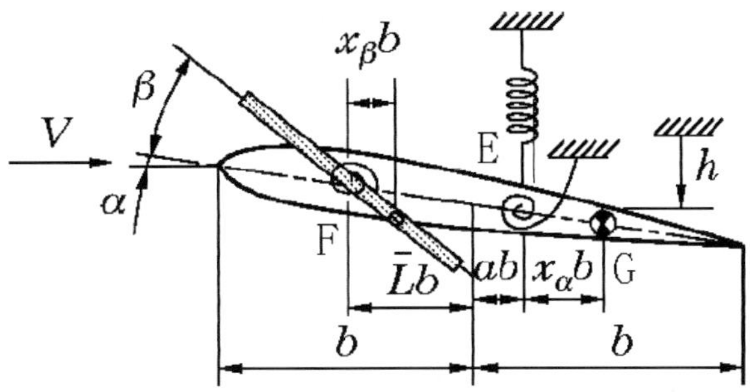

Figure 1. Schematic diagram of airfoil model with external storage

Considering the steady flow, the governing equations of the airfoil without the aerodynamics of the store can be described as

$$\begin{cases}(m+m_\beta)\ddot{h}+(S_\alpha+S_\beta-m_\beta L)\ddot{\alpha}+S_\beta\ddot{\beta}+K_h h=Q_h\\(S_\alpha+S_\beta-m_\beta L)\ddot{h}+(I_\alpha+I_\beta+m_\beta L^2-2S_\beta L)\ddot{\alpha}+(I_\beta-S_\beta L)\ddot{\beta}+K_\alpha\alpha=Q_\alpha\\S_\beta\ddot{h}+(I_\beta-S_\beta L)\ddot{\alpha}+I_\beta\ddot{\beta}+K_\beta\beta=0\end{cases} \quad (1)$$

where the superscript denotes the differentiation concerning real-time t_1. m is the airfoil mass considered per unit length and m_β is the mass of the store. The three symbols (K_h, K_α, K_β) signify the linear stiffness coefficients in 3-DOFs. the mass of inertia is $I_\alpha = mr_\alpha^2 b^2$ and $S_\alpha = mx_\alpha b$ is the static unbalance for the elastic axis. The counterparts of the external store about ***F*** are $I_\beta = m_\beta r_\beta^2 b^2$ and $S_\beta = m_\beta x_\beta b$.

In Eq. (1), $Q_h = -2\pi\rho V^2 b\alpha$ and $Q_\alpha = 2\pi\rho V^2 b^2\alpha(\bar{L}+a)$ (ρ is the air density). It is worth noting that the simple expression for stable aerodynamics should be applicable to low reduced frequencies and low wind speeds ($\boldsymbol{V / b\omega_\alpha}$). Linear stiffness in three degrees of freedom can be given as

$$K_h = m\omega_h^2,\ K_\alpha = I_\alpha\omega_\alpha^2,\ K_\beta = I_\beta\omega_\beta^2 \quad (2)$$

in which $\boldsymbol{\omega_i\left(i=h,\alpha,\beta\right)}$ is the respective angular eigen frequency of the constrained single-DOF system related to linear plunging and pitching springs. Considering the cubic nonlinear restoring forces and structural damping, and $t=\omega_\alpha t_1$ is introduced as a non-dimensional time. Eqs. (1) would be rewritten in the matrix form, as follow

$$\mathbf{M}\ddot{\mathbf{q}}+\mathbf{C}\dot{\mathbf{q}}+\mathbf{K}_Q\mathbf{q}+\mathbf{K}_3(\mathbf{q})=0,\quad \mathbf{q}=\begin{bmatrix}h_l & \alpha & \beta\end{bmatrix}^T \quad (3)$$

the superscript means the differentiation concerning the parameter *t*. Note that, the non-dimensional plunge displacement is still denoted by *h*. In system (3), the dimensionless matrix are expressed in detail as

$$\mathbf{M}=\begin{bmatrix}\mu+\mu_\beta & \mu x_\alpha+\mu_\beta x_\beta-\mu_\beta\bar{L} & \mu_\beta x_\beta \\ \mu x_\alpha+\mu_\beta x_\beta-\mu_\beta\bar{L} & \mu r_\alpha^2+\mu_\beta r_\beta^2+\mu_\beta\bar{L}^2-2\mu_\beta x_\beta\bar{L} & \mu_\beta r_\beta^2-\mu_\beta x_\beta\bar{L} \\ \mu_\beta x_\beta & \mu_\beta r_\beta^2-\mu_\beta x_\beta\bar{L} & \mu_\beta r_\beta^2\end{bmatrix},\quad \mathbf{C}=\begin{bmatrix}c_h & 0 & 0\\ 0 & c_\alpha & 0\\ 0 & 0 & c_\beta\end{bmatrix}$$

$$\mathbf{K}_Q=\begin{bmatrix}\mu(\omega_h/\omega_\alpha)^2 & 2Q & 0\\ 0 & \mu r_\alpha^2-(1+2a)Q & 0\\ 0 & 0 & \mu_\beta r_\beta^2(\omega_\beta/\omega_\alpha)^2\end{bmatrix},\quad \mathbf{K}_3(\mathbf{q})=\begin{bmatrix}k_{h3}h^3 & 0 & 0\\ 0 & k_{\alpha3}\alpha^3 & 0\\ 0 & 0 & k_{\beta3}\beta^3\end{bmatrix}$$

The mass ratio of airfoil is $\mu=\dfrac{m}{\pi\rho b^2}$, and the one of store is $\mu_\beta=\dfrac{m_\beta}{\pi\rho b^2}$. $\boldsymbol{r_\alpha}$ and r_β both are related to the mass moment of inertia mentioned above, as $I_\alpha=mr_\alpha^2b^2$ and $I_\beta=mr_\beta^2b^2$. Associated with the pitch mode, $Q=(V/b\omega_\alpha)^2$ is the non-dimensional flow velocity.

3. A MODIFIED ITERATION SCHEME

System (3) is a self-excited oscillation, which means the response of LCO's are both independent of the initial conditions, such as frequency and amplitude. Firstly, a new time scale would be introduced as

$$\tau=\omega t \quad (4)$$

in which the ω is the previously unknown angular frequency of LCOs. Then, System (3) can be expressed as

$$\omega^2\mathbf{M}\mathbf{q}''+\omega\,\mathbf{C}\mathbf{q}'+\mathbf{K}_Q\mathbf{q}+\mathbf{K}_3(\mathbf{q})=0 \quad (5)$$

now the prime denotes the differentiation concerning τ . Equation (5) also can be rewritten as

$$\omega^2\mathbf{M}\mathbf{q}''+\omega^2\mathbf{q}=\omega^2\mathbf{q}-\omega\,\mathbf{C}\mathbf{q}'-\mathbf{K}_Q\mathbf{q}-\mathbf{K}_3(\mathbf{q}):=\mathbf{H}(\omega,\mathbf{q},\mathbf{q}') \quad (6)$$

with initial conditions as

$$\mathbf{q}(0)=\alpha \quad \mathbf{q}'(0)=0 \quad (7)$$

where α is the LCO's amplitude to be determined. To obtain the LC, an iteration algorithm was presented as

$$\begin{aligned}&\boldsymbol{\omega}_{k-1}^2\mathbf{M}\mathbf{q}_k''+\boldsymbol{\omega}_{k-1}^2\mathbf{q}_k\\ &=\boldsymbol{\omega}_{k-1}^2\mathbf{q}_{k-1}-\boldsymbol{\omega}_{k-1}\mathbf{C}\mathbf{q}_{k-1}'-\mathbf{K}_Q\mathbf{q}_{k-1}-\mathbf{K}_3(\mathbf{q}_{k-1}):=\boldsymbol{G}(\boldsymbol{\omega}_{k-1},\mathbf{q}_{k-1},\boldsymbol{\omega}_{k-1}\mathbf{q}_{k-1}'),\\ &(\boldsymbol{k}=1,2,\cdots)\end{aligned} \quad (8)$$

according to initial conditions mentioned above, it could be expressed as

$$\mathbf{q}_k(0)=\alpha_k \quad \mathbf{q}_k'(0)=0 \quad (9)$$

There are existing unknown parameters, such as $\omega_{k-1}\mathbf{q}'_{k-1}$ on the right side of the Eq. Since Eq. (8) contains damping terms, ω_{k-1} is coupled to $\mathbf{q}'_{k-1}$. Therefore, ω_{k-1}^2 that arises in linear terms cannot be seen as an independent unknown. For more further simplifying the iteration scheme, the algorithm is modified as

$$\begin{aligned}
&\boldsymbol{\omega}_{k-1}^{2}\mathbf{M}\mathbf{q}_k'' + \boldsymbol{\omega}_{k-1}^{2}\mathbf{q}_k \\
&= \boldsymbol{\omega}_{k-1}^{2}\mathbf{q}_{k-1} - \boldsymbol{\omega}_{k-2}\mathbf{C}\mathbf{q}_{k-1}' - \mathbf{K}_Q\mathbf{q}_{k-1} - \mathbf{K}_3\left(\mathbf{q}_{k-1}\right) := G\left(\boldsymbol{\omega}_{k-1}, \mathbf{q}_{k-1}, \boldsymbol{\omega}_{k-2}\mathbf{q}_{k-1}'\right), \\
&(k = 2, 3, \cdots)
\end{aligned} \tag{10}$$

In the kth iteration stage, ω_{k-2} is a prior known quantity. Then, the items with unknown frequency, i.e., ω_{k-1}^2, can be treated as a dependent parameter. The starting function could be set as

$$\mathbf{q}_0(\tau) = \boldsymbol{\alpha}_0 \cos \boldsymbol{\omega}_0 t \tag{11}$$

If the sequences $\{\boldsymbol{\omega}_k, k = 1, 2, \cdots\}$ and $\{\mathbf{q}_k, k = 1, 2, \cdots\}$ are convergent, the exact solutions should be obtain. In every iteration stage, the right side of Eq. (10) could be expressed in Fourier series

$$G\left(\boldsymbol{\omega}_{k-1}, \mathbf{q}_{k-1}, \boldsymbol{\omega}_{k-2}\mathbf{q}_{k-1}'\right) = \sum_{i=1}^{\phi(k)} \left[c_{k-1,i}\left(\boldsymbol{\omega}_{k-1}^2, \boldsymbol{\alpha}_{k-1}\right)\cos\left(i\boldsymbol{\omega}_{k-1}t\right) + s_{k-1,i}\left(\boldsymbol{\omega}_{k-1}^2, \boldsymbol{\alpha}_{k-1}\right)\sin\left(i\boldsymbol{\omega}_{k-1}t\right)\right] \tag{12}$$

where $\varphi(k)$ is a positive integer, and the coefficients $c_{k-1,i}\left(\boldsymbol{\omega}_{k-1}^2, \boldsymbol{\alpha}_{k-1}\right)$ and $s_{k-1,i}\left(\boldsymbol{\omega}_{k-1}^2, \boldsymbol{\alpha}_{k-1}\right)$ both are functions of $\boldsymbol{\omega}_{k-1}^2$ and $\boldsymbol{\alpha}_{k-1}$. By eliminating the secular terms, the approximation ω_{k-1} and α_{k-1} are obtained at (k-1) th-order.

Setting

$$c_{k-1,1}\left(\boldsymbol{\omega}_{k-1}^2, \boldsymbol{\alpha}_{k-1}\right) = 0, s_{k-1,1}\left(\boldsymbol{\omega}_{k-1}^2, \boldsymbol{\alpha}_{k-1}\right) = 0, k = 1, 2, \cdots \tag{13}$$

Base on the Newton-Raphson method (NRM), these equations could be solved quickly. Different from the existing Mickens procedure, it could be considered ω_{k-1}^2 as an independent unknown parameter in Eq. (13). Then, the original Mickens procedure can be simplified by this iteration scheme.

4. NUMERICAL EXAMPLe

The parameters of the numerical example are set as follow,

$$a = -0.41,\ b = 0.118,\ x_\alpha = 0.15,\ x_\beta = 0.2,\ \bar{L} = 0.18;$$

$$c_h = 0.2,\ c_\alpha = 0.2,\ c_\beta = 0;$$

$$\mu = 12.8,\ \mu_\beta = 4.0;\ r_\alpha^2 = 0.3, r_\beta^2 = 0.89;$$

$$\omega_h = 34.6,\ \omega_\alpha = 88, \omega_\beta = 60.$$

In addition, Q is the varying parameter, and the nonlinear stiffness in each DOF is given by $k_{h3} = k_{\alpha 3} = 0$, $k_{\beta 3} = 10$. For comparison, the numerical solutions are given by the RK method. In Figure 2, Comparing the results with calculating different iterations in the pitch direction by the proposed method. The result provided by 5 iterations, denoted by dashed lines, has a relatively large discrepancy to the numerical one. As the iteration proceeds to 20, the result approaches the numerical ones closely. Since algebraic equations in every iteration stage are always linear, the solution can be attained with 10 harmonics or more. The accuracy can be further improved by taking into more harmonics.

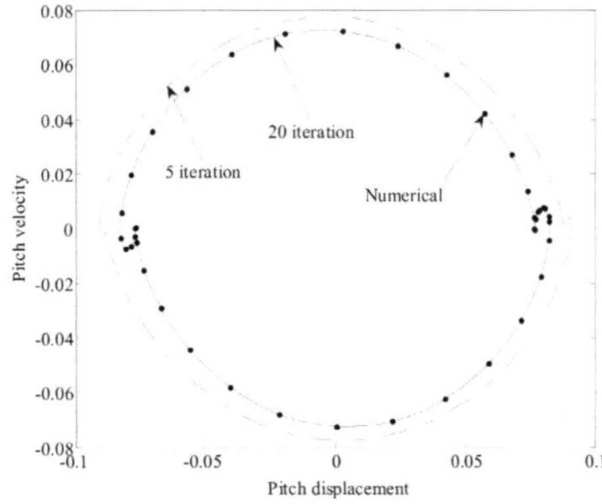

Figure 2. Comparison between the obtained results for the pitch LCO of System (3)

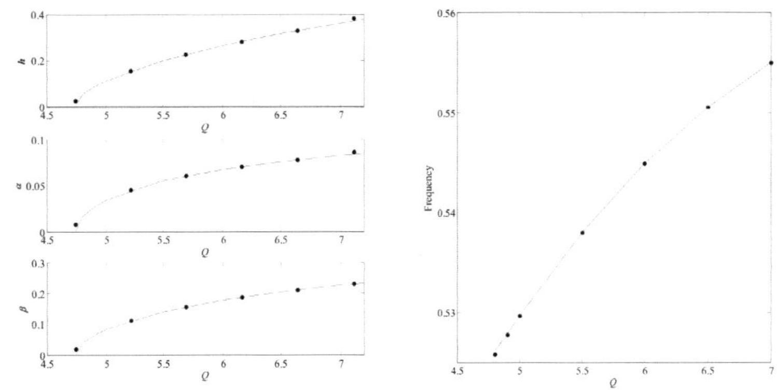

Figure 3. The LCO amplitudes and frequency for System (3) (heavy dots represent the RK solutions; solid lines denote results by the proposed method with 20 iterations.)

H is denoted the maximum value of $h(t)$. A and B represents the ones of $\alpha(t)$ and $\beta(t)$, respectively. They would be considered separately as the amplitude of LCO in the plunge, the pitch, and the external store. According to the classical flutter analysis method, the critical flutter wind speed of the considered system is located at $Q_f = 4.7429$. If the value of Q exceeds this point, the system will lose its stability and a stable LCO arises, as Figure.3 shown. As the value of Q increases, very good agreement of the results by the improved iterative process with the numerical solution can be observed.

The phase planes of the LCOs are shown in Figure 4. These diagrams suggest that the dynamic response of the studied system would be periodic and stable in region [4.7429,7.0]. The amplitude and the shape of LCOs vary with increasing Q. It implies that the motion in the plunge direction remains harmonically, while the ones in the other two DOFs evolve significantly along with the wind speed increases.

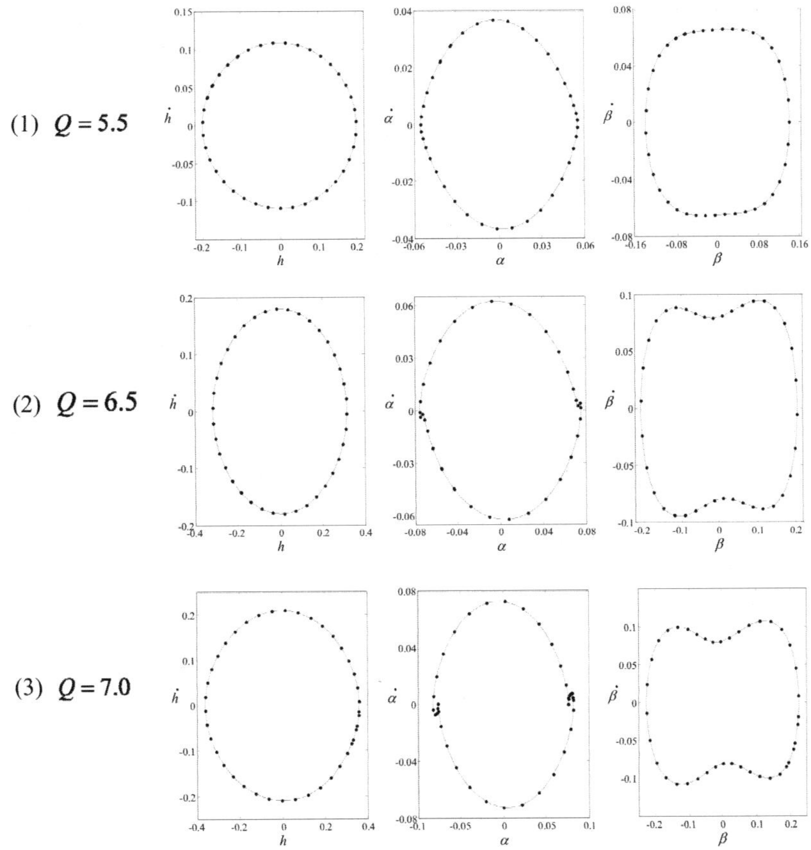

Figure 4. Phase planes of LCOs with three different Q, solid lines represent the results given by the improved process with 20 iterations and dots denote RK solutions.

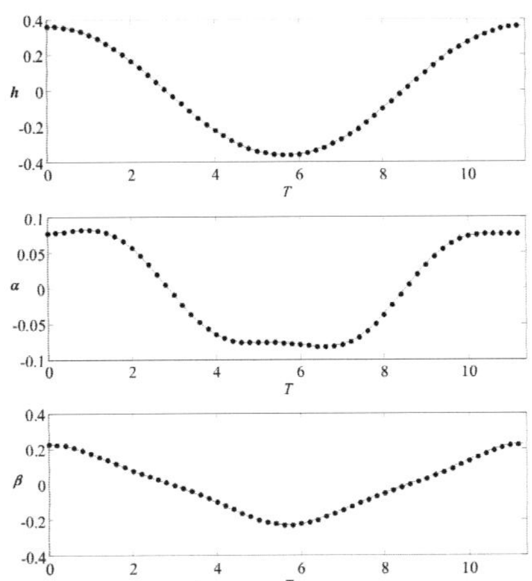

Figure.5. LCO's time histories for one period at Q = 7.0, solid lines represent the results given by the improved method, dots denote RK solutions.

According to Figure 5, the time history in plunging is like harmonic motion, while the responses of other two DOFs are quite different. This situation has also been mentioned in previous literature. One of the possible reasons is there is no nonlinearity existing in the plunge direction.

5. Conclusions

The Mickens iteration method has been improved to analyze the nonlinear airfoil with an external store. Different from the existing procedures, all algebraic equations in every iteration stage preserves linear. Because the unknown frequency is replaced by the previous iteration. Numerical example shows the LCOs given by the improved method agree well with the RK solutions, which proves the effectiveness of the proposed method. With the wind speed varying, LCOs in three directions are analyzed in detail. The high precision and effectiveness imply that the improved iterative process might be applies to some nonlinear dynamical systems with multiple DOFs.

Acknowledgment

This work is supported by Guangdong Education Department project (2022KTSCX069, 2021KTSCX123, 2019KTSCX086), the Guangdong Basic and Applied Basic Research Foundation (2020A1515010836), Guangxi Key Laboratory of Manufacturing System & Advanced Manufacturing Technology (Grant No.19-050-44-S001), Guangzhou Science and technology planning project (202102080184), and Quality Engineering project of Guangdong Polytechnic Normal University (NO. 2020zlgc13).

References

[1] Zhang, L. & Chen F.Q., "Bifurcation behavior and chaotic dynamics of a three-degree-freedom aeroelastic system. " Int. J. Non-Linear Mech. 109, 63-79 (2019).

[2] Zheng, Y.Z. & Qiu Z.P., "An efficient method for flutter stability analysis of aeroelastic systems considering uncertainties in aerodynamic and structural parameters. " Mech. Syst. Signal Proc. 126, 407-426 (2019).

[3] Niu, Y.B., Wang, Z.W., Wang, D.Q. & Liu B., "Modified Homotropy Analysis Method for Nonlinear Aeroelastic Behavior of Two Degree-of-Freedom Airfoils. " Int. J. Struct. Stab. Dyn. 16, 867-878 (2016).

[4] Liu, G., Lv, Z.R., Liu, J.K. & Chen, Y.M., "Quasi-periodic aeroelastic response analysis of an airfoil with the external store by incremental harmonic balance method " Int. J. Non-Linear Mech. 100, 10-19 (2018).

[5] Hu, H., "Solutions of the Duffing-harmonic oscillator by an iteration procedure. " J. Sound Vibr. 298, 446-452(2006).

[6] Marinca, V. & Herisanu, N., "A modified iteration perturbation method for some nonlinear oscillation problems. " Acta Mech. 184, 231-242 (2006).

[7] Mickens, R. E., "Iteration method solutions for conservative and limit-cycle x (1/3) force oscillators. " J. Sound Vibr. 292, 964-968 (2006).

[8] Gholami, A., Ganji, D.D., Rezazadeh, H., Adel, W. & Bekir, A., "Application of modified Mickens iteration procedure to a pendulum and the motion of a mass attached to a stretched elastic wire. " Int. J. Nonlinear Sci. Num. Simul. 21(2021).

Orbit and sequence planning method for space multi-target skimming observation

Yuhao Liu[1,2], Zijin Fang[1] and Bing Zhang[1]

[1] CASIC Space Engineering Development Co., Ltd., Beijing 100854, China;

[2] Graduate School, The Second Research Academy of CASIC, Beijing 100854, China

18851193526@163.com

Abstract. Based on the research of space-based multi-target skimming observation, this paper studies the observation sequence and observation orbit path planning of low Earth orbit (LEO) targets with the goal of fuel optimization. Based on the Lambert intersection algorithm and branch and bound algorithm, a two-layer branch and bound multi-target transfer observation model is established. Through the first layer of this model, the optimal observation sequence and the target location when the fuel is optimal are obtained. To obtain the global optimal solution, the target position is expanded spherically and the optimal skimming observation point of the corresponding target is determined by using the second layer of the model, and then the optimal multi-target observation orbit is found. Finally, the simulation results obtained by using the model are analyzed and compared with the Hohmann transfer speed increment to verify the reliability of the simulation results.

1. Introduction

With the continuous development of space technology, the demand for space observation is becoming increasingly significant. With the increasingly significant problem of space congestion and the continuous upgrading of space confrontation, China has put forward higher requirements for space observation capability. At present, space target detection in China mainly relies on ground-based systems, and there is a large blind area. However, space-based optical observation combines the advantages of space-based platforms and proximity optical imaging and can obtain detailed information about the target when conducting close skimming observation of the target to achieve better observation purposes, which has broad prospects for development [1].

The skimming observation of the observer to the target needs to be realized through orbit change, which will consume the fuel inside the observer. To enable the observer to observe as many targets as possible in a flight mission and improve the observation efficiency, it is particularly important to make the total fuel consumption as small as possible by reasonably planning the order of observation targets and the transition path of the observer. At present, the multi-pulse rendezvous problem is important in rendezvous, and the research on it has been relatively perfect and mature, but the research on the grazing observation of multiple targets is less. In the multi-target skimming observation problem, the determination of the observation sequence is the first problem to be solved, followed by the determination of the optimal skimming observation point and the orbital path planning.

In this paper, taking the multi-target observation problem as the background and the fuel optimization as the goal, a two-layer branch-bound multi-target transfer observation model based on the Lambert intersection algorithm and the branch-bound algorithm is established to determine the observation sequence and transfer orbit. Finally, three LEO satellites are used as observation targets for program simulation to find the optimal observation sequence and orbit transfer scheme. Finally, the simulation results are compared with the Hohmann optimal solution, and the results are analyzed.

2. Problem description

Lambert problem is a basic problem of double pulse orbit transfer in aerospace engineering, which has a very wide application in the field of orbit change and rendezvous observation in aerospace and is still a hot topic of aerospace experts and scholars in recent years. Lambert's problem with optimal fuel is described as follows. As shown in Figure 1, suppose that the position of the spacecraft before the first orbit change is r_1, the speed is v_1, the instantaneous speed after ignition is $v_1^{'}$, and the speed increment is Δv_1, and so on, the position of the spacecraft before the nth orbit change is r_n, the speed is v_n, the instantaneous speed after ignition is $v_n^{'}$, and the speed increment is Δv_n.

At this time, the problem that we set the optimization objective function to $\min \Delta V = \sum_{1}^{n} \Delta v_i$ is called the fuel optimal multi-objective Lambert problem, which is also the most basic problem to be solved in this paper.

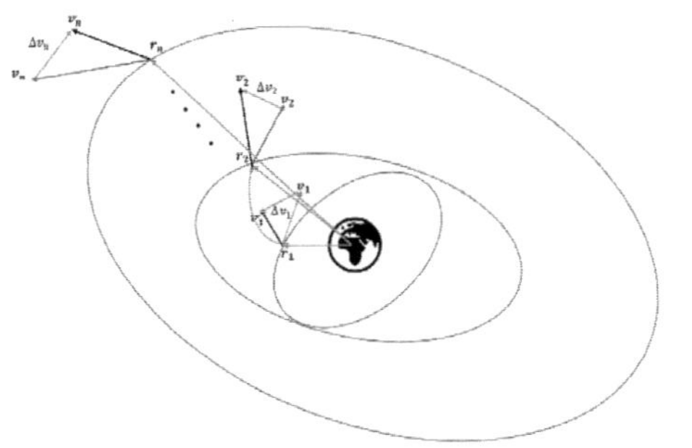

Figure 1. Schematic diagram of multi-target Lambert orbit transfer

3. Model establishment

In this section, a two-layer branch and bound multi-target transfer observation model is established. The following is the model establishment process.

3.1 Model description

The two-layer branch and bound multi-target transfer observation model in this paper is a two-layer nested model based on the Lambert intersection algorithm and branch and bound algorithm. The first layer model: determine the observation sequence. Lambert's rendezvous algorithm, time shooting method, and branch and bound method are used to traverse all possible sequences and paths of observation spacecraft and target spacecraft rendezvous, and determine the optimal fuel rendezvous sequence as the optimal observation sequence. At this time, the position of each spacecraft corresponding to the optimal sequence is fixed, which is determined by the mean anomaly corresponding to the traversed time. The second model: determine the optimal skimming observation point and transfer path. We turn the inertial coordinate system to the orbital coordinate system, take the position of each target satellite at this time as the center of the ball, expand the observation distance to a spherical area, and randomly select points in the spherical area as the skimming observation points of the target spacecraft. Based on these observation points. Finally, we use Lambert's intersection algorithm and branch and bound algorithm again to find the optimal transfer orbit. The model flow chart is as shown in Figure 2:

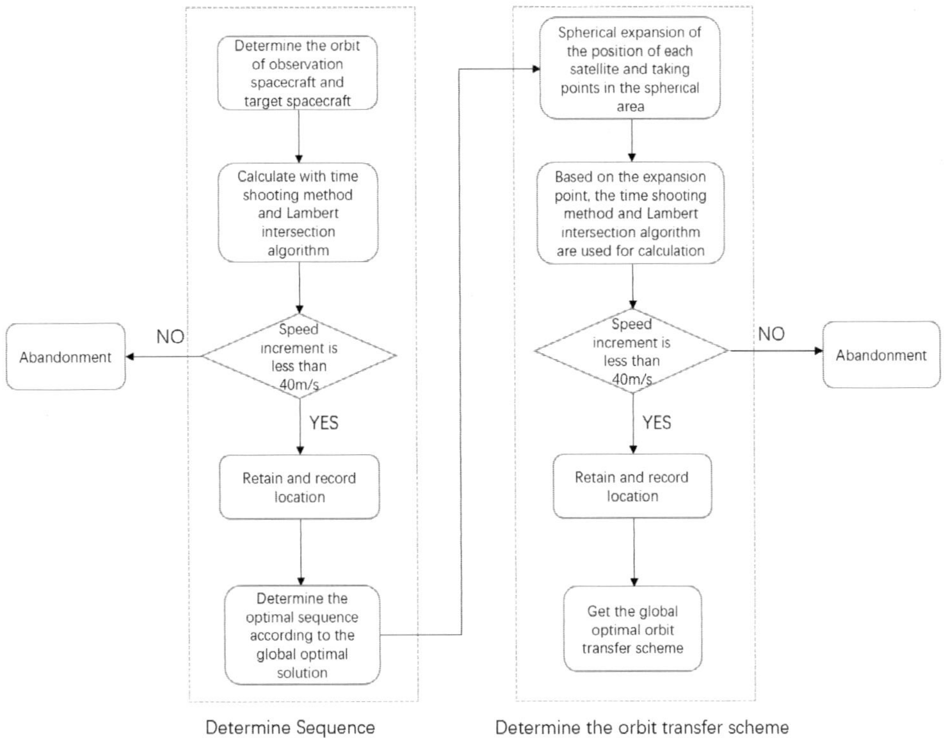

Figure 2. Model flow chart

3.2 Orbit determination based on Lambert

From r_1 and r_2, we can get the difference between the true anomaly of my spacecraft at these two positions:

$$\Delta f = \tan^{-1}\frac{|r_1 - r_2|}{r_1 \cdot r_2} \tag{1}$$

If the length of the connecting line between the two positions is c, then half of the perimeter of the triangle composed of the two positions and the geo-center is S, which meets the following formula [2]:

$$S = \frac{r_1 + r_2 + c}{2} \tag{2}$$

The minimum value of the semimajor axis of the ellipse passing through these two points is:

$$a_{\min} = S/2 \tag{3}$$

The time of spacecraft flying along the minimum energy ellipse is:

$$\Delta t_m = \sqrt{\frac{S^3}{2\mu}}\left[\frac{\pi - (\beta_m - \sin\beta_m)}{2}\right] \tag{4}$$

The calculation formula of β_m is:

$$\beta_m = 2\sin^{-1}\sqrt{K} \tag{5}$$

The calculation formula of K is:

$$K = \frac{r_1 + r_2 - c}{r_1 + r_2 + c} \tag{6}$$

According to different conditions for Δf and Δt_m, the calculation formulas of the semi-major shaft a and semi-diameter p are different.

When $\Delta f \le 180^{\circ}, \Delta t \le \Delta t_m$, a and p satisfy the following calculation formula:

$$\begin{cases} \mu^{1/2}\Delta t = a^{3/2}\left[(\alpha-\sin\alpha)-(\beta-\sin\beta)\right] \\ p = \dfrac{4a(S-r_1)(S-r_2)}{c^2}\sin^2\dfrac{\alpha+\beta}{2} \end{cases} \tag{7}$$

When $\Delta f \le 180^{\circ}, \Delta t \ge \Delta t_m$, a and p satisfy the following calculation formula:

$$\begin{cases} \mu^{1/2}\Delta t = a^{3/2}\left[2\pi-(\alpha-\sin\alpha)-(\beta-\sin\beta)\right] \\ p = \dfrac{4a(S-r_1)(S-r_2)}{c^2}\sin^2\dfrac{\alpha-\beta}{2} \end{cases} \tag{8}$$

When $\Delta f \ge 180^{\circ}, \Delta t \le \Delta t_m$, a and p satisfy the following calculation formula:

$$\begin{cases} \mu^{1/2}\Delta t = a^{3/2}\left[(\alpha-\sin\alpha)+(\beta-\sin\beta)\right] \\ p = \dfrac{4a(S-r_1)(S-r_2)}{c^2}\sin^2\dfrac{\alpha+\beta}{2} \end{cases} \tag{9}$$

When $\Delta f \ge 180^{\circ}, \Delta t \ge \Delta t_m$, a and p satisfy the following calculation formula:

$$\begin{cases} \mu^{1/2}\Delta t = a^{3/2}\left[2\pi-(\alpha-\sin\alpha)+(\beta-\sin\beta)\right] \\ p = \dfrac{4a(S-r_1)(S-r_2)}{c^2}\sin^2\dfrac{\alpha+\beta}{2} \end{cases} \tag{10}$$

The calculation formula of α and β is:

$$\begin{cases} \sin\dfrac{\alpha}{2} = (a_{\min}/a)^{1/2} \\ \sin\dfrac{\beta}{2} = (Ka_{\min}/a)^{1/2} \end{cases} \tag{11}$$

μ is the constant of gravity.

The orbital elements of the spacecraft running orbit can be calculated from the position vectors of two fixed points r_1 and r_2 as well as the semi-major axis a and semi-path are calculated above.

Eccentricity e can be calculated by the following formula:

$$e = \sqrt{1-p/a} \tag{12}$$

From the two position vectors r_1 and r_2 and the true anomaly f_1 and f_2 at the time t_1 and t_2:

$$f_i = \arccos\frac{p-r_i}{r_i e} \tag{13}$$

Then we can calculate the eccentric anomaly and the mean anomaly:

$$\begin{cases} E_1 = 2\tan^{-1}\left(\sqrt{\dfrac{1-e}{1+e}}\tan\dfrac{f_1}{2}\right) \\ M_1 = E_1 - e\sin E_1 \end{cases} \tag{14}$$

Then the time when the spacecraft passes the perigee in orbit can be obtained:

$$t_p = t_1 - \sqrt{\frac{a^3}{\mu}}\left(E_1 - e\sin E_1\right) \tag{15}$$

With this time as the benchmark, we can calculate the true anomaly, the eccentric anomaly, and the mean anomaly at zero time, so that we can get the distance r_0 from the spacecraft at zero time to the earth center by backward deduction from Equation (12).

The unit normal vector of the track surface is:

$$n = \frac{r_1 \times r_2}{|r_1||r_2|} \tag{16}$$

Orbit inclination and right ascension of ascending node can also be obtained:

$$\begin{cases} i = \cos^{-1} n_z \\ \Omega = \tan^{-1} \dfrac{n_x}{-n_y} \end{cases} \tag{17}$$

The velocity inclination and velocity of the spacecraft at zero time are:

$$\begin{cases} \theta_0 = \tan^{-1} \dfrac{e \sin f_0}{1 + e \cos f_0} \\ V_0 = \sqrt{\mu \left(\dfrac{2}{R_0} - \dfrac{1}{a} \right)} \end{cases} \tag{18}$$

At this time, the unit vector of zero-time position r_0 can be obtained by rotating the angel counterclockwise and unitizing, and then the velocity unit vector v_0 at zero time can be obtained:

$$v_0 = \cos\theta_0 (n \times r_0) + \sin\theta_0 r_0 \tag{19}$$

The eccentricity vector $\vec{e}$ is:

$$\vec{e} = \frac{R_0 V_0^2 \cos\theta_0}{\mu} (v_0 \times n) - r_0 \tag{20}$$

Perigee argument w is:

$$w = \tan^{-1} \frac{e_z}{\left(e_y \sin\Omega + e_x \cos\Omega\right) \sin i} \tag{21}$$

From the above formula, we can get six basic parameters of the orbit.

3.3 Find the speed increment

It is required to solve a multi-objective fuel optimization problem and ensure that the total velocity increment after multiple orbital transfers is minimum. Therefore, to obtain the above optimal results, it is necessary to solve the velocity increment at each step. When the spacecraft is running in the original orbit, it is assumed that it will select a speed pulse at the position on the orbit to conduct orbit change. The six orbits of the original orbit are semimajor axis a_0 , eccentricity e_0 , orbit inclination i_0 , right ascension of ascending node Ω_0 , perigee argument w_0 , and true anomaly f .

The following identity matrix can be obtained:

$$P = \begin{bmatrix} \cos w_0 \cos\Omega_0 - \sin w_0 \sin\Omega_0 \cos i_0 \\ \cos w_0 \sin\Omega_0 + \sin w_0 \cos\Omega_0 \cos i_0 \\ \sin w_0 \sin i_0 \end{bmatrix} \tag{22}$$

$$Q = \begin{bmatrix} -\sin w_0 \cos\Omega_0 - \cos w_0 \sin\Omega_0 \cos i_0 \\ -\sin w_0 \sin\Omega_0 + \cos w_0 \cos\Omega_0 \cos i_0 \\ \cos w_0 \sin i_0 \end{bmatrix} \tag{23}$$

From the above Equations (22) and (23), it can be obtained that the speed v_1 of spacecraft before orbit change is:

$$v_1 = \sqrt{\frac{\mu}{a_0(1-e_0^2)}} [-\sin f \cdot P + (e_0 + \cos f) \cdot Q] \tag{24}$$

Formula (25) can be simplified by Formula (24):

$$v_1 = \sqrt{\frac{\mu}{a_0(1-e_0^2)}} \cdot \sqrt{\sin^2 f + (e_0 + \cos f)^2} \tag{25}$$

Formula (25) is further simplified:

$$v_1 = \sqrt{\mu \left(\frac{2}{r_1} - \frac{1}{a_0}\right)} \tag{26}$$

The speed v_2 after orbit change can be obtained from the orbital elements determined by the Lambert algorithm in 3.2:

$$v_2 = \sqrt{\mu(\frac{2}{r_1} - \frac{1}{a})} \quad (27)$$

Speed pulse increment in one orbit change can be obtained:

$$\Delta v = v_2 - v_1 \quad (28)$$

3.4 Branch and bound method

The basic idea of the branch and bound method is to search all feasible solution spaces of the planning problem, divide the solution space into small branches, set a range for each small branch, continue to search for feasible solutions for the remaining branches, and repeat the above process. The whole process of branch delimitation can be divided into three steps: branching, relaxation, and delimitation [3].

Branch: decompose the overall multi-objective transfer problem into a single objective transfer problem.

Relaxation: After relaxing some conditions of the problem, we can get the relaxation problem of the problem, and it will be relatively easy to solve the relaxation problem. To find the transfer scheme with the minimum total speed increment after multiple transfers, the speed increment limit is appropriately enlarged during each transfer.

Bounding: After the relaxation problem of the subproblem is obtained, the relaxation problem should be bounded and solved. In the first layer of the model, we need to limit the speed increment of the orbit transfer. In the second layer of the model, we need to limit the transfer speed increment and also limit the transfer time. The speed increment should be within 40m/s, and the transfer time should correspond to the transfer time of the first layer of the model[4].

Compared with the general enumeration method, the branch and bound method can greatly reduce the complexity of the operation through layer-by-layer screening and comparison, and can also obtain very accurate results. It has obvious advantages in terms of the simplification of the operation and the accuracy of the results.

3.5 Coordinate System Transformation and Spherical Coordinate Extension

As shown in Figure 3, the geocentric inertial coordinate system $(O-XYZ)$ is defined as the origin O is located in the center of the earth, the axis pointing to the vernal equinox, and the axis Z pointing to the north pole of the earth. The axis X, the axis Y, and the axis Z form a right-handed coordinate system. The orbital coordinate system $(O_0 - X_0Y_0Z_0)$ is defined as: the origin O_0 is located at the satellite centroid, the axis Z_0 is collinear with the vector diameter of the satellite and points to the earth center, and the axis X_0 is located in the orbital plane, which is the same as the direction of satellite motion. And the axis Y_0, axis X_0 and axis Z_0 also form a right-handed coordinate system [5].

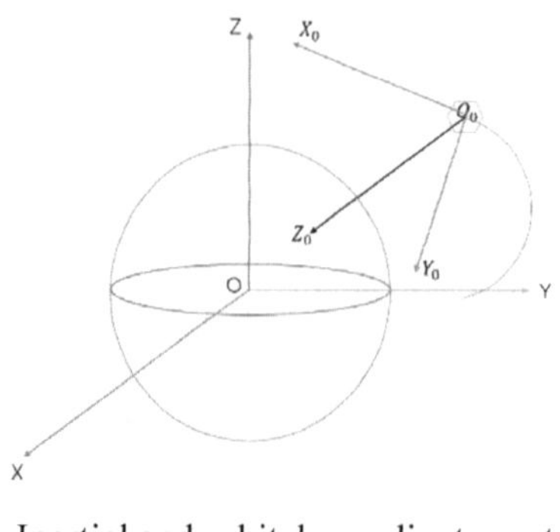

Figure 3. Inertial and orbital coordinate systems

In this paper, the indirect transformation method is used for coordinate transformation, and the attitude transformation matrix from the inertial coordinate system to the orbital coordinate system is calculated according to the satellite orbital elements, to realize the transformation of the coordinate system.

The mathematical expression of this method is as follows:

$$r_0 = L_{001} L_Z(u) L_X(i) L_Z(\Omega) r \tag{29}$$

L_X and L_Z respectively represent the rotation matrix around the axis and axis Z . u , i and Ω respectively represent the latitude argument, orbital inclination, and right ascension of the ascending node in the satellite orbital parameters, which are known in advance. L_{001} is the inverse matrix of the coordinate axis:

$$L_{001} = \begin{bmatrix} 0 & 1 & 0 \\ 0 & 0 & -1 \\ -1 & 0 & 0 \end{bmatrix} \tag{30}$$

With this method, the inertial system can be converted to the orbital system, which is ready for the spherical coordinate expansion with the satellite position as the center[6].

In the first layer of the model, we have obtained the best transfer order and the best satellite transfer position. After determining the specific position of the satellite and carrying out the coordinate transformation, in the sub-layer model, we need to take the satellite as the center of the sphere and the satellite's observation distance as the radius to establish a spherical region, and randomly select the dry points in the spherical region as the satellite skimming observation points. Based on these skimming observation points, we use the branch and bound method again and establish the best path for point-to-point Lambert transfer[7]. The schematic diagram is shown in Figure 4.

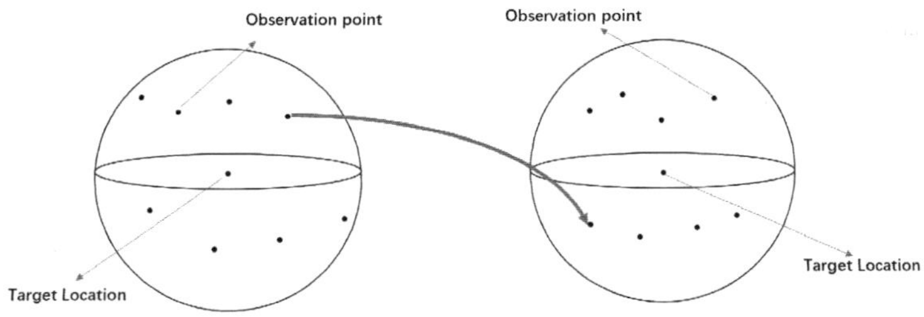

Figure 4. Point to point lambert transfer

4. Simulation Experiment and Analysis

4.1 Simulation conditions

The background of the mission is that three LEO satellites are flying in orbit, and the observer itself is flying in its orbit. The observer shall successively skim and observe the three LEO satellites through orbit change, with a maximum observation distance of 50km. An optimal observation sequence and an optimal orbit transfer path shall be determined to minimize the total speed increment. Their orbit parameters are shown in Table 1.

Table 1. Simulation conditions

Satellite name	Orbit height	Eccentricity	Orbit inclination	RAAN	Argument of perigee	True anomaly
Observer	839.5km	0.0010622	98.9095°	63.0039°	204.8205°	155.2458°
Satellite1 1	847.5km	0.0011135	98.7045°	355.2968°	167.7952°	192.3492°
Satellite2 22	845.5km	0.0011191	98.7469°	21.6404°	135.7557°	224.4513°
Satellite3 3	800.05km	0.0006727	98.3848°	229.7108°	153.3593°	206.7935°

4.2 Simulation results and analysis

Under the above simulation conditions, there are six cases in the access sequence. The optimal transfer speed increment corresponding to these six cases is calculated as shown in Table 2.

Table 2. Optimal velocity increment of six sequences

Access Sequence	**Optimal speed increment (m/s)**
Satellite 1, Satellite 2, Satellite 3	63.4863
Satellite 2, Satellite 1, Satellite 3	59.5168
Satellite 1, Satellite 3, Satellite 2	68.3541
Satellite 2, Satellite 3, Satellite 1	62.8143
Satellite 3, Satellite 1, Satellite 2	63.6193
Satellite 3, Satellite 2, Satellite 1	61.5307

According to the results above, the optimal transfer sequence is Satellite 2, Satellite 1, and Satellite 3.

At this time, record the target position corresponding to the optimal transfer orbit. Randomly select several points in the area within 50km of the center of the sphere where the observation target is located as the skimming observation points. After that, carry out point-to-point Lambert orbit transfer. Use the branch and bound method to find the optimal skimming observation Lambert transfer track. After screening and comparison, the inertial coordinates of the optimal observation point of each target and the corresponding velocity increment are shown in Table 3. The optimal transfer velocity increment is 72.9362m/s. Currently, the corresponding trajectory is the optimal Lambert transfer trajectory. From the analysis of simulation results, when conducting a transfer simulation between two orbits, the solution obtained should be obtained during coplanar orbit transfer, which is similar to Hohmann orbit transfer. The model calculates that the optimal transfer speed increment after three orbital changes is 59.5168m/s, and the average orbital change speed increment is about 20m/s, which is slightly higher than the Hohmann optimal solution, which is in line with the expected results.

The best observation points are shown in Table 3. When selecting the skimming observation point of the target, it is easier to find the optimal observation point by selecting the observation point on the cross-section of the transfer track plane and the target spherical region. After the observation point is selected, the trajectory is optimized, and the optimal velocity increment is 72.9362m/s, which is slightly larger than the original 59.5168m/s. The analysis result may be that during the transfer between two points, the observation points taken are relatively random, and some points are not taken to the observation track expansion plane, resulting in a larger velocity increment. In addition, to ensure that each target is at the originally determined optimal position when observing multiple targets, and to ensure that the observer can reach the target at the corresponding position for skimming observation at each point-to-point transfer, it is necessary to limit the transfer time, which is also the reason why the obtained solution is larger than the Hohmann solution.

Table 3. Optimum observation point and maximum velocity increment

sequence	starting point	Observation points of satellite 2	Observation points of satellite 1	Observation points of satellite 3	Optimal speed increment
Satellite 2 Satellite 1 Satellite 3	[2701818.7092, 6184974.1169, -2552361.8648]	[-814754.15309, 856028.32790, -7124414.6563]	[-673798.62929, 1145511.0065, -7095969.5068]	[-2900514.5954, -4508681.6448, -4769711.2738]	72.9362

5. Conclusion

Based on the research of space-based multi-target observation, this paper studies the observation sequence and orbit path planning of low Earth orbit (LEO) targets with the goal of fuel optimization. Based on Lambert's intersection algorithm and branch and bound algorithm, a two-layer branch, and bound multi-target transfer observation model is established, and the optimal observation sequence and optimal observation trajectory of the proposed observation target are determined. When the spacecraft is transferring between two orbits on the first layer of the model, there are countless possible situations, but only the finite region on the target orbit can meet the solution close to the Hohmann optimal solution, so the key to finding the optimal solution is to find the location of the finite region of the target orbit, generally near the intersection of the observer's initial orbit plane and the target orbit. When the optimal skimming observation point near the target is determined in the second level of the model, and the skimming observation point is selected on the plane of the initial orbit of the observer, the velocity increment obtained will be smaller, and it is easier to find the optimal solution.

The research on the optimal fuel of space-based space multi-target observation in this paper can provide some reference and theoretical basis for future space multi-target observation tasks.

References

[1] Cui, H. Z., Liu, W. L., Tang, G. S. and Han, C. (2015) Mission design and orbit determination of space-based surveillance system.

[2] Zhang, J., Wang, W., and Xi, X. N. (2008) Orbit design of approaching drifty walker constellation. *Journal of Astronautics*.

[3] Sergienko, I. V., Iemets, O. O. and Yemets, O. O. (2013) Optimization problems with interval uncertainty: branch and bound method. *Cybernetics & Systems Analysis*.

[4] Yang, X. R., Liang, J. H.(2010) Research on optimal dual-impulse rendezvous incorporating ground observations. *Journal of Spacecraft TT&C Technology*.

[5] Yang, L., Yi, D., Wang Zhengming|. (2007) Coordinate transformation methods from the inertial system to the centroid orbit system. *Aerospace Control*, 25(2), 4-8.

[6] Chai, R., Savvaris, A., Tsourdos, A., Chai, S. and Xia, Y. (2019) A review of optimization techniques in spacecraft flight trajectory design. *Progress in Aerospace Sciences*, 109(Aug.), 100543.1-100543.15.

[7] Barea, A., Urrutxua, H., and Cadarso, L. (2020) Large-scale object selection and trajectory planning for multi-target space debris removal missions. *Acta Astronautical*, 170.

Research on Travelling Gust Field Interpolation for Dynamic Response of the Aircraft

Zhiwei Jing[1], Libo Wang[1] and Chu Tang[2*]

[1] The First Aircraft Institute, Aviation Industry Corporation of China, Ltd., Xi'an, Shaanxi, 710089, China

[2] Unmanned System Research Institute, Northwestern Polytechnical University, Xi'an, Shaanxi, 710072, China

[*]Corresponding author's e-mail: tc@nwpu.edu.cn

Abstract. The traveling gust by the explosion has a strong influence on the aircraft's dynamic response. Previous research often takes the hypothesis that the aircraft is 'frozen' at the initial location and focuses on the time interpolation of the traveling gust field. This paper investigates the influence of two traveling gust filed interpolation methods on the aircraft dynamic response. The first one is the 'frozen field' interpolation method, and the other one is the 'spatial field' interpolation method which considers the time and aircraft location. Simulation results demonstrate that the 'frozen field' interpolation method increases the dynamic response of the aircraft.

1. Introduction

The explosion will produce many damage effects including overpressure and the traveling gust. Previous research has shown that the overpressure mainly damages local structures, while the traveling gust causes a significant dynamic response of the whole aircraft. Since this study focuses on the aircraft dynamic response, the effect of the traveling gust on the aircraft will be investigated.

Compared with the ordinary gust, the traveling gust by an explosion is an interesting excitation and propagates at high speed, thus the traditional aerodynamics calculation method of the aircraft in the gust cannot meet the new requirement. In the weak field, previous research has proven that the explosion wave can be treated as a traveling gust that propagates at sonic speed [1]. Then, the influence of the traveling gust on the aircraft is treated as the downwash, based on which quasi-steady and unsteady aerodynamics modeling methods have been established [2-6]. These aerodynamics modeling methods have considered the effect of the traveling gust enveloping process on the aircraft.

When the traveling gust approaches the aircraft from the tail or one wingtip, the dynamic response of the aircraft has been investigated [6-7]. Simulation results show that the traveling gust enveloping process could produce a significant effect on the aircraft's dynamic response. However, these simulation results are on the hypothesis that the aircraft is 'frozen' to the initial location and the change of its flight location is neglected.

The objective of this paper is to investigate the influence of the traveling gust filed interpolation method on the aircraft dynamic response. Two interpolation methods of the traveling gust field are presented and studied, one is called 'frozen field' which has been used in reference [6-7], and the other one is called 'spatial field' which intends to take into account the aircraft flight location in the explosion field. It should be emphasized that the traveling gust enveloping process will be ignored in this research.

A civil aircraft is selected to carry out the research on traveling gust field interpolation for aircraft dynamic response. The simulation results reveal that the 'frozen field' interpolation method increases the aircraft's dynamic response in the tail-on gust, compared to the dynamic response by the 'spatial field' interpolation method.

2. Traveling Gust Field Interpolation

2.1. Traveling gust field

In a numerical simulation of an explosion field, the large-scale explosion simulation software can be used to calculate the explosion field parameters when the explosive yield and height are given. The simulation results include spatial overpressure field, traveling gust field, and other parameters at different moments. At a certain moment, the spatial overpressure contour is shown in Figure 1 (the spatial coordinates have been transformed by dimensionless treatment). The overpressure field and traveling gust field propagate at supersonic speed along with the wavefront, and the parameters at the wavefront show a peak state.

Previous studies and flight tests have revealed that overpressure mainly affects the dynamic response characteristics of the local structure such as hatch and control surfaces, while the traveling gust affects the overall response of the aircraft. Therefore, this research focuses on the effect of traveling gust on aircraft dynamic response.

The explosion field parameters at different locations will change with time, as shown in Figure 2. It can be seen that as the distance to the explosion center increases, the field parameters decay rapidly. If the observation point is fixed at a certain location, the gust velocity is zero before the traveling gust propagates to the observation point. When the traveling gust reaches the observation point, the gust velocity will show a changing trend as shown in Figure 3, in which the traveling gust is decomposed into a vertical component U_z and a horizontal component U_x. Figure 3 depicts that the traveling gust velocity grows rapidly to the peak, and then decreases a little slowly. Subsequently, there will also be a reverse phenomenon of the gust velocity.

To ensure the flight safety of the aircraft, it is generally required that the aircraft fly away from the explosion center, which is shown in Figure 4. Then the traveling gust will chase the aircraft at supersonic speed, and the gust is called a tail-on traveling gust. When the aircraft is far from the explosion center, the chasing speed can be regarded as the speed of sound. Furthermore, it can be considered that the traveling gust field is a two-dimensional field with the horizontal traveling gust U_x and the vertical traveling gust U_z.

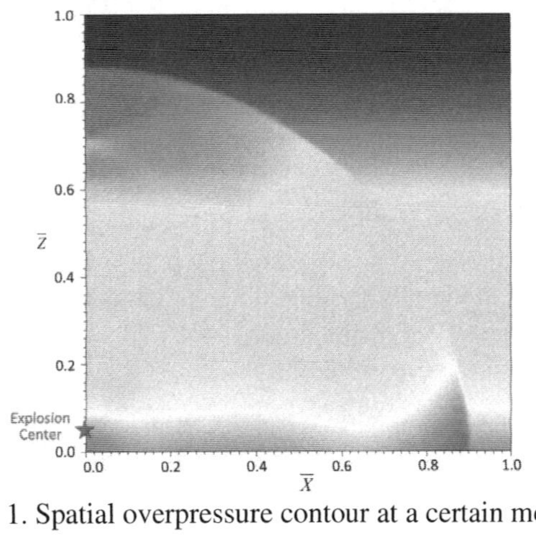

Figure 1. Spatial overpressure contour at a certain moment.

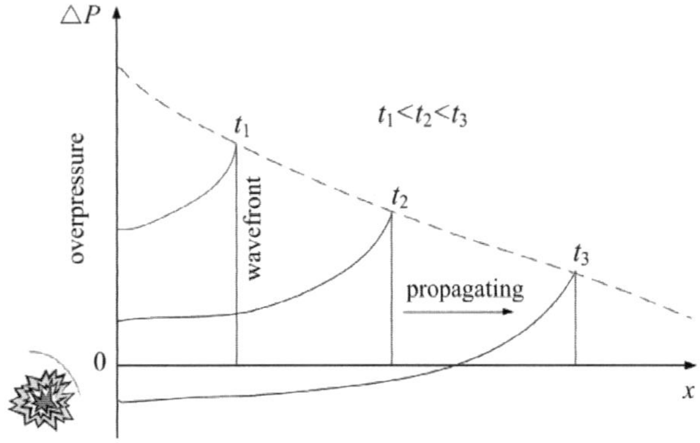

Figure 2. Overpressure curve illustration at different locations.

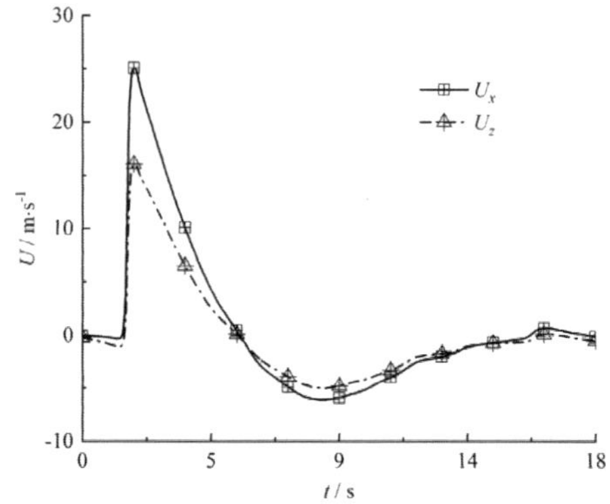

Figure 3. Velocity profiles of the traveling gust (one certain observation location).

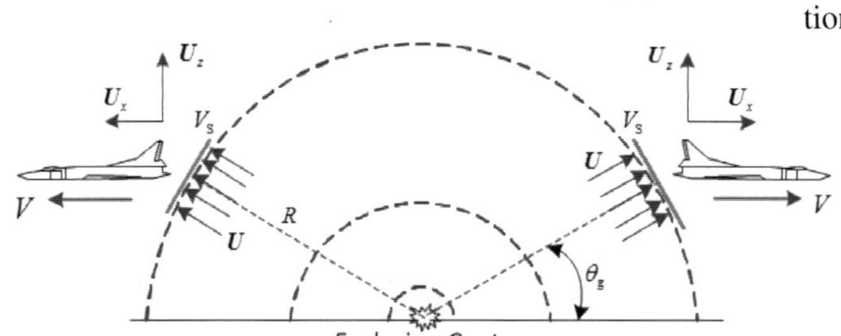

Figure 4. Illustration of the aircraft exposed to the tail-on traveling gust.

2.2. Frozen field Interpolation

The essence of the frozen field interpolation method is to 'freeze' the aircraft at the initial location by ignoring the aircraft flying away from the explosion center. No matter how the height and horizontal distance to the explosion center change, the traveling gust field at time t is interpolated by Equation (1).

$$\begin{bmatrix} U_x(t) \\ U_z(t) \end{bmatrix} = \begin{bmatrix} U_{x0}(H_0 \quad R_0 \quad t) \\ U_{z0}(H_0 \quad R_0 \quad t) \end{bmatrix} \tag{1}$$

Where H_0 is the initial flight altitude, R_0 is the initial horizontal distance to the explosion center, and U_{x0} and U_{z0} are the horizontal and vertical components of the original traveling gust velocity.

As a matter of fact, as the aircraft flies farther and farther from the explosion center, the traveling gust speed reduces gradually. Therefore, the frozen field interpolation method can produce a conservative speed for the traveling gust, and the simulation results of the dynamic response of the aircraft could also be conservative.

2.3. Spatial field Interpolation

The spatial field interpolation method not only considers the change of time, but also synthesizes the spatial position change of the aircraft, so this method is more accurate than the frozen field interpolation method. However, the spatial field interpolation method will increase the simulation cost due to the fact that there are three variables for interpolation.

The traveling gust field at time t is interpolated by employing the spatial field interpolation method as Equation (2).

$$\begin{bmatrix} U_x(t) \\ U_z(t) \end{bmatrix} = \begin{bmatrix} U_{x0}(H(t) \quad R(t) \quad t) \\ U_{z0}(H(t) \quad R(t) \quad t) \end{bmatrix} \tag{2}$$

Where $H(t)$ is the aircraft's flight altitude at time t, and $R(t)$ is the aircraft's horizontal distance to the explosion center at time t.

3. Flight Dynamics

3.1. Dynamics equations

The flight dynamics equations come from reference [8]. Since this research focuses on the longitudinal response of the aircraft in the tail-on traveling gust, the unsymmetrical degrees of freedom are neglected.

The force equations are defined as Equation (3), where q is the pitching angular velocity, θ is the angle of pitch, u and w are the body-axes components of the aircraft velocity vector, $F_{x\mathrm{A}}$ and $F_{z\mathrm{A}}$ are the components of the aerodynamic forces, T is the engine thrust and φ is the installation angle of the engine, m is the aircraft mass, g is the gravity acceleration.

$$\begin{cases} \dfrac{du}{dt} = -qw - g\sin\theta + \dfrac{T\cos\varphi + F_{x\mathrm{A}}}{m} \\ \dfrac{dw}{dt} = qu + g\cos\theta + \dfrac{T\sin\varphi + F_{z\mathrm{A}}}{m} \end{cases} \tag{3}$$

The pitching moment equation of the aircraft is simplified as Equation (4), where I_y is the aircraft pitching inertia, $M_{y\mathrm{A}}$ is the aerodynamic pitching moment, and $M_{q\mathrm{A}}$ is the aerodynamic damping moment.

$$\frac{dq}{dt} = \frac{M_{y\mathrm{A}} + M_{q\mathrm{A}}}{I_y} \tag{4}$$

One more kinematic equation is required to determine the pitching angle by Equation (5).

$$\frac{d\theta}{dt} = q \tag{5}$$

The navigation equations are required to determine the aircraft flight altitude and the horizontal distance to the explosion center shown in Equation (6).

$$\begin{cases} \dfrac{dR}{dt} = u\cos\theta + w\sin\theta \\ \dfrac{dH}{dt} = -u\sin\theta + w\cos\theta \end{cases} \tag{6}$$

To calculate $F_{x\mathrm{A}}$, $F_{z\mathrm{A}}$ and $M_{y\mathrm{A}}$, the traveling gust should be taken into account. It often treats the traveling gust as an overall downwash effect upon the aircraft. With considering traveling gust velocity, the corrected velocity vector of the aircraft can be updated by Equation (7).

$$\begin{bmatrix} u' \\ w' \end{bmatrix} = \begin{bmatrix} u \\ w \end{bmatrix} + \begin{bmatrix} -\cos\theta & -\sin\theta \\ -\sin\theta & \cos\theta \end{bmatrix} \begin{bmatrix} U_x \\ U_z \end{bmatrix} \tag{7}$$

Then the flight dynamic pressure of the aircraft is calculated by Equation (8), where ρ is the air density.

$$\bar{q} = \frac{1}{2}\rho\left[(u')^2 + (w')^2\right] \tag{8}$$

The angle of attack of the aircraft is determined by

$$\alpha = atan\left[\frac{w'}{u'}\right] \tag{9}$$

It should be emphasized that the enveloping process of the traveling gust is neglected, and this research focuses on the traveling gust field interpolation methods. Therefore, the aerodynamic characteristics can be interpolated by Equation (10).

$$\begin{cases} C_L = C_L(M,\alpha) + dC_L(M,\alpha,\delta e) \\ C_D = C_D(M,\alpha) + dC_D(M,\alpha,\delta e) \\ C_m = C_m(M,\alpha) + dC_m(M,\alpha,\delta e) \end{cases} \tag{10}$$

Where M is the flight Mach number, δe is the trimmed deflection of the elevator, dC_L, dC_D, and dC_m are induced by the elevator deflection.

Then it is routine work to calculate F_{xA}, F_{zA}, and M_{yA} by using C_L, C_D, and C_m.

3.2. Simulation flowchart

The dynamic response calculation process of the aircraft in the traveling gust is demonstrated in Figure 5. Basically, there is no big difference between this process and the traditional flight dynamics simulation process. The nonlinear ordinary differential equations, as presented in Equations (3) - (6), are solved by employing the fourth-order Runge-Kutta method.

The key study is how to interpolate the traveling gust velocity from the spatial field data, which is depicted in the dashed box of Figure 5. When using the spatial field Interpolation, the calculation cost will be considerable, and the simulation often takes a relatively long time.

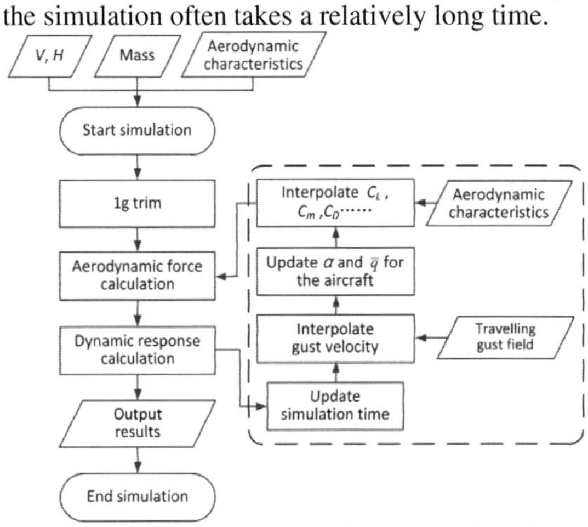

Figure 5. Dynamic response simulation flowchart.

4. Example and Results

4.1. Simulation parameters

The example aircraft is shown in Figure 6, which is a large civil transport aircraft with two engines. In the following simulation, the flight velocity M is 0.6, flight height H is 3000 m, and the initial horizontal distance to the explosion center is 7000 m.

To investigate the influence of the traveling gust field interpolation on aircraft dynamic response, two interpolation methods of the traveling gust field will be conducted,

1) The aircraft is 'frozen' to the initial location and the flight location is neglected in determining the traveling gust velocity, which is presented in chapter 2.2 and denoted with 'Frozen Field'.

2) The flight location of the aircraft is considered in determining the traveling gust velocity, which is presented in chapter 2.3 and denoted with 'Spatial Field'.

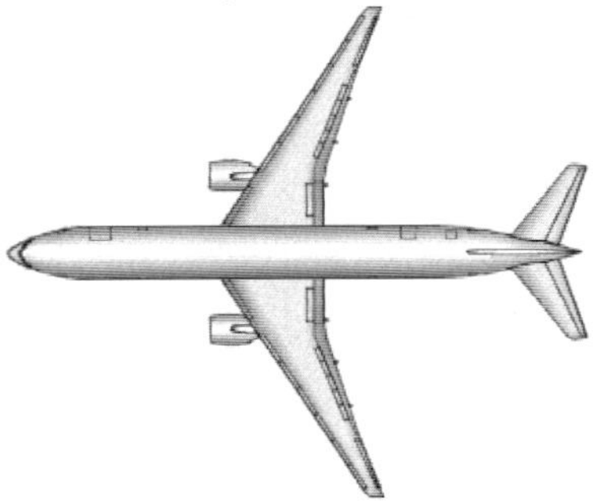

Figure 6. Illustration of the example aircraft.

4.2. Simulation results

The typical simulation results are shown in Figures 7 - 12 which present the comparison between the curves by the spatial field interpolation method and the curves by the frozen field interpolation method.

Overall, the simulation results based on the spatial traveling gust field interpolation method are smaller than the simulation results based on the frozen traveling gust field interpolation method. Taking the maximum normal load factor shown in Figure 12 as an example, the result by the frozen field interpolation method is 2.08 g, while the result by the spatial field interpolation method is 1.85 g which is reduced by about 11.1%. Moreover, compared with the simulation results denoted with 'frozen field', the peak values of the simulation results denoted with 'spatial field' appear a little later. Therefore, using the frozen traveling gust field interpolation method will produce little more conservative results which can ensure the structure safety in the engineering practice.

The main reason for the decrease of the simulation results by the spatial traveling gust field interpolation is that as the aircraft flies away from the explosion center and the flight distance increases, the speed of the spatial traveling gust decreases. Specifically, as shown in Figure 13, the frozen traveling gust field interpolation method produces the green curve denoted with 'U_z_R7000', while the spatial traveling gust field interpolation method gives the red curve denoted with 'U_z_Sim' which is interpolated from the basic explosion field related to time and location (as shown by the blue curve). When considering the flight location of the aircraft, it can be seen from the green curve and red curve that the gust speed profile by 'spatial field' has a smaller peak value and arrives at the peak value a little later.

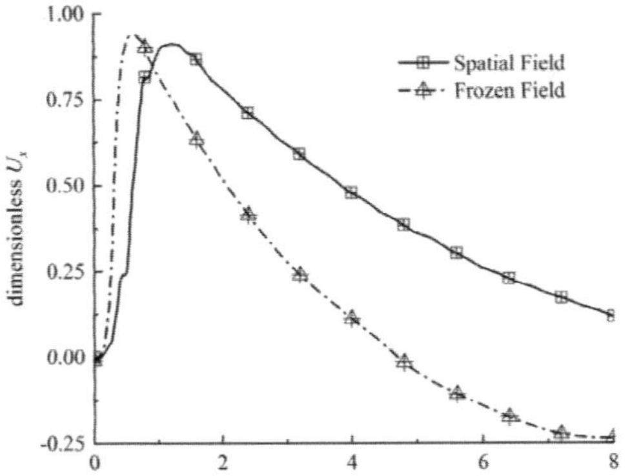

Figure 7. Dimensionless horizontal velocity profile of the traveling gust.

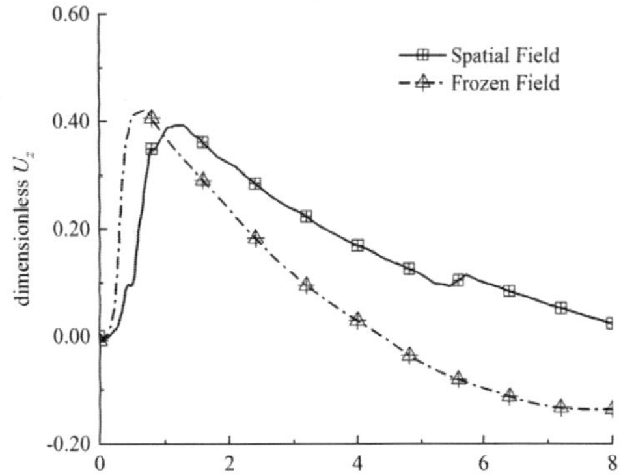

Figure 8. Dimensionless vertical velocity profile of the traveling gust.

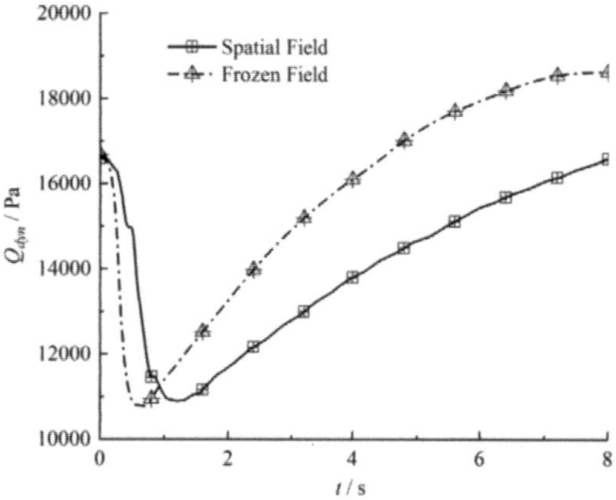

Figure 9. Time history of the flight dynamic pressure.

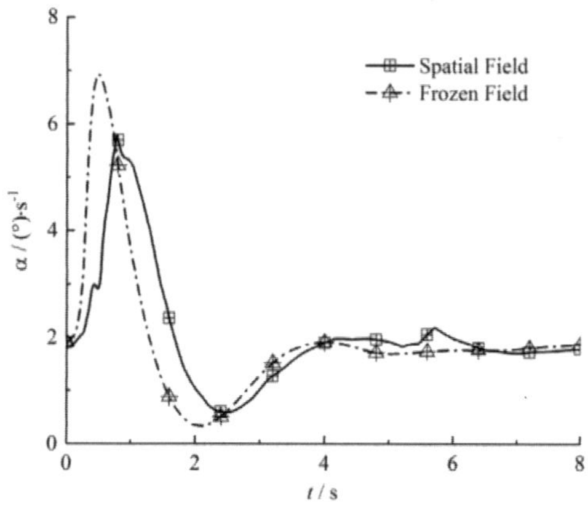

Figure 10. Time history of the angle of attack.

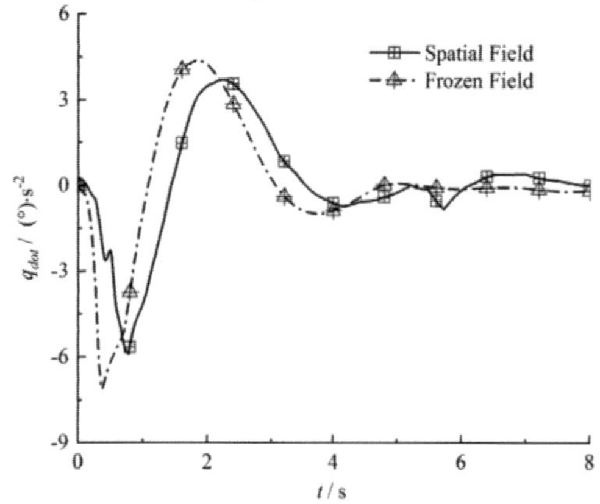

Figure 11. Time history of the pitching angular acceleration.

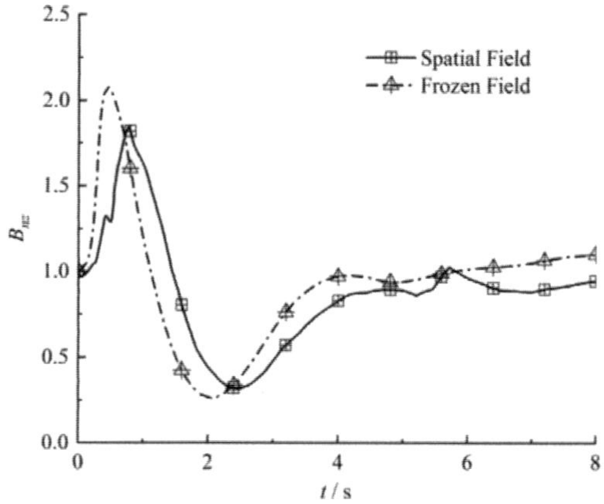

Figure 12. Time history of the normal load factor.

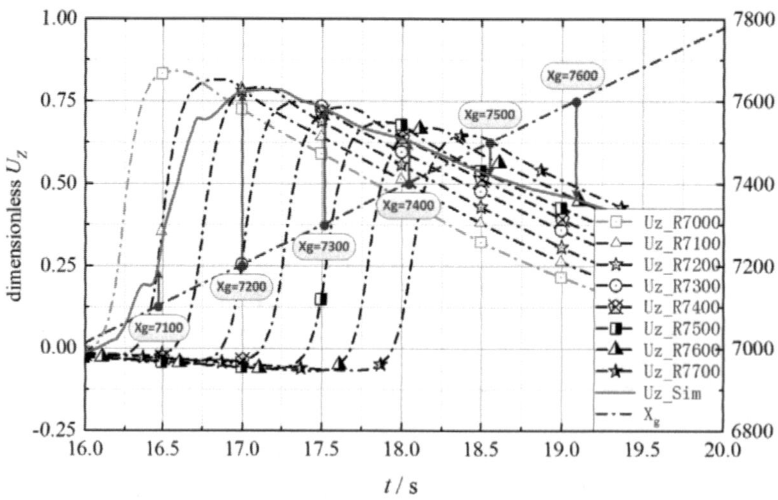

Figure 13. Determination of the spatial Uz due to the simulation time and aircraft location.

5. Conclusions

To investigate the influence of the traveling gust field interpolation method on aircraft dynamic response, two interpolation methods are presented, one is the 'frozen field' method and the other one is the 'spatial field' method.

By ignoring the enveloping process of the traveling gust, the simulation of the aircraft's dynamic response is carried out. The comparison between simulation results reveals that the 'frozen field' interpolation method increases the dynamic response, and the curve denoted with 'frozen field' arrives at its peak much earlier than that denoted with 'spatial field'.

References

[1] Goodman T R, Sargent T P. Some Evidence Justifying the Representation of Blasts as Traveling Gusts, Journal of the Aero/Space Sciences, 9 (1959): 608-608.
[2] Drischler J A, Diederich F W. Lift and Moment Response to Penetration of Sharp-Edged Traveling Gust with Application to Penetration of Weak Blast Waves, NACA TN 3956, 1957.
[3] Wen G B, Sun Z S. Numerical Computation of Unsteady Subsonic Aerodynamic Forces on Wing Exposed to Traveling Gust, Acta Aerodynamical Sinical, 1(1981): 44-52 (in Chinese).
[4] Wen G B, Sun Z S. Numerical Computation of Unsteady Subsonic Aerodynamic Forces on Wing-Body-Tail Exposed to Traveling Gust, Acta Aeronautical et Astronautical Sinical, 3(1981): 23-30 (in Chinese).
[5] Jing Z, Tang C. Modelling Technique of the Aircraft Unsteady Aerodynamics Due to the Traveling Gust, International Journal of Plant Engineering and Management, 2(2022): 86-102.
[6] Jing Z, Tang C. Dynamic Response Modelling and Simulation of Aircraft in the Traveling Gust, Flight Dynamics, 3 (2021): 8-13 (in Chinese).
[7] Jing Z, Tang C. Dynamic Response Analysis of the Aircraft Exposed to the Lateral Traveling Gust, Journal of Physics: Conference Series 2364 0120243.
[8] Stevens B L, Lewis F L. Aircraft Control and Simulation (2nd Edition), John Wiley & Sons, Inc., New Jersey, 2003, pp. 107-115.

Analysis and application of deep space exploration satellite telemetry data combined with empirical mode decomposition and ARIMA model

Di Liu[1], YunFei Cui[2*], Zhuonan Feng[3] and Yanjuan Wang[4]

[1234]Beijing Aerospace Control Center Beijing, China

13426048463@163.com

Abstract. With China's deep space exploration program constantly expanding, the number of satellites on-orbit is increasing year by year. The stable operation of satellites in orbit hinges on long-term monitoring and analysis of the telemetry data downloaded from the satellite on the ground, which enables quick detection of any irregularities. If we can grasp the changing trend and operation law of satellite telemetry data, we can monitor and control satellites and reduce losses better. Aiming at a large amount of telemetry data of deep space exploration satellites, the data presents non-stationary and nonlinear characteristics, The time series to be processed is constructed sequentially using the acquired telemetry data, and it is decomposed using the empirical mode decomposition approach. The non-stationary signal is divided into a stationary sequence and the sum of a series of stationary intrinsic mode functions. Then the decomposed sets of results are used as the input of the ARIMA model, and the ARIMA model is constructed to predict the predicted value of each decomposed sequence, and the weighted summation of the predicted values is performed to reconstruct the predicted results. Lastly, actual satellite telemetry data is used to confirm the method's viability. The experimental findings demonstrate that the model can more successfully anticipate the short-term trend of satellite telemetry data.

1. Introduction

At present, as the pace of our country's deep space exploration is getting faster and faster, the number of satellites launched every year is also increasing. Modeling and analysis of satellite telemetry data to realize automatic data interpretation, condition monitoring, failure prediction, etc. are becoming the current major challenges [1]. With the progress of our country's aerospace technology in recent years, the performance of deep space exploration satellites is getting better and better, leading to the available failure data being very limited. Due to a decreasing number of failure data, it is more difficult to use failure data as a data set for model warning and verification[2]. Hence, in order to interpret satellite telemetry data for the purpose of health management, a precise and trustworthy prediction model is required.

The threshold method is one of the most important and commonly used methods to monitor satellite status on the ground. When the parameter exceeds the alarm threshold, the ground will receive abnormal signals, to deal with the abnormality. However, some abnormal satellite state variable telemetry results do not exceed the alarm threshold, and their changing laws have changed compared with the previous historical laws, which are prone to false negatives[3][4]. Yang Jiasen[5] proposed a data-driven method for correlation knowledge discovery and traditional methods for anomaly detection from a new perspective of correlation in the telemetry branch. Based on the dynamic properties of the telemetry data, Li WeiZheng[6] carried out sampling and sequence difference operations based on the similarity of the parameter values between the cycles of the telemetry data and constructed a hybrid autoregressive moving average model to forecast the outcomes of the telemetry data. The actual value is compared to find anomalies. Yang[7] proposed an improved method based on the online least squares support vector machine algorithm to realize online anomaly detection. The LsSVM, developed by Chen Bing[8] and others, was used to identify aberrant satellite states.

This article analyzes and applies the satellite remote test data of deep-space detection with ARIMA combined with the experience modal decomposition of the experience. The telemetry test data is constructed into a time sequence, and the prediction of remote testing data converts to the change of time sequence of time sequences. The data establishes their respective self-regression moving average models after the experienced modular decomposition and finally reconstructs the final prediction result. The experimental findings demonstrate that the approach has a positive impact on the prediction of a particular satellite state, which can serve as a foundation for assessing satellite status.

2. Models and methods

The model used in this paper to analyze and predict the telemetry data of deep space exploration satellites is shown in Figure 1.

Step 1: Clean the telemetry data collected in the previous period, remove outliers, and then construct it into a time series

Step 2: Perform empirical mode decomposition on the pre-processed telemetry data to obtain multiple intrinsic mode functions and a trend term

Step 3: Build an ARIMA model for the output results in Step 2

Step 4: Perform a weighted sum of the output results in Step 3 to reconstruct the predicted value.

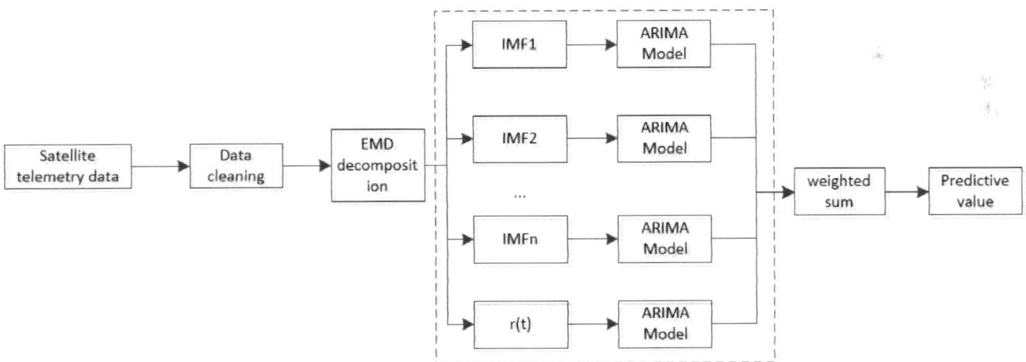

Figure 1. Example of a figure caption. *(figure caption)*

2.1. Empirial Mode Decomposition (EMD)

EMD is good in analyzing non-stationary and nonlinear data[9]. Using this approach, complicated multi-scale signals may be adaptively divided to be a total of zero-mean oscillatory components, or numerous Intrinsic Mode Functions (*IMFs*). The IMF needs to fulfill both of the following constraints:

(1) The maximum number of zero-crossing points and extreme points in the whole input signal is 1, respectively;

(2) The mean value of the envelope created by its local maximum point and minimum point is set to be zero at any given time;

$$x(t)=\sum_{i=1}^{\lambda}(r_i(t)+IMF_i(t)) \tag{1}$$

$x(t)$ is the original signal, and there are λ intrinsic mode functions $IMF(t)$, which represent the oscillation changes of the original signal at different frequency segments, and are the local characteristics of the signal; rt(t) is the original signal trend term, which reflects the slow change in the signal.

The specific steps of EMD are shown in Figure 2:

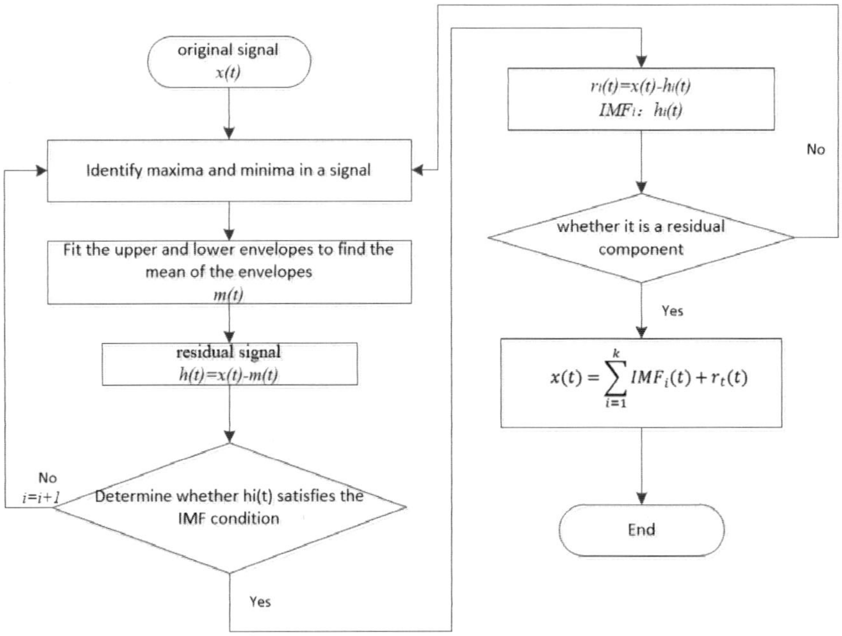

Figure 2. EMD Process

2.2. ARMIA

The most popular statistical model used for time series forecasting is known as the ARIMA (p, d, q). The model is to stabilize the non-stationary time series through d-order difference, and then use autoregression (AR(p)) and moving average process (MA(q)) to model the obtained stationary time series. In the process of establishment, the AIC and BIC criteria are generally used to determine the p and q values [10]

The ARIMA model has three parameters: p, d, and q.

p — is the AR term, which used in the forecast model stands for the lag number of the time series data.

d — signifies that the time series data becomes stable after been differentiated in numerous orders.

q — known as the MA term, stands for the quantity of lags in the forecast inaccuracy employed in the forecast model

Its mathematical form is expressed as:

$$\hat{z}_t = \mu + \varnothing_1 * z_{t-1} + \ldots + \varnothing_p * z_{t-p} + \zeta_1 * e_{t-1} + \ldots + \zeta_q * e_{t-q} \quad (2)$$

The AR model's coefficient is denoted by $\varnothing$ in the formula, while the MA model's coefficient is denoted by ζ .

3. Application Analysis

3.1. Data cleaning and time series construction

The deep space exploration satellite telemetry data collected on the ground sometimes exceeds the normal threshold range seriously due to factors such as decoding errors or equipment failures during telemetry transmission. The main difference between outliers and anomalous values is that the appearance of outliers often deviates greatly from the normal threshold range, which will not be eliminated for some time, and there are often large areas of telemetry parameters that deviate from the normal threshold range [11]. For example, a certain value of 0xAA (padding code) always appears in a transmitted parameter due to a decoding error. Outliers will decrease the developed prediction model's accuracy, hence it is vital to pre-process the satellite telemetry data to remove outliers brought on by decoding mistakes. Different telemetry parameters collected by the satellite are downloaded to the ground receiving system at a certain sampling interval. For telemetry data with a stable sampling interval, the results received from the ground are also arranged in chronological order. Telemetry data is structured as a time series as follows:

$$\{\partial_1(t), \partial_2(t), \partial_3(t), ..., \partial_N(t)\} \quad (3)$$

The data used in this article is the data sampled before the same period T15:00-T16:00 every day from October 1st to October 20th, 2020. The data will be pre-processed and arranged according to the order of reception.

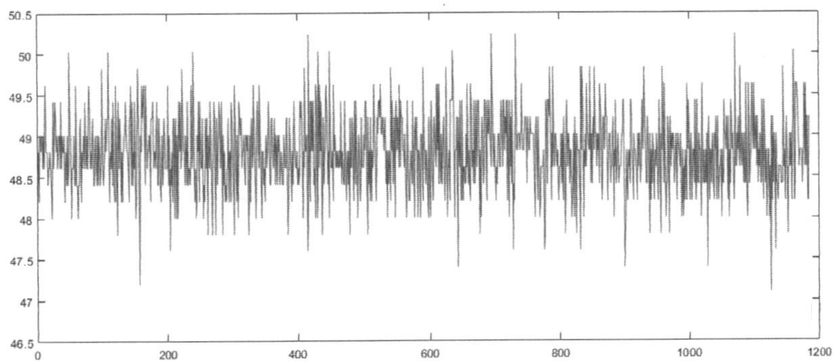

Figure 3. Time series of satellite payload power source current telemetry data

3.2. *Time series of telemetry data of a satellite payload power supply current*

Eight intrinsic mode functions and a trend term are obtained using the empirical mode decomposition technique of the time series obtained above, as shown in Figure 4.

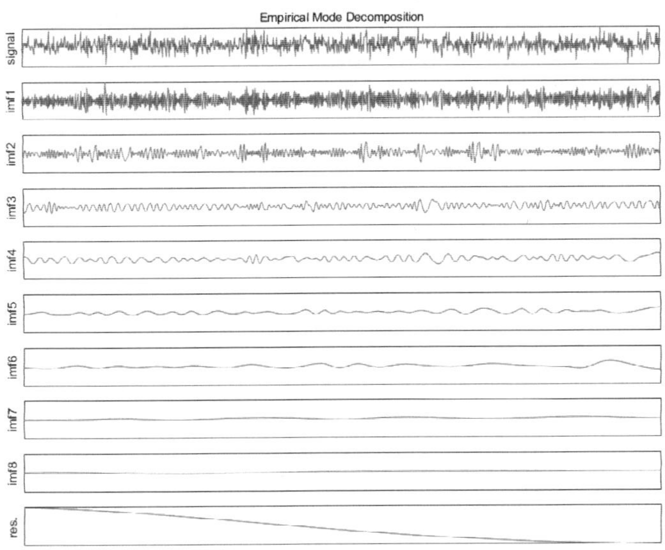

Figure 4. Decomposed telemetry time series

Each deconstructed component is given its own ARIMA model, and to calculate the p and q values, the partial autocorrelation function PACF and sample autocorrelation function ACF are, respectively, trimmed and tailed. The ARIMA model used for fitting is the one that causes the AIC to attain the minimal value, in accordance with the AIC criteria[12]. The AIC standard function is:

$$AIC = n\ln L + 2(p+q+1) \tag{4}$$

In the formula, The likelihood function is L. The model's p and q values are determined when the AIC value is the lowest.

3.3. Evaluation indicators

MSE, RMSE, and MAE are used as the accuracy evaluation indicators of model prediction, where y^ represents the predicted value of the model and y represents the true value. The better the precision of the experimental data given by the prediction model, the lower the value of the three assessment markers. The following are the calculation formulae for the three evaluation indicators:

$$MSE = \frac{1}{n}\sum_{a=1}^{n}(\hat{y}_a - y_a)^2 \tag{5}$$

$$RMSE = \sqrt{\frac{1}{n}\sum_{a=1}^{n}(\hat{y}_a - y_a)^2} \times 100\% \tag{6}$$

$$MAE = \frac{1}{n}\sum_{a=1}^{n}|\hat{y}_a - y_a| \tag{7}$$

Table 1 displays the comparison results between the EMD-ARIMA and ARIMA prediction models utilizing the aforementioned assessment indicators. In Figure 5, the prediction impacts of the two approaches are compared. The findings demonstrate that the suggested technique has a greater forecast accuracy when compared to the other two prediction methods, demonstrating its viability and efficiency.

Table 1. The comparison results of EMD-ARIMA and ARIMA prediction models

Predictive Model	MSE	RMSE	MAE
EMD-ARIMA	0.1512	0.3889	0.3118
ARIMA	0.2138	0.4624	0.3660

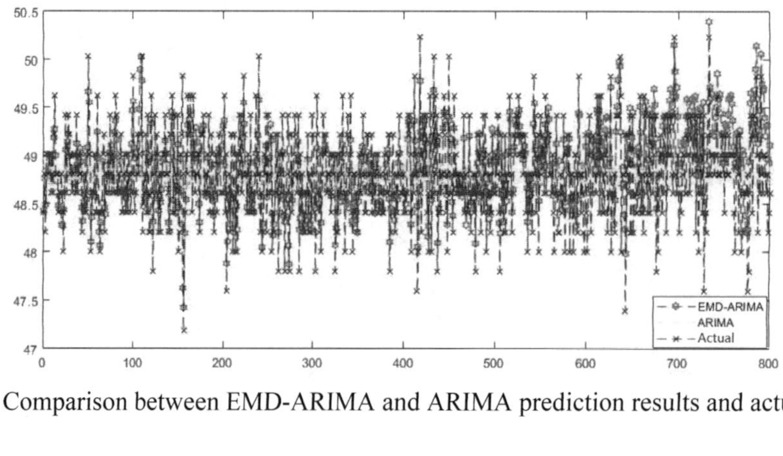

Figure 5. Comparison between EMD-ARIMA and ARIMA prediction results and actual results

4. Conclusion

Remote test data is a reflection of the operating status of the spacecraft. By analyzing the remote test data, the operation rules of the spacecraft are grasped, and accurate predictions can better monitor and control the spacecraft, avoid risks and reduce losses. In order to address the issue of nonlinearity and non-stationarity in the time series telemetry data of deep space research satellites, this study employs the approach of empirical mode decomposition to determine its overall change trend. The weighted sum approach is used to reconstruct the prediction results after the decomposed time series telemetry sequences have each been established using their own ARIMA models and the decomposed values have been forecasted. The outcomes of the experiments demonstrate that the approach has a good influence on prediction and can serve as a foundation for evaluating the state of satellites.

The predictions studied in this paper are all real-time, and the model only predicts the next step of the remote test data. The future long-term prediction of multi-dimensional remote test data is still a worthy question of in-depth research. Whether the remote test data is needed, the setting of the sampling algorithm and the establishment of a long-term remote test model requires in-depth practice and exploration.

References

[1] Kang Xu. Anomaly Detection Method for Sequential Telemetry Data[D]. Nanjing University of Aeronautics and Astronautics
[2] GU Sheng, WEI Jiao-long, PI De-chang. Particle Swarm Optimization-Fuzzy Support Vector Machine Based Prediction of Spacecraft Parameters[J]. Journal of Astronautics, 2014, 035(011):1270-1276
[3] L. X. Jiang, H. W. Li, and G. Q Yang. A survey of spacecraft fault diagnosis research[J]. Journal of Astronautics, vol.30, no.4, pp.1320-1326, 2009.
[4] Peng Xiyuan, Pang Jingyue, Peng Yu, Liu Datong. Review on anomaly detection of spacecraft telemetry data[J]. Chinese Journal of Scientific Instrument,2016,37(9):1929-1945.
[5] Yang Jiasen, Meng Xin, Chen Tuo, Zhi Jia, Li Hu. Anomaly detection of spacecraft based on the telemetry data correlation[J]. Chinese Journal of Scientific Instrument, 2018, 039(008):24-33.
[6] Li W Z, Meng Q, Fault Detection for in-orbit Satellites Using an Adaptive Prediction Model, Chinese Journal of Space Science, 2014, 34(2):201-207.
[7] Yang Y, Hou N. Data series forecasting and anomaly detection methods based on online least squares support vector machine. IEEE,2013.
[8] Chen Bing, Lu Gang, Fang Hongzheng, Ai Li. Method of Satellite Anomaly Detection Based on Least Squares Support Vector Machine[J]. Computer Measurement & Control,2014, 22(003):690-692.
[9] N. E. Huang, Z. Shen, S. R. Long, M. C. Wu, Q. Shin, H. H. Zheng, N. C. Yen, C. C. Tung, and H. H. Liu, "The Empirical Mode Decomposition and the Hilbert spectrum for nonlinear and non-stationary time series analysis, Proc. Royal Soc. London A, vol. 454, pp. 903-995, 1998.
[10] Shi D.M, Xie X.Y Applied Time Series Analysis[M]. Higher Education Press,2011.
[11] Zhong Zuhua. Research on prediction methods for telemetry time series of the satellite[D]. Nanjing University of Aeronautics and Astronautics, 2015.
[12] Akaike, Hirotsugu. A new look at the statistical model identification: IEEE Transactions on Automatic Control， 1974

AUTHOR INDEX